2nd
EDITION

CELL WALL DEFICIENT FORMS

Stealth Pathogens

2nd
EDITION

CELL WALL DEFICIENT FORMS

Stealth Pathogens

Lida H. Mattman, Ph.D.

Professor Emeritus
Department of Biology
Wayne State University
Detroit, Michigan

CRC Press

Boca Raton   Boston   London   New York   Washington, D.C.

Library of Congress Cataloging-in-Publication Data

Mattman, Lida H.
    Cell wall deficient forms: stealth pathogens/Lida H. Mattman. -
- 2nd ed.
       p. cm.
    Includes bibliographical references and index.
    ISBN 0-8493-4405-0
    1. L-form bacteria–Pathogenicity.
QR77.3.M372 1992
616′.014—dc20
                                                     92-26456
                                                      CIP

No claim to original U.S. Government works
International Standard Book Number 0-8493-4405-0
Library of Congress Card Number 92-26456
Printed in the United States of America   3 4 5 6 7 8 9 0
Printed on acid-free paper

DEDICATION

This book is dedicated to Paul E. Mattman, ever-helpful adviser, to our children Sandra and Paul, and to my students who were also my teachers.

ACKNOWLEDGMENTS OF FINANCIAL SUPPORT

The author gratefully acknowledges that her research has received financial support from Wayne State University Grants-in-Aid (National Institutes of Health), the Damon Runyon Society, the American Thoracic Society, the Michigan Heart Association, the U.S. Public Health Service, the Michigan Cancer Foundation, the American Medical Association, the Departments of Education of the State of Michigan and the U.S. Government, and from generous anonymous donors, especially two current friends.

PREFACE

This volume has a dual thrust: to describe the unrecognized omnipresent role of wall-deficient organisms in all aspects of microbe participation in life, including initiation of the food chain. Secondly to note that the majority of unexplained negative cultures concern infection with these variants.

Clandestine, almost unrecognizable, polymorphic bacterial growth occurs as often as the stereotyped classical boxcars of bacilli, pearls of cocci, and neatly sinuous *Treponema* so well known. The variant growth (termed spheroplast, protoplast, L-phase, transitional, etc.) contains less of the rigid wall components, and the shapes which result from diminution, discontinuities, or complete absence of wall are almost endlessly variegated, ranging from ultramicroscopic to syncytial. Binary fission ceases; budding is one of the common forms of reproduction.

Forms with decreased wall facilitate genetic experiments since DNA in transformation traverses only a narrow barrier. Similarly, fusion of these wall-deprived forms is common, and conjugation bridges have been clearly delineated. Interesting and unique characteristics of spore DNA have been revealed by spore spheroplasts. Biochemical analyses of the cell become easier when contamination with mural components is minimal.

Significantly, in the clinical laboratory, these organisms can be found easily, without costly time-consuming special procedures — one need only look at the direct specimen with the Acridine orange stain, or the primary culture if placed in an all-purpose "soupy" broth. These variants are critically important in septicemia, meningitis, urinary tract infection, heart valve infection, arthritides, blinding ocular inflammation, and a host of other maladies. For many syndromes at least a fourth of cases yield only these "L forms". In other conditions, such as Whipple's disease and sarcoidosis, the variants are the only organisms which ever appear in culture.

CWD forms predominate in mycobacterioses. In tuberculosis and sarcoidosis blood cultures reveal variant growth in 48 hours which gives characteristic biochemical reactions, stains with fluorescent antibody, reacts in Elisa, and is acid fast when Kinyoun's stain is buffered. Acid-fast variant colonies may be seen in sputum and in sediment of spinal or pleural fluids.

Contributions to this volume by many individuals are here noted. Alva Johnson and Sandra Augustine critiqued the manuscript. Librarians of Shiffman Medical Library, especially Linda Blakely, John Coffey, Natalie King, Lora Robbins, and Ruth Taylor gave tireless assistance. Susan McCarthy of the National Department of Agriculture made exceptional library investigations. Harris L. Coulter translated manuscripts. Susan Cook, Eve Ann Page, and Nancy Treece contributed unique laboratory studies. Repeated cooperation by Charles San Clemente is remembered. James C. Murphy donated a hard-to-find bacterium. Finally, this second edition would not have materialized without the most capable assistance of secretary-correlator, Tina Manire, and without the multifaceted support of Raymond K. Brown.

Valuable cooperation also came from other university staff members, Lila Buza, William Coloff, James Jay, Jack Petzoldt, professional photographer, Harold Rossmoore, Linda Van Thiel, and David Adamany, University President.

Lida Holmes Mattman
Wayne State University
Detroit, Michigan

THE AUTHOR

Lida Holmes Mattman, Ph.D., is currently Professor Emeritus in the Department of Biological Sciences at Wayne State University in Detroit where she is engaged in research and lecturing. She has served as President of the Michigan Branch of the American Society for Microbiology, as Chairman of the Medical Division of the Michigan Academy of Sciences, and held various offices in the local chapter of Sigma Xi.

Her studies have concerned investigating the role of surface tension depressants in immunological systems, the first complement fixation with a bacteria-free virus, the first report of wound botulism, geotrichum mycemia, nasal carriage of *Clostridium tetani*, antibiotic cure of rhinoscleroma, antibiotic sensitivity testing of *Coccidiodes immitis*, and electron microscope studies of Peptococci. The most in-depth studies relate to L Forms spontaneously occurring *in vivo* and *in vitro*.

CWD *S. pneumoniae* from blood of man caused myocarditis in mouse. Acridine orange stain of microcolony. (Photo submitted by Joseph Merline.)

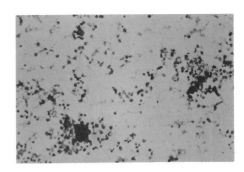

CWD streptococci in blood culture of subacute bacterial endocarditis. (Photo by R. Nativelle and M. de Paris.)

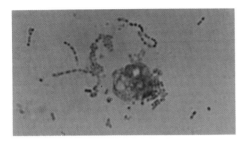

Streptococci reverting from a colony of pleomorphic growth. (Photo submitted by L. Salowich.)

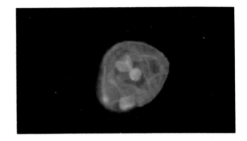

Classical streptococci or staphylococci inoculated into acridine orange form L-bodies with internal reverting cocci. (Photo submitted by Robert Ricketts.)

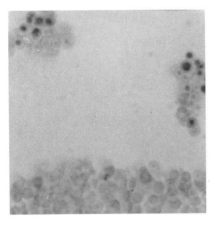

Spheres growing from blood culture. Patient also had these in urinary casts. Brown and Brenn's gram stain. Organisms reverted to streptococci. (From Domingue, G. J., Woody, H. B., Farris, K. B., and Schlegel, J. U., *Arch. Intern. Med.*, 139; 1355–1360, 1979.)

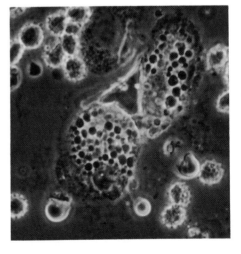

Microcolony of CWD forms in urine of idiopathic hematuria. (From Domingue, G. J., Thomas, R., Walter, F., Serrano, A., and Heidger, P. M. Jr.)

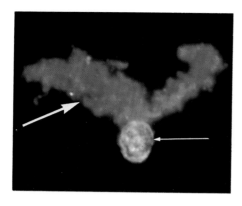

Acridine orange stains growth from erythrocyte of patient with pelvic inflammatory disease. Similar growth stained with *N. gonorrhoeae* fluorescent antibody. (Photo submitted by J. Gray.)

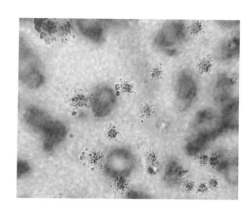

CWD *N. gonorrhoeae* in pour plate culture of blood from patient with urogenital disease. (Photo submitted by J. Gray.)

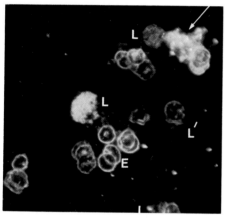

Darkfield of buffy coat of blood of gonorrhea patient shows leukocytes (L) which have engulfed organisms. (Photo submitted by J. Gray.)

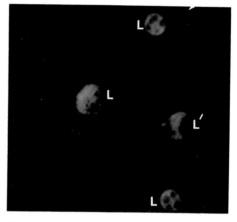

UV illumination of same field shows that FA to the gonococcus stains phagocytized CWD organisms. (Photo submitted by J. Gray.)

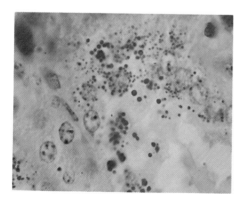

N. caviae in mouse tissue. Organism grows as gram-negative spheres. (Photo submitted by B. Beaman.)

Mouse several months after injection of L-forms of *N. caviae*. (Photo submitted by B. Beaman.)

CWD colonies of *P. aeruginosa* in pour plate culture of distilled water. Stain basic fuchsin. Orig. magn. × 1000. (Photo submitted by Stanley Susan.)

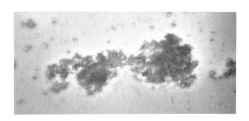

CWD colony of *P. aeruginosa* in pour plate culture of patient's blood. Orig. magn. × 540. (Photo submitted by Judith Besserer.)

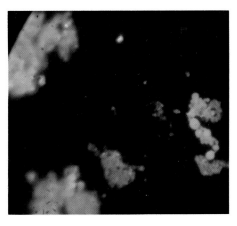

Culture of rheumatoid arthritis synovial fluid stained with fluorescein-labeled muramidase. Orig. magn. × 1000. (Photo submitted by G. Denys.)

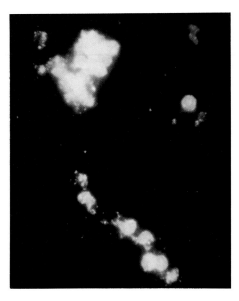

Leukocytes in rheumatoid arthritis synovial fluid have material stained with fluorescein-muramidase. (Photo submitted by F. Carlock.)

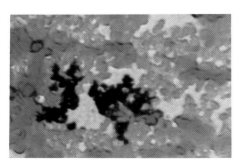

Gram stain of rheumatoid arthritis synovial fluid. Orig. magn. × 1000. (Photo submitted by L. Tunstall.)

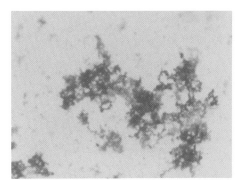

CWD *S. marcescens* in blood culture. Orig. magn. × 1000. (Photo submitted by W. D. LeBar.)

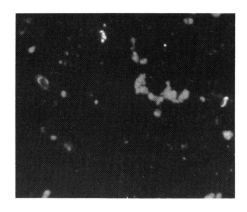

CWD staphylococci in blood culture identified with fluorescent antibody. (Photo submitted by R. Zafar.)

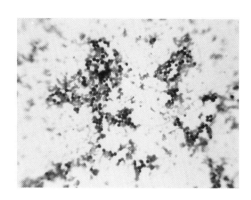

Staphylococci are reverting in CWD colonies. Orig. magn. × 1000. (Photo submitted by L. Salowich.)

Minute CWD colonies of *S. aureus* beside one classical colony. Triphenyltetrazolium chloride in pour plate of blood culture. Orig. magn. × 100. (Photo submitted by R. Zafar.)

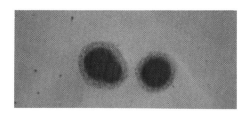

CWD colonies of *S. aureus* reduce triphenyltetrazolium chloride. Orig. magn. × 1000. (Photo submitted by M. Whalen.)

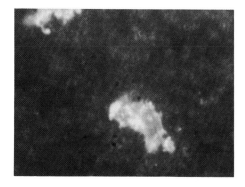

CWD *S. aureus* in "negative" blood culture stained with fluorescein-labeled muramidase. Orig. magn. × 1000. (Photo submitted by J. Jubinski.)

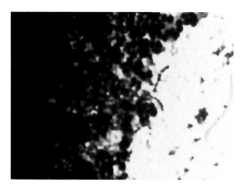

S. pneumoniae type III reverting from CWD stage. Fatal pneumonia in kidney transplant patient. (Photo submitted by L. Wickers.)

CWD organisms in direct smears of aqueous humor.
(All photographs by Carolyn Barth and Philip Hessburg.)

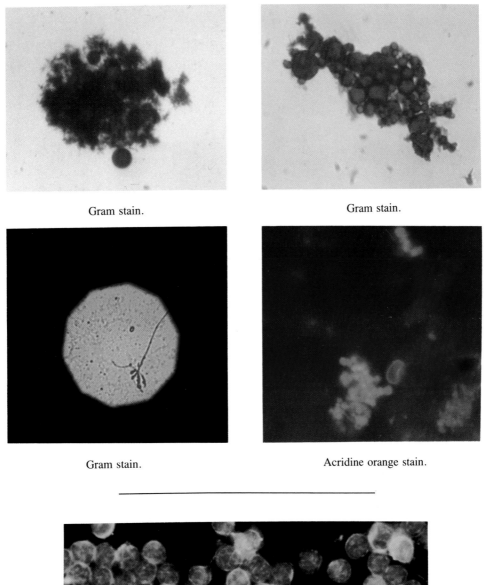

Gram stain.

Gram stain.

Gram stain.

Acridine orange stain.

Fluorescein-labeled anti-complement antibody detects immune complex within leukocytes in aqueous humor.

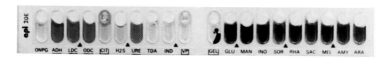

Reactions in API panel typical of CWD organisms from tuberculosis or sarcoidosis patient. (Photo submitted by M. S. Judge.)

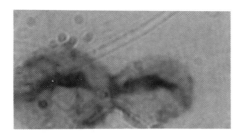

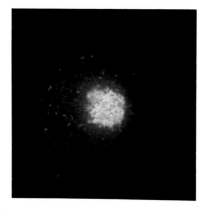

Intracellular growth has joined 2 erythrocytes in blood culture of tuberculosis patient. Sarcoid blood shows similar forms, sometimes acid fast.

Fluorescent antibody to *M. tuberculosis* stains colony from unincubated blood of tuberculosis patient. Blood was lysed and organisms concentrated by saponin.

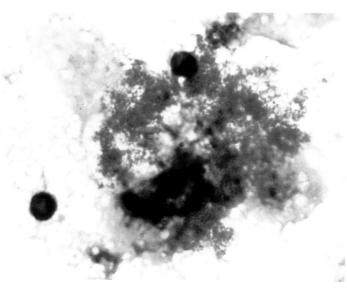

Acid-fast colony in lung of guinea pig given blood-borne CWD *M. tuberculosis* with cortisone. (Photo submitted by M. S. Judge.)

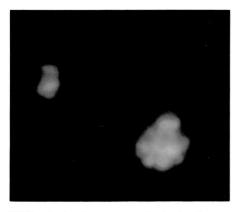

CWD colonies in sputum stained with fluorescent antibody to *M. tuberculosis*. (Photo submitted by A. Byrne.)

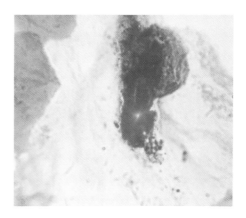

Acid-fast growth from L-body in sputum smear. Intensified Kinyoun's stain. Epithelial cells on left. (Photo submitted by C. Steffen.)

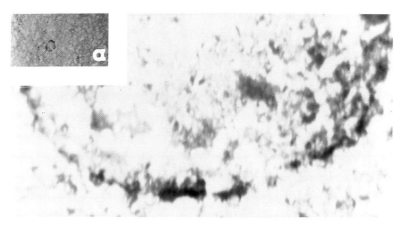

Edge of L-body (orig. magn. × 1000) in sputum. Insert shows L-body with low magnification. Victoria Blue stain. Acid fast material is blue. (Photo submitted by E. A. Page.)

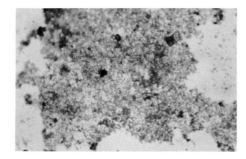

Retention of Victoria Blue dye shows acid fastness of colony from blood of tuberculosis patient. (Photo submitted by N. Thieda.)

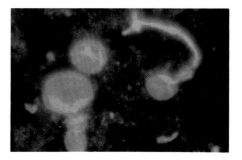

Fluorescent monoclonal antibody to *M. tuberculosis* stains L-bodies and parasitized red cells in blood culture of tuberculosis patient. (Photo submitted by A. Johnson and P. Almenoff.)

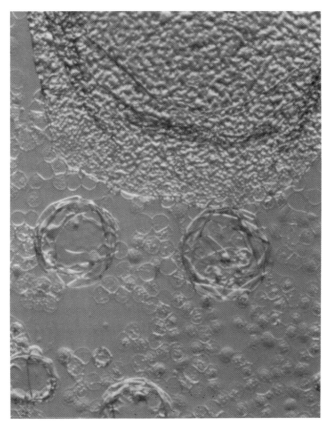

The ''phenol phenomenon''. L-bodies develop from blood in Auramine Rhodamine stain. Clinical impression of patient's disease was Ulcerative Colitis.

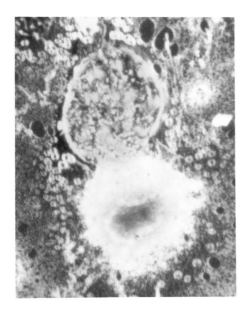

The ''phenol phenomenon'' with blood of Great Dane suffering bloat. Similar growth results from blood of some healthy canines. Is this the Clostridium Paula Sanch found in blood of dogs with bloat?

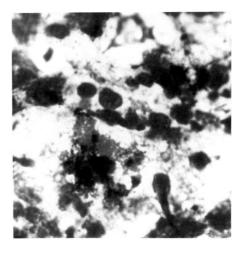

CWD growth in spleen of mouse given blood culture from sarcoid patient. Intensified Kinyoun's stain.

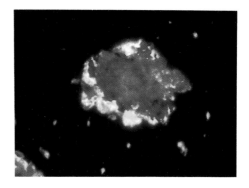

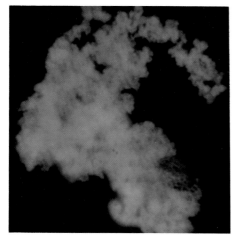

Fluorescent antibody (1–1000) made in response to sarcoid organisms stains CWD colony from blood of tuberculosis patient.

Autofluorescent CWD colony from sarcoidosis blood.

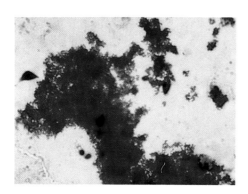

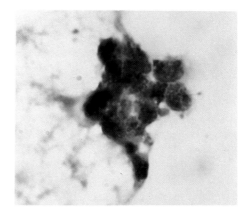

Acid-fast growth from ocular tissue of mouse which received CWD organisms from sarcoid case.

Acid-fast material in leukocytes of incubated blood culture of a sarcoidosis case.

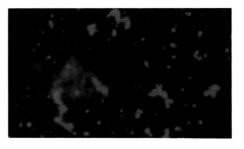

Lung tissue of mouse which received CWD organisms from the blood of a sarcoidosis case.

Acridine orange-stained microcolonies in blood culture of a Boeck's sarcoid case.

All photographs this page submitted by M. S. Judge.

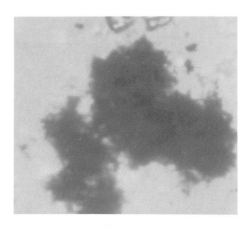

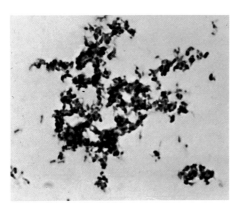

Partially acid-fast colony of CWD *M. paratuberculosis* from involved intestinal wall in Crohn's disease. (Photo submitted by R. Chiodini.)

Microcolony from blood of a Crohn's patient is acid-fast by intensified Kinyoun's stain.

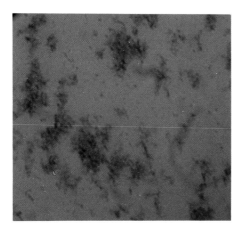

Colonies from a Crohn's disease blood culture concentrated in xylol. Intensified Kinyoun's stain.

Colony in pour plate of Crohn's patient's blood has reduced triphenyltetrazolium chloride.

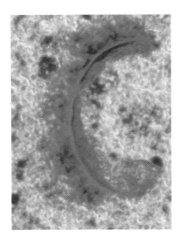

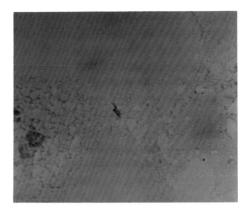

Rim of L-body with outgrowths. Crohn's patient's blood. Three days in mail. No culture medium. Intensified Kinyoun's stain.

M. avium intracellulaire colonies from patient's blood on surface of medium. Impression smear. At left, beginning acid fastness. In center, 2 reverting rods.

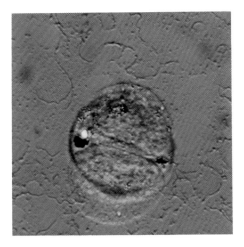

L-body in culture of *B. burdorferi* in BSK medium with no inhibitor. Orig. magn. × 400.

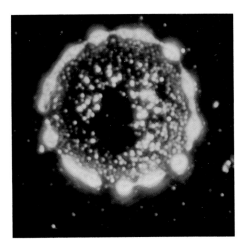

L-body in unconcentrated spinal fluid of MS case. Orig. magn. × 400.

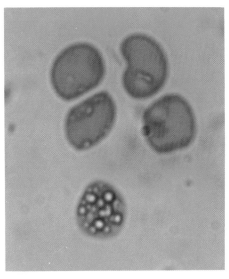

Parasitized erythrocyte of mouse infected with *C. kutcheri*. (Photo submitted by R. J. Smith, C. Zammit and D. Randall.)

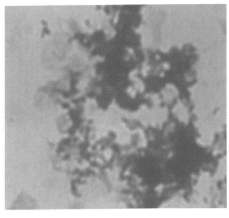

CWD. *H. influenzae* in blood of patient with *H. influenzae* meningitis. Gram stain. (Photo submitted by W. D. LeBar.)

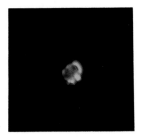

Parasitized erythrocyte of patient with CWD *C. sordelli* septicemia. Gram stain viewed under dark field.

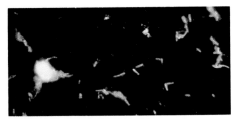

Fluorescein-labeled muramidase stains early growth of *B. cereus*. Blue is autofluorescence of organisms lacking muramic acid.

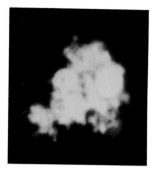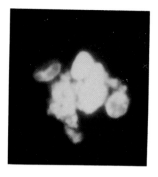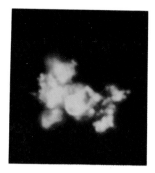

Autofluorescence in CWD colonies of *C. albicans* mycohemia. No classical organisms found. (Photo submitted by D. Swieczkowski.)

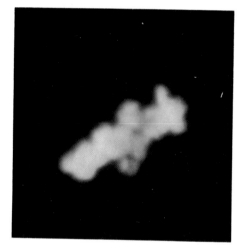

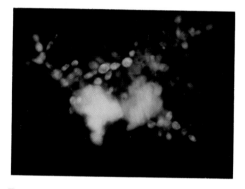

Small colony in blood culture is stained with *B. dermatitidis* fluorescent antibody. (Photo submitted by G. Senatore.)

Fungal contaminant growing in acridine orange. (Photo submitted by S. Susan.)

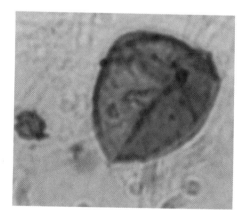

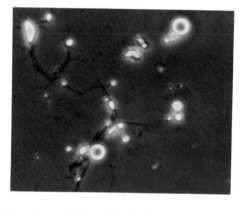

Intraerythrocytic growth in blood of a Kaposi's patient.

Fungal growth from blood of Kaposi's patient.

TABLE OF CONTENTS

Chapter 1

HISTORY

Currently bacteriology holds the belief that each species of bacteria has only a certain very simple form (coccus, bacillus, spirillum, or filament) and that it retains this form during its only mode of reproduction, which is by division into two. All colonies of bacteria form in this way from a single individual. The slight change of form which happens during growth consists essentially of elongation, or shortening, or a local swelling. More remarkable deviations in form which are observed frequently are considered involution or degenerative. In contrast, this writer, using carefully prepared pure cultures, found that bacteria pass thorugh stages with markedly different morphology.

These comments were made by Willibald Winkler, M.D., in 1899.[32]
Winkler continued,

I have to confess that the results of the studies have surprised me, so that I was tortured with doubts of many observations and I had to repeat them often. I know that in many quarters this new view will not be accepted immediately.

Zopf has written,

For example, according to Cohn, a micrococcus can produce only micrococci, not rods or spirals, and similarly, spirals can give rise to spirals only and not to rods and cocci, etc. This theory now has only historical value . . . . As strengthened by the author the new theory holds that the fission fungi, probably with some exceptions, are able to pass through different developmental stages.[34]

Zopf was publishing his findings in 1892. Yet, even earlier it had been noticed that bacteria were not merely identical units formed by an identical format by Robert Koch and an associate, the less known but nevertheless important bacteriologist, Ernst Bernhard Almquist.

Ernst B. Almquist (Figure 1) was born in Skogs-Tibble, Sweden, in 1852, of extraordinary lineage. His dynamic grandfather, Carl Jonas Love Almquist, is firmly implanted in history as one whose brilliant writing (in 21 volumes!) greatly influenced the development of Swedish literature.[8]

Almquist's training for medicine and bacteriology was arduous. He became a student in the venerable University of Uppsala when 18 years of age and only 11 years later received the license to practice medicine. The title and privileges of Doctorate of Medicine came a year later. However, during this long education period, his activities were far from pedestrian. Notably, one vacation was spent in France, working with Pasteur. Two years involved a unique adventure, serving as physician in an expedition headed for the North Pole. The sojourn is generally known as the Vega Expedition, commemorating their ship named for the brightest star in the North. Although the goal of the North Pole was not reached, the expedition yielded important finds, both geologic and geographic. Islands off the north Siberian coast still bear the name ''Almquist Islands.''

During another break in academia, Almquist worked in Koch's laboratory, where he was to return again in 1903. Thus, he was one of the few who received dual stimulation from both Pasteur and Koch.

Later he was a clinician as city physician in the large Serafiner Hospital in Stockholm. Repeatedly he was concerned with outbreaks of typhoid, cholera, and dysentery in Sweden and elsewhere. Much of his professional career was as professor in public health at Karolinksa Institute in Stockholm. His signature is preserved for posterity on the certificate of the Nobel prize awarded to von Behring. Almquist himself received decorations from Germany, France, and Switzerland. Almquist's first wife died in 1915. He must have met his second wife,

FIGURE 1.   Ernst Bernhard Almquist (1852-1946).

Gerda Troili-Petersson, in the laboratory. She was a bacteriologist capable of painstaking studies.[31] Her improved method of single cell isolation was used by her husband to assure that he was deriving pure cultures for examination.

A record of his life of 94 years is documented, especially for the professional years, not only by year but by month, recording his occupation at any one time. Perhaps it was the same meticulous care which made him stop to scrutinize what others considered merely debris or, at most, dying microbes in their cultures. As his wife recalls, for years his evenings were spent in investigative endeavors.

In 1881 Koch and Almquist together observed that the blood of typhoid fever patients contained both typical rods and oval, nonmotile bodies, the so-called Eberth-Koch forms. Almquist later concluded that Koch became so involved with multiple research goals that he lost interest in the several extraordinary bacterial forms which he, Koch, had described. "Thus, science has gone from one extreme to the other; scientists have striven to make Nature less complicated than it is."[3]

Almquist himself, however, was an exception. He was addicted to investigation of variants for the remainder of his long and fruitful career. Almquist consistently noticed nonclassical forms occurring spontaneously in disease and furthermore produced aberrant growth in a large number of species, by extremely varied techniques. After 21 years of probing, he commented, "Nobody can pretend to know the complete life cycle and all the varieties of even a single bacterial species. It would be an assumption to think so."[3]

Almquist pioneered in public health microbiology, an interest sparked by association with many epidemics of intestinal infection. In 1882 it was his responsibility to check a large outbreak of typhoid. He concluded that the epidemiology of typhoid was somehow connected with the life cycle of the pathogen. It seemed logical to him that dissemination of the infectious unit was as tiny granules rather than typical bacilli.[4] In experimental studies Almquist found that growth at 10 to 20°C favored atypical rather than classical bacterial morphology.[1,2] He hypothesized that a pathogenic bacterium in food or in water becomes an amorphous "plasmodium," which after a few hours in the body of man reverts to more classical morphology.

Almquist's observations that atypical bacterial stages resist deleterious factors are little remembered. He found certain variants of Enterobacteriaceae surprisingly heat resistant. Since pasteurization removes the threat of milk-borne typhoid, there are three possible explanations for Almquist's data: (1) the thermoduric forms are of avirulent species or are rendered avirulent by heating; (2) the variants are not heat resistant in milk; (3) pathogens never exist in atypical stages in milk, which seem unlikely since all body fluids foster growth of the atypical forms. That heat can result in loss of virulence is suggested by Almquist's finding that *V. cholerae* variants which survive heating are not agglutinated by specific antiserum.[4]

Almquist was also ahead of his time in looking for evidences of conjugation or other genetic exchange of nuclear material. Mixing the filtrates of two strains, he found evidence of recombination between *B. miserum* and his "*B. antityphosa,*" which was the odd name he gave the variant of the typhoid bacillus. (Did he, originally, think of this variant as a possible vaccinating agent?) Likewise, Almquist obtained a hybrid form by mixing cultures of *B. lineare* and *B. plasmaticum.*[4]

When Almquist defines his terms, it becomes clear that the pleomorphic growth which he studied is related to the atypical organisms inhabiting our cultures today.[3] He called an amorphous stage a plasmodium, noting that Löhnis called this a symplasm and that it was somewhat similar to the promycelium of Tulasne. Currently it is termed a syncytium. His mycelioid form was a long intricate thread without branches. His conidium was a free globule which could sprout to a rod, a spirillum, or a new globule. When these free globules had a diameter of only 0.5 μm, he called them microconidia and noted that Lohnis considered them gonidia. Almquist's sporangium was an oversized conidia which could contain several cells, globules, or chains. He noted that there were bodies smaller than the microconidia; these he termed microoidium. He found them difficult to observe with a microscope, better visible in photographs.

G. Koraen, a co-worker of Almquist, studied variation in four species of micrococci, including *S. aureus.*[19] Similar to the findings of Almquist with Gram-negative rods. Koraen found that the cocci, if dried, became tiny granules which under certain conditions could revert to typical cocci. There were also stages in which the typical cocci were connected by strands with the old amorphous growth.

Almquist's work also stimulated studies with the cholera vibrio. Examining many strains of *V. comma,* Olsson[28] found that a decrease in temperature of only 2°C could change the vibrio into nonmotile spheres which continued excellent growth. Most remarkably, when again cultured at 37°C, the return of comma configurations was accompanied by increased virulence. This reinforced the concepts of Almquist that the intestinal pathogens may spend the winter in atypical form in rivers and soil and upon the onset of summer temperature gain in virulence.

Haag subjected *Bacillus anthracis* to a complete morphological review.[10] He reported that other authors, too, had observed a variant which could certainly complicate diagnosis, the "Buchnerform," subtilis-like, in which the anthrax bacillus is motile. Another wall deficient form described by Haag is a fine Gram-negative rod, not especially resistant, requiring a heavy inoculum to initiate growth. This rod is also motile, varies in size from 0.2 to 0.4 μm, and produces a dark colony only 0.1 mm in diameter. Haag also noticed that when *B. anthracis* was cultured with products of Pseudomonas, it became entirely atypical and asporogenous.

A clandestine stage in Mycobacteria was noted as early as 1892 by Maffucci,[23] who injected *M. tuberculosis* var. *bovis* into chickens. On occasion, the fowl's intestine became studded with pinhead-sized gray granules, which showed no tubercle bacilli on smear or culture. However, these nodules caused typical tuberculosis when injected into rabbits or guinea pigs.

Very pertinent today are findings concerning culture of *T. pallidum*. A study in Leningrad by Roukavischnikoff[30] found that when syphilitic blood is placed in suitable medium, the organism changes from an invisible phase to demonstrable granules and other stainable structures. Under optimum circustances, spheroid masses of spirochetes develop.

Viruslike stages of several organisms had been implied from infectivity in the invisible stage. Friedberger and Meissner[9] demonstrated infection in guinea pigs and rabbits given the typhoid bacillus, by temperature elevation and development of specific antibody, without finding the organism in cultures or smears of the animal tissue.

Any review of morphological variation in bacteria seems but a feeble gesture compared with the analysis by Löhnis[21] of 1309 articles written between 1838 and 1918. Some of this record defies interpretation. Electron microscopy was needed but not developed. However, many drawings and micrographs were remarkably clear, and the same variations often were recorded in the same species by numerous investigators. Relying largely on drawings or photographs, which describe more exactly than words, the following data can be abstracted from Löhnis's review:

Large, budding, yeastlike forms. Oval or round budding organisms were noticed in cultures of *Clostridium chauvei* by Ghon and Sachs and by Grassberger in 1903 and by Hibler in 1908. Maasen, in 1904, noticed similar bodies in his cultures of *V. cholerae.*

Discharging cysts. Currently these would be termed L-Bodies or Giant Round Forms. These were noted by Winogradsky in Nitrosomonas in 1902 and, remarkably, in the pneumococcus by Artigalas in 1885.

Filaments which branch or develop buds. This phenomenon was noted in *B. radiciola* by Conn in 1900, in *E. coli* by Matzuschita in 1900 and by Kellerman and Scales in 1916. This fungoidal state was recorded for *Pasteurella pestis* by Albrecht and Ghon in 1900 and by Rowland in 1914.

Slender, gracefully branching filaments were especially characteristic of mycobacterial growth, as noted for *M. tuberculosis* by Metchnikoff in 1888, Crookshank in 1896, Migula in 1900, and Meirowsky in 1914. Often refractile granules developed at the sides of the filaments or terminally. Since these could serve as sites for new growth, they have frequently been termed "spores." Meirowsky (1914) found similar aberrant growth in *M. leprae,* and Kedrowski (1910) found branching hyphae, both slender and very wide, in pure cultures of the leprosy bacillus.

The same graceful filaments observed in Mycobacteria were recorded in numerous other species in Löhnis's review. This actinomyces-like stage of *Pseudomonas pseudomallei* was described by Marx and Carpano in 1913. Löhnis, who noticed that morphological variants were of almost ubiquitous natural occurrence, found the Actinomyces-like stage in *Pseudomonas fluorescens* and many other species (Löhnis, and Smith, 1916). Other bacteria recorded as assuming this thin Actinomyces-like form include *Salmonella typhi* (Matzuschita, 1900), *Salmonella enteriditis* (Meirowsky, 1914), *Spirillum rubrum* (Meirowsky, 1914; Reichenbach, 1901), *Treponema pallidum* (Meirowsky, 1914), and *Vibrio cholera* (Stamm, 1914). It is interesting that this fine-filament stage, which also characterizes some species of Mycoplasma, has been noted in so many species by men famous in microbiology for their basic contributions.

Wavy filaments with large lateral buds. This growth was noted in *Spirocheta gallinarum* by Prowazek in 1906 and in *Salmonella enteriditis* by Meirowsky in 1914, occurring with the actinomyces-like growth which he also recorded for *S. enteriditis.*

Balloon forms with rhizoid sprouting were observed in cultures of *M. leprae* by Lutz in 1886, in *V. cholera* by Fischer in 1903 and by Stamm in 1914, and in *Spirillum rubrum* by Meirowsky in 1914.

Sheets of growth. Syncytia were noted repeatedly, but the descriptions are not concise. One clear drawing shows circles developing in what was previously a structureless mass, in a culture of *V. cholera*, followed through developmental changes by Hueppe in 1885.

The summary of Löhnis cites numerous examples of nonclassical growth in root nodule and nitrogen-fixing bacteria, which are not named here because of the taxonomic changes in those genera. Löhnis documented the life cycles of Azotobacter and a total of 42 miscellaneous strains,[20] concluding that their life cycles were no less complicated than those of fungi. He concluded that since the strains he examined behaved essentially alike, analogous results might be expected with all bacterial species,[22] namely, alternate growth in an organized and in an amorphous stage. Often many cells would melt together, leaving their empty walls, forming a symplasm. Within this sheet, regenerative units developed which might become cells of normal shape. He concluded that all bacteria, in addition to forming their symplasm, multiply not only by fission but also by formation of "gonidia," which can grow to normal cells or enter the symplastic stage. Some of them are filtrable. The life cycle of each bacterial species is composed of several subcycles showing morphological and physiological differences. He noted large budding cells, easily mistaken for Torula. Löhnis commented that standard textbooks contain photographs of the variants, never noticed because "our eyes have been trained so very well not to see them."[22]

Later, two Americans, Ralph Mellon[24,25] and Arthur Kendall,[12] began to publish observations on microbial growth which did not resemble the classical parent. Kendall used a special medium to demonstrate viable filtrable stages from almost every genus tested. A third American, Hadley,[11] named and described "G"-colonies. His tiny G-colony could be comprised entirely of classical bacteria, entirely of heteromorphic units with many filtrable granules, or of a mixture of the classical and the atypical. However, today in the American vocabulary, G-Forms are classical bacteria which produce minute colonies.[33]

At about this time, *Escherichia coli* was found on occasion to exhibit the Large Bodies of the cell wall deficient cycle. Their occurrence in empyema of the gall bladder was reported by Nungester and Anderson[27] and in urine by Mellon.[26]

The history of wall deficient variation can be dated from 1852 when Perty[29] documented his observation of the Large Bodies which constitute an omnipresent part of cell wall deficient growth. As he studied the flora of Swiss lakes, he made fine-line drawings of bacterial forms which to this day remain neatly delineated and fresh in color.

Thus, records of the other life of bacteria dated back for decades. However, the methodology was disconcertingly complex, and seldom was it possible to obtain colonial growth for observation on the surface of solid medium. Research with cell wall deficient variants in America and England awaited the findings from the laboratories of Emmy Klieneberger (Figure 2) and Louis Dienes (Figure 3). At last the growth of colonies could be followed in pour plates and at times even on the surface of agar.

Klieneberger noticed that when *Streptobacillus moniliformis* was grown in serum agar, the typical colonies were often accompanied by smaller ones which in structure and component units exactly duplicated a Mycoplasma.[13,14] It was a logical conclusion that the Streptobacillus had a travelling companion from the Mycoplasma class. Because her laboratory was part of the Lister Institute, she designated these concomitants as L_1, giving rise to the term "L Form." For many years, she investigated the physical and immunological characteristics of the L Forms, often comparing them with the Mycoplasma. Her carefully compiled review articles[15-18] were crucial to the dissemination of information regarding this field. She repeatedly stated that her findings were extensions of earlier knowledge.

Dienes reinvestigated the phenomenon of L Form concomitants in *S. moniliformis*[5] and discovered the unexpected; the small colonies are variants of the larger and can revert to the typical form. He began a methodical investigation of the factors which result in L-colonies, and an investigation of their fine structure.

Without Dienes it is unlikely that recognition of wall deficient variants by U.S. investigators would have become a reality. He not only published his findings in detail, but his

FIGURE 2.   Emmy Klieneberger-Nobel in 1950.

FIGURE 3.   Louis Dienes is photographed in 1972 beside his incubator, which has supported the growth of so many cell wall deficient microbes. The dependable incubator, made over 75 years ago by a local artisan, is heated with gas.

laboratory was always open to those who wished to observe firsthand the appearance of the variant growth and technics through which it could be procured. His findings were not, at first, considered of general interest. Reputedly, he has remarked that he became known to the hospital medical staff only after a hospital artists' show displayed his watercolors. However, his dedication and unique observations have always been appreciated by his fellow microbiologists. His character is best summarized in a recent letter accepting honorary membership in the American Society for Microbiology.

It is a great satisfaction to learn that the judgment of those who work in the same field is that I did not waste the opportunity given to me. I regard it as the greatest privilege that since my earliest youth I could give the larger part of my time to scientific work and study.[7]

REFERENCES

1. **Almquist, E.**, Studien über das Verhalten einiger pathogenen Mikroorganismen bei niedriger Temperatur, *Zbl. Bakt. I Abt. Orig.*, 48, 175–186, 1908.
2. **Almquist, E. and Koraen, G.**, Studien über Biologie und Wuchsformen der Diphtherie-bakterie, *Z. Hyg.*, 85, 347–358, 1918.
3. **Almquist, E.**, Variation and life cycles of pathogenic bacteria, *J. Infect. Dis.*, 31, 483–493, 1922.
4. **Almquist, E.**, *Biologische forschungen ueber die Bakterien*, P. A. Morstedt and Soner, Stockholm, 1925.
5. **Dienes, L.**, L organisms of Klieneberger and *Streptobacillus moniliformis, J. Infect. Dis.*, 65, 24–42, 1939.
6. **Dienes, L. and Weinberger, H. J.**, The L forms of bacteria, *Bacteriol. Rev.*, 15, 245–283, 1951.
7. **Dienes, L.**, Letters in acknowledgement of election to honorary membership, *A.S.M. Newslett.*, March, 1972.
8. *Encyclopaedia Britannica*, Vol. 1, William Benton, Chicago, 1965, 660.
9. **Friedberger, E. and Meissner, G.**, Zur Pathogenese der Experimentellen Typhus-Infektion des Meer-schweinchens, *Klin. Wochenschr.*, 2, 449–450, 1923.
10. **Haag, F. E.**, Der Milzbrandbazillus, seine Kreislaufformen und Varietaten, *Arch. Hyg.*, 98, 271–318, 1927.
11. **Hadley, P.**, The instability of bacterial species with special reference to active dissociations and transmissible autolysis, *J. Infect. Dis.*, 40, 1–312, 1926.
12. **Kendall, A. I. and Walker, A. W.**, Occurrence of bacteria in the filtrable stage in active bacteriophage, *J. Infect. Dis.*, 53, 355–371, 1933.
13. **Klieneberger, E.**, The natural occurrence of pleuropneumonia-like organisms in apparent symbiosis with *Streptobacillus moniliformis* and other bacteria, *J. Pathol. Bacteriol.*, 40, 93–105, 1935.
14. **Klieneberger, E.**, Some new observations bearing on the nature of the pleuropneumonia-like organism known as L₁ associated with *Streptobacillus moniliformis, J. Hyg.*, 42, 485–496, 1942.
15. **Klieneberger-Nobel, E.**, Origin, development and significance of L-forms in bacterial cultures, *J. Gen. Microbiol.*, 3, 434–442, 1949.
16. **Klieneberger-Nobel, E.**, The L-cycle: A process of regeneration in bacteria, *J. Gen. Microbiol.*, 5, 525–530, 1951.
17. **Klieneberger-Nobel, E.**, Filterable forms of bacteria, *Bacteriol. Rev.*, 15, 77–103, 1951.
18. **Klieneberger-Nobel, E.**, L-forms of bacteria, in *The Bacteria, Vol. 1: Structure*, Gunsalus, I. C. and Stanier, R. Y., Eds., Academic Press, New York, 1960, 361.
19. **Koraen, G.**, Studien über Umformung von mikrokokken in trocknender Kultur, *Z. Hyg.*, 85, 359–365, 1918.
20. **Löhnis, F.**, Life cycles of the bacteria, *J. Agric. Res.*, 6, 675–702, 1916.
21. **Löhnis, F.**, Studies upon the life cycles of the bacteria, *Mem. Natl. Acad. Sci.*, 16, 1–335, 1921.
22. **Löhnis, F.**, Studies upon the life cycles of the bacteria. II. Life history of azotobacter, *J. Agric. Res.*, 23, 401–432, 1923.
23. **Maffucci, A.**, Die huhnertuberculose. Experimentelle untersuchungen, *Z. Hyg.*, 11, 445–486, 1892.
24. **Mellon, R. R.**, Studies in microbic heredity. VI. The infective and taxonomic significance of a newly described ascospore stage for the fungi of blastomycosis, *J. Bacteriol.*, 11, 229–252, 1926.
25. **Mellon, R. R.**, Studies in microbic heredity. XII. Microbic dissociation *in vivo* as illustrated by a case of subacute septicopyemia, *J. Bacteriol.*, 13, 99–110, 1927.
26. **Mellon, R. R.**, Studies in microbic heredity. I. Observations on a primitive form of sexuality (zygospore formation) in the colon-typhoid group, *J. Bacteriol.*, 10, 481–501, 1925.
27. **Nungester, W. J. and Anderson, S. A.**, Variation of a bacillus coli-like organism, *J. Infect. Dis.*, 49, 455–472, 1931.
28. **Olsson, P. G.**, Zur variation des Choleravirus, *Zbl. Bakt. I Abt. Orig.*, 76, 23–34, 1915.
29. **Perty, M.**, *Zur Kenntnis Kleinster Lebensformen*, Verlag von Gen & Reinert, 1852.
30. **Roukavischnikoff, E. J.**, Zur Frage dur Entwicklungsstadien des Syphiliserregers, die im Blute des infizierten Menschen und Versuchstiere zirkulieren, *Zbl. Bakt. I Abt. Orig.*, 115, 66–71, 1929.
31. **Troli-Petersson, G.**, Einzellkultur von lamgsam wachsenden Bakterienarten, speziell der Propionsäurebak-terien, *Zbl. Bakt. I Abt. Orig.*, 42, 526–528, 1915.
32. **Winkler, W.**, Untersuchungen über das Wessen der Bakterien und deren einordnung im pilzsystem, *Zbl. Bakt. II Abt. Orig.*, 5, 569–579, 1899.
33. **Wise R. and Spink, W. W.**, The influence of antibiotics on the origin of small colonies (G variants) of *Micrococcus pyogenes* var. *aureus, J. Clin. Invest.*, 33, 1611–1622, 1954.
34. **Zopf, W.**, Zur Kenntnis der Organismen des Amerikanischen Baumwollsaatmebles, *Zopf's Beitr. z. Physiol. u. Morphol. neiderer Organismen*, 1, 57–97, 1892.

Chapter 2

DEFINITIONS

In the current literature one finds "L Form", "L phase", and "spheroplasts" working as synonyms. "Protoplasts" is a term denoting complete absence of wall. "Cell wall deficient" (CWD) is an all-inclusive term encompassing protoplasts as well as the others.

It should be emphasized that CWD microorganisms, whether protoplasts or containing some wall, have interesting characteristics missing in classic organisms. These include the following:

1. Disintegration of many genera occurs if fixed to slide by heating.[2]
2. They usually require soft agar, grow subsurface, and may not grow unless autoclaved medium is aged.
3. They typically grow within erythrocytes.
4. They are often serophilic.
5. Most genera grow best in hypertonic milieu.

Definitions for these organisms are an almost amusing subject. When their existence was publicized by Klieneberger-Nobel she loyally dubbed them L Forms for her Lister Institute.[3] The earlier microbiologists, publishing through decades, gave them a variety of terms, best forgotten, which reflect the botanical training of the investigators.

A well-intentioned committee decreed that the term L Form be limited to stable variants of known parentage which grew with Mycoplasma-like colonies. This decree was almost completely ignored throughout the world. Roux has stated, "I do not know the difference between an L Form and a spheroplast."[5] In truth, these two commonly employed terms refer to pleomorphic variants of bacteria or fungi with altered cell walls, which may require hypertonicity, may be highly fastidious in growth requirements, and have atypical colonies. They may or may not revert to the classic stage. Except when produced artificially *in vitro* they rarely make Mycoplasma-like colonies.

The terms "cell wall deficient" and "cell wall defective" have the same connotation and are usually used as synonyms with L Form and spheroplast. "Cell wall deficient" was devised by Dienes as an all inclusive term when it appeared that "L Form" might be restricted in meaning. "Cell wall divergent" is more appropriate because the walls are not always thin; they usually contain substrate adsorbing lysozyme and react with antibody vs. the classic stage. Many times new or augmented lipids result in very slow adsorption of stains. Sometimes the altered wall permits longevity as in the Neisseria and Mycobacteria.

"CWD" is an exceedingly useful term widely present in the literature and is the all-inclusive designate employed in this book. However, since CWD is a clumsy term, in the future it may well be replaced entirely by the term "L Form", which became prominent in the English literature.

L-phase variant — This is a term used when one wishes to imply that the organism is capable of reverting to the classic stage. Thus, there is no uniformity of terminology for cell wall-altered microbes, with lack of agreement even among authorities. The terms used throughout this book are those of the original investigator or otherwise are the commonly used equivalent names.

Protoplasts — The definition for *protoplasts* is one where happy harmony prevails. Protoplasts are derived from bacterial or fungal cells but have completely lost the cell wall. All examples which have had complete loss of wall confirmed by electron microscopy (EM) are round. They are often larger than the original cell. As these were originally induced, it was thought they could only complete a division already initiated. There is now much evidence that protoplasts of some genera revert and can propagate in serial passage.

Spheroplasts — The term is logical in that all wall deficient variants can grow as spheres, varying in size from submicroscopic to giants, 40 to 50 μm. The term is illogical in that under certain conditions most, if not all, variants can grow as Actinomyces-like filaments, and as syncytia.

The small spheres in wall-deficient variants can often be distinguished from classic cocci under phase microscopy. Presumably due to the dense wall in the parent, they appear evenly opaque or tinted. The small spheres of the variant, in contrast, have a sharp, thin, black outline which often appears devoid of content.

Transitionals — If bacteria were politicians, these would belong to the conservative party. They have not been moved by any powerful force to become (stable) L Forms with Mycoplasma-like colonies, yet they have not put on the characteristic armor of the wall-shielded organism. To call them transitional is optimistic since, whether made in the test tube or in a vertebrate host, they frequently resist all efforts to cause their reversion. Sometimes they give a clue to their parentage, as in certain of the beautiful color micrographs of Nativelle and Deparis.[4] (See Frontispiece.) More often they deceive and are Actinomyces-like, yeastlike, or resemble either minute or overgrown cocci. Many times the terms "heteromorphic", "bizarre", and "atypical" have been applied. In our experience, CWD growth from bacteremia, meningitis, and urinary tract infection is usually of this morphology. The colonies will differ from one laboratory to another, varying with the medium and whether an antibiotic is incorporated.

L-bodies — These are the spheres which often appear empty until stained or examined in EM section. The L-bodies within a colony constitute the "vesicles" or "vacuoles" at the periphery of growth. When EM sections reveal that they actually are free of cytoplasm, they are termed "ghosts". In contrast, viable L-bodies are especially active metabolism centers. Reversion often occurs within them. In Candida, reversion may occur in the thick wall of the L-body. In the appropriate species, cytochrome oxidase and niacin may be detectable only in these globes.

Filtrable forms — Since many bacteria in the classic form pass through 0.45-μm-pore filters, the term "filtrable microbes" should be reserved for variants which pass through a porosity of 0.25 μm or less. Most CWD forms include filtrable, viable units, but this is not invariable, depending on age of culture and nutrients supplied.

Pleomorphs — For common parlance the descriptive noun pleomorphs is sometimes used to refer to heteromorphic cell wall divergent bacteria and fungi.

Induction — This is the transformation of a classic bacterium to a CWD form through *in vitro* manipulation.

Reversion — This occurs when a CWD variant returns to its typical morphology. Sometimes morphological reversion occurs, but the organism will not grow in macroscopic colonies or on routine media.

Gymnoplasts — These are fungal or bacterial cells living without encasement by any cell wall. This term is considered of interest by many because in a broad biological sense, although not in the terminology of CWD investigators, the designate "protoplast" can refer to the cell cytoplasm plus surrounding membrane, regardless of whether the wall is also present. Use of the word "gymnoplast" would avoid all possible confusion in this respect.

L-dependents — These are commonly seen in clinical variants or those induced *in vitro*. We use the term to describe bacteria and Candida cells which look entirely typical in the Gram stain but which grow only on the periphery of a CWD microcolony. They can be cultured only on media which supports the variant. Sometimes they represent a stage in true reversion of the organism.

Physiological cripples — These are not uncommon in examining clinical isolates. *Pseudomonas aeruginosa* in the CWD state, largely atypically pleomorphic, may include rods with typical morphology and motility, as viewed under dark-field illumination. The same phenomenon has been noted for several other species. No colonies develop in methods used for routine subculture. These have been repeatedly described in guinea pigs inoculated with CWD stages of *Mycobacterium tuberculosis* as typical acid-fast rods viewed microscopically but unable to grow with standard methods.

Mureinoplasts — These are variants which retain their murein layer but which have lost their typical lipoprotein and lipopolysaccharide layers.

Osmoplasts. — Another situation in which a rod retains its morphology despite a damaged wall concerns *P. aeruginosa*. The cell may become osmotically fragile without assuming a spherical shape. Such osmotically sensitive organisms which have retained the rod form have been termed "osmoplasts" by Asbell and Eagon.[1] A similar phenomenon of fragility without loss of structural rigidity has been noted in *Escherichia coli* and *P. aeruginosa* by Voss.[6]

Bdelloplasts, — These differ from other CWD forms in possessing most of their cell walls and being osmotically insensitive. They are CWD variants formed by penetration of the parasitic bacteria Bdellovibrios.

Cytoplasts — These are protoplasts prepared in a special way. Cytoplasts, usually made from algae, are obtained by pressing the cytoplasm with membrane out from cells whose surface has been cut. Cytoplasts are viable for days and are useful in fusion studies.

REFERENCES

1. **Asbell, M. S. and Eagon, R. G.,** The role of multivalent cations in the organization and structure of bacterial cell walls, *Biochem. Biophys. Res. Commun.*, 22, 664–671, 1966.
2. **Charache, P.,** Atypical bacterial forms in human disease, in *Microbial Protoplasts, Spheroplasts, and L Forms*, Guze, L. B., Ed., Williams & Wilkins, Baltimore, 1968, 484–494.
3. **Klieneberger-Nobel, E.,** Origin, development and significance of L forms in bacterial cultures, *J. Gen. Microbiol.*, 3, 434–442, 1949.
4. **Nativelle, R. and Deparis, M.,** Formes évolutives des bactéries dans les hémocultures, *Presse Med.*, 68, 571–574, 1960.
5. **Roux, J.,** In vivo study of L forms, spheroplasts, protoplasts, interaction with cells and tissues, in *Spheroplasts, Protoplasts and L forms of Bacteria*, INSERM, 65, 235–246, 1976.
6. **Voss, J. G.,** Lysozyme lysis of gram-negative bacteria without production of spheroplasts, *J. Gen. Microbiol.*, 35, 313–317, 1964.

Chapter 3

COMPARING MYCOPLASMA AND CWD FORMS

The one essential difference between Mycoplasma and CWD forms is that Mycoplasma by definition are unable to make any cell wall components. In contrast, CWD bacteria and fungi have potential for making wall components and usually possess layers varying from vestigial to substantial.

Another characteristic which clearly sets Mycoplasma apart from bacteria in any form is the G + C ratio of the Mycoplasma DNA, the mole percent of guanine plus cytosine/ adenine + thymine + guanine + cytosine. Certain of the Mycoplasma have G + C ratios below those known for any bacterium, i.e., in the range from 23 to 25%. Such species include *Mycoplasma neurolyticum*[21] and *M. mycoides* var. *capri*.[12] In contrast, the G + C base composition of bacteria may reach 70% for some Mycobacteria and does not go lower than 28 to 30% for any bacterial genus.[21] However, in general, Mycoplasma resemble CWD so closely that differentiation is difficult.

The difficulty in distinguishing Mycoplasma from CWD forms is exemplified by data from Pachas and co-investigators[23] who suggest that *Acholeplasma laidlawii* A is the L variant of a Corynebacterium. The supporting evidence is patterns of membrane proteins in gel electrophoresis and a high level of nucleotide sequence homology.

The Mycoplasma are so divergent that separate genera have been established: the Acholeplasma, free living forms not requiring serum in their growth medium; Thermoplasma, which propagate happily at 60°C, thus, probably sharing the water of your morning shower; Ureaplasma, common and sometimes pathogenic in the genital tract of many vertebrates; Anaeroplasma which grow exclusively anaerobically; Spiroplasma, which complicate the raising of corn and other vegetables; Mycoplasma, which include the agent of "virus" pneumonia and many more mammalian pathogens. In this chapter they will all be considered Mycoplasma, although taxonomically Mycoplasmatales.

SIMILARITIES RELATING MYCOPLASMA AND CWD FORMS

FINE STRUCTURE

Comparing the fine structure of CWD forms and Mycoplasma is interminable; some Mycoplasma species have markedly different morphology from others (Figure 1A).[3] The growth in Figure 1B might not be recognized as Mycoplasma.[25] Second, the resultant growth of either class of organism is dependent on culture media, comparable media not always sufficing for both Mycoplasma and CWD variants.

However, a careful study by Dienes and Bullivant[7] has shown that a Mycoplasma strain and several wall-deficient bacterial variants have almost identical fine structure when cultured in solid medium, the growth of each L Form resembling that of Mycoplasma. Irregular branches and nodular growths extended three-dimensionally from basic masses 4 to 6 μm in diameter. In addition, both Mycoplasma and the wall-deficient bacteria form elementary corpuscles, 0.25 μm or less in diameter, produced singly and in chains by budding of the larger masses. The elementary bodies in turn enlarge to become the foundation of a new colony. The tiny elementary bodies, the large burgeoning masses, and enormous L-bodies characterize all the L Forms and the Mycoplasma. There is probably no aspect which uniquely characterizes the morphology of Mycoplasma in contrast to bacterial variants.

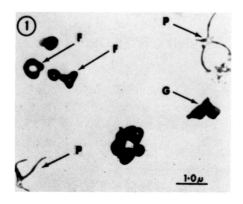
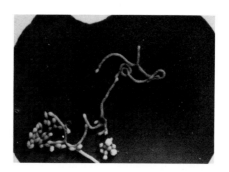

A   B

FIGURE 1.   (A) Electron micrograph of a mixed culture of *Mycoplasma gallisepticum* (G), *M. felis* (F), and *M. pneumoniae* strain 3546 (P). The three different morphologies can be clearly discerned. (Magnification × 12,500). (From Boatman, E. S., *J. Infect. Dis.*, 127 (Suppl.), S12–14, 1973. With permission.) (B) Electron micrograph of *M. mycoides* shows forms different from the spheres considered typical of Mycoplasma. (Magnification × 8600.) (From Rodwell, A. W. and Abbot, A., *J. Gen. Microbiol.*, 25, 201–214, 1961. With permission.)

Stable L Forms, like Mycoplasma, may have no cell wall, revealing only a trilaminar membrane. However, naturally occurring CWD microorganisms commonly have residual lamina of the parent cell wall.

ATMOSPHERE OPTIMUM FOR GROWTH
Another aspect shared by Mycoplasma and CWD forms is their usual preference for anaerobic conditions. The well-known Mycoplasmas of man, with the exception of *M. pneumoniae,* grow better anaerobically. CWD variants of most bacteria also prefer anaerobiosis. This refers to anaerobiosis obtained in the customary anaerobic jars. When all traces of O_2 are removed from medium, CWD of some species, e.g., Mycobacteria, will not grow. Dienes et al. early noted that unless the medium was ideal, L-variants of Salmonella grew only anaerobically.[6] Some aged strains of *Proteus mirabilis* spontaneously grow only anaerobically as CWD forms, and a staphylococcal CWD variant from a brain abscess initially grew only in anaerobic cultures.[20] However, CWD forms of Proteus and other genera under certain conditions have maximum growth aerobically.

SERUM REQUIREMENTS
Mycoplasmatales genera differ in requirements for serum or serum components. As expected, the genera from animal sources are those requiring serum. Likewise, CWD forms from vertebrates may be serum dependent or have growth improved by serum components. However, CWD forms of Proteus and many other nonfastidious genera need no serum and will grow in synthetic-type media.

RIBOSOMES
Because Mycoplasma and wall-free bacteria resemble protozoa in many respects, it is interesting that the ribosomes of Mycoplasma resemble the 70 S ribosomes of bacteria, differing from those of the protozoa.[28] This observation favors the theory that Mycoplasma and bacteria have common ancestry.

HEMADSORPTION

The tendency of Mycoplasma to adsorb leukocytes[1] and erythrocytes[5] will probably be duplicated in an equally variegated way in wall-deficient microbial colonies. Hemadsorption by CWD colonies is common, but the bacterial species involved have not been identified.[33]

SUGAR PHOSPHOTRANSFERASE

There is evidence of a relationship between Mycoplasma and bacteria in their sugar phosphotransferase system, the basic mechanism by which glucose enters the cell. Mycoplasma cytoplasm complements *Escherichia coli* membranes in this function with nearly the same specific activity as *E. coli* cytoplasm.[4]

pH REQUIREMENTS

Both CWD forms and Mycoplasma commonly prefer 7.8 to 8.0 on solid medium. Exceptions: ureaplasma needs pH 6.0. *Thermoplasma acidophila* has an optimum pH between 1.0 and 2.0. In broth with a trace of agar, CWD forms of most pathogens grow most rapidly if the pH is 5.5 to 6.0 in vigorously shaken culture.

BIOCHEMICAL PROPERTIES SHARED

Both can synthesize lysine.[26] Sphingolipid is rare in both Mycoplasma[24] and CWD variants.[15] Both use the Embden-Meyerhof pathway.[32]

FACTORS DISTINGUISHING MYCOPLASMA FROM CWD FORMS

GROWTH IN LIQUID CULTURE

The growth of CWD forms of certain species in liquid is stringy or mucoid, whereas Mycoplasma growth tends to be evenly diffuse. With either type of organism, many units may be present without clouding the medium since a microbial cell wall is the main contributor to turbidity.

COLONIAL APPEARANCE

Mycoplasma are noted for making "fried egg" colonies on agar surfaces. Most CWD forms occurring spontaneously in clinical situations and in nature make colonies different from the parent but not "fried eggs". The CWD colonies are characteristic for the species and sometimes even the strain (see Figures 2 and 3). Exceptions: *M. pneumoniae* and *Ureaplasma urealyticum* have the Mycoplasma-like colonies only after extensive incubation. Colonies of *M. incognitus* are not Mycoplasma-like (see Figure 4A).

SENSITIVITY TO PENICILLIN

Most species of Mycoplasma are not inhibited by 20,000 U/ml of penicillin,[30] whereas CWD forms may be inhibited by 2000 U/ml. Exceptions: *M. neurolyticum* is inhibited by 40 U/ml of penicillin.[11,31] Penicillin-sensitive CWD variants are known.[17] The much-studied penicillin-binding proteins (PBPs) are present on protoplasts of *P. mirabilis* and absent from *M. fermentans, M. arthritidis, M. neurolyticum, M. hominis* and *A. laidlawii*. Intriguing is the finding that PBP 4, of the seven membrane PBPs of *P. mirabilis,* is missing from the protoplasts.[18]

NUCLEIC ACID HOMOLOGY

No nucleic acid mating has been demonstrated between Mycoplasma and bacteria or their variants. DNA homology studies informatively distinguish a Mycoplasma species from

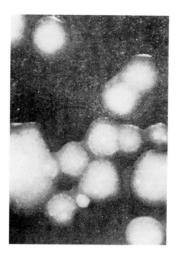

FIGURE 2.   Fourteen-day colonies of *Streptomyces hygroscopicus* L form. (Magnification × 8). (From Gumpert, J., *Z. Allg. Mikrobiol.*, 22(9), 617–627, 1982.)

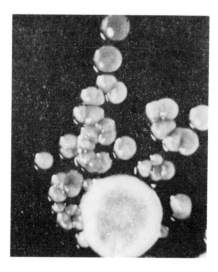

FIGURE 3.   Small 10-d colonies of *Streptomyces griseus* L Form surround single revertant colony. (From Gumpert, J., *Z. Allg. Mikrobiol.*, 22(9), 617–627, 1982.)

a bacterial variant. The genetic relationship of two species can be determined by extracting DNA from one species to serve as a template for formation of radioactive RNA from labeled nucleotide triphosphates.[27] The RNA is then mixed with thermally denatured DNA of the second species. The reaction proceeds at a temperature which allows pairing of complementary DNA and RNA base sequences. The mixture is then filtered through a nitrocellulose membrane which traps the hybrid DNA-RNA pairs. The radioactivity on the filter, therefore, corresponds to the degree of hybridization.[27] The diphtheroid which appeared among colonies of *M. hominis* type 1 has been shown to be merely an uninvited guest.[27] Similarly, no hybridization of nucleic acids was found between *M. pneumoniae* and *Streptococcus "MG"*, showing that their antigenic resemblance is coincidental. *M. gallisepticum* might be accompanied by the L phase of *Hemophilus gallinarum* but is not related by genome.[21] Thus, to

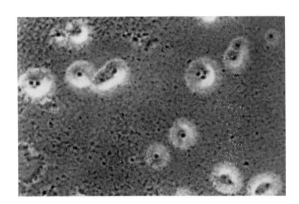
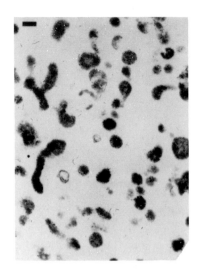

A   B

FIGURE 4. (A) Colonies of *Mycoplasma incognitus*. (B) Electron micrograph of *M. incognitus* growth. (From Lo, S. C., Shih, J. W., Newton, P. B. III, Wong, D. M., Hayes, M. M., Benish, J. R., Wear D. J., and Wang, R. Y., *Am. J. Trop. Med. Hyg.*, 41(5), 586–600, 1989. With permission.)

date, nucleic acids of Mycoplasma and bacteria have acted as distinct entities, although antisera against proteins may show cross-reactivity, as demonstrated when *M. pneumoniae* infection stimulates immunoglobulins which agglutinate *Streptococcus MG*.

BIOCHEMICAL ACTIVITIES

Zehender concluded that basic differences in metabolic pathways separated Mycoplasma from bacteria and their wall-deficient variants.[32] A marked difference is exhibited by the Mycoplasma in the citric acid cycle, being insusceptible to arsenate. Arsenate resistance indicates that phosphorylation is not used by the Mycoplasma. Furthermore, the Mycoplasma demonstrate independence of the cytochrome systems, as shown by resistance to cyanide. This is supported by their resistance to dinitrophenol.

The lysis of agar is a peculiar property of certain species in the CWD stage noted for *Staphylococcus aureus*[19] and *Clostridia welchii* type D.[14] Agar hydrolysis has not been found in the Mycoplasma.

Mycoplasma do not produce indole, do not reduce nitrate,[13] and rarely produce H_2S.[10] Mycoplasma do not use phosphorylation in the citric acid cycle. Mycoplasma produce DNAase more commonly than do CWD forms.[22] Mycoplasma synthesize chiefly mono-glycosyl diglycerides; in L Forms diglycosyl diglycerides predominate.

Metabolically, the Mycoplasma are less versatile than are microbial wall-deficient bacteria. The wall-deficient variant tends to mimic its parent classic bacterium. In contrast, Mycoplasma catabolize fewer substrates.

BACTERIOCIN SUSCEPTIBILITY

An unusual bacteriocin of *Streptococcus faecalis* inhibits the L Forms of most bacterial species, but not 33 strains of Mycoplasma.[9] This study clearly demonstrates a basic difference between the Mycoplasmal and most bacterial membranes.

FIGURE 5. Rat, 25 d old, with bilateral cataracts following intracerebral inoculation as newborn (24 h) with *Spiroplasma mirum* passaged 26 × on artificial medium. (From Tully, J. G., Whitcomb, R. F., Clark, H. F., and Williamson, D. L., *Science*, 195, 892–894, 1977. With permission.)

WALL-ASSOCIATED STRUCTURES

Structures which are part of the cell wall or attach to the wall are naturally missing in the Mycoplasma. There is no evidence that antigens common to many bacteria such as the Rantz antigen or the CA antigen are synthesized by Mycoplasma. Thus, absent are flagella and their precursor flagellin, pili, and capsules.

PATHOGENICITY

It would be hard to identify differences in pathogenicity between Mycoplasma and CWD microbes. In vertebrates, almost any organ may be infected by Mycoplasma or CWD microorganisms. For example, both can cause fatal pulmonary infections. The most recently identified Mycoplasma with this potential is *M. incognitus,* responsible for lethality in previously vigorous young adults (Figure 4 A and B)[16.] See the chapter dealing with lung disease for CWD pneumococci in a kidney transplant patient.

CWD bacteria and CWD fungi can now be recognized as the previously noncultivatable agents of uveitis (see Chapter dealing with ocular disease). Correspondingly, ocular disease of rodents is caused by *Spiroplasma mirum* (Figure 5),[29] and keratoconjunctivitis (pinkeye) of sheep and goats can be caused by *M. conjunctivae.*[2]

SUMMARY

Both mycoplasma and CWD variants cause a wide spectrum of diseases in animals and plants. Both classes of organisms usually prefer microaerophilic conditions and are serophilic. There are more resemblances than differences in fine structure. The "fried egg" colony typical of most mycoplasmatales is made by some *in vitro*-induced CWD strains. Ribosomes suggest a common ancestry. Some enzymes are produced by both.

Differences are numerous. Penicillin tolerance in mycoplasma exceeds that of CWD bacteria as does resistance to bacteriocins. Most CWD forms do not make Mycoplasma-like colonies. DNA homology studies show a marked dissimilarity.

REFERENCES

1. **Addis, S. and Meloni, G. A.,** Adsorbimento dei leucociti umani alle colonie di *Mycoplasma hominis* tipo 1, *Ann. Sclavo,* 11, 84–90, 1969.
2. **Barile, M. F., Del Giudice, R. A., and Tully, J. G.,** Isolation and characterization of *Mycoplasma conjunctivae* sp. n. from sheep and goats with keratoconjunctivitis, *Infect. Immun.,* 5, 70–76, 1972.
3. **Boatman, E. S.,** Morphologic heterogeneity of the mycoplasmatales, *J. Infect. Dis.,* 127 (Suppl.), S12–14, 1973.
4. **Cirillo, V. P.,** Complementation between membrane and cytoplasmic components of the PEP-dependent sugar phosphotransferase system (PTS) of Mycoplasmas and *E. coli, Am. Soc. Microbiol. Abstr. Annu. Meet. 1973,* p. 162, 1973.
5. **Del Giudice, R. A. and Pavia, R.,** Hemadsorption by *Mycoplasma pneumoniae* and its inhibition with sera from patients with primary atypical pneumonia, *Bacteriol. Proc. 1964,* p. 71, 1964.
6. **Dienes, L., Weinberger, H. J., and Madoff, S.,** The transformation of typhoid bacilli into L forms under various conditions, *J. Bacteriol.,* 59, 755–764, 1950.
7. **Dienes, L. and Bullivant, S.,** Morphology and reproductive processes of the L forms of bacteria. II. Comparative study of L forms and Mycoplasma with the electron microscope, *J. Bacteriol.,* 95, 672–687, 1968.
8. **Gumpert, J.,** Growth characteristics and ultrastructure of protoplast type L forms from streptomycetes, *Z. Allg. Mikrobiol.,* 22(9), 617–627, 1982.
9. **Harwick, H. J., Montgomerie, J. Z., Kalmanson, G. M., Hubert, E. G., Potter, C. S., and Guze, L. B.,** Differential action of a streptococcal bacteriocin on Mycoplasmas and microbial L forms, *Infect. Immun.,* 4, 194–198, 1971.
10. **Hayflick, L.,** Fundamental biology of the class Mollicutes, order Mycoplasmatales, in *The Mycoplasmatales and the L Phase of Bacteria,* Hayflick, L., Ed., Appleton-Century-Crofts, New York, 1969, 16–47.
11. **Hottle, G. A. and Wright, D. N.,** Growth and survival of *Mycoplasma neurolyticum* in liquid media, *J. Bacteriol.,* 91, 1834–1839, 1966.
12. **Jones, A. S. and Walker, R. T.,** Isolation and analysis of the deoxyribonucleic acid of *Mycoplasma mycoides,* var. *capri, Nature (London),* 198, 588–589, 1963.
13. **Kandler, G. and Kandler, O.,** Ernahrungs und stoffwechselphysiologische untersuchungen an pleuropneumonie-ähnlichen organismen und der L phase der bakterien, *Zentralbl. Bakteriol. Parasitenkd. Infektionskr. Hyg. Abt. 1 Orig.,* 108, 383–397, 1953.
14. **Kawatomari, T.,** Studies of the L forms of *Clostridium perfringens, J. Bacteriol.,*76, 227–232, 1958.
15. **Labach, J. P. and White, D. C.,** Identification of ceramide phosphorylethanolamine and ceramide phosphorylglycerol in the lipids of an anaerobic bacterium, *J. Lipid Res.,* 10, 528–534, 1969.
16. **Lo, S. C., Shih, J. W., Newton, P. B. III, Wong, D. M., Hayes, M. M., Benish, J. R., Wear, D. J., and Wang, R. Y.,** Virus-like infectious agent (VLIA) is a novel pathogenic mycoplasma, *Mycoplasma incognitus, Am. J. Trop. Med. Hyg.,* 41(5), 586–600, 1989.
17. **Martin, H. H.,** Biochemistry of bacterial cell walls, *Annu. Rev. Biochem.,* 35, 457–484, 1966.
18. **Martin, H. H., Schilf, W., and Schiefer, H.,** Differentiation of Mycoplasmatales from bacterial protoplast L forms by assay for penicillin binding proteins, *Arch. Microbiol.,* 127, 297–299, 1980.
19. **Mattman, L. H., Tunstall, L. H., and Rossmoore, H. W.,** Induction and characteristics of staphylococcal L forms, *Can. J. Microbiol.,*7, 705–713, 1961.
20. **Mattman, L. H.,** L forms isolated from infections, in *Microbial Protoplasts, Spheroplasts, and L Forms,* Guze, L. B., Ed., Williams & Wilkins, Baltimore, 1968, 472–483.
21. **McGee, Z. A., Rogul, M., and Wittler, R. G.,** Molecular genetic studies of relationships among Mycoplasma, L forms and bacteria, *Ann. N.Y. Acad. Sci.,* 143, 21–30, 1967.
22. **Neimark, H. C.,** Deoxyribonuclease activity from "lactic acid pleuropneumonia-like organisms", *Nature (London),* 203, 549–550, 1964.
23. **Pachas, W. N., Schor, M., and Aulakh, G. S.,** Evidence for the bacterial origin of *Acholeplasma laidlawii A, Diag. Microbiol. Infect. Dis.,* 3(4), 295–309, 1985.
24. **Plackett, P. and Smith, P. F.,** in *The Biology of Mycoplasmas,* Smith, P. F., Ed., Academic Press, New York, 1971, 31.
25. **Rodwell, A. W. and Abbot, A.,** The function of glycerol, cholesterol and long-chain fatty acids in the nutrition of *Mycoplasma mycoides, J. Gen. Microbiol.,* 25, 201–214, 1961.
26. **Smith, P. F.,** Comparative biosynthesis of ornithine and lysine by Mycoplasma and L forms, *J. Bacteriol.,* 92, 164–169, 1966.
27. **Somerson, N. L., Reich, P. R., Chanock, R. M., and Weissman, S. M.,** Genetic differentiation by nucleic acid homology. II. Relationships among Mycoplasma, L forms, and bacteria, *Ann. N. Y. Acad. Sci.,* 143, 9–20, 1967.

28. **Tourtellotte, M. E., Pollack, M. E., and Nalewaik, R. P.,** Protein synthesis in Mycoplasma, *Ann. N.Y. Acad. Sci.,* 143, 130–138, 1967.

29. **Tully, J. G., Whitcomb, R. F., Clark, H. F., and Williamson, D. L.,** Pathogenic Mycoplasmas, cultivation and vertebrate pathogenicity of a new spiroplasma, *Science,* 195, 892–894, 1977.

30. **Ward, J. R., Maddoff, S., and Dienes, L.,** *In vitro* sensitivity of some bacteria, their L forms and pleuropneumonia-like organisms to antibiotics, *Proc. Soc. Exp. Biol. Med.,* 97, 132–135, 1958.

31. **Wright, D. N.,** Nature of penicillin-induced growth inhibition of *Mycoplasma neurolyticum, J. Bacteriol.,* 93, 185–190, 1967.

32. **Zehender, C.,** Vergleichende weitere Untersuchungen über die Wirkung von Atmungsgiften auf pleuropneumonia-ähnliche Organismen, *Proteus vulgaris* und dessen stabile L-phase, *Zentralbl. Bakteriol. Parasitenkd. Infektionskr. Hyg. Abt. 2 Orig.,* 109, 22–23, 1956.

33. **Merline, J.,** personal communication.

Chapter 4

PROPERTIES & PECULIARITIES

MORPHOLOGY

THE L-CYCLE

A bacterium remembering that it is a Protistos, sharing characteristics with both protozoa and fungi, is diagrammed in Figure 1. Not all bacteria, when induced to alter their growth habits, gyrate through a complete sequence of changes and return to the classic form. Rather, the organism often stops at one variation level, or sometimes exhibits unique modifications. Although each investigator gains impressions from his own experiences, the diagram depicts stages frequently encountered.

Stages of the L-cycle have been followed carefully by many. The large vacuolated bodies, now known to be characteristic of L-colonies, were noted by Dienes and Edsall in *Streptobacillus moniliformis* from human infection.[20] It is to be noted that throughout decades, observations concerned naturally occurring variants; there were no antibiotics.

The bacterium most commonly studied in the L-cycle, *Proteus mirabilis*, is a favorite because isolates easily undergo variation.[25,95] Nelles, describing the cycles in *P. mirabilis* and Salmonella, noted atypical development originating in the smallest granules. These expanded and formed fine mycelia. Large bodies (L-bodies) formed under certain conditions.[70]

Rubio-Huertos et al. compared the L-cycle of Proteus with two species previously not examined, *Pseudomonas fluorescens* and *Rhizobium lupinus*, using both light and electron microscopy. The inducing agents, glycine and penicillin, respectively, caused all three organisms to expand into L-bodies by 3 h. Rhizobium differed only in making uniquely small L Form colonies.[83]

Schellenberg made a continuous observation of *Proteus vulgaris* L Forms with phase-contrast microscopy. He found that some large bodies propagated by division; others generated and released refractile granules which gave rise to L-type microcolonies. On penicillin-free medium, the parent bacteria were again produced within the large body as it disintegrated.[86,87] The L-bodies have been studied extensively[89] in many species[40] because of the endless variety of changes which occur within them and because of their unbelievable size. The role of L-bodies in the life of *Bacillus anthracis* was sketched by Haag (Figure 2).[39] L-bodies made under natural conditions are predominantly viable; e.g., the globules formed in response to antagonistic strains of Proteus isolates usually develop into bacterial colonies.[24]

L-bodies are seen frequently in clinical cultures of *Escherichia coli*, *Hemophilus influenzae,* and *H. parainfluenzae*, and in Streptococci. Gonococci are almost unique in spontaneously transforming into large bodies on continued incubation. All strains of *S. moniliformis* and *Bacteroides funduliformis* produce abundant large bodies.[22] Ørskov increased the yield of Proteus L-bodies by incorporating 2.0% caffeine in the culture.[74]

The rapidity with which L-bodies develop by expansion of small granules, or in some instances by fusion of several granules, is remarkable. In Group A Streptococci, Agarwal and Ganguly noticed large bodies as early as 1.5 h after subculture from liquid to fresh solid medium. Diameters varied from 7 to 10 mm. Their peripheries, stained by Giemsa's method, varied in breadth in different L-bodies. The spheres' interiors did not take the Giemsa stain.[1]

Similarly, when a minimal inoculum of *Staphylococcus aureus* is placed in a "complete

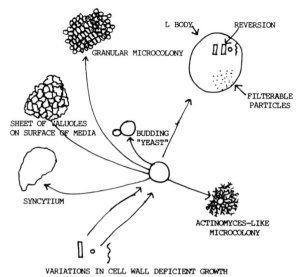

VARIATIONS IN CELL WALL DEFICIENT GROWTH

FIGURE 1. Variations in cell wall deficient growth.

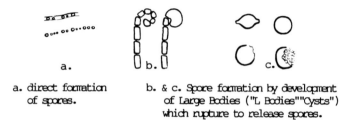

a. direct formation of spores.

b. & c. Spore formation by development of Large Bodies ("L Bodies""Cysts") which rupture to release spores.

FIGURE 2. Bacillus anthracis develops spores by two mechanisms. (From Hagg, E., *Archiv. Hyg.*, 98, 296, 1927.)

medium,'' abundant giant L-bodies appear by 8 h, constituting the only growth[61] (Figure 3). Propagation exclusively as these giant spheres occurs when the inoculated broth contains from 1 to 30,000 cocci/ml, varying with the strain of *S. aureus*. With continued incubation the giant spheres are replaced by Staphylococci.

The mechanism for L-body production varies with genera. The Morax-Axenfeld bacillus may be unique. The rods lengthen, then twist repeatedly at closely spaced intervals. The portions demarcated by twisting of the strand then swell to become spheres.[54] Thus, the fascinating L-bodies, so often the site of reversion and accelerated metabolism, may form by fusion, by ballooning of granules from segments of a twisted bacillus, or by evagination of a spiral organism as described in the chapter on Spirochaeta. Bibel and Lawson described a method for production and harvesting numerous streptococcal L-bodies.[8]

THE NUCLEOID

Many investigators have been intrigued with the behavior of the cell nucleus throughout the L-cycle. There is no single pattern for distribution of nuclear material; rather, all possible variations occur within the L-body, where it has most often been studied. The activity of the nucleus is dependent on the genus of the organism and the culture conditions. In early

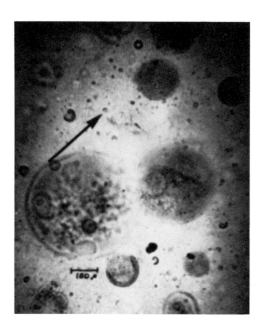

FIGURE 3.   In liquid synthetic medium at 8 h, L bodies produced earlier lyse and release classic staphylococci (arrow). (Mattman, L. H., Tunstall, L. H., and Rossmoore, H. W., *Can. J. Microbiol.,* 7, 705–713, 1961. With permission.)

growth, without cytoplasmic division, the nucleus may divide at a normal rate in Staphylococci, *E. coli, Bacillus anthracis,* Salmonella, Proteus, and Pasteurella.[94] This yields an excess of nucleic acids. In high salt media, much DNA is discarded from the cells. There may be increased degradation of RNA.

One hour after Proteus is placed in penicillin medium, the nuclear material may gather into one or two chromatin granules which gradually proliferate to form a small peripheral wreath. From this point on, a variety of possibilities exist. By 5 h the nuclear substance may appear as many small round grains; in other cells the chromatin is in rough large particles or a mixture of small and large masses; in some cells, the chromatin assumes a knit-like structure.[86]

The abundance and wide dispersion of DNA, also noted by Fitz-James,[29] must be necessary when multiple classic organisms are synthesized within one L-body. Fifty bacteria can be made within one sac, as noted in the Mycobacteria[60] and other genera.

As Hirokawa cultured *E. coli* in penicillin broth, the DNA content per cell increased more than four times in 3 h. Coincidentally, the number of bacilli reverting within a large body corresponded to the number of nuclear masses.[42]

VARIATION IN DEVELOPMENT OF UNITS

A bacterium does not always enlarge into a spherical L-body; rods, especially, may develop outcroppings resulting in so-called "rabbit ears". These bulges may extend to become mycelia.[41] Concerning lateral outgrowths, de Terra and Tatum postulate that a muralytic enzyme in Neurospora weakens wall areas, resulting in cytoplasmic flow outward and initiating a branch.[15] Thus, the rhizoid extensions in CWD forms may result from haphazard distribution of wall-eroding enzymes.

Scanning microscopy reveals that when Staphylococci are grown in methicillin, roughening of the surface corresponds with failure to lay down new wall as normal autolysis occurs.[28] The wall-deficient growth is not identical for all genera. White found that the penicillin-induced variants of *Vibrio cholerae* were not only large spheres and branching forms, but also extraordinary "starfish", some of which retained motility. Staining disclosed double or multiple chains of nuclear material in the broad part of the branches.[96]

SYNCYTIA

Of all the structures in the L-cycle, the syncytia are the most incredible. A syncytium is the structureless layer which in the older literature was designated "symplasm", consisting of numerous nuclei embedded in a sheet of cytoplasm. Naturally occurring and induced L-phase syncytia are commonly encountered. These sheets of growth, in which classic bacteria sometimes revert, leave the investigator desiring a more definitive way to identify them as microbial. Specific identification may be possible by use of specific enzymes or fluorescent antibody.

HALOS

It does not seem strange that syncytial growth occurs spontaneously from wall-deficient variants when it is noted that such webs of protoplasm are produced by *E. coli* merely by propagation at an acid pH or by exposure to sublethal concentrations of zephiran.[65]

Jones followed steps in synthesis of syncytia formed from coalescing aggregates of the large-celled Azotobacter genus.[49] As the cell walls disintegrated, the cytoplasm coalesced. The regenerative granules within this film varied from minute bodies, scarcely discernible microscopically, to larger forms readily visible when stained with Neisser's blue. The granules emerging from the symplasm grew into young Azotobacter cells, reproducing by fission.

Dienes and Edsall noted that various bacteria, especially Bacilli, tend to be surrounded by these amorphous halos.[18,19] Although visible with phase microscopy, halos are often unrecognized since they resist staining, even by crystal violet. Only silver-precipitation, or similar stains, reveals the multiple units comprising the halo.

SUBSTRATE AND MORPHOLOGY — ORDER FROM CHAOS

In the dawn of bacteriology, Cohn assigned bacteria to the class Chaos.[14] His designation still fits the L-cycle. Order will gradually appear when factors responsible for morphogenesis are defined. Data on morphogenetic factors of classic microorganisms are given below. Similar influences may be assumed to operate as CWD determinants of morphology.

AMINO ACIDS AND PEPTONES

Arthrobacter develop rods, rather than spheres, if peptone is added to the glucose-mineral salts medium.[55] In contrast, tryptone will maintain Geodermatophilus in coccoid form.[47]

The triangle form of the of the yeast *Trigonopsis variabilis* is easily maintained with such amino acids as methionine, proline, hydroxyproline, or alanine.[85] With other amino acids, ellipsoidal rather than triangles predominate, and with ammonium sulfate as the N_2 source only 11% of the cells are triangular.

The morphology of the dimorphic fungus, *Histoplasma capsulatum*, depends on both temperature and nutrition. The yeast phase requires cystine,[84] as well as biotin,[84] zinc,[76] and citrate.[76]

CARBOHYDRATES INFLUENCE MORPHOLOGY

Nickerson and Mankowski[73] found that Candida develops only budding yeast cells when glucose is the sole carbon source. With a less readily utilized carbon source, such as soluble starch, glycogen, or dextrin, there is extensive filamentation.

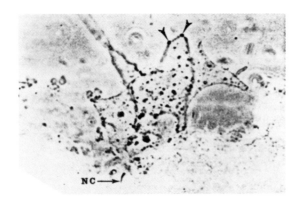

FIGURE 4. Spreading film rather than classic bacteria or colonies characterizes growth from some canine actinomycosis. (From Buchanan, A. M., *Vet. Microbiol.*,7, 587–605, 1982.)

In anaerobic culture, *Mucor rouxii* grows in shapes covering its entire dimorphic spectrum when the hexose content of the medium is manipulated. Thus, with 0.01% glucose, the hyphae are long and narrow; with higher sugar concentrations they are shorter and wider; with 8.0% glucose almost all the cells grow as isodiametric, spherical budding yeasts.[6]

IONS

The presence of simple ions can determine the shape of an organism. For some fungi, mycelia form only when Zn, Cu, and Fe cations are all present.[64] Excess zinc can prevent fungal sporulation.[77,81] *Aspergillus parasiticus* may appear hyphal or yeastlike, depending on the presence or absence of Mn ions.[16] When less than 10^3 mM of MnCl$_2$ is present, yeastlike forms replace the hyphae. The little-known round yeastlike forms of this Aspergillus are exceedingly interesting in their proclivity to produce aflatoxin. Several mutants of *Bacillus subtilis* and *B. licheniformis* grow as cocci except in high-salt medium.[82]

COLONIAL MORPHOLOGY

GROWTH AT VARYING DEPTHS IN AGAR

Razin and Oliver[80] sought to explain why L-phase colonies grow *into* the agar. They concluded that agar penetration, which occurs within a few hours after inoculation, is caused by capillary forces which draw the minute plastic organisms into the dried agar gel, together with the water surrounding them. Penetration does not occur when the agar surface is highly moist.

The tendency of CWD bacteria to slip rapidly between the fibrils of agar may explain why clinical isolates usually grow exclusively slightly subsurface. As noted with *in vitro*-made variants of Mycobacteria, the subterranean tunneling of the organisms may result in elevations which only occasionally break to the surface. Therefore, much information can be gained from sectioning subsurface colonies or from compressing small squares of the subsurface growth under a coverslip. For some unexplained reason, the subsurface tunneling often results in rectangular rather than round colonies. An unusual membranous CWD colony type, which seemed to overgrow classic colonies, developed in an Actinomyces culture.[12] (See Figure 4.)

PECULIARITIES IN *CORYNEBACTERIUM DIPHTHERIAE* COLONIES

L-colonies of *C. diphtheriae* may have unique structural peculiarities, including a predominance of extremely long cells giving the colonies a cross-banded appearance. Such

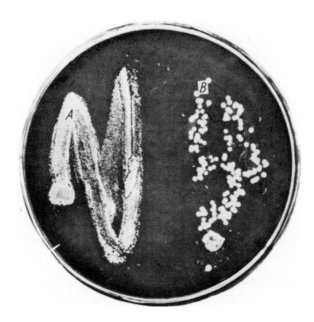

FIGURE 5. Penicillin-induced L forms of *Proteus mirabilis* have two colony types. (From Gumpert, J. and Taubeneck, U., *Z. Allg. Mikrobiol.*, 15(6), 399–410, 1975.)

colonies also have spiral and large pear-shaped bodies with a double membrane. Small granular forms show active motility.[50]

VARIATION IN CWD COLONIES, INCLUDING ROUGH AND SMOOTH

Induced under the same conditions, a strain may show more than one colony type.[38] Figure 5 shows two different colonies, neither resembling Mycoplasma. Little mention has been made of a common CWD colony type, termed "rough". Variants of many species, antibiotic induced or spontaneously occurring, yield some rough colonies with uneven surfaces and very irregular outlines.[68]

It cannot be too strongly emphasized that the smooth, the rough, and other colonies of twisted granular rods make up most CWD colonies from clinical material. Only rarely are vacuolated Mycoplasma-like colonies encountered.

PIGMENTS

WATER-SOLUBLE PIGMENTS

The effect of wall deficiency on pigmentation includes every possible change. Why would diminution of the cell wall almost invariably cause disappearance of a water-soluble diffusible pigment?

Many reports concern Pseudomonas with only nonpigmented wall-deficient variants found. The antibiotic-induced L-phase colonies studied by Edman and Mattman[26] were never pigmented. The 15 L-phase isolates of *Pseudomonas aeruginosa* induced by culture in distilled water by Stanley Susan never showed pigment. Similarly, colorless growth characterized the CWD variants in five cases of *P. aeruginosa* septicemia. Likewise, in a spectrophotometric study, Hubert and associates[45] found no trace of pigment in CWD colonies of *P. aeruginosa*. Some of these variants were stable, while others rapidly reverted. Pigmentation has always returned with reversion, although not always with original intensity.

WATER-INSOLUBLE PIGMENTS

Penicillin-induced L Forms of *Serratia marcescens* have been examined in two laboratories. Bandur and Dienes found no pigment in either colonies with dense centers or another colony type which grossly resembled classic colonies but was comprised of L-bodies.[4] Pigmentation is not lost by diffusion through the agar since the absorption spectrum for prodigiosin was missing in the agar extract.[44] Although the metabolic steps leading to pigment production are quite well known, the actual site of completion has not been found. The final synthetic step is coupling of the mono- and bipyrrole precursors.[67] The authors proposed that the cell wall is a necessary physical site, either for production of one of the precursors or for their union.

The characteristic pigment of *Staphylococcus aureus* is absent as wall deficient variants are commonly cultured. However, in certain media pigment develops rather rapidly.[88] On standard media, colonies of L-bodies may develop bright pigment, but only after incubation at room temperature for periods up to 6 weeks.

Thus, a typical characteristic of CWD variation is pigment loss, whether the pigment is diffusible or nondiffusible. An exception to this was found in *Listeria monocytogenes,* recovered from mice injected with the classic form. The growth consisted of the granular colonies usually encountered from *in vivo* situations. By casual microscopic examination, these were similar to colonies of the parent Listeria, although more granular. Remarkably, the blue-green pigment seen by reflected light, which characterizes *L. monocytogenes* was still present in the CWD stage.[11]

Rarely, variants acquire new pigmentation. An example concerns certain Enterobacteriaceae which may acquire yellow or brown pigments after reversion. A most remarkable change occurred in a carbomycin-dependent Staphylococcus which grew as large budding yeasts, exhibiting a bright pink pigment.[43] The strain may, indeed, have wall-deficient characteristics, suggested by the variation in cell sizes and its propagation by budding. Reverse mutation was demonstrated in which Staphylococcus-type cells were again recovered. Antibiotic-dependent CWD forms are not widely recognized.

RETENTION OF LUMINESCENCE AND FLUORESCENCE

The nonpathogenic *V. cholerae* var. *albensis* is characterized by luminescence. This property is very little affected when the Vibrio are converted into spheroplasts in serum containing 0.5 *M* sucrose.[5] Spheroplasts of *Anabaena variabilis* can contain chlorophyll, as shown by exposure to UV[7] (Figure 6).

INHIBITION BY CLASSIC BACTERIA

Despite the abundant data that some microbes nurture wall-deficient forms or even foster their reversion, there are many other examples where classic bacteria inhibit CWD forms. Boris et al.[10] found that strains of *S. aureus* regularly produce a substance markedly inhibitive to stable staphylococcal and streptococcal L Forms. The interfering thermostable substance diffuses through agar, but not through a dialysis membrane. Kagan and co-workers likewise found that classic Staphylococci can inhibit L Forms.[51]

Marraro noticed that a parent form of *Streptococcus faecalis* inhibited the growth of the CWD phase of the same organism occurring spontaneously in urine culture.[59] This was shown when 0.5 ml of urine from a urethritis case was cultured on serum-sucrose medium suitable for both CWD and classic organisms. The hazy CWD growth of *S. faecalis* was absent in zones around the parent streptococcal colonies. Thus, not only the inhibition, but also coexistence *in vivo* was demonstrated.

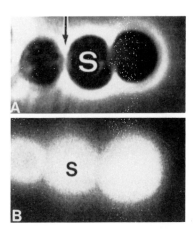

FIGURE 6. (A) Spheroplasts of *A. variabilis* formed by lysozyme; (B) the same spheroplasts examined under UV light show retained chlorophyll. (From Berliner, M. D., Neely-Fisher, D., Rosen, B. H., and Fisher, R. W., *Protoplasma* 139, 36–40, 1987. With permission.)

Fodor and Rogers found that parent *B. licheniformis* inhibited the L phase of the same species by a heat-labile, nondialyzable substance, produced only on agar, not in liquid. Surprisingly, inhibition was reciprocal. The L Forms, made with benzyl penicillin, cycloserine, or methicillin, all suppressed growth of the classic Bacillus.[30]

GENERATION TIME

Wall-deficient colonies are notoriously small. A plate dotted with a thousand colonies may be tossed aside as negative unless examined microscopically. Does this indicate a relative long generation time? Apparently it does. Gumpert and Taubeneck, examining a stable L Form of *Proteus mirabilis* in microculture,[37] found a generation time varying from 60 to 90 min, in contrast to 10 to 15 min for the parent bacterium.

The irregular sprouting, budding, and synthesis of tiny elementary bodies does not guarantee colony development on schedule. Microcolonies of CWD forms are expected in culture by 2 d for most genera, including the Mycobacteria. It must be emphasized that even the appearance of tiny colonies of CWD Mycobacteria at 2 d is phenomenal when contrasted with typical classic colonies appearing at 4 to 6 wks.

Large inocula of certain species give heavy growth by 24 h, such as occurs with penicillin-induced *P. mirabilis*. The period can, however, be exceedingly prolonged as in the 2 to 4 months needed for colonial development of *C. diphtheriae*[50] in certain media. Under some conditions, penicillin-induced L-phase colonies of *Staphylococcus aureus* appear only when 48-h incubation at 37°C is followed by 2-weeks storage at 4°C.

OSMOTIC FRAGILITY

Osmotic sensitivity is often included in the definition of wall deficient bacteria. It would be more accurate to state that the variants *may* require hypertonic environments. In this respect McQuillen may be quoted:[63] "Several workers have observed that spheroplasts of the same species prepared in different ways, e.g., penicillin, DAP-deprivation, glycine, may have very different osmotic sensitivities. Penicillin spheroplasts of *E. coli* require a sucrose concentration of about 10% w/v whereas DAP-deprived and glycine-induced spherical forms are much more robust and some can be washed with water without undergoing lysis. These

FIGURE 7. Protoplast of *B. megaterium* obtained by lysozyme treatment of cells pretreated with Polymyxin B. (From Newton, B. A., *J. Gen. Microbiol.*, 12, 226–236, 1955. With permission.)

differences are not understood." Nor is there an explanation for the cuboidal *B. megaterium* spheroplasts made with lysozyme by Newton[72] (Figure 7).

Studying *P. aeruginosa* and *E. coli,* Azuma and Witter[3] found spheroplast fragility minimal at temperatures optimum for growth. Gilby and Few[33] noted that protoplasts formed in hypotonic media resisted osmotic explosion better than those cultured at higher osmolarities. These examples suggest that the concept of osmotic fragility characterizing all wall-deficient cells be tempered, depending on species and growth milieu. By cultivating *V. comma* in 3.0% glycine, Chatterjee and Williams produced round, motile forms which needed no osmotic stabilizer.[13] Panos and Barkulis found stable L Forms of streptococci did not lyse in a wide range of sucrose concentrations. This indicated that these forms resemble mitochondria and protoplasts in rapidly adjusting to various osmotic pressures without measurable effect on viability and function.[75]

Susceptibility to osmotic changes differs with CWD variants of Gram-positive vs. Gram-negative organisms. Gram-positive cocci, for example, have an internal osmotic pressure varying from 20 to 30 atm,[66] whereas that of Gram-negative rods varies from 4 to 8 atm.[31,46] Also, differences based on the size of the elements of the L Form can be expected. The small elements of *Proteus vulgaris* L Forms resist osmotic changes better than larger elements.[93] Most populations of CWD growth contain particles diverse in size, and, thus, within the same culture, fragile and resistant elements exist.

RESISTANCE OF CWD FORMS TO PHYSICAL FACTORS

Much wall-deficient growth does not withstand pressure. James et al. found that protoplasts of *Streptococcus pyogenes* lyse when centrifuged.[48]

Larger CWD colonies do not resist drying and have a marked tendency to autolyze.[21] Some antibiotic-induced variants of Mycobacteria may lyse within 30 min following removal from an anaerobic environment. Conversely, certain phases of the L-cycle are remarkably more resistant than classic stages. Stewart and Wright[90] found unstable L-phase serogroup B *Neisseria meningitidis* and stable CWD group A β-hemolytic streptococci surviving at 4°C for approximately 6 months. Three year survival at −20°C and better at −80°C was

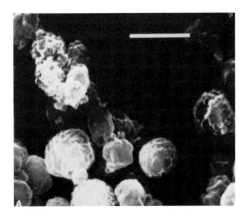
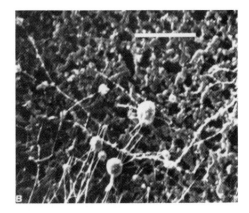

FIGURE 8. Lysolecithin finds substrate molecules in walls of Staphylococci. (A) Before lysolecithin; bar = 2 μm; (B) after lysolecithin treatment; bar = 2 μm. (From Takahashi, T., *Microbiol. Immunol.*, 24(5), 463–468, 1980. With permission.)

found for L Forms of *Staphylococcus aureus, Streptococcus pyogenes,* and *E. coli.*[92] Glycerol and dimethylsulfoxide were equally effective in improving the percentage of survival.

The L Form of *Streptobacillus moniliformis* was relatively resistant to alternate freezing and thawing,[79] much more than were bacterial protoplasts.

There is evidence that certain very tiny granules in the cytoplasm are the storage-resistant elements. Prévot, in examining anaerobes, noticed that these bacteria survived lyophilizing only if richly endowed with such minute hard granules.[78] Such particles may represent elementary bodies.

Spheroplasts of *Euglena* rapidly repair radiation-induced breaks in nuclear DNA if the cells have chloroplasts. Spheroplasts lacking chloroplasts only poorly repair such DNA damage. In this DNA repair, no difference exists between the variants and parents.[71]

EFFECT OF LYTIC AGENTS

Correlated with the susceptibility of wall-deficient forms to hypotonic solution is their susceptibility to lytic agents, usually due to membrane exposure to lipid-dissolving solvents. Razin and Argaman[79] showed that bacterial protoplasts, in many respects, are like the cell-wall-free Mycoplasma in sensitivity to lysis by surface-active agents, primary alcohols, and alkali. Spheroplasts are less sensitive than protoplasts. It has been noticed that lysis increases with the molecular weight of the alcohol used.[34]

Examination of the surface-active agent, sodium lauryl sulfate, revealed that only a fifth as much is required for lysis of spheroplasts of *Salmonella typhi* as for normal bacilli.[17] This susceptibility, along with susceptibility to lysis by pancreatic lipase, was considered to be due to the easy accessibility of the lipoprotein membrane in the variant as opposed to its protected position in the thick-walled classic bacterium.

The phospholipid, lysolecithin, present in various tissues, may be a defensive substance lysing L Forms. Takahashi[91] found that a low concentration of the phospholipid lyses Staphylococcal L growth (Figure 8). Some L-phase variants were 100-fold more susceptible to lysolecithin than were their parent bacteria.[58]

All surface-active agents do not necessarily act by the same mechanism. When protoplasts are exposed to both anionic detergents and cationic detergents, it is obvious that both agents cause disruption of the membrane, but time-sequence studies indicate that the mechanism probably differs.[35]

In some studies, lysis has been purposeful in releasing protoplast components. The nonionic detergent Brij 58 has desirable characteristics in that 30 μg can release 1 mg of protein from *E. coli*. Thus, the resultant solution consists predominantly of spheroplast constituents, only lightly contaminated with the detergent.[9]

ARE ALL WALL-DEFICIENT VARIANTS OF A STRAIN IDENTICAL?

Abundant data confirm that wall deficient variants produced under one set of circumstances do not duplicate variants made by other means. For example, spheroplasts made by EDTA-lysozyme or by penicillin can host bacteriophage φX 174, whereas serum spheroplasts are phage resistant. The morphologies of the penicillin-induced forms and the serum-induced variants also differ.[32] Diena et al.[17] found properties of *Salmonella typhi* spheroplasts varying with glucose content of the medium. Spheroplasts growing in 0.5% glucose were nonmotile and immune to a phage which lysed the motile spheroplasts produced in 1.0 to 2.0% glucose. It is a common observation that wall-deficient colonies induced by one antibiotic will have morphology and reactions completely different from colonies induced by another antibiotic.[36]

L FORMS AS PLASMID RECIPIENTS

A streptokinase shuttle plasmid produced more streptokinase when inserted in *Proteus mirabilis* L form than when given to classic *Streptococcus sanguis* or *S. lactis*. This is of interest since the streptokinase protein is still the most widely used thrombolytic agent.[57]

Another example of protoplasts accepting plasmid DNA concerns protoplast-type *P. mirabilis* accepting information for production of a proteinase indispensable in cheese manufacturing. Protoplasts of *P. mirabilis* strain L99 were interesting for their significant yield of chymosin.[52]

Another L form, *P. mirabilis* LVI, synthesized interferon alpha when transformed with recombinant plasmids.[56]

An unexplained phenomenon concerning plasmids has been noted when *Propionibacterium freudenreichii* changed from a rod to a sphere. The organism acquired several new species of DNA, presumably plasmids. Since the wall of the sphere was wider than in the bacillus, this was an example of cell wall divergence rather than deficiency.[69]

ELECTRICAL PROPERTIES

Study of the electrical properties of *Micrococcus lysodeikticus* protoplasts showed that removal of the bacterial wall decreases the dielectric constant by two orders of magnitude. Likewise, removal of the wall reduces the effective, homogeneous conductivity of the organism in $2 M$ sucrose from 0.045 mho/m to less than 0.001 mmho/m. It was concluded that the cell wall had the dominant effect on the dielectric properties of bacteria.[27]

ROLE OF CWD FORMS IN NUTRITION OF AMOEBA

Spheroplasts of a certain Bacteroides were found superior to the parent bacterium as food supply for *Entamoeba histolytica*.[2] As characteristic for CWD forms, the spheroplasts of the Bacteroides contained only 6% RNA, compared with 13% in the parent bacterium. It was suggested that spheroplast RNA was degraded, perhaps yielding labile nucleotides to serve the amoeba as coenzymes in biosynthesis.

WHY ARE SOME STRAINS HIGHLY SUSCEPTIBLE TO CWD FORM INDUCTION?

Certain bacterial strains easily assume the wall-deficient state. So far, attempts to associate this trait with other characteristics have brought negative results. An isolate of *E. colimutabile* required as much penicillin to produce L Form colonies as standard non-variegating strains of *Escherichia coli*.[53]

Salmonella strains isolated from carriers were compared with isolates from active disease. No statistically valid difference was found between the strains.[62]

SUMMARY

Bacterial life is subject to a series of diverse morphological stages, only one of which is the classic rod, coccus, spirillum, or spirochete. Granular, mycelial, and large body stages, are all capable of reverting to the parent form. Syncytial, "starfish", and "rabbit ear" forms also characterize CWD variation. The shape the variants adopt is largely determined by the presence and concentration of certain amino acids, carbohydrates, ions, and gases in the environment.

The nucleus of an L-phase organism may also assume a variety of appearances, including fine grains, meshworks, and large and small masses of nuclear material. When reverting, cytoplasmic DNA in an L-body condenses to a number of dense units from which a corresponding number of classic organisms will form.

Colonial morphology may be the vacuolated Mycoplasma type, but usually colonies have a variety of other conformations.

Striking changes accompany cell wall deficiency. One difference is the *in vivo* micro-colony formation by most species which makes them immediately detectable in culture. Because of a difference in their walls they do not disperse to appear as single units. While the generation time is known to be prolonged for all CWD forms, the time for colony development in different genera may vary from 1 d to 1 month.

Many CWD forms are delicate, being lysed by drying, centrifuging, a normotensive medium, or even aerobic atmosphere. In contrast, the small spheres of the variants may have exceptional resistance to exposure to 60°C, to prolonged storage, to alternate freezing-thawing, and to exposure to NaOH.

CWD variants induced by one agent are different from those resulting from another agent or situation. For instance, variants resulting from exposure to antibiotic do not duplicate those produced by manipulation of the medium pH.

REFERENCES

1. **Agarwal, S. C. and Ganguly, N. K.,** Reproduction in L phase of group A beta hemolytic streptococci, *Indian J. Med. Res.,* 57, 1387–1390, 1969.
2. **Albach, R. A., Shaffer, J. G., and Watson, R. H.,** Morphology, antigenicity, and nucleic acid content of the Bacteroides sp. used in the culture of *Entamoeba histolytica, J. Bacteriol.,* 90, 1045–1053, 1965.
3. **Azuma, Y. and Witter, L. D.,** Variation of the osmotic fragility of spheroplasts with growth temperature, *Can. J. Microbiol.,* 12, 422–424, 1966.
4. **Bandur, B. M. and Dienes, L.,** L forms isolated from a strain of Serratia, *J. Bacteriol.,* 86, 829–836, 1963.
5. **Barak, M., Ulitzur, S., and Merzbach, D.,** Determination of serum bactericidal activity with the aid of luminous bacteria, *J. Clin. Microbiol.,* 18, 248–253, 1983.

6. **Bartnicki-Garcia, S.,** Control of dimorphism in *Mucor* by hexoses, inhibition of hyphal morphogenesis, *J. Bacteriol.*, 96, 1586–1594, 1968.

7. **Berliner, M. D., Neely-Fisher, D., Rosen, B. H., and Fisher, R. W.,** Spheroplast induction in *Anabaena variabilis* Kütz and *A. azollae* Stras., *Protoplasma*, 139, 36–40, 1987.

8. **Bibel, D. J. and Lawson, J. W.,** Morphology and viability of large bodies of streptococcal L forms, *Infect. Immun.*, 12, 919–930, 1975.

9. **Birdsell, D. C. and Cota-Robles, E. H.,** Lysis of spheroplasts of *E. coli* by a non-ionic detergent, *Biochem. Biophys. Res. Commun.*, 31, 438–446, 1968.

10. **Boris, M., Teubner, D., and Shinefield, H.,** Bacterial interference with L forms, *J. Bacteriol.*, 100, 791–795, 1969.

11. **Brem, A. M.,** The Role of L forms in the Pathogenesis of *Listeria monocytogenes* infections, Ph.D. dissertation, University of Michigan, Ann Arbor, 1968.

12. **Buchanan, A.,** Morphological variants developing in L form cultures of two strains of *Actinomyces* spp. of canine origin, *Vet. Microbiol.*, 7, 587–605, 1982.

13. **Chatterjee, B. R. and Williams, R. P.,** Preparation of spheroplasts from *Vibrio comma*, *J. Bacteriol.*, 85, 838–841, 1963.

14. **Cohn, F.,** Untersuchungen über Bacterien, *Beitr. Biol. Pflanz.*, 1, 127–224, 1872.

15. **de Terra, N. and Tatum, E. L.,** A relationship between cell wall structure and colonial growth in *Neurospora crassa*, *Am. J. Bot.*, 50, 1066–1068, 1961.

16. **Detroy, R. W. and Ciegler, A.,** Induction of yeast like development in *Aspergillus parasiticus*, *J. Gen. Microbiol.*, 65, 259–264, 1971.

17. **Diena, B. B., Wallace, R., and Greenberg, L.,** The production and properties of *Salmonella typhi* spheroplasts, *Can. J. Microbiol.*, 10, 543–549, 1964.

18. **Dienes, L.,** Morphologic elements in the halo of subtilis colonies, *Proc. Soc. Exp. Biol. Med.*, 31, 1211–1214, 1934.

19. **Dienes, L.,** Production of amorphous extra-bacterial substances in bacterial cultures, *J. Infect. Dis.*, 57, 22–25, 1935.

20. **Dienes, L. and Edsall, G.,** Observations on the L organism of Klieneberger, *Proc. Soc. Exp. Biol. Med.*, 36, 740–746, 1937.

21. **Dienes, L.,** Isolation of L type growth from a strain of *Bacteriodes funduliformis*, *Proc. Soc. Exp. Biol. Med.*, 47, 385–387, 1941.

22. **Dienes, L.,** The significance of the large bodies and the development of L type of colonies in bacterial cultures, *J. Bacteriol.*, 44, 37–73, 1942.

23. **Dienes, L.,** Reproductive processes in Proteus cultures, *Proc. Soc. Exp. Biol. Med.*, 63, 265–270, 1946.

24. **Dienes, L.,** Further observations on the reproduction of bacilli from large bodies in Proteus cultures, *Proc. Soc. Exp. Biol. Med.*, 66, 97–98, 1947.

25. **Dienes, L.,** The development of Proteus cultures in the presence of penicillin, *J. Bacteriol.*, 57, 529–546, 1949.

26. **Edman, L. and Mattman, L. H.,** L variation in *Pseudomonas aeruginosa* induced by antibiotics, *Bacteriol. Proc. 1961*, p. 82, 1961.

27. **Einolf, C. W., Jr. and Carstensen, E. L.,** Passive electrical properties of microorganisms. IV. Studies of the protoplasts of *Micrococcus lysodeikticus*, *Biophys. J.*, 9, 634–643, 1969.

28. **Fass, R. J., Carleton, J., Watanakunakorn, C., Klainer, A. S., and Hamburger, M.,** Scanning-beam electron microscopy of cell wall defective staphylococci, *Infect. Immun.*, 2, 504–515, 1970.

29. **Fitz-James, P. C.,** Cytological and chemical studies of the growth of protoplasts of *Bacillus megaterium*, *J. Biophys. Biochem. Cytol.*, 4, 257–265, 1958.

30. **Fodor, M. and Rogers, H. J.,** Antagonism between vegetative cells and L forms of *Bacillus licheniformis* Strain 6346, *Nature (London)*, 211, 658–659, 1966.

31. **Gebicki, J. M. and James, A. M.,** The preparation and properties of spheroplasts of *Aerobacter aerogenes*, *J. Gen. Microbiol.*, 23, 9–26, 1960.

32. **Gemsa, D. and Davis, S. D.,** Failure of bacteriophage ϕX 174 to multiply in serum spheroplasts, *Nature (London)*, 215, 176–177, 1967.

33. **Gilby, A. R. and Few, A. V.,** Osmotic properties of protoplasts of *Micrococcus lysodeikticus*, *J. Gen. Microbiol.*, 20, 321–327, 1959.

34. **Gilby, A. R. and Few, A. V.,** Lysis of protoplasts of *Micrococcus lysodeikticus* by alcohols, *J. Gen. Microbiol.*, 23, 27–33, 1960.

35. **Gilby, A. R. and Few, A. V.,** Lysis of protoplasts of *Micrococcus lysodeikticus* by ionic detergents, *J. Gen. Microbiol.*, 23, 19–26, 1960.

36. **Godzeski, C. W.,** In vivo persistence of L phase bacteria, in *Microbial Protoplasts, Spheroplasts, and L forms*, Guze, L. B., Ed., Williams & Wilkins, Baltimore, 1968, 379–390.

37. **Gumpert, J. and Taubeneck, U.,** Beobachtungen über die Vermehrung der stabilen L form von *Proteus mirabilis* in flüssigen Nährmedien und die Funktion der Zellwand bei der Zellteiling, *Z. Allg. Mikrobiol.,* 6, 211–218, 1966.

38. **Gumpert, J. and Taubeneck, U.,** Characterization of a stable spheroplast type L form of *Proteus mirabilis* D 52 as cell envelope mutant, *Z. Allg. Mikrobiol.,* 15(6), 399–410, 1975.

39. **Haag, E.,** Der Milzbrandbazillus, seine Kreislaufformen und Varietaten, *Archiv. Hyg.,* 98, 296, 1927.

40. **Hadley, P.,** Microbic dissociation, the instability of bacterial species with special reference to active dissociation and transmissible autolysis, *J. Infect. Dis.,* 40, 1–312, 1927.

41. **Hahn, R. E. and Ciak, J.,** Penicillin-induced lysis of *E. coli, Science,* 125, 119–120, 1957.

42. **Hirokawa, H.,** Biochemical and cytological observations during the reversing process from spheroplasts to rod-form cells in *Escherichia coli, J. Bacteriol.,* 84, 1161–1168, 1962.

43. **Hsie, J. Y., Kotz, R., and Epstein, S.,** Carbomycin requiring mutants of *Micrococcus pyogenes* var. *aureus, J. Bacteriol.,* 74, 159–162, 1957.

44. **Hubert, E. G., Potter, C. S., Kalmanson, G. M., and Guze, L. B.,** Pigment formation in L forms of *Serratia marcescens, J. Gen. Microbiol.,* 55, 165–167, 1969.

45. **Hubert, E. G., Potter, C. S., Hensley, T. J., Cohen, M., Kalmanson, G. M., and Guze, L. B.,** L forms of *Pseudomonas aeruginosa, Infect. Immun.,* 4, 60–72, 1971.

46. **Hugo, W. B. and Russell, A. D.,** Quantitative aspects of penicillin action on *E. coli* in hypertonic medium, *J. Bacteriol.,* 80, 436–440, 1960.

47. **Ishiguro, E. E. and Wolfe, R. S.,** Control of morphogenesis in *Geodermatophilus:* ultrastructural studies, *J. Bacteriol.,* 104, 566–580, 1970.

48. **James, A. M., Hill, M. J., and Maxted, W. R.,** A comparative study of the bacterial cell wall, protoplast membrane and L form envelope of *Streptococcus pyogenes, Antonie van Leeuwenhoek,* 31, 423–432, 1965.

49. **Jones, D. H.,** Further studies on the growth cycle of Azotobacter, *J. Bacteriol.,* 5, 325–333, 1920.

50. **Kagan, G. Ya. and Savenkova, V. T.,** The technique of obtaining L forms of *C. diphtheriae* and some morphological peculiarities, *Zh. Mikrobiol. Epidemiol. Immunobiol.,* 31, 55–58, 1960.

51. **Kagan, G. Ya., Janeff, J., and Kagan, B. M.,** Inhibition of L forms of staphylococci by coccal forms of staphylococci, *Proc. Soc. Exp. Biol. Med.,* 132, 807–809, 1969.

52. **Klessen, C., Schmidt, K. H., Gumpert, J., Grosse, H., and Malke, H.,** Complete secretion of activable bovine prochymosin by genetically engineered L forms of *Proteus mirabilis, Appl. Environ. Microbiol.,* 55, 1009–1015, 1989.

53. **Klieneberger-Nobel, E.,** Origin, development and significance of L forms in bacterial cultures, *J. Gen. Microbiol.,* 3, 434–443, 1949.

54. **Klieneberger-Nobel, E.,** The L cycle, a process of regeneration in bacteria, *J. Gen. Microbiol.,* 5, 525–530, 1951.

55. **Krulwich, T. A. and Ensign, J. C.,** Activity of an autolytic N-acetylmuramidase during sphere-rod morphogenesis in *Arthrobacter crystallopoietes, J. Bacteriol.,* 96, 857–859, 1968.

56. **Laplace, F., Egerer, R., Gumpert, J., Kraft, R., Kostka, S. and Malke, H.,** Heterologous signal peptide processing in fusion interferon synthesis by engineered L forms of *Proteus mirabilis, FEMS Microbiol. Lett.,* 59, 59–64, 1989.

57. **Laplace, F., Müller, J., Gumpert, J., and Malke, H.,** Novel shuttle vectors for improved streptokinase expression in streptococci and bacterial L forms, *FEMS Microbiol. Lett.,* 65, 89–94, 1989.

58. **Mardh, P. A. and Taylor-Robinson, D.,** The susceptibility of aerobic and anaerobic bacteria L phase variants, Candida, Protozoa and viruses to lysolecithin, *Acta Pathol. Microbiol. Scand. Sect. B,* 82, 748–752, 1974.

59. **Marraro, R. V., Pfister, R. M., and Rheins, M. S.,** Growth inhibition of bacterial variants, *Can. J. Microbiol.,* 16, 640–642, 1970.

60. **Mattman, L. H., Tunstall, L. H., Mathews, W. W., and Gordon, D. L.,** L variation in *Mycobacteria, Am. Rev. Respir. Dis.,* 82, 202–211, 1960.

61. **Mattman, L. H., Tunstall, L. H., and Rossmoore, H. W.,** Induction and characteristics of staphylococcal L forms, *Can. J. Microbiol.,* 7, 705–713, 1961.

62. **Mattman, L. H., Tunstall, L. H., and Kispert, W. G.,** A survey of L variation in the salmonellae, *Zentralbl. Bakteriol. Parasitenkd. Infektionskr. Hyg. Abt. 1 Orig.,* 210, 65–74, 1969.

63. **McQuillen, K.,** Some aspects of bacterial structure and function, observations of bacterial cell walls, protoplasts and spheroplasts together with a study of the effects of linoleic acid and vitamin D_2 on *Lactobacillus casei,* 4th Int. Congr. Biochem. (Vienna), 13, 406–442, 1958.

64. **Metz, O.,** Über Wachstum und Farbstoffbildung einiger Pilze unter dem Einfluss von Eisen, Zink, und Kupfer, *Arch. Mikrobiol.,* 1, 197–251, 1930.

65. **Meyer-Rohn, J. and Tudyka, K.,** Gestaltsanderungen von *Escherichia coli* unter der Einwirkung von Antibiotica, Sulfonamiden, chemischen und physikalischen Noxen, *Arzneim. Forsch.,* 7, 151–157, 1957.

66. **Mitchell, P. and Moyle, J.,** Liberation and osmotic properties of the protoplasts of *Micrococcus lysodeikticus* and *Sarcina lutea, J. Gen. Microbiol.,* 15, 512–520, 1956.

67. **Morrison, D. A.**, Prodigiosin synthesis in mutants of *Serratia marcescens*, *J. Bacteriol.*, 91, 1599–1604, 1966.
68. **Moustardier, G., Brisou, J., and Perrey, M.**, Milieu de culture pour l'isolement des organismes L dans les urethrities amicrobiennes a inclusions, *Ann. Inst. Pasteur (Paris)*, 85, 515–520, 1953.
69. **Naud, A. I., Legault-Demare, J., and Ryter, A.**, Induction of a stable morphological change in *Propionibacterium freudenreichii*, *J. Gen. Microbiol.*, 134, 283–293, 1988.
70. **Nelles, A.**, Die Vernehrung der Stabilen L formen der Bakterien, *Arch. Hyg. Bakteriol.*, 139, 294–304, 1955.
71. **Netrawali, M. S. and Nair, K. A. S.**, Gamma radiation-induced single strand breaks in DNA and their repair in spheroplasts and nuclei of light-grown and dark-grown *Euglena* cells, *Int. J. Radiat. Biol.*, 43, 19–30, 1983.
72. **Newton, B. A.**, A fluorescent derivative of polymyxin, its preparation and use in studying the site of action of the antibiotic, *J. Gen. Microbiol.*, 12, 226–236, 1955.
73. **Nickerson, W. J. and Mankowski, Z.**, Role of nutrition in the maintenance of the yeast-shape in *Candida*, *Am. J. Bot.*, 40, 584–592, 1953.
74. **Ørskov, J.**, Some observations on aberrant bacteria morphology, *Proteus vulgaris*, *Acta Pathol. Microbiol. Scand.*, 24, 198–202, 1947.
75. **Panos, C. and Barkulis, S. S.**, Streptococcal L forms. I. Effect of osmotic change on viability, *J. Bacteriol.*, 78, 247–252, 1959.
76. **Pine, L. and Peacock, C. L.**, Studies on the growth of *Histoplasma capsulatum*. IV. Factors influencing conversion of the mycelial phase to the yeast phase, *J. Bacteriol.*, 75, 167–174, 1958.
77. **Porges, N.**, Chemical composition of *Aspergillus niger* as modified by zinc sulfate, *Bot. Gaz.*, 94, 197–205, 1932.
78. **Prévot, A.**, Étude sur la longévité des bactéries anaerobies non sporulées, *Ann. Inst. Pasteur* (Paris), 67, 471–473, 1941.
79. **Razin, S. and Argaman, M.**, Lysis of Mycoplasma, bacterial protoplasts, spheroplasts and L forms by various agents, *J. Gen. Microbiol.*, 30, 155–172, 1963.
80. **Razin, S. and Oliver, O.**, Morphogenesis of Mycoplasma and bacterial L form colonies, *J. Gen. Microbiol.*, 24, 225–237, 1961.
81. **Roberg, M.**, Weitere untersuchungen uber die Bedeutung des Zink, fur *Aspergillus niger*, *Zentralbl. Bakteriol. Parasitenkd. Infektionskr. Hyg. Abt. 1 Orig.*, 84, 196–230, 1931.
82. **Rogers, H. J., McConnell, M., and Burdett, I. D. J.**, Cell wall or membrane mutants of *Bacillus subtilis* and *Bacillus licheniformis* with grossly deformed morphology, *Nature (London)*, 219, 285–288, 1968.
83. **Rubio-Huertos, M., Küster, E., and Flaig, W.**, Licht und elektronenmikroskopische Untersuchungen an L formen bei *Pseudomonas fluorescens*, *B. proteus* und *Rhizobium*, *Zentralbl. Bakteriol. Parasitenkd. Infektionskr. Hyg. Abt. 1 Orig.*, 109, 24–31, 1955.
84. **Salvin, S. B.**, Cysteine and related compounds in the growth of the yeast-like phase of *Histoplasma capsulatum*, *J. Infect. Dis.*, 84, 275–283, 1949.
85. **Šašak, V. and Becker, G. E.**, Effect of different nitrogen sources on the cellular form of *Trigonopsis variabilis*, *J. Bacteriol.*, 99, 891–892, 1969.
86. **Schellenberg, H.**, Über die Chromatinsubstanz von *Proteus vulgaris* bei Penicillin-einwirkung, *Zentralbl. Bakteriol. Parasitenkd. Infektionskr. Hyg. Abt. 1 Orig.*, 161, 425–432, 1954.
87. **Schellenberg, H.**, Untersuchungen uber die durch Penicillin Induzierte Pleomorphie bei *Proteus vulgaris*, *Zentralbl. Bakteriol. Parasitenkd. Infektionskr. Hyg. Abt. 1 Orig.*, 161, 433–457, 1954.
88. **Smith, J. A. and Willis, A. T.**, Some physiological characteristics of L forms of *Staphylococcus aureus*, *J. Pathol. Bacteriol.*, 94, 359–365, 1967.
89. **Stempen, H. and Hutchinson, W. G.**, The formation and development of large bodies in *Proteus vulgaris* OX-19, *J. Bacteriol.*, 61, 321–355, 1951.
90. **Stewart, R. H. and Wright, D. N.**, Storage of meningococcal and streptococcal L forms, *Cryobiology*, 6, 529–532, 1970.
91. **Takahashi, T.**, Lytic susceptibility of staphylococcal L forms to lysolecithin, *Microbiol. Immunol.*, 24(5), 463–468, 1980.
92. **Takahashi, T., Nakatsuka, S., Okuda, K., and Tadokoro, I.**, Preservation of bacterial L forms by freezing, *Jpn. J. Exp. Med.*, 50(4), 313–318, 1980.
93. **Thorsson, K. G. and Weibull, C.**, Studies on the structure of bacterial L forms, protoplasts, and protoplast-like bodies, *J. Ultrastruct. Res.*, 1, 412–427, 1958.
94. **Tulasne, R.**, Existence of L forms in common bacteria and their possible importance, *Nature (London)*, 164, 876–887, 1949.
95. **Tulasne, R.**, Quelques données nouvelles sur la formation et les caractères de culture des formes L du *Proteus vulgaris*, *C. R. Soc. Biol.*, 144, 1200–1203, 1950.
96. **White, P. B.**, A note on the globular forms of *Vibrio cholerae*, *J. Gen. Microbiol.*, 4, 36–37, 1950.

Chapter 5

COMPOSITION OF WALL DEFICIENT FORMS

COMPONENTS

LOSS OF MUREIN COMPONENTS

A prominent characteristic of wall deficient forms is loss of rigidity. Thus, not unexpectedly, an L Form of Group A β-hemolytic Streptococcus lacked essential components of the murein layer, namely, glucosamine, tramnose, and muramic acid.[25] Sharp found muramic acid absent from the L Forms of both Staphylococci and Streptococci.[37] Likewise, muramic acid and ribitol were completely absent from both penicillin-induced and cycloserine-induced L Forms of *Staphylococcus aureus* analyzed by Pratt.[29]

SUBSTITUTES FOR THE RIGID MUREIN LAYER

In the absence of sufficient galactose, a strain of Bifidobacterium is unable to make a rigid murein layer, and its ramifying walls show lack of direction by a firm wall. It builds a substitute polysaccharide wall of glucose, galactose, and rhamnose, with less galactose than required for the normal wall.[5]

RETENTION OF MUREIN COMPONENTS

Most CWD forms, stable or revertible, have residua of cell wall. In a stable L Form of Proteus, Fleck found all the classic layers of the rigid wall of Gram-negative organisms: lipoprotein, lipopolysaccharide, and some glycopeptide.[6] In the glycopeptide fraction, which is rigid in a classic organism, he found that a certain peptide, glutamic acid, diaminopimelic acid (DAP), and hexosamines were all missing in the L Form. This loss of components resulted in the formation of new cross-linkages and a plastic rather than a rigid wall.

James et al.[13] detected no murein components in protoplasts but found murein in certain L Forms of streptococci. However, the mucopeptide in the derived L Form of *Streptococcus pyogenes* differed from the corresponding parental cell wall component in containing only a few peptide chains and very little reducing sugar. The mucopolysaccharide layer of spheroplasts, though structurally defective, is demonstrable by silver methionine staining of spheroplasts of *Vibrio parahaemolyticus* and *V. alginolyticus*.[12]

AMINO SUGARS IN L FORMS

The antibiotic-induced changes in the wall do not follow a set pattern, not even for a single antibiotic. Fodor,[7] applying a uniform method to make penicillin L Forms from a strain of *Staphylococcus aureus* derived three distinct variants. These differed in their *N*-acetyl amino sugar content on a dry weight basis as follows:

Group (a): Low	1 to 5	μg/mg
Group (b): Medium	5 to 20	μg/mg
Group (c): High	>20	μg/mg

Fodor concluded that the groups were inhibited at different stages of cell wall mucopeptide synthesis. The acid-insoluble amino sugars were minimal in all the variant strains, being 1/15 the quantity in the parental staphylococcus wall.

DAP IN THE CELL WALL

The cell wall building block, DAP, resides in the plastic wall of many wall deficient bacteria.[36] Its presence seems to correlate with ease of reversion of some variants.

Kandler and Zehender[15] strengthened the concept that DAP is important in reversion by finding that six L phase, revertible, cultures of *Proteus vulgaris* contained large amounts of DAP, concentrated as a gelatinous material as the organisms are heated in 1 *N* sodium hydroxide. In contrast, ten stable L Forms of this same species lacked DAP.

In all bacterial mutants which do not synthesize DAP, there is classic growth only on DAP-containing media. When placed in media lacking DAP, they initially degrade their own murein. Since lack of DAP prevents the coalescence of components necessary for new rigid wall, the organisms are consequently transformed into spheroplasts. When the mutants are again suspended in DAP-supplemented media, they make an atypical murein for some time which is not cross-linked. Cross-linking of wall components with its resultant reversion to the normal bacillary shape is a slow process.[20]

Many studies indicate that in some wall-deficient forms pieces of the cell wall are synthesized and dutifully pulled through the pores of the cell membrane but somehow lack structural detail which would permit them to link together. These organisms may for years continue to manufacture ineffective units of cell wall.[16]

One of the first clues that the muramic acid is altered in the Proteus penicillin-induced L Form was the report by Sharp that the muramic acids of parent and derived L Forms differed in solubility. Only the variant muramic acid dissolved in 4 *N* HCl.[37]

Mitchell and Moyle[23] have added another interesting aspect to considerations which may explain why a cell which has once lost its wall is unable to resynthesize this in the future. They postulate that perhaps the building blocks are sufficiently soluble to diffuse spontaneously into the culture medium rather than remain together against the wall where their union is facilitated. This would certainly explain the increased reversion which occurs in the presence of a very firm medium.

''Penicillin murein'' sometimes demonstrates a synthetic error in interconnection of peptide side chains and also in the size of the units. When spheroplast murein of Proteus is degraded with lysozyme, the product consists largely of a C6 mucopeptide lacking any peptide cross-linkage and lesser quantities of cross-linked C3 and C4 mucopeptides. In contrast, murein of the classic rod-shaped Proteus yields highly cross-linked C6 and C5 mucopeptides.[22]

In the streptococcus also, normal components of the cell wall may be present but not interlaced to make a stable deposit on top of the membrane. The rhamnose polysaccharide from L-Form membranes is devoid of amino sugars, suggesting that these might be necessary for binding the polymeric rhamnose to the streptococcal membrane.[2]

TEICHOIC ACID

The teichoic acid of a parent *Streptococcus pyogenes* contains glycerol, phosphorus, alanine, and glucose, while the L-Form teichoic acid lacks alanine. Furthermore, the length of the teichoic acid chain in its derived L Form, estimated from the phosphorus content, was approximately one half that of the coccal teichoic acid. Thus, again, the L variant is making abnormally small cell-wall precursors.[38]

LOSS OF POLYSACCHARIDES AND PROTEIN

Polysaccharide A of staphylococci is another wall component which disappears completely from L Forms of *Staphylococcus aureus*. As shown with CWD cell extracts, this compound is not detected in agar-gel diffusion tests with specific antiserum.[29]

Other malformations of spheroplast walls concern proteins. In some Enterobacteriaceae, the murein layer in the classic organism is topped by protein, visible in the electron microscope as granules evenly covering the entire murein surface. The arrangement of the protein may be necessary for normal morphogenesis of the murein since the protein granules are sparse and irregularly distributed in some Proteus L Forms.[21]

TABLE 1
Comparative Membrane Lipid in Classic Bacteria vs. Derived L Forms

Organism/investigator	Classic (%)	L Form (%)	Fatty acid content
Proteus P 18[a]			
Krembel[18]	6.0	24.0	Phosphatides 90% in
Nesbitt & Lennarz[24]	10.3	14.5	parent, 80% in de-
Rebel & Mandel[33]	12.0	21.0	rived L form; both
Smith & Rothblat[39]	4.4	8.6	contain Vitamin K
Weibull et al.[44]	—	59.0	and Coenzyme Q;[33]
Proteus sp.[42a]	6.0	24.0	proportion of lauric,
			myristic, palmitic,
			hexadecanoic,
			stearic, and octade-
			cenoic acids varies
			in classic vs. L
			Forms[18]
Staphylococcus aureus			
strain 110[43]	25.0	28.3	Much more glycolipid
strain H	24.6	29.8	in L Form than in
			classic.[43]
Streptobacillus	2.6	19.4	
moniliformis[42]	—	40.0	
Streptococcus pyogenes			
strain AED[1a]	15.3	35.6	*Cis*-vaccenic acid pre-
strain 416[13a]	16–20	28–33	dominates in parent;
strain ADA[39]	1.3	5.0	oleic acid in var-
strain GL 8[39]	0.9	4.8	iant.[1]

[a] Whole cells analyzed. Lipids in whole cell are considered primarily of membrane origin.

In a related finding, Fleck showed that the classic wall of Proteus was composed of high molecular weight glycopeptide fractions, whereas in the L Form, low molecular weight components were more numerous. He concluded that loss of peptide fractions resulted in modified repolymerization.[6]

CYTOPLASMIC MEMBRANE

TOTAL LIPID

In itself, the term "CWD" implies alterations in the constitution of the bacterial cell wall resulting from deletion and/or faulty synthesis of wall components. Unanticipated, however, is the hypertrophy of the cytoplasmic membrane, of such magnitude that it may comprise 25% of the variant's substance. This may account for many characteristics of these variants, such as the penetration of early growth into agar, the tendency to collect at an interface of medium components, and hydrocarbon solubility.

Lipids of L Forms and their parent cells are compared in Table 1. Qualitative as well as quantitative changes occurring in the membrane of *Streptococcus pyogenes* as it attains the stable L Form were delineated by Cohen and Panos.[1] The authors discuss the possibility that changes in lipid composition might be associated with the inability of the L Form to synthesize a rigid cell wall. With a few exceptions the change was quantitative rather than qualitative.[18]

In contrast to results with *S. pyogenes*, there was no qualitative difference in the neutral lipids from normal bacteria and from the derived L Forms of *P. vulgaris* P 18.[33] A great complexity of lipid material in both the stable L-variant and the parent Proteus was disclosed: quinones, hydrocarbons, fatty aldehydes, alcohols, free fatty acids, tri-, di-, and mono-glycerides. When quantitative considerations are made, however, again one finds approximately twice as much lipid in the L Form as in the parent (Table 1).

In conclusion, the lipid content of L-phase organisms is usually increased significantly compared with the parent strains. In most species this does not involve qualitative differences. Immunological evidence of retained similarity in staphylococcal L Forms was obtained when antiserum vs. membranes of the L Forms reacted with membranes of the parent Staphylococcus and its derived L Form.[29]

CHOLESTEROL IN THE MEMBRANE

The question of whether L Forms can synthesize cholesterol has been answered by several investigators, always negatively. Classic bacteria and their CWD progeny follow the same pattern: they do not synthesize cholesterol; they can grow without it, but if the sterol is in the culture medium, they deftly incorporate it into the cytoplasmic membrane.

Rebel et al. found no cholesterol in L-Form cultures of *P. mirabilis* (P 18) unless the sterol was added to the medium. When it was supplied to the culture, the L Forms absorbed and stored this compound.[32] When labeled cholesterol (4-^{14}C) is added to culture medium, both classic and CWD organisms incorporate the sterol into the cell membrane. Razin and Shafer found that protoplasts, spheroplasts, and parent bacteria behave similarly with a few quantitative exceptions: *Streptobacillus moniliformis* is outstandingly active, and *P. mirabilis* is unusual in having an L phase which takes up much more cholesterol than does the parent Proteus.[30] The membrane of a stable L phase of *S. moniliformis* grown in serum medium was found to be rich in cholesterol, which comprised almost a third of the membrane lipid.[31]

All data support the findings of Edward[3] that requiring preformed cholesterol differentiates most Mycoplasma from L-phase variants. However, CWD variants resemble Mycoplasma in incorporating cholesterol into their membranes if it is present in the culture medium. Neither can synthesize cholesterol.

MEMBRANE PROTEIN IN L FORM

The first analysis of proteins in the membrane of *Streptococcus pyogenes* and its stabilized L Form was made by Panos et al.[27] Although no qualitative differences were found, there were some provocative quantitative changes. Most remarkable was the sevenfold increase in the glucosamine associated with the membrane protein.

WHOLE CELL ANALYSES

LIPOPOLYSACCHARIDES AND LIPIDS

Whereas the lipid increases in L Forms, the lipopolysaccharide may decrease to $1/3$ or less the amount present in the parent form. Qualitatively, the lipopolysaccharide is similar in Proteus parent and L Form, containing heptose, hexosamine, phosphorus, 3-deoxyoctulosonate, glucose, galactose, and fatty acids.[24] However, quantitative differences are remarkable. The bacillary form contains approximately ten times as much glucose and slightly more fatty acid than the L Form, whereas the L Form contains slightly more galactose.

COMPOSITION CHANGES WITH SALTS IN MENSTRUUM

The content of an L Form can differ depending on the sodium chloride in the culture medium. It will be noted that physiological saline promotes the highest relative amount of saturated fatty acids[28] (Figure 1). This may have wide application if NaCl levels can influence production of many enzymes.

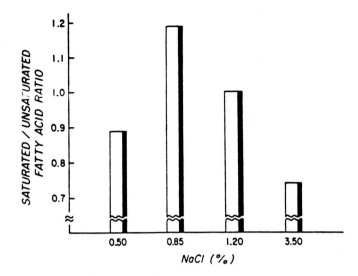

FIGURE 1. Changes in NaCl concentration results in alterations in the ratio of saturated to unsaturated fatty acids in streptococcal L Form. (From Panos, C., in *The Bacterial L-Forms,* Madoff, S., Ed., Marcel Dekker, New York, 1986. With permission.)

POLYSACCHARIDES

Decrease of carbohydrate is expected as a bacterium minimizes its cell wall. However, the complete antigenic change encountered by King et al. was extraordinary. A stable L Form of *Streptococcus faecium* contains only 1% the polysaccharide of its parent form, and there is no detectable serological relationship between the polysaccharides in the parent and in the L Form.[17]

GROUP-SPECIFIC POLYSACCHARIDE

The enterococci (Group D) peculiarly possess their group-specific antigen in some locus other than the cell wall. The absence of cell-wall components in L-phase variants of Group A streptococci is logical, but is the D antigen, which is not part of the wall, still present in the L phase? Hijmans found the D antigen absent in stable and revertible enterococci and not present in their culture medium to indicate excretion.[10] When the unstable variant was allowed to revert, it simultaneously regained the D antigen. This suggests derepression of an enzyme in cytoplasm or membrane.

TEICHOIC ACIDS

Since the M protein of Group A β-hemolytic streptococci is found in the culture fluid, Pratt tested the liquid culture of staphylococcal L Forms for teichoic acid. Its absence confirmed that staphylococcal L Forms have lost the ability to synthesize teichoic acid.[29]

RNA

The RNA in a stable L Form is characteristically less than that of the parent. In a *P. vulgaris* L Form, irreversible through 150 penicillin-free passages, the ratio of RNA to protein was 50% lower than in the parent. In contrast, the DNA-protein ratio was approximately 25% higher in the L Form.[14] In another study, the authors concluded that the RNA-DNA ratio for a Proteus L Form is the inverse of that in normal bacteria, being 2.5 for normal Proteus and 0.14 for its L Form.[42]

The change in RNA content also applies to streptococcal L Forms. Early in log-phase growth, the ratio of RNA to DNA can be 5:1 for the L Form and 10:1 for the coccus.

However, the total DNA content is similar in both organisms and remains constant throughout their respective growth cycles.[26]

The relative lowering of RNA content has an unexpected explanation. Penicillin spheroplasts of *Escherichia coli* synthesize RNA at nearly the same ratio as classic bacteria, even when the spheroplasts are no longer increasing in total mass or RNA content. Therefore, low RNA content is not due to minimal synthesis but more probably to instability of this nucleic acid in spheroplasts. Ehrenfeld and Koch traced the formation and degradation of RNA by using (^{14}C) uracil. Their studies suggest that the spheroplast overproduces RNA since only 15% of the ribosomal RNA synthesized during active growth is converted to completed ribosomes. This turnover of normally stable RNA may indicate that *E. coli* can degrade species of RNA as necessary rather than independently regulate synthesis of RNA species.[4]

RIBOSOME POSITIONING

Changes in metabolic patterns of CWD variants may be related in some instances to changes in ribosome position. During exponential growth, protoplasts of *S. faecalis* have their ribosome precursor particles firmly affixed to the cytoplasmic membrane and have sedimentation constants of 103S, 130S, 153S, and even heavier.[34]

PHOSPHORUS IN L FORMS

Total phosphorus content in Mycoplasma, in salt-requiring and nonsalt-requiring L Forms, and in their parent bacteria is equivalent, with the exception of elevation in most salt-requiring L Forms. Typical figures were 6.57 mg P for 100 mg dry wt of salt-requiring L Form vs. 0.41 mg P in nonsalt-requiring forms. Langenfeld and Smith postulated that a high salt concentration may increase the acid-soluble phosphorus content by prohibiting excretion of low molecular weight phosphorylated compounds.[19]

CAPSULAR ANTIGENS

Spheroplasts of *Hemophilus influenzae* retain antigens of the capsule as shown by rocket immunoelectrophoresis. The authors postulate that retention of this antigen in capsule-free phases may explain virulence in nontypable strains.[11]

DNA OF SPORE SPHEROPLASTS

Since spheroplasts may be produced from bacterial spores, it is interesting to examine the product. Whether the characteristic resistance of spores to traumatic physical conditions is due to the elaborate dense covering or to the spore DNA has been investigated by removal of the coat and study of the resultant spore spheroplasts. Spore spheroplasts are made by digestion with lysozyme, after a complex pretreatment.[8]

The DNA in spore spheroplasts still retains the unique heat resistance characterizing DNA of *Bacillus subtilis* spores. DNA of a spore spheroplast, boiled in the spore for 2 min, still retains 80% of its transforming activity. Transforming ability is similar in DNA extracted and rendered protein-free before exposure to heat.[35] Loss of heat resistance of the DNA occurs only as the spores germinate.

Other studies by Tanooka and Sakakibara[40,41] indicate that the DNA of spore spheroplasts is also highly resistant to ionizing and ultraviolet light irradiation, retaining the characteristics of the DNA in intact dormant spores. Loss of cortex components does affect viability, though not DNA resistance. They noted that one characteristic of spore DNA associated in all probability with this resistance was its high density, as demonstrated by Halvorson and co-workers.[9] Coincidentally, the authors confirmed that the spheroplasts lost the ability to reproduce when heated for 10 min at 80°C. Thus, spheroplast analyses have shown that much of the unique resistance of a spore lies in its DNA rather than in its multilayered tegument.

SUMMARY

A most unexpected change in a stable CWD variant is the increased lipid in the membrane, in some instances reaching four times that of the parent. Also remarkable are the endless modifications in cell-wall components which often have a complex wall too loosely integrated to be rigid. Also strange are the multiple variants resulting when a single antibiotic acts on a pure culture population. These may consist of genetically stable classes of variants which differ in their wall murein.

The concentration of DNA within a single cytoplasmic membrane shows that a variant organism is sometimes multinucleate. In contrast, less RNA is needed because of the slower growth of the variants. Excess RNA appears to be selectively degraded.

Analysis of spores treated to render them spheroplastic supplies the information that spore DNA is, indeed, unique in resistance to heat and ultraviolet radiation.

REFERENCES

1. **Cohen, M. and Panos, C.,** Membrane lipid composition of *Streptococcus pyogenes* and derived L form, *Biochemistry,* 5, 2385–2397, 1966.
2. **Cohen, M. and Panos, C.,** Cell wall polysaccharide biosynthesis by membrane fragments from *Streptococcus pyogenes* and stabilized L form, *J. Bacteriol.,* 106, 347–355, 1971.
3. **Edward, D. G. ff.,** Problems of classification - An introduction, *Ann. N.Y. Acad. Sci.,* 143, 7–8, 1967.
4. **Ehrenfeld, E. R. and Koch, A. L.,** RNA synthesis in penicillin spheroplasts of *Escherichia coli, Biochim. Biophys. Acta,* 169, 44–47, 1968.
5. **Exterkate, A. and Veerkemp, J. H.,** Biochemical changes in *Bifidobacterium bifidum* var. *pennsylvanicus* after cell wall inhibition, *Biochim. Biophys. Acta,* 231, 545–549, 1971.
6. **Fleck, J.,** Etude du glycopeptide de la ''Paroi non rigide'' d'une forme L stable de *Proteus.* I. Constitution chimique, *Ann. Inst. Pasteur (Paris),* 115, 51–55, 1968.
7. **Fodor, M.,** Studies on the cell wall mucopeptide synthesis of staphylococcal L forms, *Naturwissenschaften,* 52, 522, 1965.
8. **Gould, G. W. and Hitchins, A. D.,** Sensitization of bacterial spores to lysozyme and to hydrogen peroxide with agents which rupture disulphide bonds, *J. Gen. Microbiol.,* 33, 413–423, 1963.
9. **Halvorson, H. O., Szulmajster, J., Cohen, R., and Michelson, A. M.,** Etude del l'acide deoxyribonucleique des spores de *Bacillus subtilis, J. Mol. Biol.,* 28, 71–86, 1967.
10. **Hijmans, W.,** Absence of the group-specific and the cell wall polysaccharide antigen in L phase variants of Group D streptococci, *J. Gen. Microbiol.,* 28, 177–179, 1962.
11. **Huber, P. S. and Egwu, I. N.,** Capsular variation in experimental strains of *Haemophilus influenzae, Med. Microbiol. Immunol.,* 173, 345–353, 1985.
12. **Inoue, A., Hasegawa, K., and Takagi, A.,** Spheroplast formation in vibrios, *Jpn. J. Bacteriol.,* 24, 533–534, 1969.
13. **James, A. M., Hill, M. J., and Maxted, W. R.,** A comparative study of the bacterial cell wall, protoplast membrane and L form envelope of *Streptococcus pyogenes, Antonie van Leeuwenhoek,* 31, 423–432, 1965.
14. **Kandler, O.,** Vergleichende Untersuchungen uber den Nukleinsaure- und Atmungsstoffwechsel von *Proteus vulgaris,* dessen stabiler L phase und den pleuropneumonie-ahnlichen Organismen, *Zentralbl. Bakteriol. Parasitenkd. Infektionskr. Hyg. Abt. 2 Orig.,* 109, 335–336, 1956.
15. **Kandler, O. and Zehender, C.,** Über das Vorkommen von α,ε-Diaminopimelinsäure bei verschiedenen L-Phasentypen von *Proteus vulgaris* und bei den pleuropneumonie-ähnlichen Organismen, *Z. Naturforsch. Teil B,* 126, 725–728, 1957.
16. **Kandler, O., Hund, A., and Zehender, C.,** Cell wall composition in bacterial and L forms of *Proteus vulgaris, Nature (London),* 181, 572–573, 1958.

17. **King, J. R., Prescott, B., and Caldes, G.,** Differences in the carbohydrate content of *Streptococcus faecium* F24 and its stable L form, *Bacteriol. Proc. 1971*, p.42, 1971.

18. **Krembel, J.,** Étude des lipides de la forme L stable de Protéus P 18, identification des acides gras de la fraction acétonosoluble, *Pathol. Microbiol.*, 26, 592–600, 1963.

19. **Langenfeld, M. G. and Smith, P. F.,** Phosphorus distribution in pleuropneumonia-like and L-type organisms, *J. Bacteriol.*, 86, 1216–1219, 1963.

20. **Leutgeb, W. and Schwartz, U.,** Zur Biosynthese von Murein, *Zentralbl. Bakteriol. Parasitenkd. Infektionskr. Hyg. Abt. 1 Orig.*, 198, 76–80, 1965.

21. **Martin, H. H.,** Composition of the mucopolymer in cell walls of the unstable and stable L form of *Proteus mirabilis*, *J. Gen. Microbiol.*, 36, 441–450, 1964.

22. **Martin, H. H.,** Murein structure in cell walls of normal bacteria and L forms of *Proteus mirabilis* and the site of action of penicillin, *Folia Microbiol.* (Prague), 12, 234–239, 1967.

23. **Mitchell, P. and Moyle, J.,** Osmotic function and structure in bacteria, *6th Symp. Soc. Gen. Microbiol.*, pp. 150–180, 1956.

24. **Nesbitt, J. A., III and Lennarz, W. U.,** Comparison of lipids and lipopolysaccharide from the bacillary and L forms of Proteus P 18, *J. Bacteriol.*, 89, 1010–1025, 1965.

25. **Panos, C., Barkulis, S. S., and Hayashi, J. A.,** Streptococcal L forms. II. Chemical composition, *J. Bacteriol.*, 78, 863–867, 1959.

26. **Panos, C.,** Cellular composition during growth of a streptococcus and derived L form, *Bacteriol. Proc. 1964*, p. 71, 1964.

27. **Panos, C., Fagan, G., and Zarkadas, C. G.,** Comparative electrophoretic and amino acid analyses of isolated membranes from *Streptococcus pyogenes* and stabilized L form, *J. Bacteriol.*, 112, 285–290, 1972.

28. **Panos, C.,** Macromolecular synthesis, survival, and cytotoxicity of an L form of *Streptococcus pyogenes*, in *The Bacterial L-Form*, Madoff, S., Ed., Marcel Dekker, Inc., N. Y., 1986.

29. **Pratt, B. C.,** Cell wall deficiencies in L forms of *Staphylococcus aureus*, *J. Gen. Microbiol.*, 42, 115–122, 1966.

30. **Razin, S. and Shafer, Z.,** Incorporation of cholesterol by membranes of bacterial L phase variants (with an appendix): on the determination of the L phase parentage by electrophoretic patterns of cell proteins, *J. Gen. Microbiol.*, 58, 327–339, 1969.

31. **Razin, S. and Boschwitz, C.,** The membrane of the *Streptobacillus moniliformis* L-phase, *J. Gen. Microbiol.*, 54, 21–32, 1968.

32. **Rebel, G., Bader-Hirsch, A. M., and Mandel, P.,** Recherches sur les lipides des formes L dérivées du Proteus P 18. I. Presence et absorption du cholesterol, *Bull. Soc. Chim. Biol.*, 45, 1327–1342, 1963.

33. **Rebel, G. and Mandel, P.,** Recherches sur les lipides des formes L derivées du Proteus P 18. IV. Étude quantitative des lipides totaux, *Ann. Inst. Pasteur* (Paris), 117, 501–506, 1969.

34. **Roth, G. S. and Daneo-Moore, L.,** Intracellular localization of sites of ribosome biosynthesis in "protoplasts" of *Streptococcus faecalis*, *Bacteriol. Proc. 1971*, p. 125, 1971.

35. **Sakakibara, Y. and Ikeda, Y.,** The dormant spore-specific state of DNA and macromolecular synthesis in spheroplasts of *Bacillus subtilis* spores, *Biochim. Biophys. Acta*, 179, 429–438, 1969.

36. **Schuhmann, E. and Taubeneck, U.,** L formen von *Escherichia coli* K 12 (λ). I. Induktion und Vermehrungsweise, *Z. Allg. Mikrobiol.*, 11, 205–219, 1971.

37. **Sharp, J. T.,** Amino sugars in L forms of bacteria and pleuropneumonia-like organisms, *J. Bacteriol.*, 86, 692–701, 1963.

38. **Slabyj, B. M. and Panos, C.,** Glycerol teichoic acid from *Streptococcus pyogenes* and stabilized L form, *Bacteriol. Proc. 1971*, p. 48, 1971.

39. **Smith, P. and Rothblat, G. H.,** Comparison of lipid composition of pleuropneumonia-like and L-type organisms, *J. Bacteriol.*, 83, 500–506, 1962.

40. **Tanooka, H.,** Ultraviolet resistance of DNA in spore spheroplasts of *Bacillus subtilis* as measured by the transforming activity, *Biochim. Biophys. Acta*, 166, 581–585, 1968.

41. **Tanooka, H. and Sakakibara, Y.,** Radioresistant nature of the transforming activity of DNA in bacterial spores, *Biochim. Biophys. Acta*, 155, 130–142, 1968.

42. **Vendrely, R. and Tulasne, R.,** Premieres observations sur la constitution chimique des formes L des bacteries, *Bull. Soc. Chim. Biol.*, 34, 785–789, 1952.

43. **Ward, J. and Perkins, H.,** The chemical composition of the membranes of protoplasts and L forms of *Staphylococcus aureus*, *Biochem. J.*, 106, 391–400, 1968.

44. **Weibull, C., Bickel, W. D., Haskins, W. T., Milner, K. C., and Ribi, E.,** Chemical, biological and structural properties of stable Proteus L forms and their parent bacteria, *J. Bacteriol.*, 93, 1143–1159, 1967.

Chapter 6

DISCLOSURES BY ELECTRON MICROSCOPY

Only the electron microscope reveals the undreamed of kinship between uniformly boxed bacteria and yeasts, aspergillus, protozoa, chlamydia, and mycoplasma. Obviously, bacteria are protista, and the bacterial morphology is only one of their growth alternatives.

Nonbacterial phases of *Streptococcus pyogenes* are shown by both scanning microscopy and thin sections viewed with transmission electron microscopy. Multiple buds are evident (Figures 1 and 2).[32] The formation internally of viable units (elementary bodies) mimics spore development by fungi and reproduction in both protozoa and chlamydia (Figure 3).[32] Beautiful studies by Takahashi delineate the possible wide range of diameters of the elementary bodies (Figure 4).[40] Both bacterial CWD forms and mycoplasma have these interior minuscule spheres which can serve as reproductive units. These elementary bodies are almost omnipresent and typically numerous. Often they are more abundant than the Actinomyces-like and yeast-like stages. Three strains of *Staphylococcus aureus* have similar fine structure. The distinctions between classic bacteria and their CWD variants are numerous: colonial morphology, qualitative and quantitative antigen changes, target organs in disease, longevity, antibiotic susceptibility, and growth requirements. While not explaining differences, fine structure does show that dissimilarities between wall-deficient forms and their parents involve much more than the cell wall.

A CLOSE-UP VIEW OF THE WALL

Just how cell-wall depleted are wall deficient variants when closely scrutinized? The findings include some unanticipated results, discussed below in connection with degrees of wall deficiency.

PROTOPLASTS

When *Clostridium botulinum* was examined as an L Form produced by penicillin and lysozyme, the organisms were true protoplasts free of cell wall layers.[3] Amazingly, certain strains of these naked cells can propagate indefinitely, withstanding the same 20 atm of pressure resisted by the rigid wall of the classic form. A hypertonic milieu is usually essential.

Stable L Forms of *Escherichia coli* (B and W 1655 F+) were found by Gumpert et al. to have no cell wall.[16] These strains grew as highly pleomorphic cells in agar medium and as small granules of different sizes in liquid. Similarly, true wall-free protoplasts were found in L Forms of Proteus[21,36] and β-hemolytic streptococci.[36]

SPHEROPLASTS AND L FORMS, I.E., NOT PROTOPLASTS

Most wall-defective bacteria, whether artificially induced or occurring spontaneously, retain some wall and are, therefore, classified as spheroplasts or L Forms or L variants, the term depending on the author's choice. Sometimes the wall layers of a variant are interrupted only in patches, as substantiated by serial sections of the unstable L phase of *E. coli* K12λ.[15] Electron micrographs may show that the murein layer of the parent wall is absent in the variant. This lamella, measuring 2.0 to 2.5 nm, is absent from certain *Proteus mirabilis* variants.[21]

Certain penicillin-induced spheroplasts of *P. mirabilis* and a Providencia strain retained only the outer layer of the integuments, as noted by van Rensburg.[34] A stable L Form of *P. vulgaris* was found by Thorsson and Weibull to have a 8.0-nm wall, contrasting with the 15-nm parental wall. They deduced the absence of the inner lipopolysaccharide layer.[41]

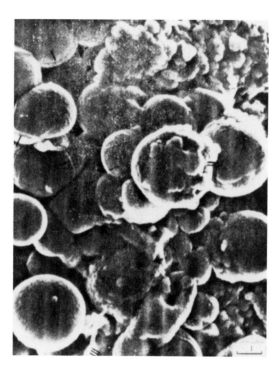

FIGURE 1. Scanning electron micrograph of *Streptococcus pyogenes* L Forms shows multiple budding from spheres. (Original magnification × 10,200.)(From Prozorovsky, S. V., Katz, L. N., and Kagan, G. Ya., *AMS USSR, H'Meditsina,* 1981.)

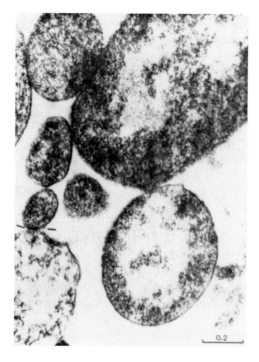

FIGURE 2. Sectioned L Form of *Streptococcus pyogenes* show propagation is by budding. (Original magnification × 100,000.)(From Prozorovsky, S. V., Katz, L. N., and Kagan, G. Ya., *AMS USSR, H'Meditsina,* 1981.)

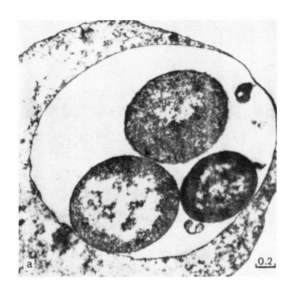

FIGURE 3. *Bacillus subtilis* L Form shows elementary bodies forming within a larger elementary body. (Original magnification × 53,000.)(From Prozorovsky, S. V., Katz, L. N., and Kagan, G. Ya., *AMS USSR, H'Meditsina*, 1981.)

The glycine-treated *Bordetella pertussis* of Milleck and associates never revealed the third mural layer evident in many of the parent rods.[26] Similarly, *E. coli, Spirillum serpens,* and Vitreoscilla sp. lost the innermost layer of their walls when spheroplasts were produced by penicillin or lysozyme. Demonstration of the layers was aided by treating the sections with uranium, thallium, or lead salts.[27,28]

Penicillin-induced spheroplasts of *Brucella abortus* clearly lose a wall stratum. A few organisms retain fragments of the missing layers; most of these revert to the parent form in antibiotic-free medium.[17]

Penicillin-induced spheroplasts of *B. suis* appear different from those of *B. abortus*. It was not apparent to Hines that cell wall is depleted, although the resulting globular forms show loss of rigidity. Surprisingly, the wall and cytoplasmic membrane are more distinct in the variant than in parent Brucella.[20]

The difficulty in inducing *in vitro* CWD forms of the tubercle bacillus is understood when one views the electron micrographs of Imaeda et al. The dense homogeneous cell wall of this bacillus may reach a thickness of 20 nm and is covered with a low-density layer, making the wall of the tubercle bacillus more than twice as thick as that of *Mycobacteria smegmatis*.[22] This probably explains why variants are more easily produced *in vitro* with *M. smegmatis* than with the pathogen.

In contrast to *in vitro* induction from the classic state, CWD Mycobacteria usually predominate *in vivo*, perhaps due to complexities of making the thick wall.

MUREINOPLASTS

A defective wall, but a wall nonetheless, characterizes bacterial forms termed "mureinoplasts".[8] The absent layers are the protein-lipid and the protein-polysaccharide portions. It is instructive that the rigid layer, which must be present due to the retained bacillary morphology, is so closely bound to the cytoplasmic membrane that it is invisible in the sectioned organism. This correlates with other data showing that the rigid layer may be of relatively minor width and closely applied to the cytoplasmic membrane. Mureinoplasts can be formed from some Gram-negative bacteria by simply washing in 0.5 *M* NaCl followed

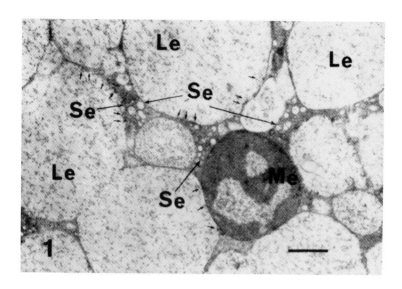

FIGURE 4. Elementary bodies of varied diameters are seen in L Form of *Staphylococcus aureus*: Large (Le), Medium (Me), and Small (Se). Sizes vary from 0.01 to 0.12 μm. (From Takahashi, T., *Jpn. J. Exp. Med.*, 50, 135–140, 1980. With permission.)

by suspension in 0.5 *M* sucrose. Such mureinoplasts are highly sensitive to EDTA and lysozyme, with resultant protoplasts.

When the same strain of organism was washed in sucrose without prewashing in NaCl, another type of wall-defective microbe resulted. The outer layers of the wall were retained but were grossly stretched and became detached from the cytoplasmic membrane except at a few points. The organism retained its rod shape, indicating an intact murein layer. The cells were easily transformed into protoplasts by the action of EDTA-lysozyme. Apparently, exposure to sucrose results in solubilizing the outer lipoprotein and lipopolysaccharide layers.

THE NUCLEOID

Although the nuclear material (*nucleoid*) of classic bacteria is not enclosed in a nuclear membrane, it is condensed in a gelatinous matrix and can be separated as a unit. Spiegelman et al. concentrated ''nuclei'' from *Bacillus megaterium* as bipartite structures composed of a nonchromatinic center surrounded by a coat of chromatin, invisible by phase microscopy.[37] DNA, invisible in phase optics, was seen as fibrous strands in electron micrographs in shadowed preparations. After digestion with DNAase, the supporting core was unchanged, but the fibrous material had either disappeared or was reduced to short lengths. The purified nuclear bodies were composed of DNA, RNA, and protein in a ratio of 1:1:3.

The ''nucleoid'' of CWD variants often differs from the centralized, compact structure of classic bacteria described above. The nuclear material of a variant is frequently more intimately intermingled with the cytoplasm than in the parent form, as noted in an unstable spheroplast of *E. coli*[15] and in variants of *Brucella suis*,[20] *C. perfringens*,[31] and *P. vulgaris*[32] (Figure 5). Dannis and Marston noted the diffuseness and lack of fibrils in the nuclear area in a stable line of staphylococcal protoplasts.[7] The extent to which changes occur is somewhat dependent on the length of time the variant stays in the aberrant stage.

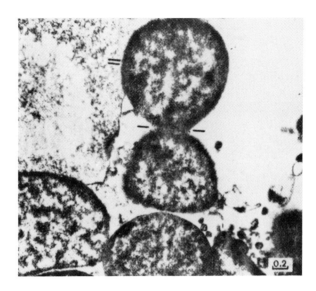

FIGURE 5. The nuclear material of the L Form of *Proteus vulgaris* is dispersed throughout the cytoplasm. (From Prozorovsky, S. V., Katz, L. N., and Kagan, G. Ya., *AMS USSR, H'Meditsina,* 1981.)

CHANGES IN THE CYTOPLASMIC MEMBRANE

MESOSOMES

As an organism becomes wall deficient, the loss of mesosomes in species which normally possess them is most striking. A mesosome is a ball of twisted cytoplasmic membrane which subtly connects the nucleoid with the encircling cell membrane.[35] When *M. smegmatis* is induced to spheroplasts with either lysozyme or glycine, some of the variants release their mesosomes; others retain them. Revertant bacteria may fail to regenerate their mesosomes and yet function normally.

Fitz-James found that a convenient way to collect mesosome vesicles is to first produce protoplasts, from which the mesosomes are extruded but still attached. Their separation from the protoplast membrane is then not difficult.[12]

EXCURSIONS OF THE CYTOPLASMIC MEMBRANE

While mesosomes disappear with protoplasting, new membranes of unknown function often appear within the cytoplasm.[1,14] Gram-negative organisms, which usually do not make mesosomes, may also develop a large number of membranous ingrowths after inductive treatment.[14,39] Such growth of multiple internal membranes is seen in Figure 6.

The converse of ingrowth may occur with *M. smegmatis:* by exfolding, the membrane of glycine-induced spheroplasts yields a number of irregularly spaced protrusions.[1]

NEW STRUCTURES

VACUOLES BETWEEN WALL AND MEMBRANE

One of the frequent changes in wall-deficient bacteria is an abnormal vacuole between the membrane and wall, often containing ribosome-like material. In penicillin-induced *B. suis* spheroplasts, such vacuoles may have a volume equaling that of the cell proper,[20] and

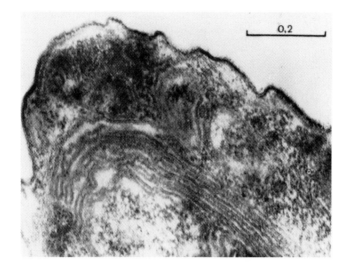

FIGURE 6. Multiple internal membranes in an L form of Vibrio strain NAG. (Original magnification ×
150,000.)(From Prozorovsky, S. V., Katz, L. N., and Kagan, G. Ya., *AMS USSR, H'Meditsina*, 1981.)

in glycine-induced variants of *Bordetella pertussis* this may sometimes cover an area 10
times greater than the portion enclosed by the cytoplasmic membrane.[26] Susan observed
them in water-cultured CWD *Pseudomonas aeruginosa*.[39]

NEW CAPSULES

Lederberg and St. Clair[25] identified their L-phase forms of *E. coli* by the refractile
capsule. Gumpert and associates similarly found a 0.15-nm capsule in *E. coli* CWD stages.[15]
In both laboratories the variants were penicillin induced, unstable forms.

IDENTIFYING-STRUCTURES IN ELECTRON MICROGRAPHS

MICROTUBULES IN GROUP D STREPTOCOCCI

A microtube traversing the entire organism characterizes Group D streptococcal L Forms
growing in and induced by penicillin. The tubules in the streptococcal variants often protrude
from the L-phase cell. They have an external diameter of about 25 nm and a hollow core
10 to 15 nm in diameter. The tubes are continuous with the cytoplasmic membrane, excluding
any possibility that they are rhapidosomes described in some other species. The inclusions
disappear after 15 months, when the L Form becomes stable.[5]

A second type of core, which is nontubular, also occurs in many strains of *Streptococcus
faecalis,* in both parent coccus and penicillin-induced protoplasts.[4] Higgins and Shockman
report that cores of either type do not occur in the streptococci in synthetic medium.[19] These
cores are similar to ones seen in a few other genera when wall deficient. (See chapter on
soil microorganisms).

INCLUSIONS IN CLOSTRIDIUM BOTULINUM L FORMS

A clue to the identification of *C. botulinum* types is given by electron microscopy, as
both types A and E have inclusions, assumed to be stored nutrient. The large, irregular
bodies in about 10% of type A protoplasts (Figure 7) consistently differ from the numerous
evenly rounded spheres in type E.[3]

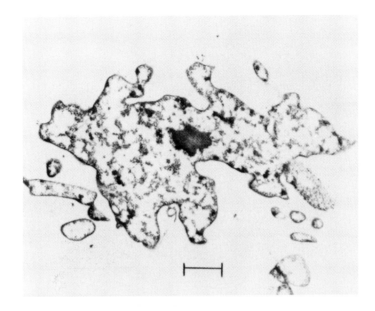

FIGURE 7.　Protoplast of *Clostridium botulinum* Type A showing characteristic irregular inclusion. Bar = 1 μm. (From Brown, G. W., Jr., Pate, J. L., and Sugiyama, H., *J. Bacteriol.*, 105, 1207–1210, 1971. With permission.)

DO ALL VARIANTS SURVIVE FIXATION FOR ELECTRON MICROSCOPY?

Methods employed in fixing CWD forms for electron microscopy are controversial. Often, standard procedures fail. The structure of plant protoplasts may be retained by immobilizing them in calcium alginate beads.[10] It has been noted that even classic bacteria may show wall damage due to osmium tetroxide.[18] Preparation for electron microscopy may remove the capsules from *Yersinia pestis*.[6,23] It is clear that whenever L-cycling is concerned, shadowed preparations should be compared with duplicates fixed for the microtome. Murray notes that the outermost, 10-nm coat of *Spirillum serpens* is not seen unless the organism is embedded in agar before fixation.[27]

FREEZE-ETCHING

In freeze-etching procedures, bacteria are solidified by freezing, rather than by chemical fixation. Freon-22 and a microtome knife cooled to −168°C are used. Unexpected findings emerge. Some of the layers apparent in the walls and membranes of chemically fixed preparations may be artifacts of fixation. In chemically fixed preparations most bacteria have a wall composed of only two layers, an outer layer and a layer directly above the cytoplasmic membrane.[33] However, exceptional bacterial walls may have only one layer,[11] others four.[29] In some species the number of wall strata remains controversial.

With freeze-etching the *cytoplasmic* membrane also seems to be less stratified than usually depicted. It appears as an integrated whole, rather than three strata, as in fixed preparations.

Freeze-etching also reveals the topography of a bacterial cell membrane, comparing its exterior and interior surfaces (Figure 8).[9] The convex surface is studded with 5- to 12-nm particles, interrupted by some particle-free patches, while particles on the concave surface

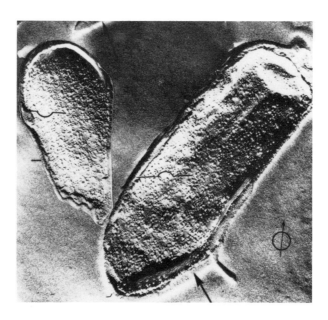

FIGURE 8.   Exterior ⌒ and interior ⌄ surfaces of protoplast cytoplasmic membrane of *Pseudomonas* sp. (From DeVoe, I. W., Costerton, J. W., and MacLeod, R. A., *J. Bacteriol.*, 106, 659–671, 1971. With permission.)

are widely separated. The membrane surfaces of bacteria and erythrocytes are remarkably similar.[38]

The value of freeze-etching was emphasized when an extremely halophilic archeobacterium was examined by both freeze-etching and routine transmission electron microscopy.[30] The flat triangular organisms appeared wall-free except in the cryostat preparations. In the future, interpretations of changes in bacterial membranes and walls will often be correlated with findings by frozen-section methods. However, new artifacts may result from the freezing procedure.[9]

An unexpected finding from frozen sections was "it has been definitely proved that mesosomes do not exist".[35] This does not negate the report that this presumed artifact behaves differently in CWD vs. the classic form.

SCANNING BEAM ELECTRON MICROSCOPE STUDIES

The scanning beam electron microscope shows how rapidly an antibiotic can alter the bacterial surface.[24] After only 5 min, lysostaphin causes marked cell wall alterations in staphylococci resembling the changes in lysostaphin-induced stable L Forms.[42]

The pleomorphism of wall-deficient bacteria is confirmed by the scanning scope. L Form colonies of Groups A and D streptococci examined by Bibel and Lawson revealed two kinds of morphology: large bodies with wrinkled exterior and granules with smooth surfaces. There was sometimes a suggestion of budding.[2]

Furutani and associates observed four steps as *Serratia marcescens* reverted from carbenicillin-induced spheroplasts to the classic stage. This proved a convenient species as the changes could be followed in liquid medium.[13]

SUMMARY

When a bacterium becomes wall deficient, the adjustment, as shown by fine structure, is often systemic, with changes throughout the cell. Biochemical alterations known to occur in the cytoplasmic membrane are also reflected in the thickness and other fine structure characteristics of this lamella which may assume new functions previously assigned to the wall. The double role assumed by the membrane of some protoplasts may explain the complex invaginations which such a membrane may make to increase its volume and surface.

New capsule-like material on the exterior of some spheroplasts may be a protective mechanism. Conceivably cytoplasmic vacuoles store essential nutrilites otherwise contributed by the wall.

The changes in the distribution and density of the nuclear material with changes in the wall are often dramatic. Physiological changes causing this response of the nucleoid remain to be explained.

REFERENCES

1. **Adámek, L., Mison, P., Mohelská, H., and Trnka, L.,** Ultrastructural organization of spheroplasts induced in *Mycobacterium* sp. *smegmatis* by lysozyme or glycine, *Arch. Mikrobiol.,* 69, 227–236, 1969.
2. **Bibel, D. J. and Lawson, J. W.,** Scanning-beam electron microscopy of L form colonies of Group A and D Streptococci, *Bacteriol. Proc. 1971,* p. 42, 1971.
3. **Brown, G. W., Jr., Pate, J. L., and Sugiyama, H.,** Fine structure of protoplasts and L forms of *Clostridium botulinum* Types A and E, *J. Bacteriol.,* 105, 1207–1210, 1971.
4. **Cohen, M., McCandless, R. G., Kalmanson, G. M., and Guze, L. B.,** Core-like structures in transitional and protoplast forms of *Streptococcus faecalis, in Microbial Protoplasts, Spheroplasts, and L Forms,* Guze, L. B., Ed., Williams & Wilkins, Baltimore, 1968, 94–109.
5. **Corfield, P. S. and Smith, D. G.,** Ultrastructural changes during propagation of a Group D streptococcal L form, *Arch. Mikrobiol.,* 75, 1–9, 1970.
6. **Crocker, T. T., Chen, T. H., and Meyer, K. F.,** Electron microscopic study of the extracellular materials of *Pasteurella pestis, J. Bacteriol.,* 72, 851–857, 1956.
7. **Dannis, D. C. and Marston, J. H.,** Fine structure of staphylococcal L forms, *Tex. Rep. Biol. Med.,* 23, 729–736, 1965.
8. **DeVoe, I. W., Thompson, J., Costerton, J. W., and MacLeod, R. A.,** Stability and comparative transport capacity of cells, mureinoplasts and true protoplasts of a gram-negative bacterium, *J. Bacteriol.,* 101, 1014–1026, 1970.
9. **DeVoe, I. W., Costerton, J. W., and MacLeod, R. A.,** Demonstration by freeze-etching of a single cleavage plane in the cell wall of a gram-negative bacterium, *J. Bacteriol.,* 106, 659–671, 1971.
10. **Draget, K. I., Myhre, S., Evjen, K., and Ostgaard, K.,** Plant protoplasts immobilized in calcium alginate, a simple method of preparing fragile cells for transmission electron microscopy, *Stain Technol.,* 63(3), 159–164, 1988.
11. **Fiil, A. and Branton, D.,** Changes in the plasma membrane of *Escherichia coli* during magnesium starvation, *J. Bacteriol.,*98, 1320–1327, 1969.
12. **Fitz-James, P.,** Fate of the mesosomes of *Bacillus megaterium* during protoplasting, *J. Bacteriol.,* 87, 1483–1491, 1964.
13. **Furutani, A., Tada, Y., and Yamaguchi, J.,** Reversion to bacillary forms of *Serratia marcescens* spheroplasts induced by carbenicillin, *Microbiol. Immunol.,* 29, 901–907, 1985.
14. **Gumpert, J., Schuhmann, E., and Taubeneck, U.,** Besonderheiten der Zellwandstruktur bei einer instabilen L form von *E. coli* K 12 (λ), *Z. Allg. Mikrobiol.,* 9, 271–281, 1969.
15. **Gumpert, J., Schuhmann, E., and Taubeneck, U.,** L formen von *Escherichia coli* K 12 (λ). II. Elektronenoptische Untersuchungen, *Z. Allg. Mikrobiol.,* 11, 283–300, 1971.
16. **Gumpert, J., Schuhmann, E., and Taubeneck, U.,** Ultrastruktur der stabilen L formen von *Escherichia coli* B und W 1655 F + [1], *Z. Allg. Mikrobiol.,* 11, 19–33, 1971.
17. **Hatten, B. A., Schulze, M. L., Huang, S. Y., and Sulkin, S. E.,** Ultrastructure of *Brucella abortus* L forms induced by penicillin in a liquid and in a semisolid medium, *J. Bacteriol.,* 99, 611–618, 1969.
18. **Heijenoort, J. van, Menjon, D., Flouret, B., Szulmajster, J., Laporte, J., and Batelier, G.,** Cell walls of a teichoic acid deficient mutant of *Bacillus subtilis, Eur. J. Biochem.,*20, 422–450, 1971.
19. **Higgins, M. L. and Shockman, G. D.,** Early changes in the ultrastructure of *Streptococcus faecalis* after amino acid starvation, *J. Bacteriol.,* 103, 244–254, 1970.

20. **Hines, W. D., Freeman, B. A., and Pearson, G. R.,** Fine structure of *Brucella suis* spheroplasts, *J. Bacteriol.,* 87, 1492–1498, 1964.

21. **Hofschneider, P. H. and Martin, H. H.,** Diversity of surface layers in L forms of *Proteus mirabilis, J. Gen. Microbiol.,* 51, 23–33, 1968.

22. **Imaeda, T., Kanetsuna, F., Rieber, M., Galindo, B., and Cesari, R. M.,** Ultrastructural characteristics of mycobacterial growth, *J. Med. Microbiol.,* 2, 181–186, 1969.

23. **Katz, L. N.,** On the submicroscopic structure of *Pasteurella pestis, Holland Zh. Mikrobiol.,* 43, 84–86, 1966.

24. **Klainer, A. S. and Perkins, R. L.,** Antibiotic-induced alterations in the surface morphology of bacterial cells, a scanning-beam electron microscope study, *J. Infect. Dis.,* 122, 323–328, 1970.

25. **Lederberg, J. and St. Clair, J.,** Protoplasts and L type growth of *Escherichia coli, J. Bacteriol.,* 75, 143–158, 1958.

26. **Milleck, J., Rockstroh, T., and Ocklitz, H. W.,** Die Feinestruktur glycininduzierter spharoplasten im vergleich zu der normaler bakterienzellen von *Bordetella pertussis, Zentralbl. Bakteriol. Parasitenkd. Infektionskr. Hyg. Abt. 1 Orig.,* 309, 37–43, 1968.

27. **Murray, R. G. E.,** On the cell wall structure of *Spirillum serpens, Can. J. Microbiol.,* 9, 381–392, 1963.

28. **Murray, R. G. E., Steed, P., and Elson, H. E.,** The location of the mucopeptide in sections of the cell wall of *Escherichia coli* and other gram-negative bacteria, *Can. J. Microbiol.,* 11, 547–560, 1965.

29. **Nanninga, N.,** Ultrastructure of the cell envelope of *Escherichia coli* B after freeze-etching, *J. Bacteriol.,* 101, 297–303, 1970.

30. **Nishiyama, Y., Sawada, K., Eda, T., Takashina, T., Usami, R., Akiba, T., and Horikoshi, K.,** Ultrastructure of archaebacterium grown under high concentration of sodium chloride, *J. Electron Microsc.,* 36(5), 307–308, 1987.

31. **Petrovici, A., Bittner, J., and Voinesco, V.,** Variabilite du *Cl. perfringens.* II. Aspects Electrono-optiques de la variante B.P.$_6$K -P.$_{1000}$, *Arch. Roum. Pathol. Exp. Microbiol.,* 22, 681–690, 1963.

32. **Prozorovsky, S. V., Katz, L. N., and Kagan, G. Ya.,** L forms of Bacteria (Mechanism of formation, structure, role in pathology), *AMS USSR, H'Meditsina,* 1981.

33. **Remsen, C. and Lundgren, D. G.,** Electron microscopy of the cell envelope of *Feffobacillus ferroxidans* prepared by freeze-etching and chemical fixation techniques, *J. Bacteriol.,* 92, 1765–1771, 1966.

34. **Rensburg, A. J. van,** Properties of *Proteus mirabilis* and providence spheroplasts, *J. Gen. Microbiol.,* 56, 257–264, 1969.

35. **Ryter, A.,** Contribution of new cryomethods to a better knowledge of bacterial anatomy, *Ann. Inst. Pasteur Microbiol.,* 139, 33–44, 1988.

36. **Shadrina, I. A.,** Comparative study of the *P. vulgaris* and *St. haemolyticus* L form substructures, *Vestn. Akad. Med. Nauk. SSSR,* 20, 17–20, 1965.

37. **Spiegelman, A., Aronson, A. I., and Fitz-James, P. C.,** Isolation and characterization of nuclear bodies from protoplasts of *Bacillus megaterium, J. Bacteriol.,* 75, 102–109, 1950.

38. **Steck, T. L., Weinstein, R. S., Straus, J. H., and Wallach, D. F. H.,** Inside-out red cell membrane vesicles, preparation and purification, *Science,* 168, 255–257, 1970.

39. **Susan, S. R.,** A Fine Structure Study of *Pseudomonas aeruginosa* in Media and Distilled Water, M.S. thesis, Wayne State University, Detroit, 1972.

40. **Takahashi, T.,** Fine structures of various elements of Staphylococcal L colonies, *Jpn. J. Exp. Med.,* 50(2), 135–140, 1980.

41. **Thorsson, K. G. and Weibull, C.,** Studies on the structure of bacterial L forms, protoplasts and protoplast-like bodies, *J. Ultrastruct. Res.,* 1, 412–427, 1958.

42. **Watanakunakorn, C., Fass, R. J., Klainer, A. S., and Hamburger, M.,** Light and scanning-beam electron microscopy of wall-defective *Staphylococcus aureus* induced by lysostaphin, *Infect. Immun.,* 4, 73–78, 1971.

Chapter 7

PUBLIC HEALTH AND NOSOCOMIAL FACETS

SALMONELLA AND SHIGELLA CARRIERS

Typhoid fever is an unsolved world health problem. Approximately 12 million cases are recognized each year.[8] Complications of the disease are not limited to the gastrointestinal tract.

Specific antibiotic therapy for typhoid can only be initiated after the diagnosis is removed from the category of "Fever of Unknown Origin". In approximately half the cases, culture of stool, urine, and blood fails to reveal the Salmonella. For example, blood cultures show the pathogen in only 40 to 70% of patients.[29] Diagnosis may be improved by identifying *Salmonella typhi* antigens in blood. See Immunology chapter.

In the earliest days of bacteriology, Almquist concluded that the pleomorphic variant of *S. typhi* was a pathogen (see History chapter). How often the CWD form alone causes disease remains to be determined. Pathogenicity of this variant for tissue culture has been demonstrated (Figure 1).[37]

Another approach to diagnosis is contributed by Prozorovsky and associates, who found the bone marrow a favorite site of residence for L Forms of *S. typhi*.[28] Not only are they present during active disease, but their recognition can aid in carrier-detection. Identification is facilitated by using antibody made vs. the L Form (Figures 2 and 3). Hauduroy demonstrated that CWD typhoid bacilli abound in the stools of convalescent patients.[15]

The possibility that rodent and other carriers of Salmonella harbor the CWD stage needs to be investigated. Ticks given classic *S. typhimurium* soon are found to be carrying only the pleomorphic variants (Figure 4).

Germ-free rats inoculated with *Shigella sonnei* soon become carriers of L Forms of the pathogen. The investigators speculate that a similar situation may occur in man.[21]

CWD FORMS IN WATER SUPPLIES

Contaminated drinking water can usually only be incriminated by epidemiological evidence since it is almost impossible to find pertinent bacteria in the water. The noncultivable infective culprit may well be CWD, undetectable by standard methods. Hauduroy found filtrable forms of typhoid bacilli in well water.[15]

Electron microscopic confirmation of wall-deficient bacteria in "potable" water may exist in the preparations of Mack et al.[25] The minute pleomorphic units in a concentrate of well water strongly resemble wall-deficient stages of bacteria. The water supplied a roadside restaurant which was a site of "food poisoning" on several occasions. No food was incriminated in repeated tests, and it became apparent that patrons who drank water were the victims. Concentrates of the water did not yield any enteropathogenic bacteria or viruses. Interesting ancillary findings included: (1) isolation of polio type II (vaccine strain), giving unsolicited vaccination to many patrons; (2) this sewage-contaminated water passed standard tests for *Escherichia coli* content. This article raises the question whether drinking water free of classic enteric bacteria may be infective because of wall-deficient bacteria.[25]

Many rivers function both for sewage disposal and city water supplies, so, not unexpectedly, such water may contain CWD stages of pathogenic bacteria. A bacterium from the Chicago river originally grew as an amorphous film. The organism, serologically related to *Shigella flexner*, required seven serial subcultures to revert.[26]

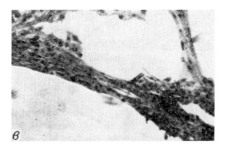

FIGURE 1. The L Form of *Salmonella typhi* has destroyed a sheet of chick embryo cells. (From Timakov, V. and Kagan, G., *L Form Bacteria and Species of Mycoplasma in Pathology,* Moscow, 1973.)

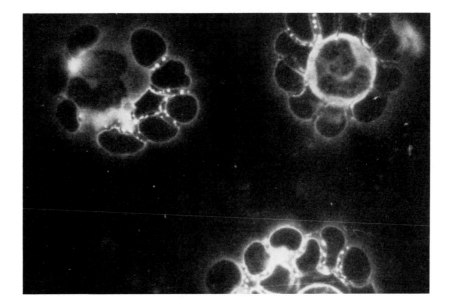

FIGURE 2. Bone marrow smear from typhoid patient. Fluorescent antibody vs. L Forms reacts with antigen on the surface of erythrocytes rosetting around monocytes and granulocytes. (Photograph furnished by I. Rakovskaya.)

A spheroplasting vibrio found in river water near Madrid is shown in Figure 5. This bacterium spontaneously develops thin-walled spheres.[40]

WALL-DEFICIENT BACTERIA IN FOODS

THE FLORA OF GROUND MEATS

Hamburger, regardless of flavor and freshness, is notoriously bacteria laden. When pour-plate cultures of hamburger are examined microscopically at 18 h, microcolonies of CWD forms predominate, as shown in Figure 6. This is usually true of hamburger of good as well as poor quality. By 48 h, when total counts are commonly made, the majority of colonies have been overgrown by the classic forms; a few have reverted. Comparing counts of microscopic colonies at 24 h with macroscopic colonies at 48 h indicates that most hamburger contains five to ten times as many CWD forms as classic bacteria. Jay finds approximately 10^7 macroscopic bacterial colonies growing from 1 g of commercial ground beef.[19] Thus, the total count of CWD plus classic bacteria is often 10^{15}/g. Culture of sausage gives results

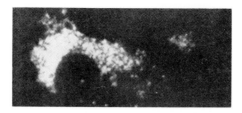

FIGURE 3. Fluorescent antibody to L Form of *Salmonella typhi* identifies organisms in bone marrow leukocyte of typhoid carrier. (Photo submitted by Vulfovich, Yu., Pogorelskaya, L., Levina, G., Prozorovsky, S., and Yliynsky, Yu.)

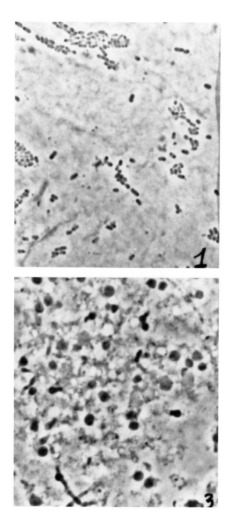

FIGURE 4. (Top) *Salmonella typhimurium* in classic state isolated 1 h after tick was injected. (Bottom) *S. typhimurium* from tick 20 d after infection. (Photo submitted by Ligangirova, N., Katz, L. N., and Konstantinova, N.)

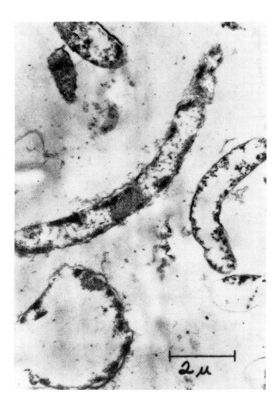

FIGURE 5. Fine structure of unidentified vibrio and its naturally forming spheroplasts. (Micrograph courtesy of M. A. Jareño and J. Pérez-Silva, Institute "Jaime Ferrán" of Microbiology, Madrid.)

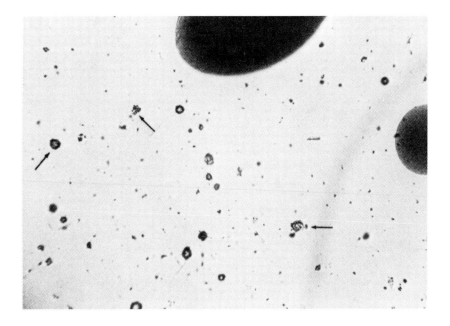

FIGURE 6. Two large classic colonies and multiple CWD colonies, of which three are indicated by arrows. Twenty-four-hour culture of hamburger. (Magnification × 200.)

similar to hamburger. It is evident that a CWD pathogen would usually be masked by "normal flora" overgrowth.

BUTTER?

Did the 230,000 lb of butter recalled from U.S. cities because of staphylococcal enterotoxin content contain L-phase variants of *Staphylococcus aureus?* It is possible, since staphylococcal wall-deficient variants are highly salt-tolerant and toxigenic. No toxigenic classic staphylococci were found.

MILK

In Germany, CWD *Streptococcus agalactiae, Staphylococcus aureus* and *Corynebacterium pyogenes* were all found to be agents of bovine mastitis by Bergmann and Böckel. They recommended that when one cultures for agents of mastitis, the medium should have stabilizing agents. Their medium included heart muscle extract, 15% inactivated horse serum, and 3.5% NaCl.[2]

Sears and associates noticed that bovines infected with *S. aureus,* after treatment with cloxacillin cease excreting the classic coccus. In contrast CWD *S. aureus* still contaminates the milk through 30 d.[31]

"STERILE" WATER IN THE HOSPITAL

Many conscientious physicians would like to see nosocomial infections decline from the current 20,000 confessed deaths per year.[23] Currently, intensive care units in the majority of large hospitals should have a sign: "ALL IS FAIR IN INTENSIVE CARE". Pre-Semmelweiss practices are followed and introduced infections brushed aside as originating in the patient. Some of these tragic and costly accidents can be avoided if the staff knows that pathogens multiply in triple-distilled water, even in the refrigerator. The pathogens are usually Gram negative rods, abundant in hospital air, which enter a bottle of water exposed even briefly to the environment. Such growth usually initiates in the CWD stage.

As early as 1976, Farmer noted that specific typing sera can identify infection sources of Pseudomonas. He emphasized the importance of *Pseudomonas cepacia* which like other pseudomonads grows even in disinfectants.[11]

P. AERUGINOSA CONTAMINATION

The world medical literature is replete with reports of serious infections caused by water thought to be sterile, as reviewed by Tomlin[38] and others.[1,4,35,39] The most common waterborne pathogen, *P. aeruginosa,* habitually requires a complex organic medium for maintenance. Paradoxically, in the CWD stage it can grow in distilled water, utilizing the few molecules which diffuse from glass and CO_2 from the air.[36]

Reports describe pneumonia of the newborn traced to bacteria in incubator humidifiers[16] or delivery room resuscitators.[12] Meningitis results from contaminated spinal anesthesia fluids[4,10,39] and panophthalmia from bacilli in eye-wetting solutions.[1] Brain abscesses follow surgery with instruments chemically sterilized and then rinsed with "sterile" water.[30] Catheterizations, both of the urinary tract[14] and of the cardiovascular system,[33] continue to cause sporadic infections; the contaminants are usually found in the distilled water used to rinse the catheters. Once-sterile water is often stored in a large glass jar with a rubber dispensing tube; as water is collected from the jar, an equal volume of room air enters. Triple-distilled and pyrogen-free water is parsimoniously stored for intermittent use, allowing proliferation of contaminants, even though refrigerated.

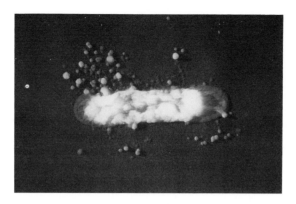

FIGURE 7. A platinum-carbon shadowed electron micrograph (EM) preparation of *Pseudomonas aeruginosa* growing in distilled water shows elementary bodies around a classic bacillus. (Sectioning in EM usually lyses the small forms.)(Micrograph provided by S. Susan).

The importance of *P. aeruginosa* in hospital infections is illustrated by the two Jarvik artificial heart recipients who survived approximately a year. At autopsy, Pseudomonas abscesses were present in almost every major organ. Apparently the possible relationship of the strains to a common source was not investigated; therefore, the possibility remains that the infections were endogenous.[6] It is estimated that over 80,000 infections occur annually in the U.S. related to implant and devices.[34]

Laboratory investigations show that "sterile" water may carry insidious CWD phase pathogens. A long-term study of "water-cultured" *P. aeruginosa* was made by Susan. L-phase growth was repeatedly demonstrated by electron micrographs and by culture.[36] The variants appear as filtrable granules, budding yeasts, and spherical "L-bodies" (Figure 7).

Spontaneous induction into the L-phase seems to occur when any strain of *P. aeruginosa* enters distilled water, e.g., strains from the urinary tract, blood, pericardial fluid, an unopened bottle of wetting solution for contact lenses, and strain 14502 of the American Type Culture Collection, a standard for disinfectant evaluation. Pseudomonas strains grow well in water, irrespective of their culture history. Strains were identified by biochemical reactions and formation of pyocyanin and fluorescein.[36]

It is important to note that water containing up to 10^6 organisms per milliliter of wall-deficient Pseudomonas appears sterile if cultured in broth or on the surface of an agar plate. A pour plate is required to reveal these variants which would be classic after a few more days of incubation in the water (Figure 8).

SIGNIFICANT FACTORS IN SUCCESSFUL CONTAMINATION WITH *P. AERUGINOSA*

Plotkin and Austrian described a series of *P. aeruginosa* infections resulting from gauze sponges soaked in zephiran.[27] This may be explained by zephiran chloride inducing wall deficient variants from Pseudomonas, and gauze remarkably stimulating their growth. Gauze and other cotton fibers which are suspended in large numbers in hospital air[24] foster the growth of Pseudomonas in any stagnant water. Rubber stoppers, like gauze, are highly nutritious for Pseudomonas wall deficient forms.

The *minimum number* of organisms required to initiate CWD growth in water, 100 to 200 bacilli placed on the surface, are often airborne in the hospital, as hospital dust may contain 300 *P. aeruginosa* per milligram,[32] and eschar shed from an infected burn patient may contain nearly 4,000,000 Pseudomonas per square centimeter.[18]

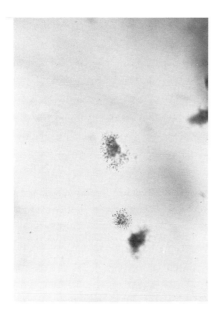 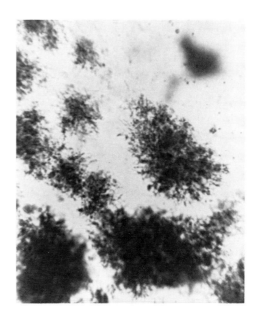

FIGURE 8. (A) Pour plate of *Pseudomonas aeruginosa* growing in distilled water shows small rhizoid colonies at bottom of the culture. (Original magnification × 100.) (B) Colonies after more extended time. (Original magnification × 1000.)(Photographs submitted by S. Susan).

DETECTING THE PSEUDOMONAS VARIANTS

Microscopic examination of water containing CWD Pseudomonas reveals abundant tiny spheres, 0.4 μm or less in diameter. Electron micrographs by Susan show tiny, membrane-bound spheres formed by budding cells and numerous minute units.[36] The stability of the variants in subculture varies; in general, when organisms are very numerous, rapid reversion occurs.

When nutrient broth subculture indicates water is sterile, as many as 8,000,000 CWD forms may be detected by pour-plate cultures in yeast dextrose agar. Mycoplasma medium containing 20% serum is a superior medium for detecting a few CWD forms of Pseudomonas.

REVERSION

Reversion in water usually begins when the Pseudomonas variants number over 1,000,000/ ml. There is much evidence that in some instances the Pseudomonas variant per se can be pathogenic. The L-stage of Pseudomonas from infections has been cultured in the laboratories of Kagan,[20] of Kenny,[22] and the author. In one instance the CWD variant of *P. aeruginosa* occurred in repeated blood cultures of a patient who expired with septicemia following cardiac catheterization.

OTHER "WATER BUGS"

It seems probable that, in addition to Pseudomonas, many other "water bugs" of nosocomial importance adapt to growth in distilled water by initially propagating in the CWD state. The variety of bacteria and fungi which can propagate in distilled water is revealed when water is cultured in yeast dextrose agar (Figure 9). Even Salmonella has been reported in the water of a pump used for cardiac surgery, suggesting that multiplication in the water had taken place.

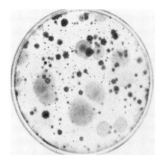
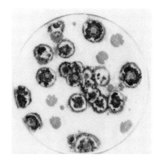
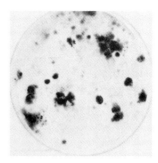

FIGURE 9. Typical findings when 0.1 ml of distilled or deionized water is incubated 2 weeks in pour plates. Initial growth is predominantly as CWD colonies.

An organism prominent as a cause of nosocomial infection is the Pseudomonad EO-1 (*Pseudomonas kingii*) found in commercial urinary catheter kits by Hardy et al.[14] The wall-deficient variant in two cases of persistent bacteremia studied in our laboratory reverted to yellow pigmented strains of this pseudomonad.

Japanese investigators suggest that *Serratia marcescens* is an especially effective nosocomial infector because it can rapidly switch from spheroplast to classic form with concomitant switch in antibiotic sensitivities.[13]

THE POSSIBILITY OF CWD STAGES IN TOXOID

No explanation has been published of why tetanus developed in Mexican children receiving a 1970 batch of imported toxoid. The possibility exists that a filtrable stage of the Clostridium regenerated sufficiently to make toxin or later became infective after storage. A precedent exists for this type of error in regeneration of filtrable forms of Group A streptococci which caused scarlet fever in persons skin-tested with Dick toxin (see chapter on Filtrable Forms).

PLAGUE-CARRYING RODENTS

When wild sand mice were tested for *Yersinia pestis,* the CWD stage was found in twice as many animals as was the classic pathogen. In 345 sick animals, the CWD variant was detected in 19, and the bacillary form in 8. Infected tissues were usually liver and spleen and, in some instances, the kidney. Identification methods for Yersinia included fluorescent antibody and reversion of the variant to the classic stage.[7]

COMMENT

It has been well established in the world literature that hospital-acquired infections, both sporadic and epidemic, are frequently due to waterborne Gram-negative rods. Contamination of distilled water by CWD forms should be considered in the epidemiology of such incidents. CWD forms of Pseudomonas vary in pathogenicity, and it is probable that pathogenic variants are important, especially in antibiotic-treated patients. We have seen *P. aeruginosa* meningitis persist after antibiotics left only the CWD variant in the spinal fluid. Such studies may document that CWD microbes per se, as they naturally propagate in water, do cause nosocomial infections. It is substantiated that water "sterile" by routine tests may contain large numbers of wall-deficient stages which, a few hours later, attain the critical number necessary for reversion to the classic pathogen. It is also known that CWD Pseudomonas can be pathogenic.[3] The cotton fibers suspended in hospital air and the rubber stoppers of bottled water are excellent stimulants of CWD growth. The early observation by Dienes et al.[5] that distilled water may cause transformation into wall deprived variants has been overlooked.

Whereas moderate numbers of Pseudomonas are unable to propagate as classic forms in distilled water, the more autotrophic CWD form can multiply, presumably with CO_2 from the air serving as the carbon source. When reaching water in small numbers, *P. aeruginosa* loses the bacillary shape and multiplies in the L phase. The L phase of *P. aeruginosa* is without pigment[9,17] or odor. Heavily contaminated water usually appears innocuously clear.

REFERENCES

1. **Ayliffe, G., Barry, D., Lowbury, E., Roper-Hall, M., and Walker, W.,** Post-operative infection with *Pseudomonas aeruginosa* in an eye hospital, *Lancet*, 1, 1113–1117, 1966.
2. **Bergmann, V. A. and Böckel, K.,** Zur diagnostik von L formen der bakterien aus Mastitiden, *Mh. Vet. Med.*, 44, 98–101, 1989.
3. **Bertolani, R., Elberg, S. S., and Ralston, D.,** Variations in properties of L forms of *Pseudomonas aeruginosa*, *Infect. Immun.*, 11, 180–192, 1975.
4. **Corbett, J. J. and Rosenstein, B. J.,** Pseudomonas meningitis related to spinal anesthesia. Report of three cases with a common source of infection, *Neurology*, 21, 946–950, 1971.
5. **Dienes, L., Weinberger, H. J., and Madoff, S.,** The transformation of typhoid bacilli into L forms under various conditions, *J. Bacteriol.*, 59, 755–764, 1950.
6. **Dobbins, J. J., Johnson, S., Kunin, C., and Devries, W.,** Postmortem microbiological findings of two total artificial heart recipients, *JAMA*, 259, 865–869, 1988.
7. **Dunaev, G. S., Zykin, L. F., Cherchenko, I. I., Klassovsky, L. N., Prozorovsky, S. V., Metlin, V. N., Rybkin, V. S., Kurilov, V. Y., et al.,** Isolation of the causative agent of plague in the L form from wild rodents in a natural focus of infection, *Zh. Mikrobiol. Epidemiol. Immunobiol.* 8, 50–54, 1982.
8. **Edelman, R. and Levine, M. M.,** Summary of a workshop on typhoid fever, *Rev. Infect. Dis.*, 8, 329–349, 1986.
9. **Edman, L. A. and Mattman, L. H.,** L variation in *Pseudomonas aeruginosa* induced by antibiotics, *Bacteriol. Proc. 1961*, p.82, 1961.
10. **Evans, R. T.,** Infection from spinal anesthesia, a warning, *Lancet*, 1, 115, 1945.
11. **Farmer, J. J., III,** Pseudomonas in the hospital, *Hosp. Practice*, 11(2), 63–70, 1976.
12. **Frierer, J., Taylor, P., and Gexon, H.,** *Pseudomonas aeruginosa* traced to delivery-room resuscitators, *N. Engl. J. Med.*, 276, 991–996, 1967.
13. **Furutani, A., Tada, Y., and Yamaguchi, J.,** Reversion to bacillary forms of *Serratia marcescens* spheroplasts induced by carbenicillin, *Microbiol. Immunol.*, 29(10), 901–907, 1985.
14. **Hardy, P. C., Ederer, G. M., and Matsen, J. M.,** Contamination of commercially packaged urinary catheter kits with the pseudomonad EO-1, *N. Engl. J. Med.*, 282, 33–35, 1970.
15. **Hauduroy, P.,** Présence des formes filtrantes du bacille d'Eberth dans le sang d'un typhique, *C. R. Soc. Biol. (Paris)*, 95, 288–289, 1926.
16. **Hoffman, M. and Finberg, L.,** Pseudomonas infections in infants, associated with high-humidity environments, *J. Pediatr.*, 46, 626–630, 1955.
17. **Hubert, E. G., Potter, C. S., Hensley, T. J., Cohen, M., Kalmanson, G. M., and Guze, L. B.,** L forms of *Pseudomonas aeruginosa*, *Infect. Immun.*, 4, 60–72, 1971.
18. **Hurst, V. and Sutter, V. L.,** Survival of *Pseudomonas aeruginosa* in the hospital environment, *J. Infect. Dis.*, 116, 151–154, 1966.
19. **Jay, J. M.,** *Modern Food Microbiology*, 4th ed., Van Nostrand Reinhold, New York, 1992.
20. **Kagan, B. M.,** Role of L forms in staphylococcal infections, in *Microbial Protoplasts, Spheroplasts, and L Forms*, Guze, L. B., Ed., Williams & Wilkins, Baltimore, 1968, 372–378.
21. **Kaminskii, G. D., Konstantinova, N. D., Popov, V. L., and Chakhava, O. V.,** Dissociation and ultrastructure of *Shigella sonnei* cells isolated from germ-free animals and obtained in vitro, *Zh. Mikrobiol. Epidemiol. Immunobiol.*, 7, 20–27, 1989.

22. **Kenny, J. F.,** Role of cell wall defective microbial variants in human infections, *South. Med. J.,* 71, 180–190, 1978.

23. LifeSpan-Plus, *Prevention,* Rodale Press, PA, 1990.

24. **Litsky, B. Y. and Litsky, W.,** Bacterial shedding during bed-stripping of reusable and disposable linens as detected by the high volume air sampler., *Health Lab. Sci.,* 8, 29–34, 1971.

25. **Mack, W. N., Lu, Y. S., and Coohon, D. B.,** Isolation of poliomyelitis virus from a contaminated well, *Health Serv. Rep.,* 87, 271–274, 1972.

26. **McDaniels, H. E. and Neal, J.,** Filter-passing bacteria in polluted water, *Proc. Soc. Exp. Biol. Med.,* 30, 115–117, 1932.

27. **Plotkin, S. and Austrian, R.,** Bacteremia caused by *Pseudomonas* sp. following the use of materials stored in solutions of a cationic surface-active agent, *Am. J. Med. Sci.,* 235, 621–627, 1958.

28. **Prozorovsky, S. V., Vulfovich, Y. V., Pogorelskaya, L. V., Levina, G. A., Ilinsky, Y. A., and Gorelov, A. L.,** Detection of the microbial variants of *Salmonella typhi* in the bone marrow of typhoid patients and carriers, *Zh. Mikrobiol. Epidemiol. Immunobiol.,* 12, 15–17, 1986.

29. **Rubin, F. A., McWhirter, P. D., Punjabi, N. H., Lane, E., Sudarmono, P., Pulungsih, S. P., Lesmana, M., Kumala, S., Kopecko, D. J., and Hoffman, S. L.,** Use of a DNA probe to detect *Salmonella typhi* in the blood of patients with typhoid fever, *J. Clin. Microbiol.,* 27, 1112–1114, 1989.

30. **Sandusky, W.,** Pseudomonas infections, sources and cultural data in a general hospital with particular reference to surgical infections, *Ann. Surg.,* 153, 996–1005, 1961.

31. **Sears, P. M., Fettinger, M., and March-Salin, J.,** Isolation of L form variants after antibiotic treatment in *Staphylococcus aureus* bovine mastitis, *J. Am. Vet. Med. Assoc.,* 191, 681–684, 1987.

32. **Selwyn, S., Maccabe, A., and Gould, J.,** Hospital infection in perspective; the importance of the gram negative bacilli, *Scot. Med. J.,* 9, 409–417, 1964.

33. **Shickman, M. D., Guze, L. B., and Pearce, M.,** Bacteremia following cardiac catheterization, report of case and studies on source, *N. Engl. J. Med.,* 260, 1164–1166, 1959.

34. **Stamm, E. E.,** Infections related to medical devices, *Ann. Int. Med.,* 89, 764–769, 1978.

35. **Stanley, M.,** *Bacillus pyocyaneus* infections, *Am. J. Med.,* 2, 253–277, 347–367, 1947.

36. **Susan, S.,** A Fine Structure Study of *Pseudomonas aeruginosa* in Media and Distilled Water, M.S. thesis, Wayne State University, Detroit, 1972.

37. **Timakov, V. and Kagan, G.,** *L Form Bacteria and Species of Mycoplasma in Pathology,* Moscow, 1973.

38. **Tomlin, C.,** Pseudomonas meningitis, *Arch. Intern. Med.,* 87, 863–867, 1951.

39. **Weinstein, L. and Perrin, T.,** Meningitis due to *Pseudomonas pyocyanea,* a report of three cases treated successfully with streptomycin and sulfadiazene, *Ann. Intern. Med.,* 29, 102–117, 1948.

40. **Jareño, M. A. and Pérez-Sila, J.,** personal communication.

Chapter 8

IMMUNOLOGY

DETECTION OF TYPHOID CASES AND CARRIERS BY THEIR SPHEROPLAST ANTIBODY OR ANTIGENEMIA

Previously it has been almost impossible to recognize a typhoid carrier. Only rarely does classic *Salmonella typhi* appear in his stools, during an intestinal upset, and attempts to detect the carrier by his antibody are unsatisfactory. He has no significant level of antibody to the classic form of *S. typhi*.

In contrast, typhoid carriers do carry significant levels of antibody to *spheroplasts* of the typhoid bacillus, detected in a simple agglutination test. Wallace et al. found that the spheroplast preparations were relatively stable when refrigerated and gave sensitive and reproducible results.[56]

Another study concerned both cases and carriers. After detection of pertinent antigens in patients' sera typhoid was diagnosed in 37 culture-negative patients. L-Form antigens were also present after the acute phase and in carriers.[17]

WALL-DEFICIENT FORMS AS VACCINES

CWD variants make superior or inferior vaccines, varying with species and mode of preparation. Milleck and Ocklitz[34] found that the wall-membrane complex from glycine-induced spheroplasts of *Bordetella pertussis* stimulates good protective antibody and is less toxic than parental walls. Encouraging results were also obtained by Mason with lysed spheroplast vaccines of Bordetella.[31] The protective antigen, histamine-sensitizing factor, toxin, and agglutinin were present in substantial amounts in the cultures. Similarly, more antibody was made by rabbits in response to *Proteus mirabilis* L Forms than after injection of the parent bacilli.[21]

A pertinent study found that membranes from stable L Forms of *Escherichia coli* stimulated an excellent anamnestic response. After a single intraperitoneal injection of membranes, a reinjection of living *E. coli* was attacked by an influx of leukocytes four to fivefold greater than if no preliminary vaccination had occurred.[22]

A new approach is worth considering in typhoid vaccination since standard typhoid vaccines are sometimes pyrogenic and have not immunized against all strains encountered. It was found that either intact spheroplasts or spheroplast lysate makes an excellent vaccine,[9] arousing five times as much protective antibody as a vaccine of classic bacilli. The toxicity of the glycine-induced variants employed, as tested in mice and rabbits, is much diminished, compared with standard vaccines. However, some pyrogenicity still remained.

Another favorable report concerns *Vibrio cholerae* immunogens.[1] In a study in India, where immunization to cholera is a lively issue, rabbits fed lysates of *V. cholera* spheroplasts produced high titers of both circulating antibody and coproantibody. Oral vaccination is logical since coproantibody is important here.[16,27]

Brucella abortus L Forms given subcutaneously results in IgM and IgG antibodies and resistance to *B. melitensis* for an observation period of 10 m.[54]

Thus, experimental data indicate that vaccines for whooping cough, typhoid, brucellosis, and cholera in the future may consist of spheroplasts. Less success was attained with CWD stages of *S. typhimurium*[49] and *B. suis*.[48]

SPHEROPLAST PRODUCTION BY LYSOZYME WITHOUT ANTIBODY

When lysozyme is employed to produce spheroplasts, EDTA is a common adjunct. While the addition of the chelating agent appears nonphysiological, perhaps it actually duplicates an *in vivo* system. The vacuoles of neutrophiles contain the chelating substance leukozyme,[42] which has EDTA-like properties. Thus, the wall-deficient bacteria common within granulocytes in exudate may be formed in these cells by synergistic action of the leukocytes, lysozyme, and leukozyme.

Species which have been induced to the spheroplast stage by lysozyme-EDTA include the halophilic vibrio, *V. parahemolyticus*,[24] and *Pseudomonas aeruginosa*.[11] In an early study, Repaske showed that lysozyme attacks many genera in a balanced solution of the enzyme, tris buffer, and EDTA (versene).[45,46] *E. coli*, *P. aeruginosa*, and *Azotobacter vinelandii* have susceptible walls under controlled conditions. Repaske concluded: "Mucopolysaccharide, the substrate for lysozyme, is a more common constituent of bacterial cell walls than was originally suspected." Furthermore, since the action of EDTA is to bind metallic ions, it was revealed that trace metals are necessary for bacterial wall integrity. The role of EDTA in assisting lysozyme disintegration of bacterial walls may be explained by participation of divalent metals in maintaining lattice structures in the bacterial wall.[12]

The presence of as little as 1.1 M NaCl in the growth medium can result in lysozyme-resistant cell walls, and KCl has a comparable effect.[5] This is a built-in effect, not removable by washing the bacteria.

Lysozyme in tissue may increase in response to mycobacterial infection, as noted when Oshima et al. analyzed granulomatous lung.[41] It is tempting to believe that this ability to stimulate lysozyme production may be responsible for the predominance of wall-deficient variants in many Mycobacteria-infected sites. Thacore and Willett formed spheroplasts from *Mycobacteria tuberculosis* by exposure to lysozyme with EDTA or monocyte extract.[53]

The enigma of why lysozyme is often lethal to bacteria, but at other times produces well-adapted viable L Forms, was elucidated by King and Gooder.[25] The best viability was obtained with much less lysozyme than that which resulted in maximum production of the round spheroplasts. Similarly, Landman[28] found that lysozyme erosion of the wall of *Bacillus subtilis* can progress to an intermediate point with resultant propagating spheroplasts.

ANTIBODY-COMPLEMENT IN INDUCTION OF VARIANTS

In an early review, Dienes et al. noted that *S. typhi* developed L Forms on exposure to antibody and complement.[10] This inducing action by antibody was found in other Salmonella species in the extensive studies of Apgar.[2] In these experiments with antibody and complement, it may be assumed that the lysozyme present (2 μg/ml) in undiluted serum is diluted beyond a level of contributory activity.[8]

In vivo lysozyme does contribute to extracellular spheroplast formation as elucidated by Muschel and co-workers. Participating factors in spheroplast formation in body fluids are antibody, complement, and lysozyme; other substances in the milieu may contribute to hypertonicity and to stability of the spheroplasts.[38-40]

Table 1, data from Crombie and Muschel,[7] clearly illustrates the varied responses of bacterial strains, even of the same species, to an antibody-complement-lysozyme system. *V. cholerae* emerges as an exceptionally supple organism which can shed its wall without requiring lysozyme.[43] Strains of *S. typhi* vary from small to large lysozyme requirements. The investigators note that the lysozyme-resistant *S. typhi* Ty2 is highly virulent and contains a large amount of Vi antigen. They conclude that there is probably a relationship between

TABLE 1
Lysozyme Requirement for
Spheroplasting

Organism	Lysozyme amount, μg*
Vibrio cholerae Inaba	0
Escherichia coli B	50
Salmonella typhi 0901	20
S. typhi Watson	100
S. typhi Ty2	500

* Lysozyme required for complete spheroplasting in the presence of adequate specific antibody and complement. Antibody requirement for killing (in absence of osmotic stabilizers) and for spheroplasting (in presence of osmotic stabilizers) were essentially the same.

Modified from Crombie, L. B. and Muschel, L. H., *Proc. Soc. Exp. Biol. Med.*, 124, 1029, 1967.

the lysozyme resistance of a bacterium and the quantitative, as well as qualitative, composition of the cell wall. Lysozyme may act after antibody and complement cause hydration of the cell wall layers with consequent exposure of the lysozyme substrate.

In summary there is evidence that the virulence armamentarium is associated with cell wall antigens not completely neutralized by the antibody-complement-lysozyme system.

Wittler noted that *Hemophilus influenzae* is more rapidly converted to L-phase variants in immune than in normal mice.[59] The abundant pleomorphic and round forms appear in lung tissue 24 to 48 h after classic bacilli are instilled intranasally.

In preparing serum spheroplasts *in vitro*, the osmotic stabilizer employed needs special consideration, as sucrose inactivates complement. As little as 0.075 M sucrose leaves only 57% of the effective complement in whole serum.[39]

SPHEROPLASTING BY LEUKOCYTE PRODUCTS OR WITHIN LEUKOCYTES

MYCOBACTERIA AND LEUKOCYTE PRODUCTS

The tubercle bacillus with its atypical wall has proved relatively lysozyme resistant, possessing only scant mucopeptide for lysozyme substrate. Furthermore, the mucopeptide is buried beneath lipoprotein and lipopolysaccharide; thus, something other than lysozyme alone is required to attack the cell wall.[58] Willett and Thacore found that two monocytic enzymes, phospholipase C and acid phosphatase, can independently bolster lysozyme attack on the tubercle bacillus, commenting that infection with the tubercle bacillus is known to increase acid phosphatase.[47,58]

IMMUNOGENS

ANTIGENS UNIQUE TO THE CWD VARIANTS

Studies of whole cells may ultimately show that some of the immunogens considered unique in a variant are present, but masked, in its parent. Nevertheless, the ability of a variant to stimulate production of an antibody not formed in response to the classic forms is of both theoretical and practical interest. A new antigen of the L-phase of *S. typhi* was detected in cell lysates. It appeared, coincidentally, with a new unidentified sugar.[23]

Unique antigens in the L Form of β-hemolytic streptococci were suggested by the thoroughness of antigen preparation by Lynn and Muellenberg.[29] In addition to whole cells, acetone-dried powders and sonic oscillator lysates were used for animal injections. However, the authors felt they had not ruled out all possibilities that the parent might contain all the L Form antigens.

Crawford found an interesting complement-fixing streptococcal antigen in L Forms, not demonstrable in the parent form. He questioned whether such antigen might not be involved in the pathogenesis of disease since its corresponding antibody appeared in convalescents.[6]

An antigen of *Candida albicans* unique to the CWD form induced with mycostatin was revealed in Ouchterlony tests by Antonina Brem, in an unpublished study.

In another instance, antigens at first erroneously appeared to be unique to the L Forms.[38] In immunoelectrophoresis of Proteus L Forms, antibody to the L Form gave two lines with the L Form extract not seen with antibody made in response to the classic organism. It was shown that the additional antibodies were specific for internal antigens of the organism.

ANTIGENS RETAINED BY CWD VARIANTS

It was noted by Ivanova and co-workers that common proteins of *Proteus mirabilis* are retained by the L Form. They concluded that electrophoretic protein patterns "may be expected to play a very important role in diagnosis of infections caused by atypical bacterial forms".[21] See also the chapter on Identification.

The H antigen has been detected in nonflagellated CWD growth by several approaches. Employing fluorescent antibody, Fleck et al.[13] detected H antigen at the periphery of the large bodies. The H antigen of CWD Proteus tended to be produced in much larger quantities in the synthetic medium of Medill rather than in horse serum milieu.[26] In CWD *S. typhi*, O, Vi, and H antigens are all demonstrable, although H is frequently diminished. The CA antigen is also present in Salmonella spheroplasts.[57]

S. typhi spheroplasts, which had only sparse flagella, were also found to contain considerable quantities of a material resembling purified flagellin of the classic strain, but slightly different when examined by immunoelectrophoresis. This was interpreted as meaning either that the extracted material was an intermediate rather than completed product or that the composition was atypical because the spheroplasts were grown in different medium.[51]

Havlíček and Havlíčková found four antigens possibly related to the P antigen present in both L Form and parent of Group A Streptococci.[20]

In contrast to loss of carbohydrate antigen, the protein responsible for type designation of *Streptococcus pyogenes* can be produced by CWD variants.[50] This M protein is usually not bound as part of the microorganism but rather released into the culture fluid.[15]

There is evidence that the cytoplasmic membrane, not wall components, is the most important fraction in disease, stimulating a necrotizing cross-reacting antibody. It could even be hypothesized that lack of some layers of the cell wall might make this critical membranous antigen more available to antibody-producing cells.

TISSUE DAMAGE FROM CWD ANTIGENS

When monkeys are given heat-killed CWD Group A Streptococci, the pathogenesis shows that their antigens per se are toxic; no multiplication with elaboration of enzymes is needed. Pancarditis and fibrotic subendocardial lesions result.[36]

Thus, studies by these investigators (Mohan and collaborators) confirm the important role of immune responses in cardiac disease and show that classic Streptococci are not required to stimulate the pertinent antibodies. Marked endocarditis, myocarditis, and pericarditis result in mice given heat killed stable L Forms of Group A Streptococci.

ANTIGENS MISSING OR MINIMIZED IN THE CWD STAGE

By definition, some components of the cell wall are absent in the wall deficient stage. Such loss was early documented in Proteus by Dienes and associates.[10] Quantitative and qualitative losses have usually been determined by biochemical analyses.

The absence of the group-specific carbohydrate of Group A Streptococci, the C substance, has been confirmed many times in analyses of CWD forms induced *in vitro*[50] or from clinical situations. Rhamnose is not detectable, as established by chromatographic studies. For Type III streptococci of Group B the type-specific polysaccharide and the R protein are minimal in the "cell wall defective" stage.[14]

INTERFERON PRODUCTION IN CHICKS

Some L Forms of Enterobacteriaceae may produce endotoxin in chick embryos. This endotoxin may, in turn, stimulate interferon production, as demonstrated by inhibition of viral propagation. Chick embryos harboring the L variants cannot be infected by infectious bronchitis or by Newcastle virus.[60]

INTERLEUKINS

Association of CWD infection with changes in interleukins is based on a case of Whipple's disease in which interleukin 2 (IL-2) activity was lowered, although the IL-2 receptors were normal. B cell growth factor could not be demonstrated. Interleukin 1 was produced in normal amounts.[18]

MITOGENS

Mitogenicity has been found in L Forms of two very different species, *Staphylococcus aureus*[30,52] and *Brucella suis*.[48] ^{3}H-thymidine incorporation in spleen microcultures was the technique used by Schmitt-Smolska and fellow investigators in studying *B. suis* L Forms.[48] Mitogenicity may be implied for streptococcal L Forms which retain, as they commonly do, these known mitogenic factors: lipoteichoic acid, streptolysin S, erythrogenic and pyrogenic toxins. The topic of mitogens is included in 95 articles concerning immunology of bacterial L Forms reviewed by Lynn.[30]

PHAGOCYTOSIS

The interactions of Group A streptococcal L Forms and mouse peritoneal macrophages were followed with electron microscopy. Active phagocytosis was demonstrated.[4]

DEPRESSION OR STIMULATION OF THE MAJOR IMMUNE BRANCHES

When CWD Listeria infect lambs, both lymphoid tissue and bone marrow progressively atrophy.[44]

In contrast to the above, when the variants are from other species, stimulation of lymphoid tissue, rather than atrophy can result from infection. The thymus-dependent zones of the spleen may be greatly stimulated by Brucella L Forms, and not by the classic bacilli.[48] A similar response was demonstrated by L Forms of Group A streptococcus.[32] The many facets of interaction between CWD organisms and host defenses has been reviewed by McGee.[33]

GRAFT VS. HOST DISEASE

Oddly, prior infection with *E. coli* can magnify the graft vs. host disease which occurs too commonly when immunocompetent cells are transferred to an immunosuppressed host. The antigens responsible for this increased response are present in wall-deficient forms, as shown by studies using a CWD mutant of *E. coli.*[37]

INCREASE IN EXUDATE CELLS

CWD *E. coli* given intraperitoneally to mice causes a four- to fivefold increase in peritoneal macrophage numbers. The macrophage bacteriocidal activity is also augmented six- to tenfold.[22]

That the increase in exudate noted above is not due to chemotaxis is suggested by a study from Harwick and associates. CWD *Streptococcus faecalis, E. coli,* and *P. mirabilis* all showed less chemotactic influence than did their parent strains.[19] In this respect they resemble the common mycoplasma of man which lack chemotactic properties.

IMMORTALIZING HUMAN LYMPHOCYTES

Human lymphocytes acquire increased longevity, though as yet not actual perpetuity, when fused with *E. coli* spheroplasts containing the entire SV40 genome. Without such fusion, the lymphocytes survive only 2 weeks; thus, transformed cells readily become apparent. The transformed lymphocytes show growth from approximately 14 d to 6 weeks. The lymphocytes remain healthy for a further 3 to 4 weeks.[3]

LOSS OF IMMUNITY IN BURNED TISSUE

Vtyurin and associates found electron microscopic evidence that burn-injured cells are sites of multiplication for L Form bacteria.[55] This adds a new facet concerning the tendency of severe infection to develop in burned tissue.

SUMMARY

In vitro studies suggest that the normally occurring CWD cycle is augmented and perpetuated *in vivo* from antibody-complement-lysozyme interaction. Without doubt, quantitative aspects are important. If the enzymatic action of complement on sensitized sites of the bacterial wall penetrates to a sufficient depth, death of the organism results, rather than viable spheroplasts or protoplasts. Similarly, excess lysozyme strips the cell wall layers so completely that bacterial death commonly results.

Monocyte enzymes can be potent contributors in promoting cell wall deficiency. Also, leukozyme C, a built-in chelating agent present in leukocyte vacuoles, contributes to formation of wall deficiency.

Differences in diseases produced by classic vs. CWD forms of the same species may be explained by quantitative or even qualitative differences in their antigens. Antigens apparently unique to the L Form have been useful in detecting typhoid carriers and in explaining persistent immune responses in such diseases as rheumatic fever. Antigenic differences must also contribute to the superiority of some CWD vaccines compared with traditional vaccines.

REFERENCES

1. **Agarwal, S. C. and Ganguly, N. K.,** Experimental oral immunization with L Forms of *Vibrio cholerae, Infect. Immun.,* 5, 31–34, 1972.
2. **Apgar, J.,** Salmonella L-Variant Production by a System of Complement Diffusion, M.S. thesis, Wayne State University, Detroit, 1957.
3. **Brzeski, H., Chambers, M., MacDonald, C., and Stimson, W. H.,** The immortalisation of human lymphocytes by spheroplast fusion, *Dev. Biol. Stand.,* 60, 105–109, 1985.
4. **Churilova, N. S., Kirillova, F. M., Neustroeva, V. V., and Vulfovich, Y. V.,** Electron microscopic study of interrelations between the L Forms of Group A beta-hemolytic streptococci and mouse peritoneal macrophages, *Zh. Mikrobiol. Epidemiol. Immunobiol.,* 12, 76–78, 1982.
5. **Chesbro, W. R.,** Lysozyme and the production of osmotic fragility in Enterococci, *Can. J. Microbiol.,* 7, 952–955, 1961.
6. **Crawford, Y. E.,** Studies of a complement-fixing antigen from Group A streptococcal L Forms. I. Preparation and preliminary test in rabbits and man, *J. Immunol.,* 84, 86–92, 1960.
7. **Crombie, L. B. and Muschel, L. H.,** Quantitative studies on spheroplast formation by the complement system and lysozyme on Gram negative bacteria, *Proc. Soc. Exp. Biol. Med.,* 124, 1029–1033, 1967.
8. **Davis, S. D., Gemsa, D., and Wedgwood, R. J.,** Kinetics of the transformation of Gram-negative rods to spheroplasts and ghosts by serum, *J. Immunol.,* 96, 570–577, 1966.
9. **Diena, B. B., Wallace, R., and Greenberg, L.,** Immunologic studies of glycine-induced spheroplasts of *Salmonella typhi, Can. J. Microbiol.,* 10, 555–560, 1964.
10. **Dienes, L., Weinberger, H. J., and Madoff, S.,** The transformation of typhoid bacilli into L Forms under various conditions, *J. Bacteriol.,* 59, 755–764, 1950.
11. **Eagon, R. G. and Carson, K. J.,** Lysis of cell walls and intact cells of *Pseudomonas aeruginosa* by ethylenediaminetetraacetic acid and by lysozyme, *Can. J. Microbiol.,* 11, 193–201, 1965.
12. **Eagon, R. G., Simmons, G. P., and Carson, K. J.,** Evidence for the presence of ash and divalent metals in the cell wall of *Pseudomonas aeruginosa, Can. J. Microbiol.,* 11, 1041–1042, 1965.
13. **Fleck, J., Minck, R., and Kirn, A.,** Etude de l'action inhibitrice spécifique des antiserums sur les cultures L des bacteries. III. Localisation de l'antigene H par la technique de l'immuno-fluorescence, *Ann. Inst. Pasteur (Paris),* 102, 243–246, 1962.
14. **Flores, A. E. and Ferrieri, P.,** The type-specific polysaccharide and the R protein antigens of the L phase from a Group B type III streptococcus, *Zentralbl. Bakteriol. Mikrobiol. Hyg. Ser. A.,* 259(2), 165–178, 1985.
15. **Freimer, E. H., Krause, R. M., and McCarty, M.,** Studies of L Forms and protoplasts of Group A Streptococci. I. Isolation, growth, and bacteriologic characteristics, *J. Exp. Med.,* 110, 853–873, 1959.
16. **Freter, R.,** Coproantibody and bacterial antagonism as protective factors in experimental enteric cholera, *J. Exp. Med.,* 104, 419–426, 1956.
17. **Gorelov, A. L., Levina, G. A., and Prozorovskii, S. V.,** Detection of antigenemia in typhoid patients at different periods in the course of infection, *Zh. Mikrobiol. Epidemiol. Immunobiol.,* 12, 33–36, 1989.
18. **Gras, C., Kaplanski, S., Farnarier, C., Bongrand, P., Chapoy, P., and Aubry, P.,** Immunological profile of Whipple disease after 17 years' evolution, *Ann. Med. Interne,* 139(1), 24–28, 1988.
19. **Harwick, H. J., Kalmanson, G. M., and Guze, L. B.,** Chemotactic activity of L Forms and Mycoplasma, *Ajebak,* 55(4), 431–434, 1977.
20. **Havlíček, J. and Havlíčková, H.,** Antigenic structure of *Streptococcus pyogenes* L Forms, *Pathol. Microbiol.,* 30, 469–480, 1967.
21. **Ivanova, E. H., Toshkov, As. S., Ivanov, B. N., and Gumpert, J. K.,** Antigenic and immunogenic properties of common proteins of the stable L Form of *Proteus mirabilis D, C. R. Acad. Bulg. Sci.,* 37, 81–84, 1984.
22. **Ivanova, E., Gumpert, J., and Popov, A.,** The lipopolysaccharide-containing cytoplasmic membranes as immunostimulants of peritoneal macrophages, *Acta Microbiol. Bulg.,* 24, 21–28, 1989.
23. **Kagan, G. Y., Koptelova, E. I., and Bitkova, A. N.,** Antigenic peculiarities of the L Form of *S. typhosa, Bull. Biol. Med.* (Russian), 1, 74–76, 1963.

24. **Kawata, T. and Takumi, K.**, The spheroplast of a vibrio and its restoration to a vibrio form, *Jpn. J. Bacteriol.*, 24, 472–473, 1969.
25. **King, J. R. and Gooder, H.**, Induction of enterococcal L Forms by the action of lysozyme, *J. Bacteriol.*, 103, 686–691, 1970.
26. **Kirn, A., Fleck, J., and Minck, R.**, Étude de l'action inhibitrice spécifique des antiserums sur les cultures L des bactéries. II. Présence d'agglutinines H dans les antiserums L, *Ann. Inst. Pasteur (Paris)*, 102, 113–117, 1962.
27. **Koshland, M. E. and Burrows, W.**, The origin of faecal antibody and its relationship to immunization with adjuvants, *J. Exp. Med.*, 65, 93–103, 1950.
28. **Landman, O. E.**, Protoplasts, spheroplasts and L Forms viewed as a genetic system, in *Microbial Protoplasts, Spheroplasts, and L-Forms*, Guze, L. B., Ed., Williams & Wilkins, Baltimore, 1968, 319–332.
29. **Lynn, R. J. and Muellenberg, M. B.**, Immunological properties of an L Form of a group A beta hemolytic streptococcus, *Antonie van Leeuwenhoek*, 31, 15–24, 1965.
30. **Lynn, R. J.**, Immunology of bacterial L Forms, in: *The Bacterial L Forms*, Madoff, S., Ed., Marcel Dekker, 1986, 263–276.
31. **Mason, M. A.**, The spheroplasts of *Bordetella pertussis*, *Can. J. Microbiol.*, 12, 539–545, 1966.
32. **Mauss, H. and Schmitt-Slomska, J.**, Changes due to streptococcal L Forms in parathymic lymph nodes and in the spleen of the mouse. *Spheroplasts, protoplasts and L Forms of bacteria, INSERM*, 65, 345–354, 1976.
33. **McGee, Z. A.**, Interaction of cell wall defective bacteria with host defenses, in *The Bacterial L Forms*, Madoff, S., Ed., Marcel Dekker, 1986, 227–285.
34. **Milleck, J. and Ocklitz, H. W.**, Protektiver and toxischer effekt der *B. pertusis* -Sphaeroplasten im tierversuch, *Dtsch. Gesundheitswes.*, 24, 379, 1969.
35. **Minck, R., Kirn, A., and Fleck, J.**, Étude de l'action inhibitrice spécifique des antiserums sur les cultures L des bactéries. I. Exclusion du rôle des anticorps O et H dans le phénomène d'inhibition spécifique des cultures L, *Ann. Inst. Pasteur (Paris)*, 101, 178–184, 1961.
36. **Mohan, C., Ganguly, N. K., and Chakravarti, R. N.**, Experimental production of cardiac injury in rhesus monkeys by L Forms of Group A streptococci, *Indian J. Med. Res.*, 86, 361–371, 1987.
37. **Moore, R. H., Lampert, I. A., Chia, Y., Aber, V. R., and Cohen, J.**, Influence of endotoxin on graft-vs.-host disease after bone marrow transplantation across major histocompatibility barriers in mice, *Transplantation* 43, 731, 1987.
38. **Muschel, L. H. and Jackson, J. E.**, The reactivity of serum against protoplasts and spheroplasts, *J. Immunol.*, 97, 46–51, 1966.
39. **Muschel, L. H. and Larsen, L. J.**, Effect of hypertonic sucrose upon the immune bactericidal reaction, *Infect. Immun.*, 1, 51–55, 1970.
40. **Osawa, E. and Muschel, L. H.**, Studies relating to the serum resistance of certain Gram-negative bacteria, *J. Exp. Med.*, 119, 41–51, 1964.
41. **Oshima, S., Myrvik, Q. N., and Leake, E.**, The demonstration of lysozyme as a dominant tuberculostatic factor in extracts of granulomatous lungs, *Br. J. Exp. Pathol.*, 42, 138–144, 1961.
42. **Ozaki, M., Higashi, Y., Amano, T., Miyama, A., and Kashiba, S.**, Nature of spheroplasting agent, leucozyme C, in guinea pig leucocytes, *Biken J.*, 175–188, 1965.
43. **Pfeiffer, R.**, Die Differentialdiagnose der Vibrionen der Cholera asiatica mit Hulfe der immunisirung, *Hyg. Infekt.*, 19, 75–100, 1895.
44. **Prosorovsky, S., Kotljarova, J., Fedotova, I., and Bakulov, I.**, Pathogenicity of listerial L Forms, in, *Spheroplasts, Protoplasts and L Forms of Bacteria, INSERM*, 65, 265–272, 1976.
45. **Repaske, R.**, Lysis of Gram-negative bacteria by lysozyme, *Biochim. Biophys. Acta*, 22, 189–191, 1956.
46. **Repaske, R.**, Lysis of Gram-negative organisms and the role of versene, *Biochim. Biophys. Acta*, 30, 225–232, 1958.
47. **Saito, K. and Suter, E.**, Lysosomal acid hydrolases in mice infected with BCG, *J. Exp. Med.*, 121, 727–738, 1965.
48. **Schmitt-Slomska, R. C., Starka, G., Mauss, H., Drach, G., and Lemaire, J.**, Immunogenic properties on mice of CWD brucella L Forms, *3rd Int. Symp. on Brucellosis, Algiers, Algeria, 1983, Dev. Biol. Stand.*, 56, 187–197, 1984.
49. **Schnauder, G.**, Tierexperimente zur Frage der epidemiologischen Bedeutung der L Formen, *Z. Hyg.*, 141, 404–410, 1955.
50. **Sharp, J. T., Hijmans, W., and Dienes, L.**, Examination of the L Forms of group A streptococci for the group-specific polysaccharide and M protein, *J. Exp. Med.*, 105, 153–159, 1957.
51. **Sparkes, G., Griffiths, B. W., Diena, B. B., and Greenberg, L.**, The presence of flagellar antigen in *Salmonella typhi* spheroplasts, immunochemical studies, *Can. J. Microbiol.*, 12, 799–805, 1966.
52. **Takada, H., Hirachi, Y., Hashizume, H., and Kontani, J.**, Mitogenic activity of cytoplasmic membranes isolated from L Forms of *Staphylococcus aureus*, *Microbiol. Immunol.*, 24, 1079–1090, 1980.

53. **Thacore, H. and Willett, H. P.,** Formation of spheroplasts of *Mycobacterium tuberculosis* by lysozyme treatment, *Proc. Soc. Exp. Biol. Med.,* 114, 43–47, 1963.

54. **Vershilova, P. A., Grekova, N. A., and Tolmacheva, T. A.,** Persistence of Brucellosis L cultures in experimental animals and immunity duration, *Vestn. Akad. Med. Nauk SSSR,* 3, 10–13, 1985.

55. **Vtyurin, B. V., Kaem, R. I., and Tumanov, V. P.,** Ultrastructural analysis of inter-relation between microbes and cells of the histohematic barriers in severe burns of the skin, *Vestn. Dermatol. Venerol.,* 6, 18–23, 1981.

56. **Wallace, R., Diena, B. B., and Greenberg, L.,** A rapid slide test for the detection of Vi antibody in human and animal sera, *Can J. Microbiol.,* 10, 551–554, 1964.

57. **Whang, H. Y. and Neter, E.,** Demonstration of common antigen in *Salmonella typhi* spheroplasts, *Can. J. Microbiol.,* 11, 598–601, 1965.

58. **Willett, H. P. and Thacore, H.,** Formation of spheroplasts of *Mycobacterium tuberculosis* by lysozyme in combination with certain enzymes of rabbit peritoneal monocytes, *Can. J. Microbiol.,* 13, 481–488, 1967.

59. **Wittler, R. G.,** The L Form of *Haemophilus pertussis* in the mouse, *J. Gen. Microbiol.,* 6, 311–317, 1952.

60. **Godzeski,** personal communication.

Chapter 9

INDUCTION BY ANTIBIOTICS, ORGANIC COMPOUNDS, AND MISCELLANEOUS FACTORS

The varied methods which have been used to induce microorganisms to lose cell wall components are a tribute both to the ingenuity of man and to the versatility of enzymes.

WHAT MECHANISMS RESULT IN CELL WALL DEFICIENCY?

When one thinks of CWD forms as resulting from an imbalance of the cell's ability to degrade and synthesize its classic thick wall, many factors in formation of these variants are explained. A change in temperature, then, can bring an alteration because autolysis or wall synthesis is favored selectively, and the same applies to the presence of cations. When gaps or thin places develop in the wall, the denuded cytoplasmic segment can diversify as other inherent factors permit. This would explain the ability of a coccus to develop Actinomyces-like branches of varying diameters. Autolysins, acting irregularly and excessively on a wall, result in weakened points through which new wall is pushed laterally by internal pressure, which may be as fantastically high as 20 atm,[49] causing the bacterium, literally, to sprout. This explains why classic bacterial morphology is rather rare, occurring only when wall synthesis and degradation are in fine balance. Otherwise, the organism may develop a syncytium, or, if a bit more restricted, become a budding yeast-like form, or otherwise express the multiple innate characteristics which are in its genome, some of which mimic protozoa.

Spontaneous autolysis is known in numerous genera. Staphylococcal growth shows especially vigorous autolysis. After 2-h incubation, many cocci escape from their walls, even in the absence of an antibiotic.[50] The chapter on reversion describes a Staphylococcus variant in a blood culture which reverted only in the presence of lysozyme.[20] A possible mechanism involved in the reversion is that the egg-white lysozyme competitively blocked excessive lysozyme made by the staphylococcus.

INDUCTION OF CWD FORMS BY ANTIBIOTICS *IN VITRO*

The finding by Pierce[52] that penicillin is a provocative agent of L Form growth initiated a pandemic of related activity, greater every year since it became possible to develop these variants at will. Earlier, Gardner[24] had noted penicillin-induced pleomorphism and Fleming's laboratory had described fantastic movement of misshapen organisms in penicillin-treated cultures[21] before penicillin-induced variant colonies had ever been observed.

The mechanisms of cell wall inhibition and, hence, L-variant growth, have been elucidated for many antibiotics. The penicillins, bacitracin, novobiocin, vancomycin, and ristocetins interrupt murein synthesis in a similar manner.[66] D-cycloserine can also produce CWD forms by interrupting cell wall structuring.[15,55]

The chemical reactions which result in killing a microbe have been defined for almost every antibiotic. As described above, when cell wall building is interrupted, wall-deficient variants are expected. Unexpected is the formation of such variants from antibiotics well known to inhibit protein synthesis. This is at least partially explained by scanning electron microscopy studies by Klainer and Perkins. They found changes in bacterial surfaces after exposure to kanamycin, tobramycin, and chloramphenicol.[36] Another possibility is production of excess autolysin which removes cell walls. Coincidentally, surface changes on *Escherichia coli* are obvious when the organism is exposed to the monobactam Aztoreonam in urine (Figure 1).[68]

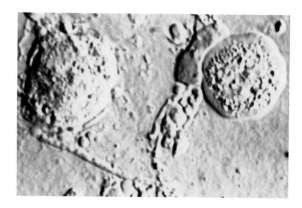

FIGURE 1. Changes in *Escherichia coli* in the presence of Aztoreonam include surface pits and blebs. (From Tsugaya, M., Washida, H., Sakagami, H., Iwase, Y., Inuzuka, K., and Takeuchi, K., *Acta Urol. Jpn.*, 32, 1883–1886, 1986. With permission.)

As one reviews the multiple effects of a single antibiotic upon varied bacterial cells, it becomes obvious that there is more than one mechanism by which each antibiotic alters bacterial metabolism. Comments of Hahn[27] and Snell and Cheng[64] which were made regarding tetracyclines can apply to other antimicrobials. The authors stated that one should not seek the mode of action of tetracyclines but rather the condition under which any or all of several mechanisms might apply.

RESPONSIVE BACTERIA

Many of the enteric bacilli induced into the L phase can continue to grow without osmotic stabilizers, and, thus, much early work concerned these bacilli. After it was realized that hypertonicity is needed to equalize the enormous internal pressure of most protoplasts, knowledge of the species of bacteria cultivable with wall deficiency increased steadily. At this date almost every bacterium of industrial, basic scientific, or medical interest has been induced into the L phase with appropriate antimicrobial compounds. The organisms induced with penicillin include Brucella[14] (Figure 2), Clostridia,[41] *E. coli*,[38] *Hemophilus influenzae*,[41] *Listeria monocytogenes*,[5,18,40] *Proteus mirabilis*,[69] *Salmonella gallinarum*,[38] *S. typhi*,[58] *Vibrio cholerae*,[48] and Vitreoscilla.[8] Penicillin-induced CWD variants of many genera are described in other chapters.

Pseudomonas aeruginosa produces stable L-colonies after exposure to Polymyxin B.[19] This organism also multiplies as L Forms when the classic bacilli enter distilled water (see chapter on public health). Fosfomycin can induce protoplasts and L Forms of *Staphylococcus aureus*.[59]

PENICILLIN INDUCTION AS A TOOL IN IDENTIFYING A CLASSIC BACTERIUM

Does penicillin induction give a clue in identification of a bacterium? In some instances the antibiotic requirement, morphology, or other characteristics of the variant colonies suggest the species or subgroup. In the Proteus genus there is a clear-cut distinction between the penicillin requirement for induction of *Proteus mirabilis* and *P. vulgaris*, which behave similarly, and *P. rettgeri*, and *P. morgani*, which have a higher antibiotic requirement.[44]

Penicillin-induced L-colonies of *Clostridium perfringens* can be distinctive for the subtypes. For example, a colony made by types B and C has many daughter colonies surrounding the initial central growth (Figure 3). Another colonial morphology appeared from types A

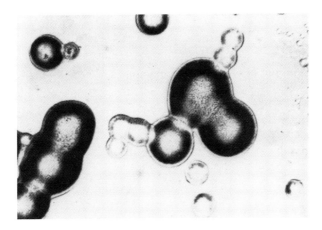

FIGURE 2. *Brucella abortus* treated with penicillin develops colonies with varied appearance. (From Christophorov, L., and Peschkov, J., *Zentralbl. Bakteriol. Parasitenkd. Infektionskr. Hyg. Abt. 1 Orig.*, 209, 497–504, 1969. With permission.)

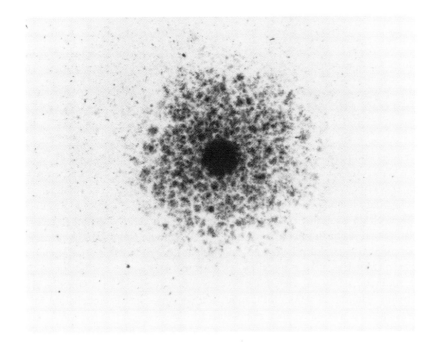

FIGURE 3. Distinctive colony from *Clostridium perfringens* type B and C. (From Kawatomari, T., *J. Bacteriol.*, 76, 227–230, 1958. With permission.)

to D inclusive but not from E or F. This colony was flat with radiating branches (Figure 4). The edges of the colony blacken after prolonged incubation. The colonies of type E are uniquely small and domed. Colonies of type F are flat with diffuse edges. Penicillin-induced L-phase growth of *C. perfringens* types A, B, C, and D, but not E, soften agar.[35]

In CWD variants of Salmonellae, the only identifying marker found is a slow rate of reproduction for *Salmonella pullorum*, corresponding to the prolonged generation time of the parent form.[45]

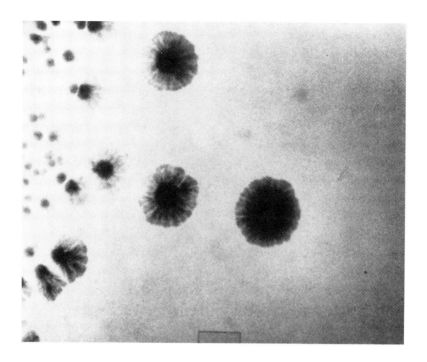

FIGURE 4. Colonies which may be formed from L-phase variants of *Clostridium perfringens* types A to D inclusive. (From Kawatomari, T., *J. Bacteriol.*, 76, 227–230, 1958. With permission.)

FORMATION OF WALL-DEFICIENT VARIANTS VS. BACTERIAL DEATH

When is penicillin bactericidal, and when does it, instead, induce wall-deficient variants? It is easily shown that the amount of substrate, i.e., the bulk of cell wall exposed to the antibiotic, is a critical facet. To induce the wall deficient stage, it is usually necessary to use a checkerboard method of exposing log-dilutions of the antibiotic to log-dilutions of the culture. Most of the antibiotic-organism mixtures will not yield variants. Excessively high levels of antibiotic result in microbial death, even if osmotic stabilizers are present. Excessively concentrated bacterial populations will remain unaffected by antimicrobial agents.

Another quantitative aspect, noted by Hijmans and Kastelein,[29] is the age of the culture when the antibiotic is added. One Enterococcus strain required an atypically brief preincubation before addition of penicillin if induction was to occur. Preincubation for only 1.5 h in growth medium was the critical requirement.

RELATIVE INDUCING ABILITY OF ANTIBIOTICS

Almost every antibiotic has induced CWD forms from some bacterial species. Why then does penicillin emerge as the most efficient inducer *in vitro* and, correspondingly, the most commonly associated with *in vivo* problems? The reason is the wide range in concentration in which penicillin induces cell wall changes. The early work of Pulvertaft[53] comparing penicillin with four other antibiotics revealed that penicillin was unique in producing "viable monsters" throughout a long series of dilutions. Similarly, Godzeski et al. found that penicillin induced L-colonies of staphylococci in a wide span of concentrations, whereas other antibiotics were effective in only one or two dilutions.[26] Additionally, Kagan et al.[33]

found that methicillin in concentrations from 1.0 to 5000 U/ml induced the L-phase of staphylococci.

CHLORAMPHENICOL

Irrefutably chloramphenicol inhibits protein synthesis.[28] Correlating with this is the fact that chloramphenicol is one of the best agents for therapy of CWD infections. However, the paradox exists that this antibiotic, which is lethal for many CWD forms, can produce CWD bacterial growth as described below. Low concentrations of the antibiotic can inhibit wall synthesis without lethality.

When chloramphenicol was incorporated in broth cultures the following organisms propagated as L-phase variants: *E. coli, Klebsiella pneumoniae, Bacillus megaterium, B. polymyxa, Serratia marcescens, Sarcina lutea, Staphylococcus aureus,* and species of Salmonella, Shigella, and Proteus.[26] The only bacterium which did not give some L-phase growth with this and many other antibiotics was *B. subtilis.* The authors remarked that perhaps this failure to convert was due to the unique lack of teichuronic acid in the *B. subtilis* cell wall. Perhaps the almost invariable success with this and many other antibiotics resulted because a wide range of antibiotic concentrations was tested in individual tubes, from 0.3 to 250 μg/ml.

Another chloramphenicol effect on walls is stimulation rather than inhibition. *Rhodotorula glutinis,* cultured in chloramphenicol, yields enormously thickened walls showing an average increase in diameter from 0.09 to 0.16 μm.[63]

ERYTHROMYCIN

Again, erythromycin is valuable as an antibiotic which may be therapeutic for infection with CWD forms. However, *in vitro,* this agent stimulates L-phase growth of *Staphylococcus aureus* and all the other organisms noted above as being inducible into the L phase by chloramphenicol.[26]

There are at least 40 macrolide antibiotics which similarly induce wall-altered variants. These include leukomycin, oleandomycin, spiramycin, and tylosin.

TETRACYCLINES

Godzeski and associates induced L-phase variants of *S. aureus, E. coli, K. pneumoniae, B. megaterium, B. polymyxa, Serratia marcescens, Sarcina lutea,* and species of Salmonella, Shigella, and Proteus with tetracycline.[26]

LYSOSTAPHIN

Watanakunakorn and associates used lysostaphin to induce staphylococcal spheroplasts and L-colonies.[71] Huber and Schuhardt found lysostaphin-induced protoplasts as metabolically active in manometric studies as untreated staphylococci.[30] Kato and associates have studied the reversion of *Staphylococcus aureus* lysostaphin protoplasts to L Forms and cocci.[34]

NYSTATIN AND AMPHOTERICIN B

Both nystatin and amphotericin B give abundant growth of CWD colonies of Candida as described in the chapter on fungi. It is enigmatic that amphotericin B can produce changes in the Candida cell wall since this antibiotic is known to affect the cytoplasmic membrane, and its receptor is considered to be a sterol.[22]

SNAIL JUICE INDUCES CELL WALL DEFICIENCY

The literature has not revealed what occurred in the mind of the man who conceived that a potent enzyme capable of dissolving cell walls of fungi might exist in the gastric juice of the snail. Was this perhaps a gentleman at a gourmet dinner who noticed that there was, as usual, sand in his escargot? Did he conclude that if the snail confidently ingested silica, the little creature must have a potent digestive weapon in its intestinal tract? Here in the snail, then, might be an enzyme more potent than those previously investigated. There had been a prolonged search for potent agents to lyse fungal walls in order to analyze wall components. The agents which dissolve bacterial walls were ineffective for yeasts and molds.

Thirty-five years after it was noticed that snail intestinal juice can dissolve the wall of yeast,[25] the finding was applied to produce viable protoplasts. The gut juice from *Heliz pomatia* dissolves the walls of Saccharomyces in a unique way, resulting in a single weakened area from which the protoplast escapes by sequential ballooning. The weakened area is often a bud scar from which a yeast cell has previously separated from the parent organism.[17]

DYES AS AGENTS WHICH PROMOTE CELL WALL DEFICIENCY OF MICROORGANISMS

Since crystal or gentian violet is commonly used to suppress bacterial growth while permitting development of mycoplasma, it should be known that these dyes efficiently produce bacterial variants which may be confused with mycoplasma. Also interesting is that CWD colonies in pour plates containing crystal violet may be blue, contrasting with the purple of the accompanying classic colonies.

EFFECTS OF PRETREATMENTS ON SUSCEPTIBILITY TO INDUCTION

RESULT OF NUMEROUS TRANSFERS *IN VITRO*

After long cultivation *in vitro,* organisms may not be inducible to cell wall deficiency by antibiotics, as early documented by Dienes.[16] Similarly, the history of the culture was important when *Cornyebacterium diphtheriae* strains were grown in penicillin-containing medium. The best production of L-type colonies resulted from freshly isolated rather than museum strains.[57]

However, many stock cultures are easily induced into the L phase. The 134 isolates of Salmonella which made penicillin-L-phase colonies were all long-established strains.[45] Old and new strains of *Brucella melitensis* and *B. suis* reacted comparably in penicillin medium, developing L-colonies and also giant cells 20 to 50 times the size of the parent organism.[14] The Catlin-Cunningham strain of *Neisseria meningitidis,* which had been in subculture for 10 years, formed antibiotic L-colonies as readily as three isolates of recent origin.[56]

INDUCING AGENTS AND CONDITIONS

Induction of CWD Forms *In Vitro*
(Excluding Early Antibiotics)

Inducing agent or conditions	Organism
Chlorpromazine (Thorazine)	*Staphylococcus aureus*[4]
Dithiothreitol	*Methanospirillum hungatti*[65]
Distilled water	*Pseudomonas aeruginosa*[67]
Fluorouracil	*Bacterium anitratum*[7]
Fosfomycin	*S. aureus*[60]
Gelatin factor (low mol wt)	*Streptococcus faecium*[2]

Inducing agent or conditions	**Organism**
Glycine	*Mycobacterium smegmatis*[1]
Imipenem (personal commum.)	*Bilophila wadsworthia* (Figure 5)
Lysin from streptococcus Group C phage	Streptococci Groups A, C, E, H form spheroplasts in 5 min.[54]
Lysoamidase	*Staphylococcus aureus*[51]
Lysozyme plus EDTA caused protoplast formation	*Escherichia coli*[39a] and *P. aeruginosa*[39a]
Lysozyme plus freeze-thaw	*Vibrio comma*[13] *E. coli*[37]
Mitomycin C & degranol	Pasteurella and Yersinia[6]
Nonhypertonic medium	Halobacteria[32]
Prereduced medium	*Methanobacterium bryantii*[31]
Progesterone	*Brucella abortus, B. suis, B. melitensis*[46]
Radioactive phosphorus, [32]P	Proteus[43]
Radioactive sodium and sulfur	Salmonella (Figure 6)[45]
7% Sodium chloride	*Bacillus subtilis*[10,11]
Streptomyces GM enzyme	Yeasts[23]
Sulfonamides (sulfathiazole, sulfadiazine, sulfamerazine, sulfamethoxypyridazine, gantrisin, and elkosin behave similarly.)	*Bacterium anitratum*[7] *Herellea vaginicola*[7] *Staphylococcus aureus*[26]
Zephiran	*E. coli*[47]

[a] Protoplasts of these two species were fused with polyethylene glycol to make the first intergenus bacterial hybrid. For dyes and snail juice see text. For induction of CWD forms by brilliant green and crystal violet, see the chapter on fungi.

IN VIVO INDUCTION OF CWD MICROBES

It should be emphasized that most pathogens injected into vertebrates initiate growth in the pleomorphic wall-deficient state. This includes such diverse organisms as *Clostridium tetani* and *V. cholerae*, a development which may contribute to a carrier state.[42] It was noticed that after L Forms of *E. coli* were antibiotic-induced in a patient with urinary tract infection, these variants soon disappeared from urine but continued to be seen in phagocytes. The authors suggested that examining the leukocytes might be a guide to the precise amount of antibiotic needed for effective therapy.[70]

Another example is spontaneous formation of the CWD stage after classic Actinomyces were injected into mice. Disease which appeared after a long interval seemed to result only after CWD formation.[3]

The classes of antibiotics which induce CWD forms *in vitro* can be shown to also produce variants *in vivo* as described in the chapters dealing with pathogens. A typical example concerns mice infected with *Brucella suis* which have residual L Forms of the Brucella in their spleens.[61]

SUMMARY

While many points remain obscure, cell wall deficiency and variation are clarified when one views classic growth as perfect cooperation between wall autolysis and replacement. Aberrant forms result whenever there is imperfect balance between construction and destruction.

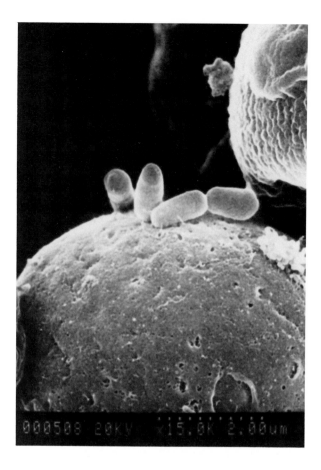

FIGURE 5.   Imipenem gradient plate has induced spheroplasts (L bodies) in *Bilophila wadsworthia*. Classic anaerobic rods sit on the variants. (Photo submitted by Hannah Wexler.)

Time-lapse photographs show that many agents and conditions cause a bacterium to alter its morphology, exchanging a rigid format for a pleomorphic outline. It is also well known that wall-deficient forms are usually part of every bacterial population, and agents loosely termed "inducers" actually may be only allowing selective growth of the variants.

By mechanisms not always defined, it seems that any antimicrobial substance can, in certain conditions, foster L-phase variations in bacteria and fungi. When the inducing occurs *in vitro*, the resultant L Forms and protoplasts are useful tools for studying genetic exchange. The first intergenus hybrid of bacteria has been made using bacterial protoplasts made with lysozyme plus EDTA.

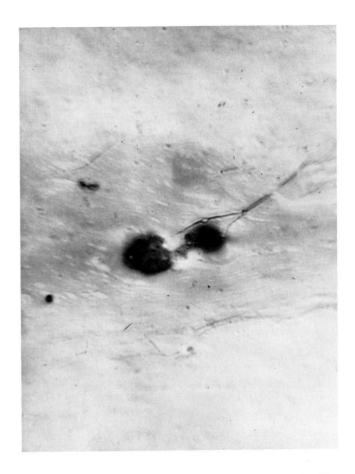

FIGURE 6. *Salmonella typhimurium* growing with a radioactive substrate. Very fine filaments have emerged from L bodies. (From Mattman, L. H., Tunstall, L. H., and Kispert, W. G., *Zentralbl. Bakteriol. Parasitenkd. Infektionskr. Hyg. Abt. 1 Orig.*, 210, 65–74, 1969. With permission.)

REFERENCES

1. **Adámek, L., Mio, P., Mohelská, H., and Trnka, L.,** Ultrastructural organization of spheroplasts induced in *Mycobacterium smegmatis* by lysozyme or glycine, *Arch. Mikrobiol.*, 69, 277, 1969.
2. **Albrewczynski, D. J. and Lawson, J. W.,** Effect of a low molecular weight component of gelatin on cAMP and the morphology of L Forms, *ASM Abstr. Annu. Meet.*, 1983, p. 100.
3. **Beaman, B. L. and Scates, S. M.,** Role of L Forms of *Nocardia caviae* in the development of chronic mycetomas in normal and immunodeficient murine models, *Infect. Immun.*, 33, 893–907, 1981.
4. **Bourdon, J. L.,** Contribution à l'étude des propriétés antibiotiques de la chlorpromazine Ou 4560 RP, *Ann. Inst. Pasteur (Paris)*, 101, 876–886, 1961.
5. **Brem, A. M. and Eveland, W. C.,** Inducing L-forms of *Listeria monocytogenes* types 1 through 7, *Appl. Microbiol.*, 15, 1510, 1967.
6. **Brzin, B.,** The influence of antibiotics and cytostatics on the morphology of *Pasteurella pseudotuberculosis* and *Yersinia enterocolitica*, Int. Symp. Pseudotuberculosis (Paris), *Symp. Ser. Immunobiol. Stand.*, 9, 207–210, 1968.
7. **Brzin, B.,** Synonymy of *Herellea vaginicola* and *Bacterium anitratum*, *Zentralbl. Bakteriol. Parasitenkd. Infektionskr. Hyg. Abt. 1 Orig.*, 209, 409–412, 1968.

8. **Brzin, B.,** An effect of penicillin on the morphology of vitreoscilla, *Zentralbl. Bakteriol. Parasitenkd. Infektionskr. Hyg. Abt. 1 Orig.,* 204, 393–398, 1967.

9. **Brzin, B.,** Morphological alterations induced in bacteria by a cytostatical agent, *Zentralbl. Bakteriol. Parasitenkd. Infektionskr. Hyg. Abt. 1 Orig.,* 205, 510–512, 1967.

10. **Burmeister, H. R. and Hesseltine, C. W.,** Selection and characterization of a pleomorphic form and an L form of *Bacillus pumilus* NRRL B-3275, *Can. J. Microbiol.,* 17, 1057–1060, 1971.

11. **Burmeister, H. R. and Hesseltine, C. W.,** Induction and propagation of a *Bacillus subtilis* L Form in natural and synthetic media, *J. Bacteriol.,* 95, 1857–1861, 1968.

12. **Calandra, G. B., Nugent, K. M., and Cole, R. M.,** Preparation of protoplasts of Group H streptococci *(Streptococci sanguis), Appl. Microbiol.,* 29(1), 90–93, 1975.

13. **Chatterjee, B. R. and Williams, R. P.,** Preparation of spheroplasts from *Vibrio comma, J. Bacteriol.,* 85, 838–841, 1963.

14. **Christophorov, L. and Peschkov, J.,** Untersuchungen über L-Formen bei den Brucellen. I. Versuche zur Gewinnung von Brucellen-L-Formen, *Zentralbl. Bakteriol. Parasitenkd. Infektionskr. Hyg. Abt. 1 Orig.,* 209, 497–504, 1969.

15. **Curtiss, R., Charamella, L. J., Berg, C. M., and Harris, P. E.,** Kinetic and genetic analyses of d-cycloserine inhibition and resistance in *Escherichia coli, J. Bacteriol.,* 90, 1238–1250, 1965.

16. **Dienes, L.,** Isolation of L type colonies from typhoid bacilli with the aid of penicillin, *Proc. Soc. Exp. Biol. Med.,* 68, 589–590, 1948.

17. **Eddy, A. A. and Williamson, D. H.,** A method for isolating protoplasts from yeast, *Nature (London),* 179, 1252–1253, 1957.

18. **Edman, D. C., Pollock, M. B., and Hall, E. R.,** *Listeria monocytogenes* L Forms. I. Induction, maintenance, and biological characteristics, *J. Bacteriol.,* 96, 352–357, 1968.

19. **Edman, L. A. and Mattman, L. H.,** L variation in *Pseudomonas aeruginosa* induced by antibiotics, *Bacteriol. Proc.,* 1961.

20. **Edwards, J. H. and Mattman, L. H.,** Lysozyme-produced variation in a staphylococcal isolate, *Bacteriol. Proc. 1972,* p.20, 1972.

21. **Fleming, A., Voureka, A., Kramer, I. R. H., and Hughes, W. H.,** The morphology and motility of *Proteus vulgaris* and other organisms cultured in the presence of penicillin, *J. Gen. Microbiol.,* 4, 257–269, 1949.

22. **Gale, G. R.,** Cytology of *Candida albicans* as influenced by drugs acting on the cytoplasmic membrane, *J. Bacteriol.,* 86, 151–157, 1963.

23. **Garcia-Mendoza, C. and Villanueva, J. R.,** Stages in the formation of yeast protoplasts with strepzyme, *Nature (London),* 202, 1241–1242, 1964.

24. **Gardner, A. D.,** Morphological effects of penicillin on bacteria, *Nature (London),* 146, 837–838, 1940.

25. **Giaja, J.,** Emploi des ferments dans les études de physiologie cellulaire, Le globule de levure dépouillé de sa membrane, *C. R. Soc. Biol. (Montpellier),* 82, 719–720, 1919.

26. **Godzeski, C. W., Brier, G., and Pavey, D. E.,** L-phase growth induction as a general characteristic of antibiotic-bacterial interaction in the presence of serum, *Antimicrob. Agents Chemother.,* pp. 843–853, 1962.

27. **Hahn, F. E.,** Modes of action of antibiotics, *Proc. 4th Int. Congr. Biochem. (Vienna),* 5, 104–124, 1958.

28. **Hahn, F. E.,** Chloramphenicol, in *Antibiotics. I. Mechanism of Action,* Gottlieb, D. and Shaw, P. D., Eds., Springer-Verlag, New York, 1967, 308–330.

29. **Hijmans, W. and Kastelein, M. J. W.,** The production of L Forms of Enterococci, *Ann. N.Y. Acad. Sci.,* 79, 371–373, 1960.

30. **Huber, T. W. and Schuhardt, V. T.,** Lysostaphin-induced, osmotically fragile *Staphylococcus aureus* cells, *J. Bacteriol.,* 103, 116–119, 1970.

31. **Jarrell, K. F., Colvin, J. R., and Sprott, G. D.,** Spontaneous protoplast formation in *Methanobacterium bryantii, J. Bacteriol.,* 149, 346–353, 1982.

32. **Jarrell, K. F. and Sprott, G. D.,** Formation and regeneration of *Halobacterium* spheroplasts, *Curr. Microbiol.,* 10, 147–152, 1984.

33. **Kagan, B. M., Molander, C. W., and Weinberger, H. J.,** Induction and cultivation of staphylococcal L Forms in the presence of methicillin, *J. Bacteriol.,* 83, 1162–1163, 1962.

34. **Kato, Y., Hirachi, Y., Toda, Y., Takemasa, N., and Kotani, S.,** Effect of the composition of reversion medium on change of *Staphylococcus aureus* lysostaphin protoplasts to coccal forms and L Forms, *Biken J.,* 29, 39–44, 1986.

35. **Kawatomari, T.,** Studies of the L Forms of *Clostridium perfringens.* I. Relationship of colony morphology and reversibility, *J. Bacteriol.,* 76, 227–232, 1958.

36. **Klainer, A. S. and Perkins, E. L.,** Surface manifestation of antibiotic-induced alterations in proteins synthesis in bacterial cells, *Antimicrob. Agents Chemother.,* 1, 164, 1972.

37. **Kohn, A.,** Lysis of frozen and thawed cells of *Escherichia coli* by lysozyme, and their conversion into spheroplasts, *J. Bacteriol.,* 79, 697–706, 1960.

38. **Kudlai, D. G., Soloviev, N. N., and Prozorovsky, S. V.,** Penicillin protoplasts of intestinal bacteria, *Zh. Mikrobiol. Epidemiol. Immunobiol.,* 32, 22–28, 1961.
39. **Kurono, M., Hirachi, Y., Kato, Y., Toda, Y., Takemasa, N., Kotani, S., Takahashi, T., and Takokoro, I.,** Intergenus cell fusion between L Form cells of *P. aeruginosa and E. coli., Biken J.,* 26, 103–111, 1983.
40. **Kotylarova, G. A., Bakulov, I. A., and Prozorovsky, S. V.,** Obtaining of L Forms of *Listeria monocytogenes, Zh. Mikrobiol. Epidemiol. Immunobiol.,* 45, 96–99, 1968.
41. **Lapinski, E. M. and Flakas, E. D.,** Effects of antibiotics on induction, viability and reversion potential of L-forms of *Haemophilus influenzae, J. Bacteriol.,* 98, 749–755, 1969.
42. **Lomov, Y. M., Tynkevich, N. K., Buchul, K. G., and Golubkova, L. A.,** Features of formation persistence and reversion of *Vibrio cholerae* L Forms in experimental animals, *Mikrobiol. Zh.,* 44(6), 79–84, 1982.
43. **Mandel, P., Sensenbrenner, M., and Vincendon, G.,** Action of beta radiation on the Proteus P 18 bacillus, *Nature (London),* 182, 674–675, 1958.
44. **Mattman, L. H., Burgess, A., and Farkas, M. E.,** Evaluation of antibiotic diffusion in L variant production by Proteus species, *J. Bacteriol.,* 76, 333, 1958.
45. **Mattman, L. H., Tunstall, L. H., and Kispert, W. G.,** A survey of L variation in the salmonellae, *Zentralbl. Bakteriol. Parasitenkd. Infektionskr. Hyg. Abt. 1 Orig.,* 210, 65–74, 1969.
46. **Meyer, M. E.,** Evolution and taxonomy in the genus *Brucella,* steroid hormone induction of filterable forms with altered characteristics after reversion, *Am. J. Vet. Res.,* 37(2), 207–210, 1976.
47. **Meyer-Rohn, J. and Tudyka, K.,** Gestaltsanderungun von *Escherichia coli* unter der Einwirkung von Antibiotica, Sulfonamiden, chemischen und physikalischen Noxen, *Arzneim. Forsch.,* 7, 151–157, 1957.
48. **Minck, R. and Minck, A.,** Obtention de formes naines (formes L) à partir d'une souche de vibrion cholérique soumise à l'action de la pénicilline, *C. R. Soc. Biol.* (Paris), 145, 927–929, 1951.
49. **Mitchell, P. and Moyle, J.,** Osmotic function in structure in bacteria, *Symp. Soc. Gen. Microbiol.,* 6, 150–180, 1956.
50. **Mitchell, P. and Moyle, J.,** Autolytic release and osmotic properties of "protoplasts" from *Staphylococcus aureus, J. Gen. Microbiol.,* 16, 184–194, 1957.
51. **Petrov, V. V., Ratner, E. N., Severin, A. I., Fikhte, B. A., and Kulaev, I. S.,** Isolation of *Staphylococcus aureus* protoplasts using lysoamidase, *Prikladnaia Biokhim. Mikrobiol.,* 26(3), 413–421, 1990.
52. **Pierce, C. H.,** *Streptobacillus moniliformis,* its associated L_1 form and other pleuropneumonia-like organisms, *J. Bacteriol.,* 43, 780, 1942.
53. **Pulvertaft, R. J. V.,** The effect of antibiotics on growing cultures of *Bacterium coli, J. Pathol. Bacteriol.,* 64, 75–89, 1952.
54. **Raina, J. L.,** Purification of *Streptococcus* Group C bacteriophage lysin, *J. Bacteriol.,* 145(1), 661–663, 1981.
55. **Reynolds, P. E.,** Antibiotics affecting cell-wall synthesis, *Symp. Soc. Gen. Microbiol.,* 16, 47–69, 1966.
56. **Roberts, R. B. and Wittler, R. G.,** The L Form of *Neisseria meningitidis, J. Gen. Microbiol.,* 44, 139–148, 1966.
57. **Savenkova, V. T.,** Some peculiarities in the formation of L-forms of *C. diphtheriae, Zh. Mikrobiol. Epidemiol. Immunobiol.,* 33, 91–93, 1962.
58. **Schegolev, A. G.,** On certain general regularities of individual response of typhoid fever bacteria strains to L-transforming effect of penicillin, *Antibiotiki,* 15, 240–243, 1970.
59. **Schmid, E. N.,** Fosfomycin-induced protoplasts and L Forms of *Staphylococcus aureus, Chemotherapia,* 30, 35–39, 1984.
60. **Schmid, E. N.,** Bacteriophages in L Form of *Staphylococcus aureus, J. Bacteriol.,* 164, 397–400, 1985.
61. **Schmitt-Slomska, J., Caravano, R., Anoal, M., Gay, B., and Roux, J.,** Isolation of L Forms from the spleens of *Brucella suis*-infected, penicillin-treated mice, *Ann. Microbiol. Inst. Pasteur (Paris),* 132A, 253–265, 1981.
62. **Schwarz, U. and Weidel, W.,** Zum Wirkungsmechanismus von penicillin. II. Nachweis eines penicillin-induzierten hydrolytischen Abbaus von Murein und Mureinvorstufen in *E. coli* B, *Z. Naturforsch. (B),* 20, 153–157, 1965.
63. **Smith, D. G. and Marchant, R.,** Unbalanced cell-wall synthesis in chloramphenicol-grown *Rhodotorula glutinis, Antonie van Leeuwenhoek,* 35, 113–119, 1969.
64. **Snell, J. F. and Cheng, L.,** Studies on modes of action of tetracyclines (II), *Dev. Ind. Microbiol.,* 2, 107–132, 1961.
65. **Sprott, G. D., Colvin, J. R., and McKellar, R. C.,** Spheroplasts of *Methanospirillum hungatii* formed upon treatment with dithiothreitol, *Can. J. Microbiol.,* 25, 730–738, 1979.
66. **Strominger, J. L., Izaki, K., Matsuhashi, M., and Tipper, D. J.,** Peptidoglycan transpeptidase and D-alanine carboxypeptidase penicillin sensitive enzymatic reaction, *Fed. Proc.,* 26, 9–22, 1967.
67. **Susan, S. R.,** A Fine Structure Study of *Pseudomonas aeruginosa* in Media and Distilled Water, M.S. thesis, Wayne State University, Detroit, 1972.

68. **Tsugaya, M., Washida, H., Sakagami, H., Iwase, Y., Inuzuka, K., and Takeuchi, K.,** Morphological studies of *Escherichia coli* in the urine of patients with acute simple cystitis treated with aztoreonam, *Acta Urol. Jpn.,* 32, 1883–1886, 1986.

69. **Tulasne, R.,** Cytologie des Proteus et ses Modifications sous l'effet de la penicilline, *C. R. Soc. Biol. (Paris),* 143, 286–288, 1949.

70. **Ubukata, K., Konno, M., and Fujii, R.,** Filamentous shape of *E. coli* treated with penicillin or cephalosporin C analogues, and its clinical significance. III, *J. Jpn. Assoc. Infect. Dis.,* 44(3), 146–155, 1970.

71. **Watanakunakorn, C., Goldberg, L. M., Carleton, J., and Hamburger, M.,** Staphylococcal spheroplasts and L-colonies. III. Induction by lysostaphin, *J. Infect. Dis.,* 119, 67–74, 1969.

Chapter 10

REVERSION AND REVERTANTS

"L Forms should revert since they contain all the machinery" has been the conclusion of Panos and Barkulis.[64] However, to the clinical microbiologist, reversion is the most frustrating word in Webster's lexicon. Many studies concern *in vitro*-made CWD forms ("Penicillin L's") which revert easily compared with the variants occurring spontaneously *in vivo*. The ensuing data will refer to both naturally occurring variants and some made by growth with antibiotics. When protoplasts initiate reversion, they first form short pieces of wall, gradually adding components for wall completion. The stages were followed with ferritin-labeled antibody by Wyrick and Gooder.[90]

TO REVERT OR NOT TO REVERT?

In general, CWD variants of Gram-positive organisms tend to be stable and derivatives of Gram-negative bacteria more easily reverted. It is reported that only a rare strain of *Proteus mirabilis* can be stabilized in the L-phase.[74] This species commonly reverts within a few hours, on the same medium used for penicillin induction, when the antibiotic is removed.

There are, however, many exceptions to correlating Gram reaction with reversibility; e.g., two types of penicillin-induced wall-deficient variants of *Serratia marcescens* were immediately stable.[2] Likewise, the parent strains from L-type cultures of Shigella are not always regained.[89] Also, while *P. vulgaris* characteristically forms very unstable CWD forms, Kandler and Kandler[37] noticed that even in this species the small granular colonies could be stabilized after two to seven passages in penicillin medium. Furthermore, the small L-type colonies of *Salmonella typhi* and other Salmonella species tend to be stable as soon as produced, although the larger mycoplasma-like colony may rapidly revert. However, it is not always possible to judge strain stability by colonial appearance, as shown by Figure 1. Three colony types are exhibited by stable strains of CWD *Escherichia coli*.[73]

Exceptions to the generality that Gram-positive bacteria make stable CWD forms are also numerous. Gram-positive cocci growing from lysed blood of kidney disease patients frequently revert. Such a coccus, making a septum for binary fission and simultaneously budding to show its variant tendency, is shown in Figure 2.[21] A nontypable Group A β-hemolytic streptococcus, the Richards strain, was successfully stabilized as an L Form only after 40 transfers on penicillin agar.[80] All of 19 penicillin-induced L-phase variants of enterococci remained unstable through 12 transfers[32] and L-phase variants of *Bacillus licheniformis* were stabilized only after 30 transfers on penicillin-containing medium.[25]

Again referring to reversion of a Gram-positive genus, Brigidi and co-investigators found great variation in reversibility of Bifidobacterium species. Easily reverting were *Bifidobacterium bifidum* var. *pennsylvanicus*, *B. thermophilum*, and *B. boum*.[7]

Organisms filtered from the sputum of tuberculosis cases revert to the classic mycobacterium, showing that these tiny forms in sputum differ from the nonreverting variants in circulating blood.[11]

In general, reversion is rapid in wall deficient variants made *in vitro,* in contrast to the stubborn irreversibility of CWD forms cultured from infection.

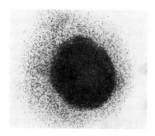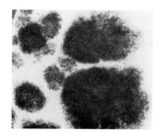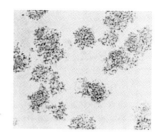

FIGURE 1.   Three colony types appearing in stable L colonies of *Escherichia coli.* (Schuhmann, E. and Taube-neck, U., *Z. Allg. Mikrobiol.*, 9, 297–313, 1969.)

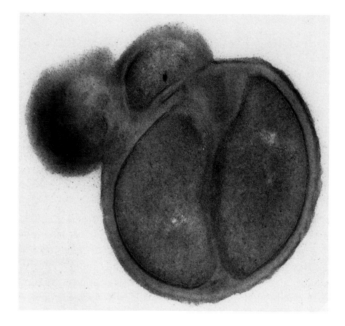

FIGURE 2.   Reverting coccus in subcultures of blood culture of kidney disease patients has both dividing septum and emerging buds. (From Domingue, G. J. and Smith, W. T., in *Cell Wall Deficient Bacteria*, Addison-Wesley, Reading, MA, 1982. With permission.)

IN VITRO REVERSION TECHNIQUES

OMISSION OF THE INDUCING AGENT

The difficulty of obtaining reversion of clinical isolates has been discussed by Taube-neck.[85] Kagan has stated that his L Forms seem to revert "when they want to".[35] This is also true in our laboratory, where the reversion rate, usually spontaneous, approximates 16%. Calderón et al., however, experienced an enviable reversion of over 50% in variants from infections.[9] Their procedure is presented in the chapter on methods. Also in Domingue's laboratory, reversion usually occurs with repeated transfers.[21]

Many L-phase variants induced *in vitro* rapidly revert when transferred to medium lacking the inducer. Thus, *P. mirabilis,* induced into cell wall deficiency by culture in a filtrate of an antagonistic strain of the same species, will revert to normal morphology and character-istics merely by growth in a medium lacking this filtrate.[23] The simple expedient of trans-ferring antibiotic-induced variants to antibiotic-free medium is widely used.

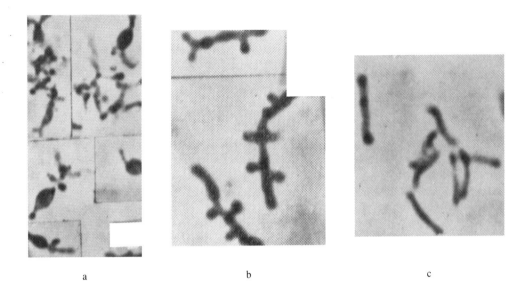

a   b   c

FIGURE 3. Showing the need for α,β-methyl-*N*-acetyl-D-glucosaminide to obtain rods from *Lactobacillus bifidus*. In the absence of the factor, thinned areas of wall result in ballooning or lateral branching (a). Increments of the growth factor decrease pleomorphism (b, c). (From Glick, M. C., Sall, T., Zilliken, F., and Mudd, S., *Biochim. Biophys. Acta,* 37, 361–363, 1960.)

Reversibility is related to the agent used for *in vitro* induction, and one agent may give relatively stable spheroplasts, while another yields unstable spheroplasts from the same organism.[77] It is well known that the longer an organism stays in the wall-deficient stage, the more difficult is the reversion.

EXPOSURE TO AN ANTIBIOTIC

Strangely, antibiotics long known for inducing wall deficiency, can sometimes foster reversion. LeBar[47] found that the CWD staphylococci from acute arthritis reverted in the diffusion zone around a gentamycin disk. The patient's severe infection with a CWD variant may have been related to her underlying disease, lupus erythematosis.

CHANGES IN NUTRITION
Role of Serum, Yeast, Vitamins

It has been noticed by many[9,70] that yeast extract fosters reversion, e.g., by Marston studying staphylococcal L-phase organisms.[51] Crawford et al. found that not only omission of serum, but also addition of 4% yeast extract permitted reversion of many L-phase streptococci after 1-month incubation.[16] Yeast extract combined with the filtrate of *Staphylococcus epidermidis* facilitated reversion of an interesting strain of *Corynebacterium acnes* from endocarditis.[92] One of the several critical aspects responsible for the high reversion rate in Calderón's laboratory may be that he employs a yeast extract which is not autoclaved.

Vitamin E was found by Fitz-James[24] to be an additive which fosters reversion of *Bacillus megaterium* protoplasts. The vitamin seems to stabilize the membrane.

Addition of Amino Sugars

Lactobacillus bifidus var. *pennsylvanicus* is a mutant growing as a uniform rod only in medium containing human milk or, more specifically, *N*-acetyl-D-glucosamine-containing saccharides. In medium containing 400 μg/ml of the amino sugar slightly curved, faintly granular rods are seen. If the essential wall precursor is present in lesser amounts, the organism is increasingly pleomorphic, as shown in the accompanying Figure 3.[28]

Salowich, using *N*-acetylglucosamine, obtained complete reversion of penicillin-induced staphylococcal CWD forms and partial reversion of *in vivo*-formed pneumococcal variants, which would not colonize on the surface of media.[70] The number of passages required, and in some instances lack of complete reversion, make this a less than ideal method.

Spermine

Lapinski and Flakas[45] describe the reversion of *Hemophilus influenzae* in the presence of spermine. They note that spermine is increased in the mammalian body in pneumonia, tuberculosis, carcinoma, and other diseased tissue, having approximately twice the spermine of the corresponding normal tissue. Spermine levels might explain why some individuals have classic bacteria in infections and others only L-phase organisms.

Exclusion of All Large Molecules

Some bacteria, suspended only in saline within cellophane sacs, grow luxuriously, metabolizing small molecules which filter through the membrane from the exterior. Czop and Bergdoll found that growth within such sacs fosters the reversion.[17]

PHYSICAL FACTORS
Support by Firm Agar or Gelatin

Changing the physical environment can foster reversion. For example, gelatin is a useful agent providing support and stimulating reversion, with a concentration of 25% exhibiting maximal effect.[38,44,83] Protoplasts of *Saccharomyces cerevisiae* and *S. fragilis* revert in gelatin medium. Treating the medium with gelatinase removes the reverting property, indicating that reversion accrues from the physical state provided by this protein, rather than from its components.

Similarly, hard agar fosters reversion of certain variants.[59] Svobodá found 2% agar ideal for reversion of *S. carlsbergensis, Candida utilis,* and *S. cerevisiae* protoplasts.[82] With less agar, the protoplasts grew temporarily but died after 8 to 10 h of incubation. All sources of agar tested were satisfactory, but he found that prolonged boiling or freeze-thawing the agar lessens the reversion. Therefore, Svobodá decontaminated agar only by boiling, not autoclaving.

Support by Cellulose Filter Pads

Membranous filters may foster reversion, again, by providing rigidity at the variant's surface. Eighteen types of filters proved totally incapable of supporting L-colony growth. However, a number of membrane filters from Millipore, Oxoid, and Schleicher and Schuell not only helped growth but also permitted reversion of wall-less forms of *Bacillus subtilis*. The composition of filters fostering reversion include nitrocellulose, cellulose-acetate, diacetate, and triacetate products.[15]

Atmosphere

Although induction into the L phase, even for aerobes, is usually best under anaerobic conditions, Freundt found that reversion occurs most rapidly aerobically.

Perhaps oxygen tension is involved in the reversion ritual devised by Fukui and associates. They report that protoplasts of *Geotrichum candidum* revert only if cultured in an agar layer less than 1 mm thick.[27] Similarly, aeration may be the critical factor in reversion occurring in agitated culture, as noted in the L phase of staphylococci shaken in liquid medium for several weeks.[75]

Temperature

The temperature needed for reversion in some instances can be critical, as found by Landman et al. for *B. subtilis*.[43] Conversely, reversion often occurs throughout the optimum growth range of the organism, e.g., 20 to 30C for *G. candidum*.[27]

Alternation of Culture Menstrua

Carter and Greig[10] obtained reversion of a diphtheroid contaminant by culturing the L-phase organism first on serum agar, then transferring a block of colonies to serum broth. After repeating this several times, a diphtheroid was consistently obtained. The contaminant was from bovine serum used as tissue culture enrichment.

Mucin

Bisset's group[4] explored reversion techniques for *B. licheniformis* isolates from normal blood. Hog mucin increased reversion about 25-fold, and diaminopimelic acid 20-fold. The majority of revertants appeared by 7 d, but some developed in the ensuing 3 weeks. Mucin, streaked over the surface of plates, produced reversion of one colony in 8 plates. The revertants were stable.

REVERSION STIMULATED BY PRODUCTS FROM MICROBES

Filtrates and other products from several genera have sponsored reversion of certain CWD isolates. Reversion studies have concerned fostering reversion with whole organisms or factors from bacilli,[12,19] staphylococci,[20,50] *Sarcina lutea*,[76] and *Neisseria perflava*.[49]

Clive and Landman also found that reversion increases after incorporating the walls of bacteria into the medium or into the membrane filter. The effect is entirely nonspecific; walls could be harvested from other strains of *Bacillus, E. coli, P. mirabilis, Salmonella typhimurium, Pseudomonas aeruginosa, Staphylococcus aureus,* and *Saccharomyces cerevisiae*. With some preparations, as high as 42% of the organisms reverted. Boiled walls foster reversion better than chemically decontaminated walls. It was concluded that heat eliminated undesirable enzymes which interfere with growth of the revertants.[15]

Attempts to foster reversion of *Salmonella paratyphi* with bacteriophage or with crude bacterial lysates prepared by exposure to penicillin, sonication, and virulent phage were unsuccessful.[42]

MYCOBACTERIAL REVERSION

Reversion of CWD mycobacteria produced by antibiotics *in vitro* is consistently reproducible. The rapidly multiplying species revert relatively soon after *in vitro* induction.[53]

Reversion of wall-deficient variants of *Mycobacterium tuberculosis* growing *in vivo* is exceedingly more difficult. Culturing with Freundt's adjuvant produces reversion of filtrable variants of the tubercle bacillus from sputum.[68] This is not successful when applied to the CWD forms found in the blood. A study by Trece and Teodecki[93] compared acetone extracts with Freundt's adjuvant as reversion stimulants of CWD blood-borne variants from tuberculosis and sarcoidosis cases. Morphological reversion to granular bacilli was obtained for both types of variants with acetone extracts but not with Freundt's adjuvant. In no isolates did macroscopic colonies grow on the surface of media.

Similarly, prolonged storage or incubation has not facilitated reversion of blood-borne *M. tuberculosis* variants. Fourteen variants of the tubercle bacillus and four from sarcoidosis were stable at 4 to 8°C for 10 to 12 years without reversion.

In summary, wall-deficient *M. hominis, M. bovis* and the variant mycobacteria from sarcoidosis[34] and from Crohn's disease[13] revert belatedly or not at all.

A MURALYTIC ENZYME

Reversion had been achieved enzymatically.[22] The reverting strain was in the blood of a drug addict hospitalized with a temperature of 105°F, a massive purulent pleural effusion and septicemia with a coagulase-positive staphylococcus. The cultures studied by Edwards were taken after penicillin therapy, when the patient's blood no longer yielded classic organisms, although his condition was little improved. The addition of muramidase to subcultures in a concentration of 2 μg/ml caused reversion of a wall deficient variant to a coagulase-positive staphylococcus. The reversion occurred by 48 h in a variety of media with commercial muramidase from two sources. The mechanism involved remains to be elucidated. Conceivably, since muramidases are not identical, the commercial product may compete with the bacterial autolysins,[65] thus inhibiting catabolism which otherwise exceeds wall restructuring.

The staphylococcus revertant of Edwards is far from a typical organism. It grows first as a filamentous, pleomorphic, Gram-negative rod which does not grow in media containing inhibitors for Gram-positive organisms.

Muramidase has not fostered reversion of CWD variants which appear to be strepto-coccal, suggested by the type of hemolysis. The resistance of the cell walls of Group A streptococci to muramidase is discussed by Barkulis et al.[3]

AGING WITH OR WITHOUT TRANSFERS

Hauduroy and Lesbre's reversion procedure,[31] essentially blind passage, has been used successfully in several countries. The method, which involves multiple transfers of inapparent growth to fresh media, cannot be termed rapid but may eventually result in classic growth.

Sometimes reversion has been the reward of mere patience. "Ripening" at room temperature or incubator temperature for a period of at least 4 months has been recommended by Trofimova.[86]

CONCENTRATED POPULATIONS

It is the general experience that reversion is more successful in cultures with high populations. Rosner recommends plating a veritable microbial paste on blood agar and subjecting this to prolonged incubation.[94] Svobodá obtained reversion of yeast protoplasts only in concentrates of 10^4 to 10^6 cells.[82] In Calderón's reversion series, he transferred a large inoculum when the inverse microscope showed many colonies. Susan found reversion of *P. aeruginosa* in distilled water when the CWD forms numbered 10^6 organism per milliliter.[81]

MUTAGENS

Reversion of stable L-cultures was produced by the mutagens: UV light, hydrolamines, and acridine. Reversion occurred several hours after application of the mutagenic agents to *S. typhi* which had been induced into the L Form 8 years earlier. However, reversion was realized with low frequency in only 3 of 45 trials.[48]

STABILIZING INDUCED L FORMS

When culturing L Forms from infections, the challenge is how to obtain reversion to facilitate identification. In contrast, when L Forms are induced *in vitro*, the reverse is the problem: how to stabilize the L Forms for genetic or other studies. Continued subculture in the presence of lysozyme can stabilize L Form *B. subtilis* which can be freeze dried or stored at −70°C in glycerol.[1]

L Forms of *Listeria monocytogenes* have been stabilized by successive passage in penicillin-containing medium, with accompanying morphological and physiological changes.[41]

TABLE 1
Characteristics Which May Be Changed in Revertants

Organism	Lost or Altered	Gained
Brucella melitensis[55]	Did not oxidize i-erythritol or glucose	Mucoid colony oxidized adonitol, D-asparagine, & D-glutamate
Erysipelothrix rhusopathiae[63]	Penicillin sensitivity changed	
Erythromycin producers		Antibiotic production increased 20–60%[91]
Klebsiella pneumoniae[29]	Vigorous growth and mouse virulence	
Mycobacterium tuberculosis[39]	Pronounced cord-factor Virulence decreased or lost	Faster, softer growth on solid media
Neisseria gonorrhoeae[69]		Made L-phase colonies without antibiotic
N. meningitidis[40]	Virulence lost Serological response changed	Became penicillin resistant
Nocardia rubra[66]	Lost red pigment	
Oleandomycin producers		110% increase in antibiotic production[46]
Proteus mirabilis[71]	Lost ability to ferment glucose	
P. vulgaris strain 52[5]	Ability to swarm & susceptibility to phage types c, d, e, g	Resistance to 80°C
Pseudomonas[88]		Revertants of carbenicillin-induced spheroplasts had 4 × increased carbenicillin resistance
Saccharomyces cerevisiae[59]	Typical morphology lost in some strains; one became polynuclear Fermentation rate & alcohol formation changed	
Shigella flexneri	Grew only at 20–28°C Loss of some fermentative enzymes	Agglutinated by antibody to: *S. boydi, S. newcastle, S. sonnei*
Shigella sp.		Brown and yellow pigments acquired
Staphylococcus aureus	Often failed to make penicillinase; lost phage sensitivity[72] Antibiotic sensitivities changed[72] 3 of 8 strains became phage untypable[51]	
Streptococci, group A[31,36,61,87] These concern revertants in blood of rheumatic fever.	Hemolysins	Resistance to heat and increased resistance to certain antibiotics
Vibrio cholerae[62]		Virulence increased up to 10 ×

L variants of *Proteus vulgaris, P. mirabilis, P. rettgeri, P. morgani, S. typhimurium* and *Shigella alkalescens* could all be propagated without reversion by adding sodium azide to the medium.[52]

CHARACTERISTICS OF REVERTANTS

B. subtilis strains are right- or left-handed, i.e., fibers in the wall curve in one direction. When spheroplasts revert to classic forms, the fibers in the bacilli are again right- or left-handed following the parental genotypes.[6]

Characteristics of reverted streptococci and other bacteria are given in Table 1. Changes in β-hemolytic streptococci are especially notable.

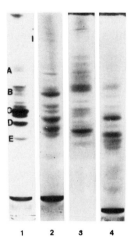

FIGURE 4. Polyacrylamide gel studies show revertant of *Staphylococcus aureus* (1) has proteins resembling those of CWD *S. aureus* in blood cultures 2–4. (From Motwani, N., Ph.D. dissertation, Wayne State University, Detroit, 1976.)

Revertants of two strains of streptococci appeared identical to the parent cocci until their fine structure was examined. In both strains the revertant wall was much thicker than in the parent.[61] The converse of this situation was found in a revertant of *E. coli*. The reverted cells had thinner walls, and wall strata never exceeded three, in contrast to the three to five layers of the original cells.[27] The protein patterns of *Staphylococcus aureus* in polyacrylamide gel can very much resemble the parent electrophorogram. This is also true of the revertant[57] (Figures 4 and 5).

INHIBITION OF REVERSION

INHIBITION OF REVERSION OF YEAST PROTOPLASTS BY 2-DEOXYGLUCOSE
2-Deoxyglucose interferes with the regeneration of yeast protoplasts which normally occurs in glucose-gelatin. Reversion is completely inhibited even if the ratio of glucose to 2-deoxyglucose is 20:1. The cell wall at the surface of nonregenerating protoplasts clearly shows fibrils, indicating reduced matrix.[84]

PRESENCE OF SERUM
Many investigators have noted that serum in a medium inhibits reversion of a wall-deficient variant.[16,27,30] On the other hand, certain sera foster reversion. In one study, it was only with combined serum and 5% chicken plasma that any remarkable steps toward reversion of *Salmonella enteritidis* occurred.[33] In another study, Vadász and Juhász were able to follow in serum-medium the complete morphological reversion of penicillin-induced *S. enteritidis*.

When numerous species of *Salmonella* were induced into vacuolated colonies by penicillin, enrichment could be either horse or guinea pig serum. Reversion occurred rapidly when guinea pig serum was in the medium, but never with horse serum.[54] The subject of serum components is discussed by Broome[8] in reviewing the work of Clementi, who found that L-asparaginase was present in the blood of only one species surveyed, the guinea pig.[14] Perhaps a similar enzyme fosters the reversion of the Salmonella.

An enigma is presented by the reversion inhibition characteristic of most sera. If deletion of serum is needed for reversion, how can this be done with the majority of wall-deficient

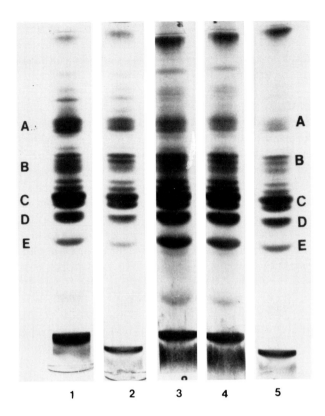

FIGURE 5. Revertant of *S. aureus* from blood culture (5) resembles classic isolates from (1) urine, (2) sputum, (3) blood culture, and (4) wound. (From Motwani, N., Ph.D. dissertation, Wayne State University, Detroit, 1976).

bacteria which are serophilic? Calderón's reversion procedure adapts to serum omission by successive steps. Other solutions are constantly sought by investigators.

DESTROYING THE REVERSION INHIBITOR

An interesting reversion inhibitor factor (RIF) has been studied.[18] The inhibitor is non-dialyzable and sensitive to trypsin and heat. A protocol for efficient reversion of *B. subtilis* includes trypsin-treatment to destroy the inhibitor.

MORPHOLOGY OF THE REVERTING ORGANISM

Prozorovsky notices that as L-phase cultures of Staphylococci revert, they exhibit many differences in morphology which are inherent peculiarities of the strains.[67] Perhaps such morphological distinctions can be used in epidemiology, as reversion of a penicillin-induced Staphylococcus can be performed more simply than phage typing. These observations are intriguing in view of the similarity exhibited by all species of antibiotic-induced Salmonella[54] and Candida.

IN VIVO MANIPULATIONS

CHICK EMBRYO INOCULATION

Passage in chick embryo produces reversion of some wall-deficient variants found in blood cultures.[58] This approach is not successful in all cases. Inoculation of chick embryo amniotic cavity or allantoic cavity did not result in reversion of any of 60 isolates of Peptococci which were wall deficient after lyophilization and prolonged storage. If reversion is not obtained by the third serial passage in the embryo, it is probable that adaptation toward a strict intracellular virus-like stage is occurring.

INJECTIONS IN MICE

Mice are frequently used to accelerate reversion of unstable L-phase strains. Reversion can be further facilitated by pretreatment of mice with mucin, as found by Godzeski working with Klebsiella variants.[95] It should be noted that mucins of different origins can vary greatly.[56]

Commonly, injections to obtain reversion are made intraperitoneally in mice. However, intracerebral inoculation in mice brought reversion of penicillin-induced *Corynebacterium diphtheriae*.[78]

Injection of mice, even when immunosuppressed, does not lead to reversion when the CWD strains are from uveitis, or when they come from tuberculosis or sarcoid blood cultures. In these cases disease occurs without reversion.

SUMMARY

Factors essential for reversion have been defined for only a few saprophytic organisms which have lost their walls by specific *in vitro* treatments. Techniques which may foster reversion include:

1. Omitting inducing antibiotic
2. Concentrating populations
3. Adding a new antibiotic, or use of a diffusion gradient
4. Aging at 4 or 37°C
5. Adding or omitting serum from varied mammals
6. Addition of yeast extract, Vitamin E, *N*-acetyl-D-glucosamine, spermine, trypsin, hog mucin, diaminopimelic acid, muramidase, chloramphenicol
7. Adding products or walls of other microbes: *Staphylococcus epidermidis;* Bacillus species; factors from *Neisseria perflava,*[9]
8. Omission of large molecules from growth medium
9. Use of mutagens: UV light, hydrolamines and acridine
10. Support by firm agar (agar better not autoclaved)
11. Support by firm gelatin
12. Freundt's adjuvant *in vitro* for mycobacteria
13. Acetone extract of mycobacteria given *in vivo* for mycobacteria
14. Inoculation of chick embryos or mice

Investigation of a thermolabile protein required for division of *E. coli* might be informative.[79]

The revertants returned to the classic stage from the CWD form often display remarkable differences when compared with the parent strain. Thus, passage through a stage with aberrant walls may be one of the important instigators of variation within a species.

REFERENCES

1. **Allen, E. J.,** Induction and cultivation of a stable L Form of *Bacillus subtilis, J. Appl. Bacteriol.,* 70, 339–343, 1991.
2. **Bandur, B. M. and Dienes, L.,** L Forms isolated from a strain of Serratia, *J. Bacteriol.,* 86, 829–836, 1963.
3. **Barkulis, S. S., Smith, C., Boltralik, J. J., and Heymann, H.,** Structure of streptococcal cell walls. IV. Purification and properties of streptococcal phage muralysin, *J. Biol. Chem.,* 239, 4027–4033, 1964.
4. **Bisset, K. A. and Bartlett, R.,** The isolation and characters of L Forms and reversions of *Bacillus licheniformis* var. *Endoparasiticus* (Benedek) associated with the erythrocytes of clinically normal humans, *J. Med. Microbiol.,* 11, 335, 1978.
5. **Bloss-Bender, L.,** Über modifikative und mutative L Formen von *Proteus vulgaris, Arch. Mikrobiol.,* 34, 16–29, 1959.
6. **Briehl, M. M. and Mendelson, N. H.,** Helix hand fidelity in *Bacillus subtilis* macrofibers after spheroplast regeneration, *J. Bacteriol.,* 169, 5838–5840, 1987.
7. **Brigidi, P., Matteuzzi, D., and Crociani, F.,** Protoplast formation and regeneration in Bifidobacterium, *Microbiologica,* 9(2), 243–248, 1986.
8. **Broome, J. D.,** L-Asparaginase, the evolution of a new tumor inhibitory agent, *Trans. N.Y. Acad. Sci.,* 30, 690–704, 1968.
9. **Calderón, N., Albuerne, A., Gonzalez, S., and Winkler, L.,** Cell wall deficient bacteria (L Forms) in meningitis, *Lat. Am. Microbiol.,* 13, 95–100, 1971.
10. **Carter, G. R. and Greig, A. S.,** The recovery of diphtheroids from L type organisms contaminating tissue cultures, *Can. J. Microbiol.,* 9, 317–320, 1963.
11. **Chandrasekhar, S.,** Cell wall deficient forms of tubercle bacilli, *Indian J. Chest Dis. Allied Sci.,* 22, 114–122, 1980.
12. **Charache, P.,** Atypical bacterial forms in human disease, in *Microbial Protoplasts, Spheroplasts, and L Forms,* Guze, L. B., Ed., Williams & Wilkins, Baltimore, 1968, 484–494.
13. **Chiodini, R. J., Van Kruiningen, H. J., Thayer, W. R., and Coutu, J. A.,** Spheroplastic phase of mycobacteria isolated from patients with Crohn's disease, *J. Clin. Microbiol.,* 24, 357–363, 1986.
14. **Clementi, A.,** Desaminidation enzymatique de l'asparagine, *Arch. Int. Physiol.,* 19, 369–373, 1922.
15. **Clive, D. and Landman, O. E.,** Reversion of *Bacillus subtilis* protoplasts to the bacillary form induced by exogenous cell wall, bacteria and by growth in membrane filters, *J. Gen. Microbiol.,* 61, 223–243, 1970.
16. **Crawford, Y. E., Frank, P. F., and Sullivan, B.,** Isolation and reversion of L Forms of beta-hemolytic streptococci, *J. Infect. Dis.,* 102, 44–52, 1958.
17. **Czop, J. K. and Bergdoll. M. S.,** Synthesis of enterotoxin by L Forms of *Staphylococcus aureus, Infect. Immun.,* 1, 169–173, 1970.
18. **DeCastro-Costa, M. R. and Landman, O. E.,** Inhibitory protein controls the reversion of protoplasts and L Forms of *Bacillus subtilis* to the walled state, *J. Bacteriol.,* 129, 678–689, 1977.
19. **Dienes, L.,** Influence of a bacillus on the reversion of *Hemophilus influenzae* L Forms to the bacterial form, *Proc. Soc. Exp. Biol. Med.,* 135, 392–394, 1970.
20. **Dienes, L. and Pachas, W. N.,** Observations suggesting the development of Streptococci from pleomorphic filamentous Gram negative bacteria, *Yale J. Biol. Med.,* 43, 337–350, 1971.
21. **Domingue, G. J. and Smith, W. T.,** in, *Cell Wall Deficient Bacteria,* Addison-Wesley, Reading, MA, 1982.
22. **Edwards, J. H. and Mattman, L. H.,** Lysozyme-produced variation in a staphylococcal isolate, *Am. Soc. Microbiol. Abstr. Annu. Meet. 1972,* 20, 1972.
23. **Falkow, S.,** Forms of Proteus induced by filtrates of antagonistic strains, *J. Bacteriol.,* 73, 443–444, 1957.
24. **Fitz-James, P. C.,** The response of growing protoplasts to various doses of vitamins A and E, in *Microbial Protoplasts, Spheroplasts, and L Forms,* Guze, L. B., Ed., Williams & Wilkins, Baltimore, 1968, 239–247.
25. **Fodor, M. and Rogers, H. J.,** Antagonism between vegetative cells and L Forms of *Bacillus licheniformis* Strain 6346, *Nature (London),* 211, 658–659, 1966.
26. **Freundt, E. A.,** Observations on Dienes' L type growth of bacteria, *Acta Pathol.,* 27, 159–171, 1950.
27. **Fukui, K., Sagara, Y., Yoshida, N., and Matsuoka, T.,** Analytical studies on regeneration of protoplasts of *Geotrichum candidum* by quantitative thin-layer agar plating, *J. Bacteriol.,* 98, 256–263, 1969.
28. **Glick, M. C., Sall, T., Zilliken, F., and Mudd, S.,** Morphological changes of *Lactobacillus bifidus* var *pennsylvanicus* produced by a cell-wall precursor, *Biochim. Biophys. Acta,* 37, 361–363, 1960.
29. **Guze, L. B., Harwick, H. J., and Kalmanson, G. M.,** Klebsiella L Forms, effect of growth as L Form on virulence of reverted *Klebsiella pneumoniae, J. Infect. Dis.,* 133(3), 245, 1976.
30. **Hamburger, M. and Carleton, J.,** Staphylococcal spheroplasts and L colonies. II. Conditions conducive to reversion of spheroplasts to vegetative Staphylococci, *J. Infect. Dis.,* 116, 544–550, 1966.
31. **Hauduroy, P. and Lesbre, P.,** Les formes filtrantes des Streptocoques, *C. R. Soc. Biol.,* 97, 1394–1395, 1927.

32. **Hijmans, W. and Kastelein, M. J. W.,** The production of L Forms of *Enterococci, Ann. N.Y. Acad. Sci.,* 79, 371–373, 1960.

33. **Juhász, I. and Vadász, J.,** Regeneration of the filtrable forms of *Salmonella enteritidis* in media containing blood plasma (fibrin structure), *Nature (London),* 176, 208–209, 1955.

34. **Judge, M. S. and Mattman, L. H.,** Cell wall deficient mycobacteria in tuberculosis, sarcoidosis, and leprosy, in *Cell Wall Deficient Bacteria,* Domingue, G. J., Ed., Addison-Wesley, Reading, MA, 1982, 257–298.

35. **Kagan, B. M.,** Staphylococcal L Forms — ecologic perspectives, *Ann. N.Y. Acad. Sci.,* 128, 81–91, 1965.

36. **Kagan, G. Y.,** Some aspects of investigations of the pathogenic potentialities of L Forms of bacteria, in *Microbial Protoplasts, Spheroplasts, and L Forms,* Guze, L. B., Ed., Williams & Wilkins, Baltimore, 1968, 422–443.

37. **Kandler, O. and Kandler, G.,** Trennung und Charakterisierung verschiedener L phasen Typen, *Z. Naturforsch.(B),* 11, 252–259, 1956.

38. **King, J. R. and Gooder, H.,** Reversion to the streptococcal state of enterococcal protoplasts, spheroplasts, and L Forms, *J. Bacteriol.,* 103, 692–696, 1970.

39. **Kochemasova, Z. N., Dukhno, M. I., Kassirskaya, N. G., and Bakanova, D. Ya.,** Biological properties of revertants obtained from the L Forms of *Mycobacterium tuberculosis, Probl. Tuberk.,* 11, 65–68, 1969.

40. **Koptelova, E. I. and Mironova, T. K.,** Stabilization and reversion of L Forms of meningococcus, *Zh. Mikrobiol. Epidemiol. Immunobiol.,* 47, 16–19, 1970.

41. **Kotylarova, G. A., Bakulov, I. A., and Prozorovsky, S. V.,** Obtaining of L Forms of *Listeria monocytogenes, Zh. Mikrobiol. Epidemiol. Immunobiol.,* 45, 96–99, 1968.

42. **Landman, O. E. and Ginoza, H. S.,** Genetic nature of stable L Forms of *Salmonella paratyphi, J. Bacteriol.,* 81, 875–886, 1961.

43. **Landman, O. E., Ryter, A., and Frehel, C.,** Gelatin-induced reversion of protoplasts of *Bacillus subtilis* to the bacillary form, electron-microscopic and physical study, *J. Bacteriol.,* 96, 2154–2170, 1968.

44. **Landman, O. E. and Forman, A.,** Gelatin-induced reversion of protoplasts of *Bacillus subtilis* to the bacillary form, biosynthesis of macromolecules and wall during successive steps, *J. Bacteriol.,* 99, 576–589, 1969.

45. **Lapinski, E. M. and Flakas, E. D.,** Reversal of penicillin-induced L phase growth of *Haemophilus influenzae* by spermine and its effects on antibiotic susceptibility, *Infect. Immun.,* 1, 474–478, 1970.

46. **Lavrent'eva, G. I. and Ovsianko, A. V.,** The use of the protoplast method for the selection of oleandomycin producers, *Antibiot. Khimiother.,* 35(4), 17–20, 1990.

47. **LeBar, W. D.,** Isolation and Identification of Cell Wall Deficient Microorganisms from Infection, M.S. thesis, Wayne State University, Detroit, 1977.

48. **Levashev, V. S.,** Genetic mechanisms of the L Form bacteria. Formation and prospects for further research, *Vestn. Akad. Med. Nauk SSSR,* 20, 32–37, 1965.

49. **Madoff, S.,** L Forms of *Haemophilus influenzae,* Morphology and ultrastructure, in *Spheroplasts, Protoplasts, and L Forms of Bacteria, INSERM,* 65, 15–26, 1976.

50. **Madoff, S.,** Introduction to the Bacterial L Forms, in *The Bacterial L Forms,* Madoff, S., Ed., Marcel Dekker, New York, 1986.

51. **Marston, J.,** Observations on L Forms of Staphylococci, *J. Infect. Dis.,* 108, 75–84, 1961.

52. **Mattman, L. H. and Starnes, R. W.,** Variants and the use of sodium azide in their propagation, *Yale J. Biology Med.,* 31, 294–302, 1959.

53. **Mattman, L. H., Tunstall, L. H., and Lenzini, G.,** Stages in L variation of *Mycobacterium phlei, Bacteriol. Proc.,* p.121, 1960.

54. **Mattman, L. H., Tunstall, L. H., and Kispert, W. G.,** A survey of L variation in the Salmonellae, *Zentralbl. Bakteriol. Parasitenkd. Infektionskr. Hyg. Abt. 1 Orig.,* 210, 65–74, 1969.

55. **Meyer, M. E.,** Evolution and taxonomy in the genus *Brucella,* Steroid hormone induction of filterable forms with altered characteristics after reversion, *Am. J. Vet. Res.,* 37(2), 207–210, 1976.

56. **Moore, R. W.,** *Mycoplasma hyoarthinosa,* in *Proc. Conf. Relationship of Mycoplasma to Rheumatoid Arthritis and Related Diseases* (Chicago), Decker, J. L., Ed., U.S. Public Health Serv. Publ. No. 1523, Washington, D.C., 1966, 88–96.

57. **Motwani, N.,** Identification of Cell Wall Deficient Bacteria by Immunofluorescence and Polyacrylamide Gel Electrophoresis, Ph.D. dissertation, Wayne State University, Detroit, 1976.

58. **Nativelle, R. and Deparis, M.,** Formes evolutives des bacteries dans les hemocultures, *Presse Med.,* 68, 571–574, 1960.

59. **Necas, O.,** Regeneration of yeast cells from naked protoplasts, *Nature (London),* 177, 898–899, 1956.

60. **Necas, O.,** The mechanism of regeneration of yeast protoplasts. I. Physical conditions, *Folia Biol.,* 8, 256–262, 1962.

61. **Nicholls, E. E.,** A study of the organisms recovered from filtrates of cultures of hemolytic *Streptococci, J. Infect. Dis.,* 62, 300–306, 1938.

62. **Olsson, P. G.,** Zur Variation des Choleravirus, *Zentralbl. Bakteriol. Parasitenkd. Infektionskr. Hyg. Abt. 1 Orig.,* 76, 23–34, 1915.

63. **Pachas, W. N. and Currid, V. R.,** L Form induction, morphology,and development in two related strains of *Erysipelothrix rhusiopathiae, J. Bacteriol.,* 119, 576–582, 1974.

64. **Panos, C. and Barkulis, S. S.,** Streptococcal L Forms. I. Effect of osmotic change on viability, *J. Bacteriol.,* 78, 247–252, 1959.

65. **Pooley, H. M. and Shockman, G. D.,** Relationship between the location of autolysin, cell wall synthesis, and the development of resistance of cellular autolysis in *Streptococcus faecalis* after inhibition of protein synthesis, *J. Bacteriol.,* 103, 457–466, 1970.

66. **Prasad, I. and Bradley, S. G.,** Cell wall defective variant of *Nocardia rubra, J. Gen. Microbiol.,* 70, 571–572, 1972.

67. **Prozorovsky, S. V.,** Reversion of the stabilized cultures of the L Forms of pathogenic staphylococci. I. Peculiarities of the reversion process, *Biol. Med.,* 4, 87–90, 1959.

68. **Ratnam, S. and Chandrasekhar, S.,** The pathogenicity of spheroplasts of *Mycobacterium tuberculosis, Am. Rev. Respir. Dis.,* 114, 549–554, 1976.

69. **Roberts, R. B.,** The in vitro production, cultivation and properties of L Forms of pathogenic Neisseriae, in *Microbial Protoplasts, Spheroplasts, and L Forms,* Guze, L. B., Ed., Williams & Wilkins, Baltimore, 1968.

70. **Salowich, L. G.,** Reversion of Stable Staphylococcal L Forms, M.S. thesis, Wayne State University, Detroit, 1965.

71. **Schmid, E. N.,** Unstable L Form of *Proteus mirabilis* induced by Fosfomycin, *Chemotherapia* 31, 286–291, 1985.

72. **Schönfeld, J. K.,** "L" forms of Staphylococci; their reversibility; changes in the sensitivity pattern after several intermediary passages in the L phase, *Antonie van Leeuwenhoek,* 25, 325–331, 1959.

73. **Schuhmann, E. and Taubeneck, U.,** Stabile L Formen verschiedener *Escherichia coli*-Stämme, *Z. Allg. Mikrobiol.,* 9(4), 297–313, 1969.

74. **Silberstein, J. K.,** Observations on the L Forms of Proteus and Salmonella, *Z. Allg. Pathol. Bakteriol.,* 16, 739–755, 1953.

75. **Simon, H. J. and Yin, E. J.,** Penicillinase studies on L phase variants, G phase variants, and reverted strains of *Staphylococcal aureus, Infect. Immun.,* 2, 644–654, 1970.

76. **Sinkovics, J.,** Pleuropneumonieahnliche eigenheiten in alten *E. coli*-Kulturen, *Acta Microbiol. Acad. Sci. Hung.,* 4, 59–75, 1957.

77. **Smarda, J.,** Microscopic picture of reversion of glycine spheroplasts into rods in *Escherichia coli, Nature (London),* 201, 927–928, 1965.

78. **Smirnovskaia, A. S.,** Comparative evaluation of different methods for regeneration of diphtheria bacilli from their filterable forms, *Zh. Mikrobiol. Epidemiol. Immunobiol.,* 28, 177–180, 1957.

79. **Smith, H. S. and Pardee, A. B.,** Accumulation of a protein required for division during the cell cycle of *Escherichia coli, J. Bacteriol.,* 101, 901–909, 1970.

80. **Stewart, R. H. and Wright, D. N.,** Storage of meningococcal and streptococcal L-forms, *Cryobiology,* 6, 529–532, 1970.

81. **Susan, S. R.,** A Fine Structure Study of *Pseudomonas aeruginosa* in Media and Distilled Water, M.S. thesis, Wayne State University, Detroit, 1972.

82. **Svobodá, A.,** Regeneration of yeast protoplasts in agar gels, *Exp. Cell Res.,* 44, 640–642, 1966.

83. **Svobodá, A. and Necas, O.,** Mechanism of regeneration of yeast protoplasts. VI. An experimental blocking of regeneration of protoplasts, *Folia Biol. (Prague),* 14, 390–397, l968.

84. **Svobodá, A., Farkas, V., and Bauer, S.,** Response of yeast protoplasts to 2-deoxy-glucose, *Antonie van Leeuwenhoek,* 35, B11–12, 1969.

85. **Taubeneck, U.,** Zur Frage der Reversion der stabilen L Form in ihre Ausgangsbakterien, *Zentralbl. Bakteriol. Parasitnkd. Infektionskr. Hyg. Abt. 1 Orig.,* 184, 290–293, 1962.

86. **Trofimova, N. D.,** Studies of the conditions for regeneration of filterable forms of Staphylococci, *Mikrobiol. J. Acad. Sci. Ukrain. RSR,* 21, 45–47, 1959.

87. **Vulfovich, I. V., Krasilnikova, O., Konstantinova, N. D., and Gamova, N. A.,** Antibiotic resistance of revertant cultures of Streptococcus group A and of unclassified streptococci isolated from the blood of rheumatism patients, *Antibiot. Med. Biotekhnol.,* 31(8), 628–633, 1986.

88. **Watanakunakorn, C., Phair, J. P., and Hamburger, M.,** Increased resistance of *Pseudomonas aeruginosa* to carbenicillin after reversion from spheroplast to rod form, *Infect. Immun.,* 1, 427–430, 1970.

89. **Weinberger, H. J., Madoff, S., and Dienes, L.,** The properties of L Forms isolated from Salmonella and the isolation of L Forms from Shigella, *J. Bacteriol.,* 59, 765–775, 1950.

90. **Wyrick, P. B. and Gooder, H.,** Reversion of *Streptococcus faecium* cell wall defective variants to the intact bacterial state, in *Spheroplasts, Protoplasts and L Forms of Bacteria, INSERM,* 65, 59–88, 1976.

91. **Zakharova, G. M., Belova, T. S., and Danilenko, V. N.,** Study of the effect of protoplast formation on the antibiotic activity of erythromycin producers, *Antibiot. Khimioter.,* 35(1), 3–4, 1990.

92. **Zierdt, C. H. and Wertlake, P. T.,** Transitional forms of *Corynebacterium acnes* in disease, *J. Bacteriol.,* 97, 799–805, 1969.
93. **Trece, N. and Teodecki, S.,** unpublished.
94. **Rosner, R.,** personal communication.
95. **Godzeski, C.,** personal communication.

Chapter 11

SEPTICEMIA AND CARDIOPATHIES

ENDOCARDITIS

Endocarditis contributes the most unfortunate false negatives in the hospital laboratory. In subacute bacterial endocarditis (SBE), routine cultures often fail. This was true before antibiotic therapy, even with the best media, and when tested blood is arterial. Currently, there has been a rise in culture-negative endocarditis cases.[31] Recent figures state the cases from which no organism is recovered may reach 40%.[9] In culture-negative SBE cases, the diagnosis is delayed, and "calculated guess" therapy often fails. When an appropriate antibiotic is found, complications may have already occurred. Culture-negative patients have longer illnesses, more frequent anemia and more embolic phenomena than culture-positive patients.[14] As is well known, without specific antimicrobial therapy, SBE is almost invariably fatal.

SUBACUTE BACTERIAL ENDOCARDITIS

The first reports of SBE blood cultures containing only CWD forms were from Nativelle and Deparis[24] illustrated with excellent color micrographs (see Frontispiece). Their series included both streptococci and staphylococci. It has been reassuring to learn that the variants in Detroit look much as they do in France. On solid medium the colonies are often composed simply of granules. (see Frontispiece) Sometimes they grow as uniform spheres which later revert to streptococci (Figure 1). Some of the early isolates of nonclassic growth from endocarditis were reported by Tanret who used chick embryo for culture of the variants and for antibiograms. His two patients, infected with streptococcal L Forms, did not respond to penicillin. One patient was successfully treated with chloramphenicol and spiramycin, and one needed a fourfold regime of streptomycin, tetracycline, tifomycine, and aureomycin.[33]

Smith[37] described isolations of L-phase streptococci from blood cultures in his laboratory. The organisms were subcultured from blood culture sediment onto sheep blood agar, using trypticase agar base, and incubated at 35 to 37°C for 48 h, in air enriched with 5% CO_2. On agar, growth always appeared as pitted colonies, difficult to pick, and subcultured onto other solid media only by removing and pushing a block of agar inverted along a new surface. Reversion to parent form was never obtained on solid media. Reversion occurred after 10 daily subcultures in brain heart infusion broth (BBL) with no supplements. When reverted, classic colonies appeared on the blood plate among L-phase colonies. Without exception, the revertants were *Streptococcus faecalis*.

S. faecalis is probably the most commonly encountered species in SBE septicemia with wall-deficient variants. It can enter the patient's circulation from his intestinal tract or oral cavity, and, thus, infections can follow surgical procedures in either area. Neu and Goldreyer[25] treated a case of CWD *S. faecalis* endocarditis which finally required incision of the infected valve. When CWD streptococci and variants of many other species are placed on serum agar, they usually do not colonize but often may be detected in impression smears (Figure 2).

Another case history of SBE follows. The pathogen was probably *S. sanguis*. The patient, who had chronic rheumatic heart disease with mitral stenosis and insufficiency without cardiac decompensation, felt reasonably well until December, when he developed malaise and chills. This presumably upper respiratory infection was treated with antibiotics, with temporary improvement. Shortly thereafter he redeveloped a low-grade temperature and

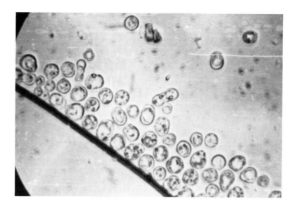

FIGURE 1. Growth from a penicillin-resistant case of SBE. These reverted to streptococci. (Magnification ×
1000.)

became progressively more anemic. His physician suspected subacute bacterial endocarditis
and hospitalized him on March 19. He had no petechiae; his lungs were clear; his heart not
enlarged. There was a grade 3 systolic murmur, followed by a grade 2 diastolic rumble.
Liver and spleen were not felt. Extremities showed no clubbing. Urinalysis and blood
chemistries were within normal limits.

Because of 18 negative blood cultures, the impression of SBE was discarded and a
malignancy considered, suggested by a gastric ulcer and persistent anemia. The patient,
however, refused an exploratory operation. New blood cultures were taken in the late
afternoon, coinciding with his daily temperature elevation. One blood sample was placed
in Medill-O'Kane medium made hypertonic with 10% sucrose and another 10 ml of blood
placed in tryptose phosphate broth containing a trace of agar and para-aminobenzoic acid.
In both media, variants developed by 48 h which rapidly reverted to α-hemolytic streptococci
unstable in subcultures. Six blood cultures taken previously in routine media were also found
to contain the same pleomorphic bacteria. A bone marrow biopsy was cultured for classic
bacteria and wall-deficient variants. Again, a variant was found which reverted to α-hemolytic streptococcus.

At the date of bone marrow biopsy there was more convincing clinical evidence of
endocarditis: systolic and diastolic murmurs as well as mitral murmurs, some splenomegaly,
and a petechia in the left fundus. The patient was started on massive doses of penicillin, 30
million U/d, reinforced with benemid to preserve penicillin blood levels. Three days of
administration did not affect the temperature excursions. Vancomycin was then administered
for 2 weeks without clinical response. Erythromycin was given for 48 h, and the patient's
temperature dropped to normal and subnormal levels for the first time in the treatment
regime. Because of an allergic reaction, erythromycin was replaced by keflin which maintained the improvement, permitting the discharge of the patient after 3 weeks of keflin
therapy.

The classic state of the bacterium had the exquisite sensitivity to penicillin and vancomycin which characterizes most oral strains of streptococci, yet the patient's condition
was not improved by these antibiotics. In contrast, the variant was very sensitive to erythromycin, which gave a dramatic clinical response. Sensitivity tests showed that the variant
was resistant to penicillin, vancomycin, tetracycline, aureomycin, declomycin, streptomycin,
and methicillin. It was never possible to determine the sensitivity to keflin, but this patient's
improvement indicated sensitivity.

Erythrophagocytosis was found when the buffy coat leukocytes of the patient were
stained by Wright's method. Phagocytosis of erythrocytes *in vivo* is known to occur in

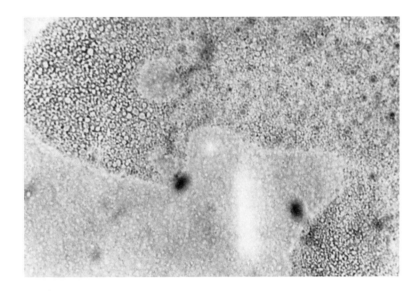

FIGURE 2. Impression smear of surface growth of a "negative" blood culture on serum agar. (Magnification × 1000.)

clostridial and *Salmonella typhi* infections. The phagocytized erythrocytes did not have detectable parasitism, and it is reasonable to assume that antigen on the surface of a red cell can make it vulnerable to phagocytic attack.

SBE WITH CWD PROPIONIBACTERIUM

In another case, Propionibacterium in a wall-deficient form was implicated. A 35-year-old woman had not felt well since the birth of her first child 16 years previously. Since then she had episodes of temperature elevations varying from 102 to 105°F. Her history, which includes rheumatic fever at age 9 years, well illustrates the complexities of diagnosing SBE when blood cultures are negative. Clinical impressions included collagen disease, which seemed improbable in view of no improvement from prednisone administration. Her spinal fluid showed 54 white cells/mm³, with 58% polymorphonuclear cells and 42% mononuclear. This fluid was negative on culture and revealed no cryptococcus with India ink staining. Her pleural fluid contained 53,800 leukocytes/mm³ with 66% polymorphonuclear and 34% mononuclear cells. Pleural fluid cultures for bacteria and fungi were negative, and no disease was produced in inoculated guinea pigs. The urine colony count was within normal limits. Biopsies of nodes, liver, skin, and muscle were not informative. Negative tests included febrile agglutinins for typhoid, paratyphoid, *Brucella abortus,* and proteus O_{xk}, O_{x19}, and O_{x2}. Complement fixation tests were negative for adenovirus, cytomegalovirus, herpes simplex, measles, *Mycoplasma pneumoniae,* and Varicella.

After unsuccessful administration of streptomycin and penicillin, more blood cultures were drawn during her evening temperature elevation (104°F). CWD forms were isolated from duplicate blood cultures taken at half-hour intervals, and antibiotic sensitivities were read. Only after 9 months of incubation did reversion to the classic bacterium occur in both of the original blood cultures continuously incubated. The isolates were identified as *Propionibacterium acnes* by biochemical tests and by gas chromatography studies by James Stewart in the laboratory of Thomas Neblett.

The patient responded to combined therapy with penicillin, tetracycline, and streptomycin; finally 500 mg four times daily of tetracycline alone was effective. She was discharged on maintenance lincomycin, to which the variant was sensitive, and was well a year later. The isolates from this patient resembled those described by Wittler et al.[34] and by Zierdt and Wertlake[35] in being unstable and giving more than one colony type. The strain from this case differed from most *P. acnes* from normal throats and from acute septicemia in that the colonies remained minute, pigment was intense, and propagation required strict anaerobiosis. On some media tiny, red, raised colonies were a variation of a flat, colorless colony.

Another detailed study concerned *P. acnes* variants in blood cultures and on the heart valves of a fatal SBE case. Zierdt and Wertlake found two forms of the organism growing from the blood of the patient.[35] The round body stage was entirely Gram-negative. The second type consisted of the spherical bodies plus numerous filaments, greatly resembling *Streptobacillus moniliformis*. Latex agglutination showed the relationship of these two variants to the parent propionibacterium, obtained as variant colonies reverted. Biochemical tests were confusing, as a large inoculum of the filamentous forms produced gas in AC medium (Difco), which is not characteristic of this species. Fluorescent antibody made to the forms growing on the heart valve and to the transitional forms in the blood culture reacted strongly with the organisms in smears and sections of the heart valves. No classic bacteria were ever found *in vivo*.

Licht et al.[19] describe an unusual case of endocarditis in a 25-year-old male. His valvular disease was complicated by a coronary embolus, causing cardiac arrest. Six blood cultures taken before antibiotic administration yielded CWD forms which reverted to α-hemolytic streptococci. The microbes in primary culture grew in isotonic as well as hypertonic medium, but required hypertonicity for subculture. The patient was resuscitated and successfully treated with combined penicillin and streptomycin, followed by erythromycin.

Wittler and coinvestigators found CWD variants which reverted to diphtheroid morphology in both the bone marrow and blood of three cases of suspected endocarditis. Apparently, classic bacteria were never found in the cultures of these patients.[34]

Wittler and associates[34] documented that a case of Propionibacterium endocarditis had the classic stages of the organism circulating when the disease was active and CWD stages in the several periods of quiescence.

ENDOCARDITIS DUE TO CWD PSEUDOMONAS

Endocarditis due to the CWD stage of *Pseudomonas aeruginosa* occurred in a 3-year-old German shepherd. Shallow respiration, moist rales, and pyrexia were among the recurrent symptoms. Intermittent lameness was later thought to be due to emboli to the skeletal muscles. Originally, blood cultures were positive for the classic form of the bacillus, but after tetracycline therapy classic bacilli never reappeared. In this chronic stage a blood culture studied by Sylvia Coleman revealed the L phase of *P. aeruginosa*. It seemed impossible to prevent relapses, and after 8 months the animal was sacrificed. The veterinarian, W. J. Bone, believes that canine bacterial endocarditis is not reported frequently because the clinical signs resemble those of such common diseases as distemper and hepatitis.[2]

ACUTE ENDOCARDITIS

Hannoun and associates[16] made the first report of staphylococcal endocarditis caused by the organism in a granular state. The infection, which followed a compound fracture of the finger, only initially responded to penicillin and was refractory to cathomycin therapy. Autopsy confirmed the diagnosis of ulcerovegetative endocarditis of the aortic valve.[1,16] Other findings of CWD forms in acute endocarditis are numerous, including *Acinetobacter calcoaceticus*.[36]

Edwards[10] had unique observations as he studied a case of staphylococcal endocarditis in a drug abuser (ASM'72). Initial blood cultures grew a coagulase-positive staphylococcus.

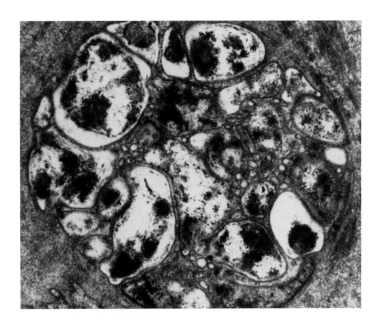

FIGURE 3. Vegetations on cardiac valves examined by electron microscopy show predominantly CWD bacteria. (Original magnification × 44,000). (From Piepkorn, M. W. and Reichenbach, D. D., *Hum. Pathol.*, 9, 163–173, 1978. With permission.)

Penicillin brought the patient's temperature from 105 to 103°F, and classic staphylococci disappeared from his blood. However, he remained febrile and toxic. The wall-deficient variants grown from his blood cultures at this time reverted to staphylococci when cultured with lysozyme. In the reversion stages the organisms morphologically were temporarily bacilli, i.e., Gram-positive rods. This reversion sequence could be made to occur repeatedly. This strange morphology as an intermediate reversion step was also reported by Charache in an early study.[6]

The question might be asked: Is a defective wall of organisms in endocarditis blood cultures a result of the unnatural environment, i.e., the culture medium? Piepkorn and Reichenbach[26] addressed this question, studying four cases of infective endocarditis in which no classic bacteria were found in repeated blood cultures taken prior to antibiotic administration. When cardiac valvular vegetations were examined by electron microscopy, classic bacteria were found in each specimen, but CWD variants predominated in three cases (Figure 3). It was concluded, "The evidence indirectly suggests a pathogenic role for these aberrant bacteria in human disease."

In two cases of culture-negative endocarditis studied in our laboratory, CWD organisms in multiple blood cultures were identified as *Staphylococcus aureus* in direct smears of the cultures stained with fluorescent antibody.

Predilection of the CWD form for vascular tissue is also demonstrated by variants of Brucella. Grekova et al.[13] found that the tissues most vulnerable to CWD forms of *B. abortus* were blood vessels of the liver, resulting in decomposition of liver trabeculae in inoculated guinea pigs. Johnson and associates found that unidentified CWD bacteria from human ocular disease can cause inflammation of mouse cardiac valves.[17]

CWD bacteria are also a threat from heterograft heart valves (Figure 4). A much used source has been the pig. Ferrans and co-investigators[11] made an electron microscopic examination of four infected porcine valve heterografts. One valve showed only thin-walled "ghosts", no classic bacteria.

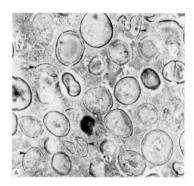

FIGURE 4. A porcine heart valve was infected with a CWD bacterium. (From Ferrans, V. J., Boyce, S. W., Billingham, M. E., Spray, T. L., and Roberts, W. C., *Am. J. Cardiol.*, 43, 1123–1136, 1979. With permission.)

ADMINISTRATION OF ANTIBIOTIC BEFORE TAKING BLOOD CULTURES

Wall-deficient stages of a pathogen may persist following antibiotic administration. Sapico et al.[29] found a single intravenous dose of penicillin can convert classic infecting organisms to osmotically fragile forms. The disease may not be controlled by an empirically chosen antibiotic, and the condition only worsens. Two thirds of the endocarditis patients in a culture-negative group reported by Hampton and Harrison had received antibiotic before blood cultures were drawn.[14]

FOLLOWING ADMINISTRATION OF AN ANTIBIOTIC INHIBITORY TO THE CLASSIC PATHOGEN

In other cases the classic bacterium has been isolated, but antibiotics to which it is sensitive fail to control the variant. Castleman and McNeely describe the typical evolution of endocarditis when administered antibiotics do not control CWD variant growth.[4] The history is typical, although cultures for CWD forms were not made. Blood cultures initially revealed *S. aureus,* sensitive to 2 U/ml of penicillin. Despite 31 d of penicillin administration, increased to a daily dosage of 40 million U and the addition of methicillin, the patient's condition progressed. Serosanguineous fluid from the chest, containing 61% neutrophils, 29% lymphocytes, and 10% monocytes, was negative by culture and smear. After therapy was initiated, repeated blood cultures were "negative" in penicillinase-containing medium. Findings at autopsy revealed acute bacterial endocarditis, involving tricuspid and aortic valves and emboli in renal, pulmonary, and coronary arteries. A Gram stain of an impression smear of the valves did not give any definite evidence of staphylococci.

INFECTIONS FOLLOWING INSERTION OF PROSTHETIC CARDIAC VALVES

Occult variants may perpetuate an infection initiated when an artificial valve is inserted. The variants have been cultured in some cases and suggested by associated facts in others. The bloodstream does not yield classic bacteria until 25 d or more after surgery, and the bacteria are usually sensitive *in vitro* to the antibiotic administered prophylactically after surgery.[28] *S. albus* was isolated by Charache from a patient who had an atrial septal defect repaired with a Teflon patch 2 years previously. After ineffectual penicillin administration, a successful response was achieved with methicillin.[6] This suggests *persisters* sequestered in tissue or in clots. These are generally accepted as representing an antibiotic-resistant CWD state. A constant feeding from a nidus of CWD growth might explain why it is often impossible to cure these patients with antibiotics. It is often necessary to replace the infected prosthesis.[30]

NUTRITIONALLY VARIANT STREPTOCOCCI

Another group of streptococci are important in endocarditis. These are the well-known nutritionally variant streptococci which require vitamin B_6 or L-cysteine, commonly supplied by growing them with *S. aureus.* These nutritionally variant streptococci are found to have both abnormal walls and atypical septation.[15]

Frenkel and Hirsch, studying thiol-requiring streptococcal mutants in cases of bacterial endocarditis and otitis media, showed that in some instances they could grow as L-phase colonies.[12] Cayeux et al.[5] found similar thiol-requiring streptococci in three cases of endocarditis and reported that their sulfur requirement could be met by a combination of L-cysteine and thioglycolate.[4] Although they did not succeed in growing the strains as L Forms, the possibility that they existed *in vivo* in a wall-deficient stage is heightened by the fact that one case failed to respond to daily doses of 50 million U of penicillin. In the other cases, combined antibiotics were effective. However, most wall-defective streptococci are not thiol-requiring mutants. For a discussion of fungal endocarditis see the chapter on Fungi.

SEPTICEMIA

As in endocarditis, CWD streptococci are the variant most commonly occurring in septicemia without accompaniment by the classic bacteria. However, septicemia or bacteremia with almost every genus in the CWD state has been reported. In a series of septicemia cases in Mexico City, CWD *S. epidermidis* was also prominent.[18] From septicemia cases Charache isolated CWD aerobic and anaerobic Corynebacteria, *Escherichia coli,* Herella, Klebsiella, and Proteus.[6] There is provocative suggestion that whenever an infected patient is pyrexic, the pathogen is circulating in the blood in a CWD stage, regardless of the site of infection. The blood is a relatively unpopulated medium from which to culture the invader.

Two patients with septicemia due to CWD streptococci expired after belated therapy.[21] Other reported septicemia cases have concerned streptococci,[6] unidentified Corynebacteria,[20] varied fungi,[20,27,32] Actinomyces,[23] and Pasteurella.[10]

CWD staphylococci in the blood of a febrile patient were identified in our laboratories by both fluorescent antibody and fluorescent lysostaphin. When treated for CWD staphylococci his temperature fell from 105 to 103°F, but 1 d later he expired. Autopsy showed that the underlying disease was tuberculosis. Reexamination of smears and cultures showed that CWD microcolonies, acid fast by an intensified method, were also present. Other cases with CWD staphylococci in septicemia have been identified rapidly in our laboratory with fluorescent antibody.

Chmel made a first report of probable transfer of a CWD streptococcus from a pet dog to cause septicemia in the owner.[7] The organism, which reverted to *Streptococcus sanguis,* was cultured from the hemodialysis patient's blood, the vascular access site, and the dog's mouth. (The collie had the habit of affectionately licking the wound of the patient.)

Considering the importance of infection in burn patients, it might be profitable to examine for CWD bacteria and fungi. In burn cases there is a clinical paradox of patients dying with septicemia with negative blood cultures.[8]

Many of the numerous mysterious culture-negative endocarditis and septicemia cases become culture positive if the acridine orange (AO)-stained smear of the blood culture is carefully examined. AO staining has generally replaced the Gram stain. The inexperienced believe the masses of staining units are artifacts since they fail to grow on the surface of blood plates in macroscopic colonies. The CWD colonies grow in pour plates containing 10% horse or swine serum and 1% separately autoclaved yeast extract. The developing microscopic colonies may all be subsurface.

The microcolonies may grow on the surface of chocolate agar if separately autoclaved yeast extract is in the medium. Their presence can be revealed by impression smears. A sensitivity test may be done by growing the variant in 2 ml of veal infusion broth containing 0.1% purified agar and 1.% separately autoclaved yeast extract. A commercial antibiotic-containing disk is added to this.

MYOCARDITIS AND COMBINED MYOCARDITIS AND ENDOCARDITIS

The etiology of human myocarditis, the common disease which may require heart replacement, remains largely unelucidated. Such cases as familial virus infections necessitating cardiac transplants in sisters have been reported. In addition, the "virus" stage of bacteria must be considered, evident from the study by Merline and associates.[22] Both pneumococcal and streptococcal variants in the Merline study caused murine myocarditis. The streptococcal wall-deficient form produced not only myocarditis, but also endocarditis. The test organisms were variants cultured from infections in man.

Both muscle and valve damage resulted in mice injected with unidentified CWD isolates from human uveitis. In some mice the pulmonary and coronary arteries were also diseased.[17]

Additionally, it is well recognized that in rheumatic fever the damage is pancarditis, involvement of myocardium, valves and pericardium. Evidence grows that CWD streptococci play a major role in all aspects of rheumatic fever (see Rheumatic Fever chapter).

HYPERTONIC MEDIA

Rosner[27] gave the first data showing the benefits from hypertonic medium which probably permits growth and reversion of CWD microbes (see chapter on Media). Many studies substantiate the value of hypertonicity in culture media. A comparative study in Mexico City found only 45% positive cultures in isotonic vs. 65% positive in hypertonic conditions.[3]

WHERE ARE POSITIVE "NEGATIVE" CULTURES TO BE FOUND?

Anyone desiring an L-type culture to study need only inspect some discards before they reach the autoclave. Any hospital with a large infectious disease unit offers a daily selection. The brightly hemolyzed "negative" cultures may contain the variants of *Staphylococcus aureus,* β-hemolytic streptococci, or, rarely, Clostridia or Salmonella. The brown-tinged blood suggests variants of Proteus, α-hemolytic streptococci, or pneumococci. Nonhemolyzed bottles may be truly negative, come from the tuberculosis ward, and, thus, probably contain CWD mycobacteria, or contain the variants of any nonhemolytic pathogen. Repeated cultures on the same patient usually show the same type of hemolysis until the patient responds to treatment.

SKIP CULTURES IN SEPTICEMIA

It is a common phenomenon when working with CWD variants in septicemia to encounter "skip cultures". These blood cultures have only variants but have been both preceded and followed by cultures which contain a classic bacterium. Similarly, CWD forms are found in blood cultures in an individual whose previous or later cultures contain a classic pathogen. Skip cultures constitute an excellent material for those who wish to become acquainted with these bizarre organisms. We have observed skip cultures in bacteremias with *Proteus mirabilis, Pseudomonas aeruginosa, Streptococcus faecalis, Staphylococcus aureus,* and *Fusobacterium fusiforme.* Such variants are highly unstable and often revert after only a few transfers in thioglycolate broth.

ALLERGY AND OTHER CONDITIONS ASSOCIATED WITH CWD BACTEREMIA

It must be noted that bacteria in the circulating blood do not always imply significant infection. We repeatedly studied an individual who had CWD *Streptococcus mutans* in his blood associated with episodes of incapacitating fever and intestinal symptoms. The actual diagnosis was made after a weekend when the family ran out of milk. He has remained symptom-free as long as he avoids milk and dairy products. These findings support the concept that bacteria from an inflamed bowel may enter the blood in a wall-deficient stage and explain such phenomena as the Pseudomonas species isolated in Crohn's disease.

SUMMARY

Negative blood cultures which characterize 20 to 30% of subacute bacterial endocarditis cases usually contain CWD microcolonies, easily detected in Acridine Orange-stained smears. The CWD microbes in the "negative" blood cultures include various species of staphylococci, streptococci, corynebacterium, propionibacterium, pseudomonas, streptobacillus, actinomyces, acinetobacter, and fungi. Similar species in CWD stages are found in septicemia.

The typical person needing a heart transplant has cardiac muscle damage, which is common to some degree in most individuals over 50 years of age. The cause of this myocarditis remains largely unknown, although viruses have been implicated in some instances. Murine models show that streptococci, pneumococci and other genera in CWD form can be etiological agents.

REFERENCES

1. **Bertrand-Fontaine, Th. and Schneider, J.,** La rôle de l'évolution biologique du germe dans la résistance aux traitements antibiotiques. A propos d'une observation de septicémie à staphylocoques, *Acad. Natl. Med. Bull.,* 143, 342–345, 1959.
2. **Bone, W. J.,** L Form of *Pseudomonas aeruginosa,* the etiologic agent of bacterial endocarditis in a dog, *Vet. Med. Small Anim. Clin.,* 65, 224–226, 1970.
3. **Calderon, N., Albuerne, A., Gonzalez, S., and Winkler, L.,** Cell wall deficient bacteria (L Forms) in meningitis, *Lat. Am. Microbiol.,* 13, 95–100, 1971.
4. **Castleman, B. and McNeely, B. U.,** Case records of the Massachusetts General Hospital, weekly clinicopathological exercises, Case 31–1964, *N. Engl. J. Med.,* 270, 1415–1422, 1964.
5. **Cayeux, P., Acar, J. F., and Chabbert, Y. A.,** Bacterial persistence in streptococcal endocarditis due to thiol-requiring mutants, *J. Infect. Dis.,* 124, 247–254, 1971.
6. **Charache, P.,** Atypical bacterial forms in human disease, in, *Microbial Protoplasts, Spheroplasts and L Forms,* Guze, L. B., Ed., Williams & Wilkins, Baltimore, 1968, 484–494.
7. **Chmel, H.,** Graft infection and bacteremia with a tolerant L Form of *Streptococcus sanguis* in a patient receiving hemodialysis, *J. Clin. Microbiol.,* 24, 294–295, 1986.
8. **Deitch, E. A.,** The management of burns, *N. Engl. J. Med.,* 323, 1249–1253, 1990.
9. **Dressler, F. A. and Roberts, W. C.,** Infective endocarditis in opiate addicts, analysis of 80 cases studied at necropsy, *Am. J. Cardiol.,* 63, 1240–1257, 1989.
10. **Edwards, J. H. and Mattman, L. H.,** Lysozyme-produced variation in a staphylococcal isolate, *Am. Soc. Microbiol. Abstr. Annu. Meet. 1972,* p.20, 1972.

11. **Ferrans, V. J., Boyce, S. W., Billingham, M. E., Spray, T. L., and Roberts, W. C.,** Infection of glutaraldehyde-preserved porcine valve heterografts, *Am. J. Cardiol.,* 43, 1123–1136, 1979.

12. **Frenkel, A. and Hirsch, W.,** Spontaneous development of L Forms of streptococci requiring secretions of other bacteria or sulphydryl compounds for normal growth, *Nature (London),* 191, 728–729, 1961.

13. **Grekova, N. A., Tolmacheva, T. A., and Vershilova, P. A.,** On the pathogenicity of nonstable *Brucella* L Forms and their revertants, *J. Hyg. Epidemiol. Microbiol. Immunol.,* 23, 129–134, 1979.

14. **Hampton, J. R. and Harrison, M. J. G.,** Sterile blood cultures in bacterial endocarditis, *Q. J. Med.,* 36, 167–174, 1967.

15. **Handrick, W., Köhler, W., Spencker, F. B., and Schneider, P.,** Endocarditis due to nutritionally variant streptococci, *Infection* 16, 371–372, 1988.

16. **Hannoun, C., Vigouroux, J., and Schneider, J.,** Isolement de formes granulaires de bactéries dans deux cas d'endocardite maligne à hémoculture "négative ou négativée," *Presse Med.,* 65, 1608–1611, 1957.

17. **Johnson, L., Wirostko, E., and Wirostko, W.,** Mouse lethal cardiovascular disease, induction by human leukocyte intracellular Mollicutes, *Br. J. Exp. Pathol.,* 69, 265–279, 1988.

18. **Larracilla-Alegre, J., Quijano-Arjona, J. L., and Paz, G.,** Atypical bacterial forms in septicemia, *Arch. Invest. Med.,* 12(4) (Suppl. 1), 27–28, 1981.

19. **Licht, A., Yeivin, R., and Levy, M.,** Successful resuscitation in coronary embolism occurring as a complication of bacterial endocarditis caused by cell wall defective *Streptococcus viridans, Isr. J. Med. Sci.,* 9, 1009–1013, 1973.

20. **Louria, D. B., Kaminski, T., Grieco, M., and Singer, J.,** Aberrant forms of bacteria and fungi found in blood or cerebrospinal fluid, *Arch. Intern. Med.,* 124, 39–48, 1969.

21. **Mattman, L. H. and Mattman, P. E.,** L Forms of *Streptococcus fecalis* in septicemia, *Arch. Intern. Med.,* 115, 315–321, 1965.

22. **Merline, J. R., Golden, A., and Mattman, L. H.,** Cell wall bacterial variants from man in experimental cardiopathy, *Am. J. Clin. Pathol.,* 55, 212–220, 1971.

23. **Merline, J. R. and Mattman, L. H.,** Cell wall deficient forms of Actinomyces complicating two cases of leukemia, *Mich. Acad.,* 3, 113–121, 1971.

24. **Nativelle, R. and Deparis, M.,** Formes évolutives des bactéries dans les hémocultures, *Presse Med.,* 68, 571–574, 1960.

25. **Neu, H. C. and Goldreyer, B.,** Isolation of protoplasts in a case of enterococcal endocarditis, *Am. J. Med.,* 45, 784–788, 1968.

26. **Piepkorn, M. W. and Reichenbach, D. D.,** Infective endocarditis associated with CWD bacteria, electron microscopic findings in four cases, *Hum. Pathol.,* 9, 163–173, 1978.

27. **Rosner, R.,** Isolation of Candida protoplasts from a case of Candida endocarditis, *J. Bacteriol.,* 91, 1320–1326, 1966.

28. **Sande, M. A., Johnson, W. D., Hook, E. W., and Kaye, D.,** Sustained bacteremia in patients with prosthetic cardiac valves, *N. Engl. J. Med.,* 286, 1067–1070, 1972.

29. **Sapico, F. L., Keys, T. F., and Hewitt, W. L.,** Experimental enterococcal endocarditis. II. Study of in vivo synergism of penicillin and streptomycin, *Am. J. Med. Sci.,* 263, 129–135, 1972.

30. **Sarot, I. A., Weber, D., and Schechter, D. C.,** Cardiac surgery in active, primary infective endocarditis, *Chest,* 57, 58–64, 1970.

31. **Smyllie, J. H., Sutherland, G. R., and Roelandt, J.,** The changing role of echocardiography in the diagnosis and management of infective endocarditis, *Int. J. Cardiol.,* 23, 291–301, 1989.

32. **Swieczkowski, D. M., Mattman, L. H., Truant, J. P., and Wilner, F. M.,** Cell wall deficient forms of *Candida albicans* in mycohemia, *Lab. Med.,* 1, 41–42, 1970.

33. **Tanret, P. and Solignac, H.,** Sur deux cas de maladie d'Osler à hemo-ovocultures positives, *Soc. Med. Hop. Paris Bull. Mem.,* 113, 247–251, 1963.

34. **Wittler, R. G., Malizia, W. F., Kramer, P. E., Tuckett, J. D., Pritchard, H. N., and Baker, H. J.,** Isolation of a Corynebacterium and its transitional forms from a case of subacute bacterial endocarditis treated with antibiotics, *J. Gen. Microbiol.,* 23, 315–333, 1960.

35. **Zierdt, C. H. and Wertlake, P. T.,** Transitional forms of *Corynebacterium acnes* in disease, *J. Bacteriol.,* 97, 799–805, 1969.

36. **Zubkov, M. N., Bogomolov, B. P., Gugutsidze, E. N., and Levina, G. A.,** Isolation of the L Forms of *Acinetobacter calcoaceticus,* Var. *lwoffii,* in infectious endocarditis, *Zh. Mikrobiol. Epidamiol. Immunobiol.,* 6, 27–29, 1984.

37. **Smith, M. R.,** personal communication.

Chapter 12

INTRACELLULAR GROWTH OF CWD FORMS

INTRAERYTHROCYTIC CWD FORMS

As well established, white blood cells rapidly scavenge undesirable particles, bacteria or otherwise, and haul them away for disposal. It has never been surprising to see neutrophiles transporting bacteria and fungi. Erythrocytes, on the other hand, were thought not to participate in moving, destroying, or housing foreign invaders.

This view of red cells as inert bodies was rudely dispelled by faculty members of the University of Camerino Medical School. Seeking to learn whether erythrocyte parasites were involved in hemolytic anemias, they found microbial life in red cells of all individuals examined.[48] (See Frontispiece.)

At Camerino the physiologists, microbiologists, and electron microscopists united to learn whether microbes in erythrocytes are viable and metabolizing. Since erythrocytes per se are unable to synthesize the bases of the purine nucleotides, cultures of erythrocytes were tested to find whether radioactive glycine and formate would be incorporated into purine nucleotides. In erythrocyte cultures, both compounds appeared in new purine nucleotides, indicating activity unassociated with the metabolism of the erythrocyte itself (Figure 1). Autoradiograms of the blood cells indicate that radioactivity from labeled substrates occurs in a variable number of erythrocytes whose morphological changes suggest degrees of lysis[48] (Figure 2).

The intraerythrocyte forms in *healthy* persons have been identified by Tedeschi's group as common staphylococci.[48] Bisset and Bartlett have concluded that *Bacillus licheniformis* resides in red cells of 30% of healthy individuals.[5]

Heat fixation distorts the outlines of most CWD forms into vague masses.[9] This is also true of the erythrocyte parasites of healthy and diseased persons: the outline of the microbes becomes unrecognizable; any Gram-positiveness is lost, and the forms are indistinguishable from the Gram-negative hemoglobin. Methanol fixation reduces this problem.

ERYTHROCYTE PARASITISM ASSOCIATED WITH DISEASE

The findings of Tedeschi's group were extended by Pohlod et al.[41] who confirmed the microbial identity of the inclusions by muramidase staining. Comparisons were made between college students, nursing home inmates and hospital patients. A bonus finding was that a presumptive *diagnosis of bacteremia* can be made in many instances by the high percentage of freshly drawn red cells with structures staining with fluorescent muramidase. While agreeing with the other studies that normal people have erythrocyte-incorporated CWD variants,[38] they found a striking quantitative difference between normal individuals and patients with symptoms of septicemia. Thus, there is now available a 10-min test to detect a probability of septicemia.

In the marked parasitism associated with active infection, it becomes obvious that the intracellular pattern differs between genera. The genus is not defined, but clues exist. The Frontispiece shows incubated red cells of a patient whose revertant was *Clostridium sordelli*.[31] The red blood cells of all five blood cultures showed the coarse, heavily Gram positive bodies in unusual patterns.

A different picture is given by staphylococcal variants and by variants in erythrocytes of patients with untreated pulmonary tuberculosis (see Frontispiece). The acid-fast characteristic of sarcoidosis mycobacterial growth within erythrocytes was first shown by Judge.[21]

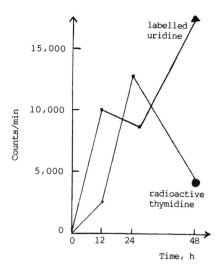

FIGURE 1. Erythrocytes of a normal person incorporating radioactive nucleosides. (From Tedeschi, G. G., Amici, D., and Paparelli, M., *Haematologia,* 4, 27–47, 1970. With permission.)

A parasitized erythrocyte from a Kaposi patient is shown with a fungus which grew from the blood, plasma, and involved tissue of the patient (see Frontispiece).

Red cell parasitism is detected by darkfield examination, by staining with acridine orange and with rhodamine-labeled muramidase,[41] indicating that considerable murein of the cell wall backbone is retained in these variants *in vivo*. Specificity of the staining with the fluorescent muramidase can be demonstrated by blocking the staining by preexposure of the parasitized erythrocytes to unlabeled muramidase.

An in depth study of red cell parasitism in nephropathies was made by Domingue et al. at Tulane University.[11] After culturing lysed erythrocytes from 489 patients, they concluded that the CWD forms in red cells were the same species causing the urinary tract infection. Many reverted organisms became available for examination and for comparison with reverted organisms from urine.

Domingue's group found different genera of pathogens in erythrocytes in different urinary tract syndromes. The lysed filtered blood of 34 of 35 patients with idiopathic hematuria grew CWD forms which reverted to streptococcal-like bacteria. In contrast, the red cells from nephrotic syndrome in children yielded variants more often reverting to staphylococci.

The red blood cells of individuals with the enigmatic disease systemic lupus erythematosis also yielded CWD variants. Thus, from varied approaches, the Tulane studies indicate that CWD bacteria associated with renal disease circulate in the patients' erythrocytes.

Residence within the red cell is confirmed by electron microscopic examination of sectioned erythrocytes. Early studies by Tedeschi's group have been greatly extended by the laboratory of Domingue. Even in smears examined with phase microscopy, the red cell membrane is occasionally distorted to accommodate a developing variant colony; commonly, the red cells enlarge and show diminished hemoglobin, similar to the fate of red cells hosting malarial Plasmodia.

How do variants penetrate erythrocytes? Malarial plasmodia enter red cells by modified phagocytosis,[24] and it seems logical that the same process results in engulfment of small microbial variants. It is possible that minute microbial variants attach to specific receptors on the red cell.

CWD FORMS AS TISSUE CULTURE CONTAMINANTS

Apparently some of the contaminants in tissue culture are wall-deficient bacteria, rather than Mycoplasma. MacPherson and Allner[29] found that prolonged treatment with 200 U/ml of neomycin was needed to clear a tissue culture of the CWD phase of a Corynebacterium. The concentration required was unexpected since less than 10 U/ml of neomycin killed both the parent bacterium and the variant in agar medium without tissue cells.

GROWTH OF BACTERIA WITHIN FUNGI

A "tracer" strain of *Candida albicans* has been employed by Mankiewicz to indicate growth of the tubercle bacillus. This yeast grows only when accompanied by the mycobacterium, hence when the yeast is added to culture medium simultaneously inoculated with digested sputum, the rapidly growing Candida indicates that multiplication of mycobacteria is also occurring. Noticeably, some of the acid fast rods are growing within the Candida. This is especially pertinent here because certain other pathogenic bacteria which tend to grow predominantly in CWD stages may assume classic form within accompanying fungal contaminants.

SUMMARY

Parasitism of circulating erythrocytes, occurring in a minimal degree as it seems to in every person, probably represents the escape of bacteria from the heavily populated intestine, oral cavity, skin, and lymph nodes of man into regions usually considered sterile. In contrast, heavy parasitism has been found in acute bacteremias, kidney infections, and other disease states. The extent of parasitism can aid in distinguishing infectious from noninfectious disease. The parasitizing organism can be identified with fluorescent antibody (F.A.) and by other methods.

In septicemia and even some chronic infections, the pathogen can best be found in the blood's buffy coat by seeking CWD as well as classic forms. Identifiable CWD variants in neutrophiles of the buffy coat may expedite diagnosis of many infections, including gonorrhea and sarcoidosis. The most sensitive stain for detecting all species of CWD microbes is fluorochrome-labeled muramidase.

Bacteria with varying degrees of cell wall loss find an ideal growth medium within certain tissue cells. For some pathogens this is related to their tissue tropisms; e.g., the typhoid bacillus grows well as a wall-deficient variant within monocytes and not within many other cells, corresponding to the residence *in vivo* of *S. typhi* in the lymph tissue of Peyer's patches.

Growth of variant bacteria within tissue cells may result in inclusion bodies which mimic parasitism by viruses or Chlamydia. At other times the invaded cells show no evidence that they harbor bacteria. Spirochetae grow mainly as small granules within host tissue or within ova of carrier ticks. Therefore, not surprisingly, wall deficiency can be produced by growing an organism in tissue culture. It follows that CWD organisms, in addition to Mycoplasma, are found as tissue culture contaminants.

REFERENCES

1. **Arduino, M. J., Bland, L. A., Tipple, M. A., Aguero, S. M., Favero, M. S., and Jarvis, W. R.,** Growth and endotoxin production of *Yersinia enterocolitica* and *Enterobacter agglomerans* in packed erythrocytes, *J. Clin. Microbiol.,* 27, 1483–1485, 1989.

2. **Baker, R. F.,** The fine structure of stromalytic forms produced by osmotic hemolysis of red blood cells, *J. Ultrastruct. Res.,* 11, 494–507, 1964.

3. **Beale, L. S.,** Observations upon the nature of the red blood corpuscle, *Trans. Microsc. Soc. London,* 12, 37, 1864.

4. **Begue, R., Dennehy, P. H., and Peter, Georges,** Hemolytic uremic syndrome associated with *Streptococcus pneumoniae, N. Engl. J. Med.,* 325(2), 133–134, 1991.

5. **Bisset, K. A. and Bartlett, R.,** The isolation and characters of L-forms and reversions of *Bacillus licheniformis* var. *endoparasiticus* (Benedek) associated with the erythrocytes of clinically normal persons, *J. Med. Microbiol.,* 11, 335–349, 1978.

6. **Braude, A. I.,** Production of bladder stones by L-forms, *Ann. N.Y. Acad. Sci.,* 174, 896–902, 1970.

7. **Burgdorfer, W., Barbour, A. G., Hayes, S. F., Benach, J. L., Grunwaldt, E., and Davis, J. P.,** Lyme disease — a tick-borne spirochetosis?, *Science,* 216, 1317–1319, 1982.

8. **Carrère, L., Roux, J., and Mandin, J.,** Culture des organismes L sur membrane chorio-allantoide d'embryon de poulet, *Montpellier Med.,* 68, 438–445, 1955.

9. **Charache, P.,** Atypical bacterial forms in human disease, in *Microbial Protoplasts, Spheroplasts, and L-Forms,* Guze, L. B., Ed., Williams & Wilkins, Baltimore, 1968, 484–494.

10. **Cleary, T. G.,** Cytotoxin-producing *Escherichia coli* and the hemolytic uremic syndrome, *Pediatr. Clin. North Am.,* 35, 485–501, 1988.

11. **Domingue, G. J., Ed.,** *Cell Wall-Deficient Bacteria, Basic Principles and Clinical Significance,* Addison-Wesley, Reading, MA, 1982.

12. **Dmochowski, L., Grey, C. E., Padgett, F., Langford, P. L., and Burmester, B. R.,** Submicroscopic morphology of avian neoplasma, *Tex. Rep. Biol. Med.,* 22, 20–60, 1964.

13. **Fitzgerald, T. J., Johnson, R. C., Miller, J. N., and Sykes, J. A.,** Characterization of the attachment of *Treponema pallidum* (Nichols strain) to cultured mammalian cells and the potential relationship of attachment to pathogenicity, *Infect. Immun.,* 18(2), 467–478, 1977.

14. **Fogh, J. and Fohg, H.,** A method for direct demonstration of pleuropneumonia-like organisms in cultured cells, *Proc. Soc. Exp. Biol. Med.,* 115, 899–901, 1964.

15. **Freeman, B. A. and Rumack, B. H.,** Cytopathogenic effect of *Brucella* spheroplasts on monocytes in tissue culture, *J. Bacteriol.,* 88, 1310–1315, 1964.

16. **Gray, J. M.,** The In Vitro Stability of Betalactamase Production by *Neisseria gonorrhoeae* Strains, and Improved Diagnosis by Detection of Cell Wall Deficient Forms, Ph.D. dissertation, Wayne State University, Detroit, 1980.

17. **Gray, J. C. and Mattman, L. H.,** Cell wall deficient *Neisseria gonorrhoeae,* in *Cell Wall-Deficient Bacteria,* Domingue, G. J., Ed., Addison-Wesley, Reading, MA, 1982, 409–425.

18. **Hamburger, M. and Carleton, J.,** Staphylococcal spheroplasts as persisters, in *Microbial Protoplasts, Spheroplasts, and L-Forms,* Guze, L. B., Ed., Williams & Wilkins, Baltimore, 221–229, 1968.

19. **Hatten, B. A. and Sulkin, S. E.,** Intracellular production of *Brucella* L Forms. I. Recovery of L Forms from tissue culture cells infected with *Brucella abortus, J. Bacteriol.,* 91, 285–296, 1966.

20. **Hindle, E.,** On the life-cycle of *Spirochaeta gallinarum, Parasitology,* 4, 463–477, 1911.

21. **Judge, M. S.,** Evidence Implicating a Mycobacterium as the Causative Agent of Sarcoidosis, and Comparison of This Organism with the Blood-Borne Mycobacterium of Tuberculosis, Ph.D. dissertation, Wayne State University, Detroit, 1979.

22. **Kagan, G. Ya. and Rakovskaya, I. V.,** The cytogenic action of the L-forms of certain pathogenic bacterial species in tissue cultures, *Bull. Exp. Biol. Med. (Russian),* 6, 69–73, 1964.

23. **Kourany, M. and Kendrick, P. L.,** Interaction between a human monocytic cell line and *Salmonella typhosa, J. Infect. Dis.,* 116, 495–513, 1966.

24. **Ladda, R., Aikawa, M., and Sprinz, H.,** Penetration of erythrocytes by merozoites of mammalian and avian malarial parasites, *J. Parasitol.,* 55, 633–644, 1969.

25. **Lavillaureix, J.,** Action de formes L pathogènes sur des cultures de cellules cancéreuses, *C. R. Acad. Sci.,* 244, 1098–1101, 1957.

26. **Lawrence, C., Brown, S. T., and Freundlich, L. F.,** Peripheral blood smear bacillemia, *Am. J. Med.,* 85, 111–113, 1988.

27. **Leishman, W. B.,** The mechanism of infection in tick fever and on the hereditary transmission of *Spirochaeta duttoni* in the tick, *Lancet,* 1, 11–14, 1910.

28. **Levaditi, C., Vaisman, A., and Chaigneau, H.,** Culture de *Spirochaeta duttoni* dans l'oeuf fécondé de poule, *Ann. Inst. Pasteur* (Paris), 80, 9–20, 1951.

29. **MacPherson, I. A. and Allner, K.,** L Forms of bacteria as contaminants in tissue culture, *Nature (London)*, 186, 992, 1960.

30. **Magath, T. B.,** Spirochetes in the blood, *Am. J. Clin. Pathol.*, 23, 691–693, 1953.

31. **Mattman, L. H., Dowell, V. R., and Neblett, T. R.,** Intraerythrocytic forms in a case with *Clostridium sordelli* bacteremia, *Bacteriol. Proc.* p.69, 1971.

32. **McCune, R. M., Jr., Tompsett, R., and McDermott, W.,** The fate of *Mycobacterium tuberculosis* in mouse tissues as determined by the microbial enumeration technique. II. The conversion of tuberculous infection to the latent state by the administration of pyrazinamide and a companion drug, *J. Exp. Med.*, 104, 763–802, 1956.

33. **McDermott, W.,** Microbial persistence, *Yale J. Biol. Med.*, 30, 257–291, 1958.

34. **Memminger, J. J., III,** Rapid Methods for the Detection of Septicemia, Ph.D. dissertation, Wayne State University, Detroit, 1983.

35. **Mettler, N. E.,** Isolation of a microtatobiote from patients with hemolytic-uremic syndrome and thrombotic thrombocytopenic purpura and from mites in the United States, *N. Engl. J. Med.*, 281, 1023–1027, 1969.

36. **Ovchinnikov, N. M. and Delektorsky, V. V.,** *Treponema pallidum* ultrastructure and mechanisms of cellular protection before and during syphilis treatment, *Vest. Dermatol. Venerol.*, 12, 37–40, 1981.

37. **Pease, P.,** *L-Forms, Episomes and Auto-Immune Disease,* E. & S. Livingstone, Edinburgh, 1965.

38. **Pease, P.,** Bacterial L-forms in the blood and joint fluids of arthritic subjects, *Ann. Rheum. Dis.*, 28, 270–274, 1969.

39. **Pickett, M. J. and Nelson, E. L.,** Observations on the problem of Brucella blood cultures, *J. Bacteriol.*, 61, 229–237, 1951.

40. **Pierce-Chase, C. A., Fauve, R. M., and Dubos, R. J.,** *Corynebacteria pseudotuberculosis* in mice. I. Comparative susceptibility of mouse strains to experimental infection with *Corynebacterium kutscheri, J. Exp. Med.*, 120, 267–281, 1964.

41. **Pohlod, D. J., Mattman, L. H., and Tunstall, L.,** Structures suggesting CWD forms detected in circulating erythrocytes by fluorochrome staining, *Appl. Microbiol.*, 23, 262–267, 1972.

42. **Ryter, A.,** Relationship between ultrastructure and specific functions of macrophages, *Comp. Immunol. Microbiol. Infect. Dis.*, 8(2), 119–133, 1985.

43. **Sansonetti, P. J., Ryter, A., Clerc, P., Maurelli, A. T., and Mounier, J.,** Multiplication of *Shigella flexneri* within HeLa cells, lysis of the phagocytic vacuole and plasmid-mediated contact hemolysis, *Infect. Immun.*, 51(2), 461–469, 1986.

44. **Schmitt-Slomska, J., Boué, A., and Caravano, R.,** Infection chronique des cellules diplöides humaines en culture par les formes L du streptocoque du groupe A, *C. R. Acad. Sci.(Paris)*, 263, 1653–1655, 1966.

45. **Sidorov, V. E.,** The body cavity of argasid ticks as habitat of spirochaetes and Brucella, *Zh. Mikrobiol. Epidemiol. Immunobiol.*, 31, 91–97, 1960.

46. **Smadel, J. E.,** Intracellular infections, *Bull. N.Y. Acad. Med.*, 39, 158–172, 1963.

47. **Tedeschi, G. G., Amici, D., and Paparelli, M.,** Incorporation of nucleosides and amino-acids in human erythrocyte suspensions, possible relation with a diffuse infection of Mycoplasms or bacteria in the L Form, *Nature (London)*, 222, 1285–1286, 1969.

48. **Tedeschi, G. G., Amici, D., and Paparelli, M.,** The uptake of radioactivity of thymidine, uridine, formate, glycine, and lysine into cultures of blood of normal human subjects, Relationships with Mycoplasma infection, *Haematologia*, 4, 27–47, 1970.

49. **Tedeschi, G. G., Amici, D., Sprovieri, G., and Vecchi, A.,** *Staphylococcus epidermidis* in the circulating blood of normal and thrombocytopenic human subjects: immunological data, *Experientia*, 32, 1600–1602, 1976.

50. **Tedeschi, G. G. and Santarelli, I.,** Electron microscopical evidence of the presence of unstable L Forms of *Staphylococcus epidermidis* in human platelets, in *Spheroplasts, Protoplasts and L-Forms of Bacteria, INSERM*, Roux, J., Ed., 65, 341–344, 1976.

51. **Wilkie, B. N. and Winter, A. J.,** Location of *Vibrio fetus* var. *venerealis* within the endometrium of the cow, *Infect. Immun.*, 3, 854–856, 1971.

52. **Wittler, R. G.,** The L-form of *Haemophilus pertussis* in the mouse, *J. Gen. Microbiol.*, 6, 311–317, 1952.

Chapter 13

L FORMS IN THROMBI

L FORMS IN SURGERY AND CONTRACEPTION

Intravascular clotting is now known to follow surgical procedures. Altemeier's group cultured "L Forms" from the thrombi of 50 patients, whereas 41 persons without thromboembolic disease gave only negative cultures.[2] Five patients with thromboembolic attack gave both clinical and laboratory evidence of encephalomeningitis. L Forms circulating in the blood were demonstrated in several general classes of thromboembolic disease, a common type being pulmonary emboli occurring within 8 to 12 d after surgery. Many patients experienced recurrent episodes of thrombi formation in veins or arteries.[1]

Some of the thrombi-associated organisms demonstrated *in vitro* interference with the anticoagulant ability of heparin. This would explain why patients may need high doses of heparin to lengthen the coagulation time of the blood to a normal level, and why antibiotic administration can reduce the need for heparin.

Altemeier noticed that Enovid (Norethnodrel and Mestranol) is a growth factor for certain wall-deficient bacteria. Thus, an explanation exists for thrombi which have been reported in relatively young women taking birth control pills.

Strengthening the thesis of an association between bacteria and thrombi, Altemeier et al. noticed that certain bacteria, particularly Bacteroides sp., produce heparinase.[1] In a related finding, Gesner and Jenkins found a strain of Bacteroides from intestinal flora which grew only on media supplemented with heparin.[3] The organism made several enzymes which, together, were responsible for heparin cleavage.

An earlier report by Hill and Lewis[5] described an "L Form" from a patient with thrombophlebitis which surprisingly was found in direct mounts of spinal fluid as well as in cultures of spinal fluid, the thrombus, and circulating blood. Immunofluorescence indicated the "L Form" from the thrombus was serologically related to *Sphaerophorus necrophorus* as was the transitional "L Form" isolated from the patient's blood culture. The variants persisted in the bloodstream of this patient for at least 5 months after the initial episode of thrombophlebitis.

In a personal communication Altemeier stated that many additional findings in his laboratory remain unpublished. He found it easier to be awarded an honorary degree from the University of Edinburgh than to convince his peers of an association between L Forms and blood clots. These stimulating findings immediately suggest new areas of investigation. In early studies it has been noted that CWD variants of streptococci cause clotting within the blood vessels of the chick embryo,[4] but the phenomenon in man was unforeseen.

Mycoplasma have also been implicated in thrombophlebitis. A case report and literature review have been published by Martinez et al.[7]

CHLAMYDIA IN CORONARY THROMBOSIS

In a broad sense of the term, Chlamydia are CWD microbes since in part of their well-known life cycle the small units are bound by only one membrane. It has been surprising that they have been implicated in the common myocardial infarction which follows a clot formed in a coronary artery.

An association has been shown between *Chlamydia pneumoniae* (originally called TWAR) and both chronic coronary heart disease and episodes of coronary thrombosis. In the blood

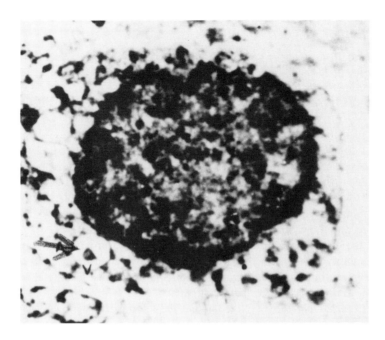

FIGURE 1.   Pericardial fluid cell of a myocardial infarction patient has inclusions near nucleus. Arrow points to vacuole containing an agent. (From Rostam-abadi sofla, H., M.S. thesis, Wayne State University, Detroit, 1976.)

of these patients a circulating immune complex containing antigen of the Chlamydia has been demonstrated in the laboratory of Leinonen.[6]

Findings in our laboratory at Wayne State University also support the conclusion that Chlamydia often may be involved in myocardial infarction. Our interest was piqued by an unidentified agent in pleural fluid of a myocardial infarction (MI) case. The agent, which was nonbacterial, could be serially transferred in guinea pigs, causing pleurisy and pneumonia. Studies by Mandeville,[11] using a sensitive complement fixation technique with convalescent serum, showed that a pertinent microbe probably did not reside in the involved coronary vessels. Examination by Menard[8] and Tunstall[10] revealed a nonbacterial agent transmissible in tissue culture and chick embryo from series of MI cases. The agent(s) produced a potent hemagglutinin with titers up to 1 to 250. Most of the isolates could be neutralized by convalescent serum from other cases. The agent was not a virus of the influenza class. Cooperating laboratories, specializing in virus isolation, reported "no growth". In retrospect, it may be concluded that antibiotics in their tissue cultures eliminated the agent. The hemagglutinin in CT sera was investigated by Mendoza and Nucci.[13] Tunstall [10] had previously shown it was unrelated to known viral hemagglutinins.

Another examination of pericardial fluids from "heart attack" patients for viruses was made by Wandzel[12] with the cooperation of Walter N. Mack of Michigan State University. The approach was to examine density gradients obtained by centrifugation. The conclusion was that no known virus was present.

Examination of coronary thrombosis (CT) pericardial fluids by Rostam-abadi sofla[9] showed two new aspects of the agent(s). It changed the leukocyte differential count in mice, and, more intriguingly, it produced inclusions resembling Chlamydia in cells of the patients' pericardial fluid and chick embryos (Figures 1 and 2). At that time *C. pneumoniae* was unknown. It should not be too difficult to repeat the work and test for identity of the Chlamydia. Preliminary studies suggested that the agent might be airborne from man to man

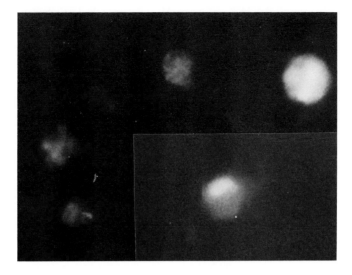

FIGURE 2. Inclusion bodies in pericardial infusion cells of a myocardial infarction (MI) patient react with fluorescent antibody made vs. agent from another MI case. (From Rostam-abadi sofla, H., M.S. thesis, Wayne State University, Detroit, 1976.)

with a long incubation period. There was also possible evidence that a carrier state could temporarily exist in a nonsusceptible individual.

COMMENT

Altemeier's unexpected finding of CWD bacterial variants in thromboembolic phenomena of varied situations, including contraceptive medication, seems well substantiated. The hormones of contraception are growth-stimulating for some CWD variants, which, in turn, can synthesize heparinase. When a thrombus forms, it may be merely a local exacerbation by a circulating, cultivable wall-deficient form.

The role of Chlamydia in coronary artery disease and acute episodes is also an important field for continuing investigation since Chlamydia antigen has been identified in the blood of CT patients. Additionally, a transmissible agent has grown from pleural and pericardial fluids of CT patients. This agent, which has been antigenically related in the majority of cases resembles a Chlamydia species by electron microscopy.

REFERENCES

1. **Altemeier, W. A., Hill, E. O., and Fullen, W. D.,** Acute and recurrent thromboembolic disease: a new concept of etiology, *Ann. Surg.,* 170, 547–558, 1969.
2. **Altemeier, W. A.,** quoted in, L Form of bacteria may cause thrombi, *Infect. Dis.,* 2, 1, March 1972.
3. **Gesner, B. M. and Jenkins, C. R.,** Production of heparinase by Bacteroides, *J. Bacteriol.,* 81, 595–604, 1961.
4. **Hannoun, C., Vigouroux, J., Levaditi, J., and Nazimoff, O.,** Étude des formes L des bactéries apparues spontanément *in vivo*. III. Histopathologie comparée des lésions provoquées par la bactérie normale et par ses formes modifiées, *Ann. Inst. Pasteur (Paris),* 92, 231–238, 1957.
5. **Hill, E. O. and Lewis, S.,** "L Forms" of bacteria isolated from surgical infections, *Bacteriol. Proc.,* p. 48, 1964.

6. **Leinonen, M., Linnanmäki, Mattila, K., Nieminen, M. S., Valtonen, V., Leirisalo-Repo, M., and Saikku, P.,** Circulating immune complexes containing chlamydial lipopolysaccharide in acute myocardial infarction, *Microb. Pathogenesis,* 9, 67–73, 1990.

7. **Martinez, O. V., Chan, J., Cleary, T., and Cassell, G. H.,** Mycoplasma hominis septic thrombophlebitis in a patient with multiple trauma: a case report and literature review, *Diag. Microbiol. Infect. Dis.,* 12(2), 193–196, 1989.

8. **Menard, R. R.,** Tissue Culture Studies of an Unidentified Agent, M.S. thesis, Wayne State University, Detroit, 1958.

9. **Rostam-abadi sofla, H.,** Animal Inoculation and Transmission Electron Microscope Studies of Pericardial Fluids, M.S. thesis, Wayne State University, Detroit, 1976.

10. **Tunstall, L. H.,** Microbiological Studies on Pleural Fluids of Coronary Thrombosis Patients, Ph.D. dissertation, Wayne State University, Detroit, 1963.

11. **Mandeville, R.,** unpublished study.

12. **Wandzel, R.,** unpublished study.

13. **Mendoza, R. A. and Nucci, E. P.,** unpublished study.

Chapter 14

URINARY TRACT INFECTIONS

DISEASE IN MAN

CHRONIC PYELONEPHRITIS AS A DISEASE WITH OCCULT ORGANISMS

The literature shows that chronic pyelonephritis is frequently of occult etiology. Angell and associates found that of 20 cases of histologically proven active chronic pyelonephritis, 12 patients had been asymptomatic, and urine cultures had been negative. Their data indicated that chronic pyelonephritis without bacteriuria is not rare and can progress without symptoms of urinary tract infection.[2]

A similar report comes from one long interested in kidney disease, Kimmelstiel, reporting in a paper with Pawlowski and Bloxdorf.[28] In an extensive autopsy series, only in 14 instances was chronic pyelonephritis accompanied by bacteriuria. In the majority of cases, 112, which were histologically classified as chronic pyelonephritis, no significant number of bacteria was found in the kidney or the urine.

The two series described above point to the conclusion that the agents of "active chronic pyelonephritis" are often bacteria which fail to grow by routine methods. One clue is given by the stimulating study of Aoki et al.[3] Bacterial antigen was detected in renal sections of six of seven patients with "abacterial" pyelonephritis. The unique tool was a fluorescent antibody prepared to an antigen native to most Enterobacteriaceae. This antigen, termed CA, occurs so consistently that it can be used to identify the common urinary tract pathogens, with the exception of Pseudomonas species. Clusters of amorphous granules observed in this study within macrophages and in interstitial tissue could well be CWD organisms.

Nishimoto and associates also concluded that chronic pyelonephritis could be recognized by detecting bacterial antigen in "aseptic" urine, testing for the CA antigen by the Elisa method. In a canine model, they found that "bacterial L-forms are inflammatory latent factors leading to pyelonephritis relapse".[27]

Chronic pyelonephritis, while often free of classic bacteria may have marked pyuria, as in a case which progressed for 3 years as both kidneys atrophied. L Forms of *Streptococcus faecalis* were implicated. The infection was eliminated with ampicillin plus erythromycia.[9]

Fernandes and Panos[16] studied a CWD streptococcus from a diseased kidney (Figure 1). The colony is the whorl-colony type, also noted in variants of *Nocardia caviae*.[4]

TREATED AND UNTREATED CASES SHOW VARIANTS

The foremost detectives in the world for solving urinary tract infections must be in Tulane University, the Veterans Administration and the Charity Hospitals of New Orleans. Domingue and Schlegel found L-phase organisms in 105 of 502 patients (21%), whose cultures were negative for classic organisms.[13] Preadministration of antibiotic could not be responsible for every incidence of the variants; sometimes no therapeutic agent had been administered. The study included finding L Forms in the urine of a patient with a 2-year history of repeated pyelonephritis and the subsequent isolation of the identical L Form from the cortex, medulla and papillae of the involved kidney.

Domingue and Schlegel noted that oil-immersion phase microscopy of the urine always revealed spheres, budding forms, cigar-shaped cells, and long filaments if CWD forms were to grow in culture. Heidger and associates[21] noted that CWD forms which were similar could be isolated from blood and urine of culture-negative kidney disease patients (Figure 2).[13]

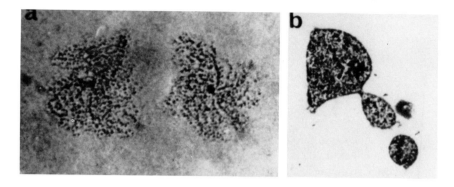

FIGURE 1.   Wall-deficient organism from biopsy of human kidney, acute renal failure. (A) Colony; (B) organism. (Magnification × 14,000.) (From Fernandes, P. B. and Panos, C., *J. Clin. Microbiol.*, 5, 106–107, 1977. With permission.)

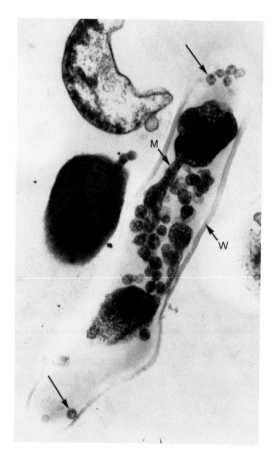

FIGURE 2.   Dense bodies within elongated forms are found in growth from both urine and blood of "culture-negative" kidney disease. A membrane (M) encloses some particles. (From Domingue, G. J., *Cell Wall Deficient Bacteria*, Addison-Wesley, Reading, MA, 1982.)

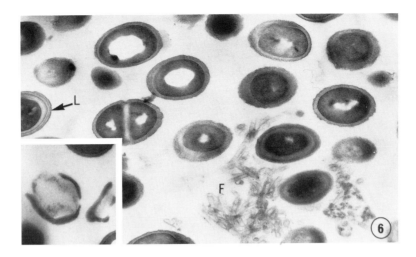

FIGURE 3. Revertant cocci in cultures of blood have atypical laminated walls. Insert shows cocci with incomplete walls. (Magnification × 30,000.) (From Domingue, G. J., *Cell Wall Deficient Bacteria*, Addison-Wesley, Reading, MA, 1982.)

Much of the work from the New Orleans group has concerned investigation of the blood-borne variants which they found were the same species causing the patients' urinary tract disease. An especially interesting micrograph of minute organisms undetectable with light microscopy shows the small spheres surrounded by revertant cocci. All stages in reversion could be followed (Figure 3).

Another unexpected finding concerns the content of "debris". A culture of lysed blood seemed to have only structureless material until subjected to high magnification (Figure 4).

Another informative series cultured by Conner et al.[8] in Bay Pines, FL, showed that among 115 patients with chronic or acute urinary tract infections, 26 had bacterial L Forms in the urine which reverted to Pseudomonas sp., *Proteus vulgaris, P. mirabilis, P. rettgeri, Escherichia coli,* the Aerobacter group, beta-hemolytic streptococci, *Staphylococcus aureus,* or *S. epidermidis.*

RENAL FANCONI SYNDROME

A child with renal fanconi syndrome was treated in a urinary clinic for $2\frac{1}{2}$ years before expiring. Her urinary casts were notable for containing 2- to 8-μm granules with dense cores. These inclusions, observed during incubation on a slide, extruded mycelia-like filaments, some of which segmented into streptococcal-like forms. A CWD strain growing from her urine was used as antigen to prepare specific antibody in rabbits. The fluorescent antibody thus prepared reacted with the urinary cast inclusions, with hepatocytes, and with CWD forms from her blood cultures as well as with the reverted streptococcus.[12]

IDIOPATHIC HEMATURIA

Recently, Domingue's laboratory made detailed studies on a woman with idiopathic hematuria whose infection was polymicrobic with CWD bacteria. Although repeated cultures for classic bacteria were negative, microcolonies were visible in the catheterized urine (see Frontispiece). Both the staphylococci and streptococci present initially grew only as variant colonies containing typical "vacuoles". Of 15 antimicrobials tested, macrodantin elicited the best *in vitro* response. Correspondingly, in 4 d, clinical improvement was definite, and

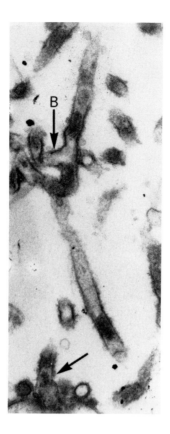

FIGURE 4. Culture of lysed blood from kidney disease patient shows elongated branching forms at 60,000 × magnification. At 1000 × magnification, the material only suggested debris. (From Domingue, G. J., *Cell Wall Deficient Bacteria,* Addison-Wesley Reading, MA, 1982.)

the patient has remained symptom free for more than a year. Growth from a colony with beginning reversion is shown in Figure 5.

In an earlier study of 35 patients with idiopathic hematuria, each had a single organism growing in urine cultures in hypertonic broth. The organism usually reverted to an alpha-hemolytic streptococcus.[13]

GLOMERULONEPHRITIS

The laboratory of Panos found evidence that L Forms of β-hemolytic streptococci may be involved in Bright's disease. Acute and chronic glomerulonephritis, as is well known, usually follow infection with Type 12 Group A β-hemolytic streptococcus. An L Form of Type 12 retained some lipoteichoic acid (LAT).[30] This compound was then shown to cause collagen deposit in glomeruli in tissue culture.[31] Interestingly, bacitracin decreased the variants' content of LAT without killing the organism.[30]

L Forms were found in blood or urine of 11 of 14 children with glomerulonephritis in remission or active stages. The three organisms which reverted were streptococci. It required only a single examination to find the L Forms.[24]

SYSTEMIC LUPUS ERYTHEMATOSIS

Two patients with systemic lupus erythematosis had only CWD forms in their urinary cultures. In both instances, the organisms were so unstable, reverting back and forth from classic to wall deficient, that classification could not be made.[13]

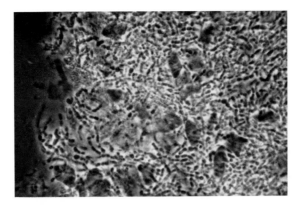

FIGURE 5. Reverted cocci appear in colony from urine of idiopathic hematuria. (Magnification × 1000.) (Photo submitted by Domingue, G. J.)

NEPHROTIC SYNDROME

Woody et al. have discovered an ignored participant in the nephrotic syndrome.[35] In this condition, serum antibodies are lowered as marked proteinuria occurs. Enigmatically, lipids are elevated in both serum and urine. Fatty deposits seen in renal cells have been thought to result from absorption by the renal cells. On the contrary, the investigators found lipid masses invariably associated with wall-deficient bacteria, which are known to absorb lipid. The authors comment that the role of the variants in the infections which plague nephrotic syndrome patients is an intriguing subject for continued investigation. One of the lipid-containing forms, termed an oval fat body, is shown in Figure 6. Extensive electron microscopic studies were made of the associated CWD forms.

LITHIASIS

Miliary tuberculosis may follow shock-wave lithotripsy with smears of urine, sputum, and gastric aspirate all negative for classic tubercle bacilli. In a recent case, diagnosis waited a month until cultures for classic bacteria became positive.[15]

THERAPY

CWD variants in urinary tract disease are usually sensitive to erythromycin and/or kanamycin. The response to kanamycin is not surprising since this agent has a rather broad spectrum. Sensitivity of Gram-negative rods to erythromycin is unexpected since for *classic bacteria* the antibiotic has a spectrum duplicating penicillin with few exceptions. However, studies of Montgomerie et al.,[25] Kagan et al.[22] Guze and Kalmanson,[20] and many others indicate that often CWD forms are more sensitive than parent bacteria to certain antibiotics. The treatment schedule of Domingue's group for five patients is given in Table 1.

VARIANTS IN RELAPSES

Wall-deficient organisms in the urine of a "cured" case quite accurately forecast a relapse. A University of Washington study indicates that bacterial variants present during antimicrobial therapy greatly increase the likelihood that the condition will relapse.[18]

Conner and co-workers followed a case whose pyelonephritis with *P. mirabilis* repeatedly relapsed as long as therapy was only chloramphenicol, to which the parent bacillus was sensitive.[8] When the L phase was eradicated with erythromycin, no relapse occurred. It is noteworthy that the classic bacterium was not sensitive to erythromycin, so this is one of many cases in which treatment had to be directed to parent and variant, respectively.

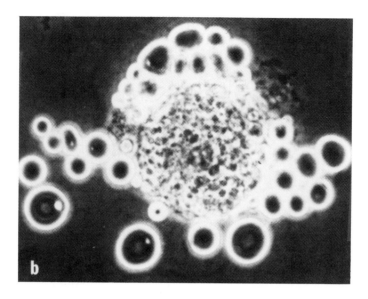

FIGURE 6.   Escape of lipid from an oval fat body reveals a giant renal epithelial cell containing many CWD organisms. (Phase contrast, × 1500.) (From Woody, H. B., Walker, P. D., and Domingue, G. J., in *Cell Wall Deficient Bacteria*, Addison-Wesley, Reading, MA, 1982.)

Patients treated with chloramphenicol can harbor CWD stages which will cause recurrence of urinary tract infection. Voureka found bacterial variants, including giant bodies, in the urine after chloramphenicol therapy and connected their occurrence with clinical relapse in some cases.[33]

One of the early studies of L variants in pyelonephritis, made by Boman, concerned a patient who had tenderness in the kidney region for 19 years. Organisms in the urine reverted to *P. mirabilis*. The case was successfully treated with chloramphenicol and ampicillin.[5]

Turck et al. noted that recurrent pyelonephritis is not always due to persisting CWD bacteria. They emphasize as an example that structural defects in the urinary tract may be involved.[32]

ANIMAL MODELS

FORMATION OF SPHEROPLASTS IN THE RAT AND MOUSE KIDNEY

Animal models demonstrate the role of wall deficient forms in urinary tract infections. Guze and Kalmanson found that rats given penicillin injections following an inoculum of *Streptococcus faecalis* harbor protoplasts in the kidney for at least 13 weeks.[19]

In a delightfully ingenious study, Demonty showed that *E. coli* infecting the rat kidney gradually consists more of spheroplasts than of bacilli.[10] This study employed no antibiotic. Twenty-one days after injecting the kidney-massaged animals, more than 50% of the organisms in the kidney are CWD spheres. Their spheroplastic stage was confirmed by resistance to osmotic shock and, second, by fluorescent antibody staining, with serum directed specifically against the spheroplasts.

Nakao and associates have published informative scanning and transmission electron micrographs of *Pseudomonas aeruginosa* as it changes to spheroplasts in the urinary bladder of untreated mice.[26]

TABLE 1
Treatment of Patients Whose Urinary Tract
Infection Yielded Only CWD Forms

Pt.	Reverted microorganism	Treatment	Repeated routine and L Form cultures
1	*Proteus mirabilis*	Ampicillin 500 mg. 5d; erythromycin 1 g/d for 10 d	Negative through 6 m
2	*P. mirabilis*	Nalidixic acid 2 g/ d for 10 d	Negative through 6 m
3	*Alcaligenes faecalis*	Nitrofurantoin 50 mg/3 × d for 24 d	Negative through 9 m
4	Klebsiella	Erythromycin 2 g/d for 10 d	Negative through 7 m
5	*Staphylococcus aureus,* coagulase +	Erythromycin 1 g/d for 10 d	Negative through 4 m

From Domingue, G. J. and Schlegel, J. U., *J. Urol.,* 104, 790–797, 1970. With permission.

BLADDER STONES IN RATS

Braude described lithiasis of the rat's urinary bladder caused by persistence of CWD variants of Proteus. It is fascinating that formation of the stones coincided with passage of the L Form into a stage not growing in the culture medium.[7]

GLOMERULONEPHRITIS IN GUINEA PIGS

Another rodent study concerns the pathogenesis of CWD variants of *Mycobacterium tuberculosis.*[36] All the early work showed that variants of this genus exhibit little pathogenesis for guinea pigs. Xalabarder likewise found no disease in the spleens, livers, or lungs where expected, but found severe disease in the kidney glomeruli.[36] Since media for L Forms may cause some damage to the kidney,[34] it was important to note that Xalabarder's work had multiple controls: pigs receiving culture medium only, others given penicillin-induced L Forms of *E. coli,* and others receiving only a 1:10 dilution of tuberculin every 2 months throughout a year. The tuberculin was given to show that dead antigenic material from tubercle bacilli did not produce the same pathogenesis as the living wall-deficient mycobacterium.

THE KIDNEY MEDULLA AS A PROTECTIVE ENVIRONMENT

The usual situation — Irrespective of pathogenicity, CWD bacteria find a protective milieu in the kidney. Variants of *E. coli* survive better in the rat kidney medulla than in the liver and spleen.[34] Similarly, *E. coli* spheroplasts which are lysed in rabbit dermis multiply in the kidney medulla of this animal.[1] The association of CWD forms with chronic and recurrent pyelonephritis is explained.

NEGATING THE PROTECTIVE HYPERTONICITY OF THE MEDULLA

Spheroplasts can be successfully lysed away if the infected rats are given glucose-water, which they imbibe in large amounts (Figure 7). Rats were injected with *S. fecalis* and treated with penicillin resulting in a persisting spheroplast infection of the kidney.[14] Cultures for

FIGURE 7. Rats under therapy for kidney CWD forms. (From Eastridge, R. R., Jr. and Farrar, W. E., Jr., *Proc. Soc. Exp. Biol. Med.*, 128, 1193–1196, 1968. With permission.)

spheroplasts were made on rats sacrificed on weeks 4, 6, and 8. When diuresis was forced by high fluid intake, the spheroplasts were almost entirely eliminated by the fourth week. The rationale for forcing fluids in patients with urinary tract infections is elucidated.

THE KIDNEY AS A SITE FOSTERING REVERSION

One study indicates that the kidney may not only protect, but also foster reversion of some CWD strains. An *E. coli* strain which was stable *in vitro* gave a high reversion rate when injected into the rat kidney medulla.[34]

LOW URINARY GLUCOSE AND LATENT INFECTION

An ingenious way to detect urinary tract infection applies to both classic and wall-deficient bacteria. Mårdh et al.[23] utilized a sensitive test paper[29] which detects the slight glucose present in normal urine. Less than the normal 1.5 mg% indicates urinary tract infection. This test detected a urinary tract infection which showed only L-phase variants of *E. coli* in culture.

CWD BACTERIA AS NORMAL FLORA IN URINE

Normal urine with a specific gravity of 1.030 may contain filtrable stages of such parasites as *Staphylococcus epidermidis* and *Streptococcus fecalis*. These can be filtered through cellophane membranes and grown in serum-enriched hypertonic medium as Actinomyces-like colonies which will revert after 2 week incubation. Normal flora variants indicate that bacteria in the urinary tract are probably in a constant state of flux between orthodox and aberrant stages.

Braude et al. noticed that spheroplasts may survive as well in hypertonic urine as in hypertonic sucrose.[6] The normal acidity of urine also preserves some spheroplasts. Gnarpe and Edebo found that spheroplast-like forms of *Proteus vulgaris* and *E. coli* remain unlysed and viable in urine which has a pH of 5.0 to 5.5.[17]

SUMMARY

The otherwise inexplicable "no growths" occurring in urinary tract infections are often explained when hypertonic enriched medium reveal variants. The annoying tendency of these infections to recur is also related to the persistence of CWD forms. Sometimes therapy needs to be directed against both classic and variant. Sometimes the infection is polymicrobial with two species in a wall-deficient stage.

Bladder stone formation noted to result from rodent infection with stable L Forms will probably stimulate search for variants in renal calculi of man, a condition considered to be a solely metabolic disease.

The tubercle bacillus in the L-stage may be responsible for some urinary tract lithiasis as suggested by a case in which stones were disrupted by laser; miliary tuberculosis rapidly developed. No classic bacilli were seen in the urine although cultures became positive after a month.

REFERENCES

1. **Alderman, M. H. and Freedman, L. R.,** Experimental pyelonephritis. X. The direct injection of *E. coli* protoplasts into the medulla of the rabbit kidney, *Yale J. Biol. Med.,* 36, 157–164, 1964.
2. **Angell, M. E., Relman, A. S., and Robbins, S. L.,** "Active" chronic pyelonephritis without evidence of bacterial infection, *N. Engl. J. Med.,* 278, 1307–1308, 1968.
3. **Aoki, S., Imamura, S., Aoki, M., and McCabe, W. R.,** "Abacterial" and bacterial pyelonephritis. Immunofluorescent localization of bacterial antigen, *N. Engl. J. Med.,* 281, 1375–1382, 1969.
4. **Beaman, B. L. and Scates, S. M.,** Role of L-forms of *Nocardia caviae* in the development of chronic mycetomas in normal and immunodeficient murine models, *Infect. Immun.,* 33, 893–907, 1981.
5. **Boman, I.,** as quoted in New attack on chronic infection, *World Med. News,* 7, 63–70, 1966.
6. **Braude, A. I., Siemienski, J., and Jacobs, I.,** Protoplast formation in human urine, *Trans. Assoc. Am. Physicians,* 74, 234–245, 1961.
7. **Braude, A. I.,** Production of bladder stones by L-forms, *Ann. N.Y. Acad. Sci.,* 174, 896–902, 1970.
8. **Conner, J. F., Coleman, S. E., Davis, J. L., and McGaughey, F. A.,** Bacterial L-forms from urinary-tract infections in a veterans hospital population, *J. Am. Geriatr. Soc.,* 16, 893–900, 1968.
9. **Davies, A. G., McLachlan, M. S. F., and Asscher, A. W.,** Progressive kidney damage after non-obstructive urinary tract infection, *Br. Med. J.,* 4, 406–407, 1972.
10. **Demonty, J.,** Experimental *Escherichia coli* pyelonephritis, formation of spheroplasts in the kidney of the untreated rat, *Antonie van Leeuwenhoek,* 36, 273–284, 1970.
11. **Domingue, G. J. and Schlegel, J. U.,** The possible role of microbial L-forms in pyelonephritis, *J. Urol.,* 104, 790–797, 1970.
12. **Domingue, G. J., Woody, H. B., Farris, K. B., and Schlegel, J. U.,** Bacterial variants in urinary casts and renal epithelial cells, *Arch. Intern. Med.,* 139, 1355–1360, 1979.
13. **Domingue, G. J., Ed.,** *Cell Wall Deficient Bacteria,* Addison-Wesley, Reading, MA, 1982.
14. **Eastridge, R. R., Jr. and Farrar, W. E., Jr.,** L Form infection of the rat kidney, effect of water diuresis, *Proc. Soc. Exp. Biol. Med.,* 128, 1193–1196, 1968.
15. **Federmann, M. and Kley, H. K.,** Miliary tuberculosis after extracorporeal shock-wave lithotripsy, *N. Engl. J. Med.,*323, 1212, 1990.
16. **Fernandes, P. B. and Panos, C.,** Wall-less microbial isolate from a human renal biopsy, *J. Clin. Microbiol.,* 5, 106–107, 1977.
17. **Gnarpe, H. and Edebo, L.,** Conditions affecting the viability of spheroplasts in urine, *Acta Pathol. Microbiol. Scand.,* 187, 33, 1967.
18. **Gutman, L. T., Turck, M., Petersdorf, R. G., and Wedgwood, R. J.,** Significance of bacterial variants in urine of patients with chronic bacteriuria, *J. Clin. Invest.,* 44, 1945–1952, 1965.

19. **Guze, L. B. and Kalmanson, G. M.**, Persistence of bacteria in "protoplasts" form after apparent cure of pyelonephritis in rats, *Science,* 143, 1340–1341, 1964.

20. **Guze, L. B. and Kalmanson, G. M.**, Action of erythromycin on "protoplasts" in vivo, *Science,* 146, 1299–1300, 1964.

21. **Heidger, P. F., Jr., Domingue, G. J., Larsen, W. J., and Smith, T. W., Jr.**, Fine structural studies of CWD bacteria isolated from human blood and urine, in *Cell Wall Deficient Bacteria,* Domingue, G. J., Ed., Addison-Wesley, Reading, MA, 1982, 149–187.

22. **Kagan, B. M., Molander, C. W., Zolla, S., Heimlich, E. M., Weinberger, H. J., Busser, R., and Liepnieks, S.**, Antibiotic sensitivity and pathogenicity of L-phase variants of staphylococci, *Antimicrob. Agents Chemother.,* pp. 517–521, 1963.

23. **Mårdh, P. A., Fritz, H., Köhler, L., and Scherstén, B.**, Hypoglucosuria and L Forms of *Escherichia coli* in the urine, *Acta Med. Scand.,* 186, 549–552, 1969.

24. **Mashkov, A. V. and Dumnova, A. G.**, Preliminary results of bacteriological examination of the blood and urine of children suffering from nephritis, *Zh. Mikrobiol. Epidemiol. Immunol.,* 1, 127–131, 1977.

25. **Montgomerie, J. Z., Kalmanson, G. M., and Guze, L. B.**, The effects of antibiotics on the protoplast and bacterial forms of *Streptococcus faecalis, J. Lab. Clin. Med.,* 68, 543–551, 1966.

26. **Nakao, M., Kondo, M., Imada, A., and Tsuchiya, K.**, An electron microscope study of pathogenesis of urinary tract infection caused by *Pseudomonas aeruginosa* P-9 in mice, *Zentralbl. Bakteriol. Mikrobiol. Hyg. Ser. A,* 260(3), 369–378, 1985.

27. **Nishimoto, T., Miyazaki, Y., Sasaki, S., Takahashi, Y., Moriyama, M., Morita, T., Kudo, K., and Tsuchida, S.**, Significance of bacterial parent form, L-form and bacterial antigen in experimental pyelonephritis, *J. Urol.,* 137, 321A, 1987.

28. **Pawlowski, J. M., Bloxdorf, J. W., and Kimmelstiel, P.**, Chronic pyelonephritis, A morphologic and bacteriologic study, *N. Engl. J. Med.,* 268, 965–969, 1963.

29. **Scherstén, B., Dahlqvist, A., Fritz, H., Köhler, L., and Westlund, L.**, Screening for bacteriuria with a test paper for glucose, *JAMA,* 204, 113–116, 1968.

30. **Slabyj, B. M. and Panos, C.**, Membrane lipoteichoic acid of *Streptococcus pyogenes* and its stabilized L Form and the effect of two antibiotics upon its cellular content, *J. Bacteriol.,* 127, 855–862, 1976.

31. **Tomlinson, K., Leon, O., and Panos, C.**, Morphological changes and pathology of mouse glomeruli infected with a streptococcal L Form or exposed to lipoteichoic acid, *Infect. Immun.,* 42, 1144–1151, 1983.

32. **Turck, M., Gutman, L. T., Wedgwood, R. J., and Petersdorf, R. F.**, Significance of bacterial variants in urinary tract infections, in *Microbial Protoplasts, Spheroplasts, and L-Forms,* Guze, L. B., Ed., Williams & Wilkins, Baltimore, 1968, 415–421.

33. **Voureka, A.**, Bacterial variants in patients treated with chloramphenicol, *Lancet,* 1, 27–28, 1951.

34. **Winterbauer, R. H., Gutman, L. T., Turck, M., Wedgwood, R. J., and Petersdorf, R. G.**, The role of penicillin-induced bacterial variants in experimental pyelonephritis, *J. Exp. Med.,* 125, 607–618, 1967.

35. **Woody, H. B., Walker, P. D., and Domingue, G. J.**, in *Cell Wall Deficient Bacteria,* Domingue, G. J., Ed., Addison-Wesley, Reading, MA, 1982.

36. **Xalabarder, C.**, L Forms of Mycobacteria and chronic nephritis, *Publ. Inst. Antituberc. Suppl.,* 7, 1970.

Chapter 15

LISTERIA MONOCYTOGENES STUDIES

Listeriosis appears to be increasing in Europe and North America[28] with a mortality of 30 to 40%. New findings include the notation that the organism may be carried in a product in which it obviously cannot multiply, namely alfalfa tablets.[13] Electrophoretic typing has indicated strain identity in major epidemics and association of types with certain foods.[29] In Bratislava, 350 strains of *Listeria monocytogenes*, isolated in Slovakia, were characterized.[11]

Ghosh and Murray[14] detected a marked difference in cell walls of *L. monocytogenes* strains, a difference unrelated to serological type. Differences in cell wall composition were shown by susceptibility to lysozyme and to lipase. Accordingly, one might conclude that some strains are especially prone to wall deficiency in natural situations.

L. monocytogenes is noteworthy as an organism suspected of producing neurological disease in the CWD state because of the difficulty with which the bacterium is cultured from infected brain. It has long been known that domestic or forest animals dead of listerial encephalitis may appear to have a diseased but noninfected central nervous system until the tissue is again cultured after months of storage at 4°C. Such technique has not been applied to diseased brain of man.

LISTERIOSIS IN CHILDREN

In neonatal illness, listeriosis may be no less important than Rh incompatibility. There is also unexplained association of fetal Rh incompatibility and listeriosis.[1,8]

The data of Lang suggest that perhaps brain tissue of human infants is an important target of the bacterium, causing infections difficult to diagnose. If the information of Table 1 carries the import which it suggests, there is a dire need for improved detection of listeriosis in the young.[21]

Lang found low Listeria antibody titers in children with cerebral palsy and disorders of established etiology, including congenital central nervous system abnormalities. In contrast, almost half of the children with cerebral disorders of *unknown* etiology had elevated serum titers to Listeria. These children had conditions including spastic paresis, arrested development, athetoid and choreiform syndromes, convulsions susceptibility, and hydrocephalus. Hydrocephalus as a postinfection complication of listeriosis has been reported.[20,38] Seeliger commented that early childhood listeriosis may leave cerebral scar tissue, responsible for such abnormalities.[34] Localization of Listeria in mouse brain is shown in Figure 1.[5]

Seeliger concludes that the clinical symptoms in infants dying from Listeria infection are usually not distinctive and are ascribed to such general categories as alimentary disorders or pylorospasm. The intestinal disorders can be explained by the rapidity with which the organism destroys living cells. Twenty four hours after infection mouse hepatocytes may show almost complete destruction of cytoplasmic organelles. Intracellular growth of the organism in the L-stage seems to play a stellar role in the cytopathogenesis.[39] In children the infectious cause is usually not recognized.[34] When infection is recognized, it is difficult to ascertain the route of infection. Transplacental passage of the organisms is often involved, although aspiration of infected amniotic fluid is evident in some cases.[32]

With listerial infections of the central nervous system, the antibody titer in the spinal fluid may be suggestively high when the serum antibody titer is only borderline.[34] However, since spinal fluid is seldom tested for antibody, the chance is slight of diagnosing Listeria

TABLE 1
Distribution of O and H titers against *L.*
***monocytogenes* among Sick Children and Healthy**
Controls[21]

	Total figure	O or H titers in serum dilutions of 1:60 or higher
Healthy controls	56	3 (= 5%)
Children with cerebral disorders		
Etiology established	27	4 (= 14%)
Etiology probably established	14	1 (= 7%)
Gross congenital abnormalities	9	—
Etiology unknown	87	43 (= 49%)
Sick children with other diseases	256	59 (= 23%)

encephalitis when the bacterium is not detected in the spinal fluid. Nevertheless, the etiology could probably be recognized through the indirect methods discussed under the next section.

Symptoms in the woman transmitting the bacterium to her unborn child are usually too mild to serve as a warning. Thirty-seven cases of listeriosis in the newborn were found by Patočka et al. to have resulted from infections in the mother which were either not apparent or minimal, although maternal serum reactions could be demonstrated.[27]

LISTERIOSIS IN ADULTS

Clinicians who see Listeria infections in adults note characteristics of the disease suggesting that infection may be much more common than is realized and the diagnosis frequently missed because the organism is masked in a clandestine state.[35] It is well recognized that infection of the central nervous system may be either meningeal or encephalitic. When the infection is in the meninges, presumably diagnosis can easily result from symptomatology and routine culture. This is not true. Spinal fluid cultures may be negative and the usual indications of meningitis missing. Louria[23] has commented that there is often no significant increase in spinal fluid leukocytes. The usual symptom of meningitis, the typical stiff neck, is not exhibited. Sometimes the only symptoms are headache and slight lethargy. Louria suggests that it might be profitable to culture spinal fluid routinely for the L Form of *L. monocytogenes* in such patients.

The diagnosis is even more difficult in the patient with Listeria encephalitis.[16] The cell count in the spinal fluid may not be greatly elevated, and cultures are usually negative. Only rarely is the blood culture positive. From experience in our laboratory with acute and severe chronic infections, it would be anticipated that wall deficient forms of the variant will grow from the blood in routine blood culture medium in 48 h. Identification of the variants with specific serum would be the final identification step.

An intriguing new aspect of pathogenicity of CWD Listeria has emerged. When mice are simultaneously infected with Rausch leukemia virus and Listeria L Forms, their plasma contains 10,000 × as much erythroleukemia-producing factor as when mice are given only the virus. Immunosuppression by the L Form of Listeria seems obvious.[18] This raises the question: when HIV positive patients are simultaneously infected with Listeria, what is the role of each pathogen?[24,33]

FIGURE 1. *Listeria monocytogenes* CWD organisms localized in brain of mouse. Impression smear. (Magnification × 680.) (Photograph submitted by A. Brem.)

Reichertz and Seeliger reported encouraging findings with skin testing antigens.[31] Skin tests are applicable to infections in inaccessible sites of the body and should detect infection with either the classic or wall-deficient stage.

Chronic disease may result from inadequately treated cases. Penicillin alone may be totally ineffective.[25,26] Louria et al. found that penicillin or tetracycline given to Listeria-infected mice often failed to eradicate the bacteria.[22] At 7 d the classic organisms were still present in lung and spleen, although the disease was in remission. Comparable phenomena may occur in man.

LABORATORY STUDIES ON WALL-DEFICIENT VARIANTS

Many people who work with listeriosis have a keen interest in L-phase variation. A significant advance was made when Brem and Eveland obtained L-phase variants of ten strains of Listeria, representing different serotypes.[6,7] Extracts of normal tissue induced L variation. When penicillin was used as an inducing agent, there was great variation in the antibiotic requirement.

Patočka et al.[27] and Suchanova and Patočka[37] obtained CWD growth of Listeria by combining penicillin and glycine. Prosorovsky and associates induced the variants by using penicillin in medium stabilized with salts.[30]

Beson[2] and Beson and Baugh[3] found a remarkably simple method to produce CWD Listeria. They noticed that the variants appeared exclusively in the early growth of the umbrella which appears subsurface after classic bacilli are stabbed into a tube of semisolid medium. This finding of wall-deficient growth was confirmed by electron microscopy. When 0.005% 2,3,5-triphenyl tetrazolium chloride was incorporated into Difco's motility medium, the dye was reduced only by the classic bacteria.[2,3]

PATHOGENICITY OF WALL-DEFICIENT VARIANTS

The isolation of *L. monocytogenes* in the L phase from a cranial lesion in man[9] has convincingly demonstrated that pathogenicity is a reality. Pathogenicity has long been suggested by the prolonged cold-storage enrichment needed to grow Listeria from the brains of forest animals dead of encephalitis, as mentioned above. To our knowledge no one has examined such tissues for CWD Listeria.

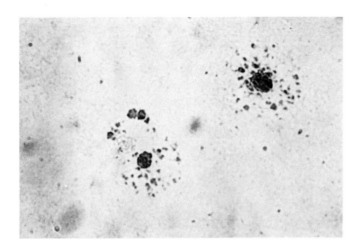

FIGURE 2.   CWD colony of *L. monocytogenes* in fetus of a mother infected with type 4b Listeria. (Magnification × 170.) (Photograph submitted by A. Brem).

L Forms of Listeria in studies by Prosorovsky and associates[30] produced the marked pathogenicity in rabbits which characterizes infection with classic Listeria, namely, paralysis of extremities and ear muscles and hepatic granulomas. Monocytosis also occurred. These variants were pathogenic for some cell lines, but not for chick embryo, and they did not give the anticipated conjunctival reaction. Similar rabbit pathogenicity was reported earlier.[19]

Other *in vitro*-produced Listeria L Forms have been lethal to chick embryo.[17] Mice have great natural resistance to L Forms of Listeria; however, the organisms can persist in the brains of mice for as long as 6 weeks[7] and can pass transplacentally in mice[5] (Figure 2).

Interesting is the spontaneous formation of wall-deficient variants in rabbits.[37] Tiny granular stages of Listeria appeared in muscles and pulmonary histiocytes after classic organisms were injected. Sheep with naturally occurring listeriosis were found to harbor L Forms of the organisms as well as classic Listeria.[30]

EPIDEMIOLOGY OF LISTERIA IN MAN

Epidemiological studies in Listeria cases are difficult due to commonly unsuccessful attempts to isolate the pathogen in suspected food or other sources. Soft cheese has often been implicated. Seeliger[36] noted that if milk contains a high concentration of Listeria, flash pasteurization, as commonly practiced, does not eliminate the pathogen. It was found that 16% of cows which had experienced a Listeria-associated abortion excreted the organism in their milk.[10]

Two characteristics of the L Form of Listeria mentioned above may help the epidemiologist. The growth of the variant in the subsurface "umbrella" in motility medium shows a need for microaerophilic atmosphere.[3] Secondly, variants differ from classical in not reducing triphenyl tetrazolium chloride.

Again, difficulties in defining the mode of spread of this pathogen may be related to an occult phase. When attempts are made to locate the source of a case, too often only an unrelated Listeria type is isolated from the contact.[4] Eveland and Baublis[12] found evidence in two cases that *L. monocytogenes* might have been acquired from pet dogs.

Many times the infection is endogenous, surfacing only after cortisone therapy. Probably also endogenous was a fatal case of Listeria meningoencephalitis occurring 2 d after tooth extraction, as observed by Gram[15] and discussed by Seeliger.[34]

LISTERIA IN LIVESTOCK

Listeriosis in domestic animals mimics the ramifications of human disease in many respects. Veterinary medicine makes significant contributions resulting from the numerous cases encountered and the autopsies and cultures which can be made under controlled conditions.

In Friesland it was concluded that listeriosis caused 46% of the bovine encephalitis. In 80% of the farms with infection, *L. monocytogenes* was found in the grass silage. Abortion in cattle was occasionally due to Listeria and could be caused by either types 1 or 4b. Correlated with the difficulty of finding this organism after human abortion, it is interesting that after bovine abortion the Listeria organism could be isolated from the vagina for about a fortnight only. Thus, it is suggested that in the bovine as in the human species, pregnancy temporarily creates a favorable *milieu* for propagation of the classic stage of Listeria. Individuals who have repeated abortions may well harbor the organism in the L phase between pregnancies.

It is interesting to note the culture methods used by Dijkstra as his laboratory investigated over 35,000 fetuses and afterbirth specimens in cattle. His extensive experience did not suggest an easy method of isolating this organism. For instance, of 12 samples of feces infected with Listeria, only 4 samples revealed the bacteria by immediate culture; in 8 instances the organism grew only after refrigeration for 7.5 months.[10] Culturing for L-phase variants might be less tedious.

SUMMARY

Serological tests indicate that *L. monocytogenes* is widely disseminated in the human population throughout Europe and the U.S., yet attempts to grow the organism from suspected cases or carriers meet with infrequent success. Most intriguing is a study suggesting that permanent cerebral damage can occur in children from undiagnosed meningitis or encephalitis. It has been suggested by clinicians that looking for the wall-deficient form may improve diagnosis. Means of culturing the CWD variant are now known.

REFERENCES

1. **Alex, R.,** Die approximative Häufigkeit der Listerioseinfektion in der Schwangerschaft, *Arch. Gynaekol.,* 186, 381–384, 1955.
2. **Beson, C. A.,** Cell Wall Deficient forms of *Listeria monocytogenes* As a Natural Phenomenon, M.S. thesis, Texas Tech University, 1984.
3. **Beson, C. A. and Baugh, C. L.,** Cell wall deficient forms of *Listeria monocytogenes* as a natural phenomenon, *Abstr. Annu. Meet. ASM,* 1983.
4. **Bojsen-Møller, J. and Jessen, O.,** Occurrence of *Listeria monocytogenes* in human faeces, epidemiological and pathogenic aspects, *Proc. 3rd Int. Symp. Listeriosis* (Bilthoven), 1966, 415–421.
5. **Brem, A. M.,** The role of L Forms in the pathogenesis of *Listeria monocytogenes* infections, Ph.D. dissertation, University of Michigan, Ann Arbor, 1968.
6. **Brem, A. M. and Eveland, W. C.,** Inducing L Forms in *Listeria monocytogenes* types 1 through 7, *Appl. Microbiol.,* 15, 1510, 1967.
7. **Brem, A. M. and Eveland, W. C.,** L Forms of *Listeria monocytogenes*, in vitro induction and propagation of L Forms in all serotypes, *J. Infect. Dis.,* 118, 181–187, 1968.
8. **Cech, J.,** The co-existence of foetal erythroblastosis and toxoplasmosis or another chronic infection, *Proc. 10th Czech. Congr. Microbiol. Epidemiol.* (Prague), 1958, 233.

9. **Charache, P.,** Cell wall defective bacterial variants in human disease, *Ann. N.Y. Acad. Sci.,* 174, 903–911, 1970.

10. **Dijkstra, R. G.,** Epidemiological investigations on Listeriosis in cattle, *Proc. 3rd Int. Symp. Listeriosis* (Bilthoven), 1966, 215–224.

11. **Elischerova, K. and Cupkova, E.,** Differentiation of Listeriae isolated in Slovakia according to newly proposed classification, *Folia Microbiol.,* 32(6), 516, 1988.

12. **Eveland, W. C. and Baublis, J. V.,** Two case reports of the assocation of human and canine listeriosis, *Proc. 3rd Int. Symp. Listeriosis* (Bilthoven), 1966, 193–195.

13. **Farber, J. M., Carter, A. O., Varughese, P. V., Ashton, F. E., and Ewan, E. P.,** Listeriosis traced to the consumption of alfalfa tablets and soft cheese, *N. Engl. J. Med.,* 322, 338, 1990.

14. **Ghosh, B. K. and Murray, R. G. E.,** Fine structure of *Listeria monocytogenes* in relation to protoplast formation, *J. Bacteriol.,* 93, 411–426, 1967.

15. **Gram, H. G.,** quoted by Seeliger, H. P. R., in *Listeriosis,* Hafner, New York, 1961, 177.

16. **Jessen, O. and Bojsen-Møller, J.,** The encephalitic type of listeriosis in man, *Proc. 3rd Int. Symp. Listeriosis* (Bilthoven), 1966, 443–445.

17. **Kalvodova, D., Jirasek, A., and Duskova, M.,** Pathogenicity of *Listeria monocytogenes* L Forms under experimental conditions, *Folia Microbiol.,* 32(6), 516, 1988.

18. **Kotlyarova, G. A., Biryulina, T. I., Snegiryova, A. E., Kokorin, I. N., Konstantinova, N. D., et al.,** Mixed infection of BALB/c mice with Rauscher leukaemia virus and L Forms of Listeria, *Eksp. Naya Onkologiya,* 4(1), 21–25, 1982.

19. **Kotlyarova, G. A., Prozorovsky, S. V., Bakulov, I. A., and Chevelev, S. F.,** Biological properties of *Listeria monocytogenes* L Forms, *Vestn. Akad. Med. Nauk SSSR,* 24, 84–90, 1969.

20. **Lang, K.,** Zur Listeriose beim Kleinkind, *Z. Kinderheilkd.,* 76, 17–21, 1955.

21. **Lang, K.,** Listeria infektion als moglich Ursache fruh erworbener Cerebral-schaden, *Z. Kinderheilkd.,* 76, 328–339, 1955.

22. **Louria, D. B., Armstrong, D., Hensle, T., and Blevins, A.,** Listeriosis in patients with neoplastic disease, *Proc. 3rd Int. Symp. Listeriosis* (Bilthoven), 1966, 387–402.

23. **Louria, D. B.,** Discussion, *Proc. 3rd Inst. Symp. Listeriosis* (Bilthoven), 1966, 446.

24. **Mascola, L., Lieb, L., Chiu, J., Fannin, S. L., and Linnan, M. J.,** Listeriosis, an uncommon opportunistic infection in patients with acquired immunodeficiency syndrome, *Am. J. Med.,* 84, 162–164, 1988.

25. **Özgen, H.,** Über die *Listeria monocytogenes,* Inaugural Diss., Gieben, 1951.

26. **Özgen, H.,** Sulfonamide und Antibiotika in Ihrer Bakteriostatischen Wirkung gegen *Listeria monocytogenes, Dtsch. Tierarztl. Wochenschr.,* 59, 339, 1952.

27. **Patočka, F., et al.,** quoted by Seeliger, H., in *Listeriosis,* Hafner, New York, 1961, 16.

28. **PHLS Communicable Disease Surveillance Centre,** Communicable Disease Report January to March 1989, *Community Med.,* 11, 255–259, 1989.

29. **Piffaretti, J. C., Kressebuch, H., Aeschbacher, M., Bille, J., Bannerman, E., Musser, J. M., Selander, R. K., and Rocourt, J.,** Genetic characterization of clones of the bacterium *Listeria monocytogenes* causing epidemic disease, *Proc. Natl. Acad. Sci.,* 86, 3818–3822, 1989.

30. **Prozrovsky, S., Kotljarova, J., Fedotova, I., and Bakulov, I.,** Pathogenicity of Listerial L Forms., in *Spheroplasts, Protoplasts and L Forms of Bacteria, INSERM,* 65, 265–272, 1976.

31. **Reichertz, P. and Seeliger, H. P. R.,** Results of skin tests with different antigens in suspected and proved listeric infections, *Proc. 3rd Int. Symp. Listeriosis* (Bilthoven), 1966, 407–414.

32. **Reiss, H. J., Potel, J., and Krebs, A.,** Granulomatosis infantiseptica (Eine Allgemeininfektion bei Neugeborenen und Sauglingen mit miliaren Granulomen), *Z. Inn. Med.,* 6, 451–457, 1951.

33. **Riancho, J. A., Echevarria, S., Napal, J., Duran, R. M., and Macias, J. G.,** Endocarditis due to *Listeria monocytogenes* and human immunodeficiency virus infection, *Am. J. Med.,* 85, 737, 1988.

34. **Seeliger, H. P. R.,** *Listeriosis,* Hafner, New York, 1961, 139.

35. **Seeliger, H. P. R.,** Opening lecture, *Proc. 3rd Int. Symp. Listeriosis* (Bilthoven), 1966, 5–11.

36. **Seeliger, H. P. R.,** Discussion, *Proc. 3rd Int. Symp. Listeriosis* (Bilthoven), 1966, 291–293.

37. **Suchanova, M. and Patočka, F.,** Pokus o dosazeni L forem *Listeria monocytogenes, Cs. Epidemiol. Mikrobiol. Immunol.,* 6, 133–139, 1957.

38. **Van Driest, G. W.,** Meningitis and sepsis in newborn due to *Listerella monocytogenes, Mschr. Kindergeneesk.,* 16, 101–109, 1948.

39. **Beson, C.,** personal communication.

Chapter 16

LATENCY AND PERSISTENCE

Latency has long been an intriguing facet of virology. The legend examples include the prevalent genital herpes, shingles, the disastrous postmeasles encephalitis, and recently unanticipated long-incubated cases of rabies.[19]

In contrast to persisting viruses, little is known of persisting bacteria and fungi. The data indicate that microbial wall deficiency frequently contributes to *in vivo* longevity. In animal studies the bacteria have often been viable at the terminus of the investigation, with their total survival capacity untested. Persistence in a nonculturable stage, later giving growth, has been reported for decades, as reviewed by McDermott,[12] who called the phenomenon "playing dead".

STAPHYLOCOCCI

Patients who seem cured of staphylococcal skin lesions may become surprisingly reinfected. From where do the staphylococci emerge to cause recurrent boils? Godzeski found L-phase staphylococci in the circulating blood of such persons, the resident site of the organism remaining unidentified.[5]

How is recurrent toxic shock syndrome explained? Sautter, with fluorescent antibody, demonstrated CWD staphylococci in the circulating blood of a woman in whom the condition was recurrent (Figure 1). Again, the site from which the cocci are fed into the blood remains unidentified.[18]

ENTERIC BACILLI

Shigella flexneri L Forms can persist in mice for 75 d, taking residence in liver, kidney, spleen, and lung.[4] In a similar study, L Forms of *Escherichia coli,* injected intraperitoneally into mice, multiplied for 295 d. Their growth in heart, liver, and other tissues was confirmed by culture and electron microscopy.[22]

Are L Forms related to the carrier state in cholera? This was investigated by injecting L Forms of El tor vibrios into rabbit gallbladders to test their persistence, which proved to be 120 d compared with carriage of the classic form of the organism for 35 d.[10]

YERSINIA

A final interpretation awaits the careful long-time study of Granfors and associates.[6] They were not able to grow CWD bacteria from the joints of arthritic individuals who previously had *Yersinia enterocolitica* intestinal infection, yet they did conclusively demonstrate with fluorescent antibody and electrophoresis that Yersinia antigen was in the synovial fluids. Yersinia antigen was demonstrated for 6 months after the initial infection, being present in 1 to 10% of the leukocytes, mostly in the polymorphonuclear cells. Considering the turnover of leukocytes and the constant elimination of antigenic materials, it seems probable that culture for CWD bacteria somehow failed. *Yersinia pestis* can survive long term *in vivo*. Zykin and associates detected L Forms of this species in ticks 3 years after their experimental infection with the classic bacteria.[23] Techniques for growing these fastidious variants have not been perfected. Were pour plates made? Was the yeast extract autoclaved separately? Was the medium aged at least 1 d before use? Methods published as suitable for *in vitro*-made CWD variants do not often suffice for those formed *in vivo*. Were checks for growth made with acridine orange? Did they add a trace of agar or gelatin to broth? Were any anaerobic cultures made? Were the variants serophilic, and if so, was heat-inactivated swine serum used?

FIGURE 1.   Fluorescent antibody shows CWD *Staphylococcus aureus* in blood culture of a patient who experienced recurrent toxic shock syndrome. (Photograph furnished by Robert Sautter.)

ERYSIPELOTHRIX RHUSIOPATHIAE

A long period of persistence was recorded for *Erysipelothrix rhusiopathiae* in experimental infection in rats. In the 8th month CWD forms resided in blood, liver, and bone. At 15 months they were still in liver and bone marrow. The bacillary forms which had been injected disappeared much earlier (Figure 2).[21] This can explain the chronic joint disease which is a major problem in the livestock industry.

GLASSER'S DISEASE OF SWINE

This disease involves multiple systems in the animals, causing lameness, peritonitis, and pericarditis. Neil and co-workers[14] found that the disease persists in animals whose tissues show only round, spheroplast-like forms of *Hemophilus suis*. Identification of the round forms in sacrificed animals was made with fluorescent antibody.

PSEUDOMONAS PSEUDOMALLEI

Kishimoto and Eveland made a finding which may explain the relapses or chronicity complicating meliodosis. Alveolar macrophages may not destroy the pathogen but, rather, continue to harbor the organisms in a wall defective state. Their experimental animal was the rabbit.[9]

RICKETTSIA PROWAZEKII

Spheroplast-like organisms developed from the virulent Breinl rickettsial strain after phagocytosis by guinea pig macrophages. Since the variants were found 2 months after infection, it suggests a mode to maintain latency.[17]

ACTINOMYCES AND NOCARDIA

The ''sulfur granules'' characteristic of actinomycosis resulted from persisting CWD *Nocardia caviae*, rather than from the classic stage (see Frontispiece).[1] Their conclusions were based on examining more than 1000 mice.

STREPTOCOCCI

In a Russian study, stable L Forms of Group A streptococci given as a single i.p. injection to mice found long-term residence in many tissues. The antigen, minimal in the liver at 10 weeks, increased thereafter. Necrosis of both hepatic parenchyma cells and blood vessels occurred. By 13 weeks, membranous glomerulonephritis was found. In the heart, L

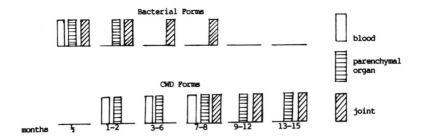

FIGURE 2. Persistence and isolation of *Erysipelothrix rhusiopathiae*, bacterial and CWD forms after experimental infection of rats. Identity of CWD forms was made by precipitin arcs with hyperimmune serum. (Toshkov, A., Mihailova, L., Cherepova, N., and Gulubov, S., *Zentralbl. Bakteriol. Parasitenkd. Infektionskr. Hyg., Abt. 1 Orig. A.,* 233, 370–375, 1975.)

Form antigen reached a maximum in 13 weeks and did not decrease during the remainder of 52 weeks. At that time, antigen was in pericardium and myocardium, most often perivascularly. In the spleen, the L Forms initially were within the Blast cells (Figure 3), and later predominately extracellular. Morphological changes accompanied proclivity for intracellular vs. extracellular growth.[8]

In a similar study in India, stable streptococcal L Forms were given intravenously to monkeys. The organisms persisted in the circulating blood for as long as 10 weeks and in the bone marrow for 4 months.[13] One animal developed a massive myocardial lesion. L Forms of Group A streptococci resist immune attack of mouse peritoneal macrophages for at least 6 to 7 d.[15]

MICROBIAL PERSISTENCE RELATED TO INTRACELLULAR SURVIVAL OF CWD FORMS

Currently, microbial persisters are considered CWD variants resting within body cells. This remarkable situation involves disappearance of a pathogen, with cultures negative even when entire organs of sacrificed animals are cultured. After a variable interval the pathogen reappears, often to cause fulminating disease. Before disappearing, the organism can be antibiotic sensitive and will exhibit the same antibiotic sensitivity after its reincarnation. Such microbial persisters can survive concentrations of antibiotics to which the bacterium is highly sensitive. As reviewed by McDermott,[12] this survival cannot be explained by lack of penetration of the antibiotic into fluids, exudates, or cells. Hibernation as an antibiotic-tolerant CWD stage appears to be a logical explanation. The phenomenon has been studied experimentally for the staphylococci,[7] for the tubercle bacillus,[11] Group A streptococci,[24] and pseudotuberculosis.[16] When reinitiating classic growth, the microbe is susceptible to low concentrations of an antibiotic it recently resisted in a 100 times greater concentration.

The interesting data of Braude[2] seem related to microbial persistence in a highly fastidious state. When Braude inoculated a stable L Form of *Proteus mirabilis* into rats, bladder stones resulted only when the organisms had apparently disappeared from the animal; i.e., no growth developed in cultures. L Forms could be recovered from kidney, bladder, and urine for only 1 week; stones first appeared at 2 weeks, becoming more numerous over a period of 2 months. The L Form had demonstrated *in vitro* sensitivity to chloramphenicol, and when this agent was given 1 week after infection, formation of stones was almost completely prevented. The conclusion seems justified that the antibiotic was acting on viable organisms which could not be cultured. Other careful experiments showed that the stones were not the result of urease in dead organisms.

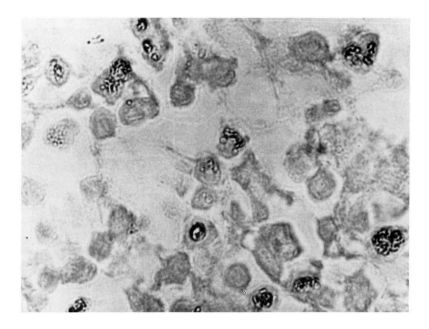

FIGURE 3.   Fluorescent antibody revealing L Forms of Group A streptococci 13 weeks after given intraperitoneally to mouse. (Magnification × 1200.)(Photographs furnished by Y. U. Vulfovich and B. Gusman.)

LATENCY IN TUBERCULOSIS

When the lungs of 44 persons who spontaneously recovered from tuberculosis were examined to test for viable tubercle bacilli, L Forms of *Mycobacterium tuberculosis* were cultured from 85% of the individuals.[20]

SUMMARY

CWD stages can persist *in vivo* for longer periods than the classic stage and can be responsible for lesions considered characteristic for the microbe. Survival may be in blood, liver, gallbladder, peritoneal macrophages, or bone marrow. Many data have been collected on streptococci, staphylococci, the enteric bacilli, and *E. rhusiopathiae* in arthritis.

REFERENCES

1. **Beaman, B. L. and Scates, S. M.,** Role of L Forms of *Nocardia caviae* in the development of chronic mycetomas in normal and immunodeficient murine models, *Infect. Immun.,* 33, 893–907, 1981.
2. **Braude, A. I.,** Production of bladder stones by L Forms, *Ann. N.Y. Acad. Sci.,* 174, 896–902, 1970.
3. **Carrère, L., Roux, J., and Mandin, J.,** Culture des organismes L sur membrane chorio-allantoïde d'embryon de poulet, *Montpellier Med.,* 68, 438–445, 1955.
4. **Fedorova, Z. F.,** In vivo study of the properties of *Shigella flexneri* L Forms biological features, *Zh. Mikrobiol. Epidemiol. Immunobiol.,* 6, 113–114, 1986.
5. **Godzeski, C. W.,** In vivo persistence of L phase bacteria, in *Microbial Protoplasts, Spheroplasts and L Forms,* Guze, L. B., Ed., Williams & Wilkins, Baltimore, 1968, 379–390.
6. **Granfors, K., Jalkanen, S., Essen, R. V., Lahesmaa-Rantala, R., Isomaki, O., Pekkola-Heino, K., Merilahti-Palo, R., Saario, R., Isomaki, H., and Toivanen, A.,** Yersinia antigens in synovial-fluid cells from patients with reactive arthritis, *N. Engl. J. Med.,* 320, 216–220, 1989.

7. **Hamburger, M. and Carleton, J.,** Staphylococcal spheroplasts as persisters, in *Microbial Protoplasts, Spheroplasts, and L Forms,* Guze, L. B., Ed., Williams & Wilkins, Baltimore, 1968, 221–229.

8. **Kagan, G., Vulfovitch, Y., Gusman, B., and Raskova, T.,** Persistence and pathological effect of strepto-coccal L Forms in vivo, in *Spheroplasts, Protoplasts and L Forms of Bacteria, INSERM,* 65, 247–258, 1976.

9. **Kishimoto, R. A. and Eveland, W. C.,** Induction of microbial variants of *Pseudomonas pseudomallei* in cultured rabbit alveolar macrophages, *Can. J. Microbiol.,* 21, 2112–2115, 1975.

10. **Lomov, Y. M., Tynkevich, N. K., Buchul, K. G., and Golubkova, L. A.,** Features of formation persistence and reversion of *Vibrio cholerae* L Forms in experimental animals, *Mikrobiol. Zh.,* 44(6), 79–84, 1982.

11. **McCune, R. M., Jr., Tompsett, R., and McDermott, W.,** The fate of *Mycobacterium tuberculosis* in mouse tissues as determined by the microbial enumeration technique. II. The conversion of tuberculous infection to the latent state by the administration of pyrazinamide and a companion drug, *J. Exp. Med.,* 104, 763–802, 1956.

12. **McDermott, W.,** Microbial persistence, *Yale J. Biol. Med.,* 30, 257–291, 1958.

13. **Mohan, C., Ganguly, N. K., and Chakravarti, R. N.,** Experimental production of cardiac injury in Rhesus monkeys by L Forms of Group A streptococci, *Indian J. Med. Res.,* 86, 361–371, 1987.

14. **Neil, D. H., McKay, K. A., L'Ecuyer, C., and Corner, A. H.,** Glasser's disease of swine produced by the intracheal inoculation of *Haemophilus suis, Can. J. Comp. Med.,* 33, 187–193, 1969.

15. **Neustroeva, V. V., Vulfovich, Y. V., and Churilova, N. S.,** Study of the survival of the L Forms of Group A streptococci in peritoneal macrophages of mice., *Zh. Mikrobiol. Epidemiol. Immunol.,* 12, 45–46, 1983.

16. **Pierce-Chase, C. A., Fauve, R. M., and Dubos, R. J.,** *Corynebacteria pseudotuberculosis* in mice. I. Comparative susceptibility of mouse strains to experimental infection with *Corynebacterium kutscheri, J. Exp. Med.,*120, 267–281, 1964.

17. **Popov, V. L., Prozorovsky, S. V., Vovk, D. A., Kekcheeva, N. K., Smirnova, N. S., and Barkhatova, O. I.,** Electron microscopic analysis of in vitro interaction of *Rickettsia prowazekii* with guinea pig macro-phages. I. Macrophages from nonimmune animals, *Acta Virol.,* 31(1), 53–58, 1987.

18. **Sautter, R.,** Vaginal Ecology in Relation to Toxic Shock Syndrome, Ph.D. dissertation, Wayne State University, Detroit, 1983.

19. **Smith, J. S., Fishbein, D. B., Rupprecht, C. E., and Clark, K.,** Unexplained rabies in three immigrants in the United States, *N. Engl. J. Med.,* 324, 205, 1991.

20. **Solovyeva, I. P., Dorozhkova, I. R., and Biron, M. G.,** Clinicoanatomical and microbiological investigation of residual lesions in spontaneously recovered tuberculosis patients, *Probl. Tuberk.,*8, 58–61, 1986.

21. **Toshkov, A., Mihailova, L., Cherepova, N., and Gulubov, S.,** Persistence and isolation of L cycle forms of *Erysipelothrix rhusiopathiae* in experimental infection, *Zentralbl. Bakteriol. Parasitenkd. Infektionskr. Hyg., Abt. 1 Orig. A,* 233, 370–375, 1975.

22. **Toshkov, A., Ivanova, E., Mikhailova, L., and Gumpert, I.,** Persistence of L Forms of *E. coli* in mice, *Acta Microbiol. Bulg.,* 11, 78–85, 1982.

23. **Zykin, L. F., Dunaev, G. S., Sayamov, S. R., and Sokolov, P. S.,** *Yersinia pestis* L Forms in rodents and ectoparasites, *Zh. Mikrobiol. Epidemiol. Immunobiol.,* 2, 36–40, 1989.

24. **Godzeski, C. W.,** personal communication.

Chapter 17

MENINGITIS AND ASSOCIATED CONDITIONS

It has been reported that 18% of meningitis spinal fluids are culture negative in untreated cases and 30% if the patient has received therapy.[8] The elevated lysozyme levels characteristic of meningitis spinal fluid may explain the atypical growth.[7] That the organism is not isolated to permit early specific therapy in many cases is shown by the permanent brain damage resulting in 38% of the 10,000 children in the U.S. who annually suffer meningitis.[2] Most of the sequelae are prevented by early identification of the pathogen.

There is every reason to believe that stains, including fluorescent antibody and tests for microbial antigens, namely, countercurrent electrophoresis and latex agglutination,[16] will identify CWD microbes as well as classic in spinal fluids.

MENINGITIS SERIES

Three meningitis series have recorded a significant percentage of CWD strains. In a large series cultured by Kagan et al.,[9] spinal fluid from 60 patients yielded wall-deficient growth exclusively. Conflicting with current concepts, all the CWD strains in this series were not penicillin resistant. Eight patients with only CWD variants in spinal fluid before therapy responded to massive doses of penicillin, 200,000 U/kg of body weight. Seven other patients had variants perhaps antibiotic induced, since pretreatment cultures had classic bacteria. The variants in five of these cases did not respond to penicillin. Brain abscesses which ruptured into the subarachnoid space commonly yielded only CWD organisms.

Reverting variants of the tubercle bacilli have been followed in spinal fluid by Alexander-Jackson[1] and others. Intensified modifications of acid-fast stains reveal small colonies of acid-fast variants in tuberculous meningitis, as they do in pleural fluids and sputa.

Calderón and associates[5] studied 89 cases of clinically typical meningitis, some of which had received antibiotic prior to hospitalization. In 21 instances, growth occurred only in hypertonic medium.

Investigators need more clues for detection of microbes. Cells may be scarce in the spinal fluid of meningitis caused by classic bacteria and fungi;[19] hence, leukocytes cannot inevitably be expected in infection by wall-deficient microbes. The levels of sugar and of protein in spinal fluid are not entirely reliable to differentiate bacterial from viral infection. There is need for a sensitive biochemical test to detect bacterial components, such as diaminopimelic acid or muramic acid.

Timakov's laboratory isolated L Forms in 56 cases of meningitis. Of these only ten reverted.[17]

SPECIES WALL-DEFICIENT IN MENINGITIS

While many CWD forms isolated in meningitis have failed to revert, others have been identified by reversion or by fluorescent antibody or had their identity suggested by finding the classic organism in the patient's blood. The species involved are described below.

STAPHYLOCOCCI

Staphylococcus aureus in the CWD stage is prominent among strains cultured from spinal fluid.[5,9,14,17] These were identified by reversion. An organism in one fatal case had unusual characteristics. Initially, growth was only under anaerobic conditions, extending

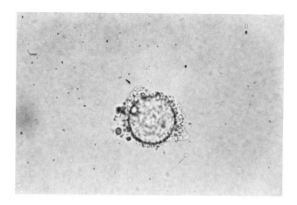

FIGURE 1.   Beginning growth of a CWD colony from spinal fluid of a fatal brain abscess. The revertant was a coagulase-positive staphylococcus.

around L-bodies (Figure 1). Reversion to a coagulase-positive staphylococcus occurred after many transfers. Autopsy revealed an abscess typical of staphylococcal infection, although routine cultures of the abscess revealed no growth, and smears were negative for classic organisms.

STREPTOCOCCI

Group A streptococci, as expected, are among the CWD inhabitants of meningitis.[9,14] The pneumococcus, wall deficient, is conspicuous in meningitis. One study indicated that CWD pneumococci can be transferred to cause infection in a second individual. A man and his relative were hospitalized within a 48-h period. Neither had received any antibiotic before spinal fluid withdrawal, and in both cases the disease was fulminant as long as penicillin alone was administered. An additional interesting point is that the infection was not controlled with chloramphenicol alone, but also required massive doses of penicillin. These two cases, which required dual antibiotics, support the philosophy of Nativelle and Deparis that "therapy must be directed against the whole organism," i.e., the classic and variant stages.[15]

A third case of variant pneumococcal meningitis was also not antibiotic induced. A woman hospitalized for arteriosclerotic heart disease developed both pneumonia and meningitis, and, as in the cases above, her infection was not controlled with heavy doses of penicillin but did respond to administration of chloramphenicol. In all three cases, primary cultures on blood and chocolate agar and in semisolid agar with ascitic fluid showed only wall deficient colonies. Transition to classic pneumococci occurred with subculturing.

Pneumococcal CWD forms can vary in penicillin sensitivity, as shown by transitional forms in another case found only after penicillin administration. This patient recovered with penicillin therapy alone.[4]

PSEUDOMONAS

Wall-deficient *Pseudomonas aeruginosa* is more frequently encountered in spinal fluid than might be anticipated.[5,9] One patient with culture-negative spinal fluid and blood actually suffered from meningitis and septicemia. This young man, during an altercation, had a screwdriver thrust through his cheek, connecting the nasal cavity and subarachnoid space. Because of the danger of meningitis from oral and nasal organisms draining into his spinal fluid, he was placed on a wide-spectrum antibiotic. However, on the fourth day, he was hospitalized with a temperature of 105°F. His spinal fluid cells numbered 2200/mm,[3] although cultures remained negative. Duplicate blood cultures were also negative for classic organisms. A prolonged examination of the Gram stain of the spinal fluid sediment showed not only pleomorphic forms, but also rare typical clusters of staphylococci, expected considering the direct connection between his nasal sinuses and spinal canal. Cultures in hypertonic medium of both blood and spinal fluid grew forms which reverted to *P. aeruginosa*. Disk antibiotic sensitivity tests with both the classic and variant growth on the surface of serum agar indicated that colymycin was a suitable antibiotic for inhibition of both the classic and the variant stage. Although it was never possible by culture to show a staphylococcal variant, it was necessary to maintain the patient on both keflin and colymycin during a prolonged hospitalization with multiple surgical procedures.

MYCOBACTERIA

Even with the auramine rhodamine stain, in many cases of tuberculous meningitis the tubercle bacillus is not detected. Alexander-Jackson[1] was one of the first to follow reversion of any organism from clinical material, observing pleomorphic masses differentiate into acid fast rods in spinal fluid of tuberculous meningitis.

NOCARDIA

CWD *Nocardia steroides* was implicated by Beaman in a fatal brain abscess afflicting a man who had received steroid therapy.[3] The variant organisms in his spinal fluid numbered 10^4. Pal studied a similar case where a Nocardia species was wall deficient in repeated samples of spinal fluid.[20]

NEISSERIA MENINGITIDIS

The meningococcus, as expected, is an important variant in meningitis. Timakov found this species not only wall deficient in disease of man, but also showed that these variants produce a pronounced encephalopathy in rabbits.[17] This species spontaneously produces budding forms, giant and microcells when growing in human tissue culture.[18]

Other organisms found wall deficient in meningitis include *Proteus mirabilis*,[5] *Klebsiella*,[5,21] α-hemolytic streptococci,[5] *Salmonella typhimurium*,[5] and *Escherichia coli*.[5] One reported case was polymicrobic with CWD *E. coli* and streptococci.[5] Fungi and Listeria are also implicated (see related chapters). CWD *Hemophilus influenzae* can appear alone in meningitis.[9,13] The variant usually also accompanies the classic stage when it is present, as found by Karris[10] studying spinal fluids collected before therapy. It was apparent that the variants in these cases were susceptible to therapy appropriate for the bacilli, as the clinical response was good in each case.

A unique report thus far comes from India. Pal[20] studied repeated samples of spinal fluid from a meningitis case which suggested that a symbiosis flourished between a CWD micrococcus and an unidentified mucoid Gram-negative rod.

POSTOPERATIVE INFECTIONS IN THE BRAIN

A series of five cases in which CWD organisms occurred exclusively concerned removal of cysts or correction of hydrocephalus in children. The revertants were staphylococcus or pseudomonas.[11] From a similar case we isolated a fungus which reacted with antibody to *Histoplasma capsulatum*. Similarly, Charache found a CWD infection in an atrioventricular shunt.[6] A streptococcus was the revertant in another case where routine culture of a subdermal hematoma was negative (Figure 2).[12]

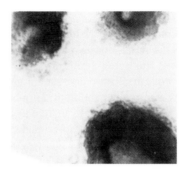

FIGURE 2.   L-colonies of streptococci from subdural fluid. (Magnification × 170.) (From Kenny, J. F., *South. Med. J.*, 71, 180–190, 1978. With permission.)

SUMMARY

The CWD variants of many bacterial genera, including Escherichia, Neisseria, Nocardia, Proteus, Pseudomonas, Staphylococcus, and Streptococcus, have been found in cases of meningitis, both before and despite antibiotic therapy. Similarly, fungal meningitis shows only CWD stages of the organisms.

Because of the difficulty of cultivating these variants from spinal fluid, these "culture-negative" cases may prompt an initial diagnosis of "aseptic meningitis", yet ultimately present acute symptoms and pathology of the more lethal infectious disease. Dual antibiotic administration is often required to control both the classic and atypical forms of the pathogen.

REFERENCES

1. **Alexander-Jackson, E.**, A hitherto undemonstrated form of *Mycobacterium tuberculosis, Ann. N.Y. Acad. Sci.*, 46, 127–152, 1945.
2. Changes Pending for Pertussis, Haemophilus Vaccines, *ASM News*, Vol. 56(12), 634–635, 1990.
3. **Beaman, B. L.**, Nocardiosis: role of the CWD state of Nocardia, in *Cell Wall Deficient Bacteria,* Domingue, G. J., Ed., Addison-Wesley, Reading, MA, 1982, 231–255.
4. **Brzin, B.**, Spheroplast-like cells of pneumococci, *Experientia,* 20, 135, 1964.
5. **Calderón, E., Albuerne, A., González, S., and Winkler, L.**, Cell wall deficient bacteria (L Forms) in meningitis, *Rev. Lat. Am. Microbiol.,* 13, 95–100, 1971.
6. **Charache, P.**, Atypical bacterial forms in human disease, in *Microbial Protoplasts, Spheroplasts and L Forms,* Guze, L. B., Ed., Williams & Wilkins, Baltimore, 1968, 484–494.
7. **Grekova, N. A., Tolmacheva, T. A., and Vershilova, P. A.**, On the pathogenicity of nonstable *Brucella* L Forms and their revertants, *J. Hyg. Epidemiol. Microbiol. Immunol.,* 23, 129–134, 1979.
8. **Jonsson, M. and Alvin, A.**, A 12-year review of acute bacterial meningitis in Stockholm, *Scand. J. Infect. Dis.,* 3, 141–150, 1971.
9. **Kagan, G. Yu., Koptelova, E. I., and Pokrovsky, B. M.**, Isolation of pleuropneumonia-like organisms, L Forms and heteromorphous growth of bacteria from the cerebrospinal fluid of patients with septic meningitis, *J. Hyg. Epidemiol. Microbiol. Immunol. (Russian),* 9, 310–317, 1965.
10. **Karris, G.**, as quoted by Mattman, L. H., L Forms isolated from infections, in *Microbial Protoplasts, Spheroplasts, and L Forms,* Guze, L. B., Ed., Williams & Wilkins, Baltimore, 1968, 472–483.
11. **Kenny, J. F.**, Bacterial variants in central nervous system infections in infants and children, *J. Pediatr.,* 83, 531–542, 1973.
12. **Kenny, J. F.**, Role of cell wall defective microbial variants in human infections, *South. Med. J.,* 71, 180–190, 1978.

13. **LeBar, W. D., Mattman, L. H., and Nathan, L. E.,** L phase *Hemophilus influenzae* in a case of meningitis, *Bacteriol. Proc. Am. Soc. Microbiol.,* 1977, p. G1.

14. **Mattman, L. H.,** L Forms isolated from infections, in *Microbial Protoplasts, Spheroplasts and L Forms,* Guze, L. B., Ed., Williams & Wilkins, Baltimore, 1968, 472–483.

15. **Nativelle, R. and Deparis, M.,** Formes évolutives des bactéries dans les hémocultures, *Presse Med.,* 68, 571–574, 1960.

16. **Robinson, R. O. and Roberts, H.,** Acute bacterial meningitis. I. Diagnosis, development, *Med. Child Neurol.,* 32, 79–86, 1990.

17. **Timakov, V.,** Bacterial L Forms in pathology, in *Spheroplasts, Protoplasts and L Forms of Bacteria,* Roux, J., Ed., INSERM, Paris, 1976, pp. 357–364.

18. **Vysotskii, V. V., Ruzal, G. I., Galeeva, O. P., and Smirnova-Mutusheva, M. A.,** Ultrastructure of meningococcal cells during incubation in continuous human amnion cell culture FL, *Zh. Mikrobiol. Epidemiol. Immunobiol.,* 1, 40–44, 1986.

19. **Wolf, R. W. and Birgara, C. A.,** Meningococcal infections at an Army training center, *Am. J. Med.,* 44, 243–256, 1968.

20. **Pal, S. R.,** communication to B. L. Beaman.

21. **Rostam-Abadi Sofla, H.,** personal communication.

Chapter 18

RHEUMATIC FEVER AND ERYSIPELAS

CROSS-REACTING ANTIBODY

Significant aspects of the enigma of rheumatic fever have been elucidated, and bacteria with incomplete walls seem to participate as much or more than classic streptococci. The appearance of pancarditis 2 to 3 weeks after a throat infection with Group A streptococci is usually associated with a rise in streptococcal antibody.[20,29] Precise work by Kaplan et al.,[18,19] Freimer and Zabriskie,[8] and Zabriskie and co-workers[40] long ago showed that antibody against the streptococcal membrane unfortunately cross-reacts with the membrane of cardiac cells. Endocarditis has been produced in mice just from the cytoplasmic membranes of streptococcal L Forms.[26]

In additional studies incriminating immunoglobulins, Lannigan and Zaki used ferritin-labeled antibody and electron microscopy.[22] Ferritin-tagged antibody to human gamma globulin attaches to the endocardium of rheumatic fever patients, revealing localized antibody.

Cross-reacting antibody of still a different type was shown for the RNA of streptococci and cardiac cells by Wilhelm and Dippell.[38] Guinea pigs injected with streptococcal RNA develop antibody strongly reactive with cardiac tissue. Lastly, Goldstein et al.[10] found an immunological reaction between the structural glycoproteins of heart valves and the specific carbohydrate of Group A streptococci. Therefore, many facets indicate that antibody produced in response to a streptococcal infection combines with cardiac tissue and is responsible for both the initial damage and the recrudescence of symptoms following reinfection with Group A streptococci.

ROLE OF PERSISTING COCCI AND CWD VARIANTS IN ANTIBODY STIMULATION

It seemed reasonable to many individuals who studied rheumatic fever that the patient's copious antibody results from bacterial persistence after the individual appears well.[2,6,24,27] Active infection must progress for at least 10 d, although the symptoms subside.[32] The probability of complications following a respiratory infection with Group A streptococci increases markedly if the patient is treated inadequately.[4] Evidence grows that continued multiplication of the streptococcus, predominately as a variant, is responsible.

PERTINENT ORGANISMS IN CIRCULATING BLOOD

CWD streptococci were found early in the blood in scarlet fever by Wittler et al.[39] and by Klodnitskaya.[21] Klodnitskaya noted that penicillin therapy did not prevent the development of the circulating variant, persisting up to the seventh day after the onset of illness. Timakov and Kagan cultured the blood of 19 rheumatic fever patients,[35] isolating L Forms in 12 instances and classic streptococci once. Sixteen control cultures yielded no growth. Electron micrograph studies also reveal classic and L Forms of streptococci in the plasma and leukocytes of rheumatic fever patients.[1]

Earlier workers,[5,14] incubating cultures for long periods, reported that streptococcal bacteremia characterizes scarlet fever. The belated growth of the cocci in culture suggests that they circulate as CWD stages. Streptococci have been found in cardiac lesions per se in small numbers in some cases of rheumatic fever.[5,14,33]

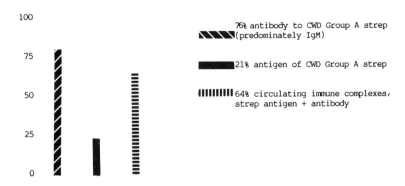

FIGURE 1. Both antigen and antibody of CWD Group A streptococci are found when rheumatic fever symptoms persist. (From Goncharova, S. A., Gorina, L. G., Vulfovich, Yu, V., Zheverzheeva, I. V., and Labinskaya, A. S., *Vestn. Akad. Med. Nauk. SSSR,* 6, 34–36, 1989. With permission.)

In studying 1343 rheumatic fever patients, Goncharova and associates[11] detected high titers of antibody to streptococcus L Forms in 76% of cases during acute disease (Figure 1). In continuing illness, antigens of both the L Form and classic stage persisted. Earlier studies had shown that the major antigen in the circulating immune complex in rheumatic fever is Group A L Form antigen.[12]

CWD STREPTOCOCCI IN CARDIAC TISSUES

Physical evidence incriminating CWD variants in rheumatic endocarditis was found by Starnes.[31] In several cases, sections of rheumatic heart valves contained pleomorphic forms reacting specifically with fluorescent antibody to Group A streptococci (Figure 2). In most of the reacting areas, the organisms had morphology characteristic of CWD variants, and their reaction with specific antibody indicated the possession of some cell wall. Morphologically, the fluorescent antibody staining bodies were indistinguishable from colonies of penicillin-induced L-phase Group A streptococci.

Starnes also found that fluorescent-antibody staining spheres were associated with the characteristic Aschoff body of rheumatic fever. Lannigan and Zaki,[22] using the electron microscope, believed the structures associated with the rheumatic heart Aschoff bodies could be colonies of streptococcal variants.

Tissue culture studies show that CWD Group A streptococci can invade and destroy cardiac cells (Figure 3).[23] Similarly, the ability of nonreverting CWD forms of Group A streptococci to damage mammalian cardiac tissue was explored in a long-term Russian study. A single intraperitoneal injection of the variants into 250 mice produced pancarditis leading to cellular death and sclerosis (Figure 4).[17] The antigen of the streptococcus could be detected by immunofluorescence throughout the 12 months of the experiment.[15] CWD Group A streptococci in the cardiac muscle of the mouse are shown in Figure 5.[15] In a later study the antigen of L Form Group A streptococci persisted in mice, rabbits, and rats for at least 2 years.[36]

Investigators also use the monkey to demonstrate rheumatic fever complications. Cardiomegaly can be produced in macaque species (Figure 6).[34]

Another study was beautifully designed by Wedum,[37] who grew 36 pieces of atrium and ventricle from rheumatic hearts as tissue cultures. In the antibiotic-free culture medium,

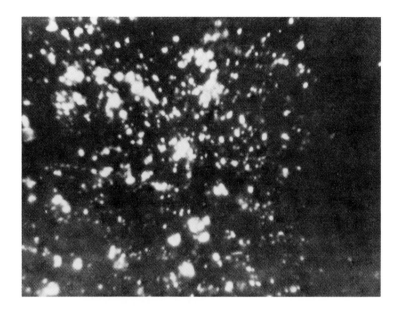

FIGURE 2. Fluorescence with antiserum to Group A streptococci in section of rheumatic heart counterstained with chelated azo dye to quench nonspecific fluorescence. (Micrograph provided by R. Starnes.)

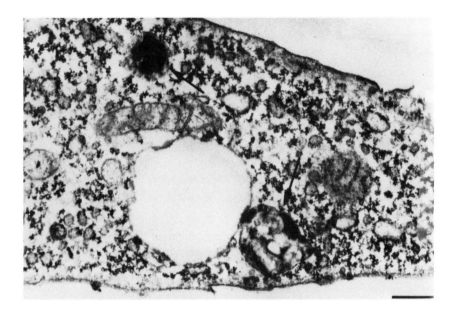

FIGURE 3. A stable L Form of Group A streptococci has invaded a cardiac cell in tissue culture. (From Leon, O. and Panos, C., *Infect. Immun.*, 13, 252–262, 1976. With permission.)

no recognized colonies developed. However, cytotoxicity was observed in tissues, in which no virus was demonstrable. It would be interesting to repeat this study, with immunological methods, testing for streptococcal antigens in the virus-like particles observed. Using fluorescent muramidase staining, one could seek tiny units of intracellular and extracellular CWD forms. Minute forms as well as giant spheres characterize L growth of Group A β-hemolytic streptococci (Figure 7).[28]

FIGURE 4. Sclerosis of mouse cardiac connective tissue follows one injection of CWD Group A streptococci. Silver impregnation stain. (Magnification × 360.)(From Kagan, G., Gusman, B., Vulfovitch, Yu., and Besouglova, T., *INSERM*, 65, 259–264, 1976. With permission.)

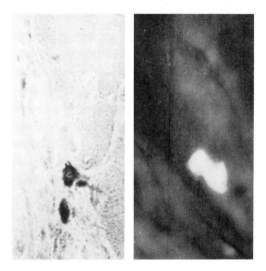

FIGURE 5. Microcolony of CWD Group A streptococcus in myocardium of white mouse. (Left) Histological staining; (right) fluorescent antibody staining. (Magnification × 500.)(From Gusman, B. S., Vulfovitch, Yu. V., and Kagan, G. Ya., *Arch. Pathol.*, 12, 26–32, 1980. With permission.)

A factor which consistently complicates recognition of β-hemolytic streptococcal L Forms from infection is that the reverted cocci are seldom typical β-hemolytic organisms. Revertants are usually α-hemolytic and vary serologically from the parent type.[16,30]

VARIANTS IN SYNOVIAL FLUIDS OF RHEUMATIC FEVER PATIENTS

The inflamed joints of seven children with rheumatic fever were found to harbor CWD β-hemolytic streptococci. The variants were isolated in a semiliquid medium containing 20% horse serum and 0.075% agar.[25]

ERYSIPELAS AND RECURRENT ERYSIPELAS

Erysipelas has a tendency to recur in the same area,[7] making recurrent erysipelas a mystery. Obviously the patient harbors the streptococcus in some tissue, but where, and in the classic or wall-deficient stage? Investigation reveals that most of these individuals (88%)

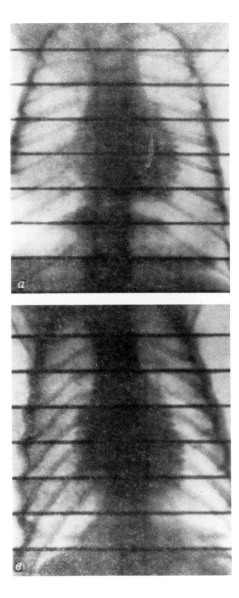

FIGURE 6. Changes in heart size of a monkey given L Forms of Group A streptococci. No reversion of the organism to classic streptococci occurred. (From Timakov, B. and Kagan, G. Ya., *L Forms Bacteria and Species of Mycoplasma,* Moscow, 1973. With permission.)

have antibody to the L Form of Group A hemolytic streptococci in their blood, usually bound in an immune complex. Only one of 21 blood donor controls had this antibody.[13]

Another study implicating streptococcal L Forms in recurrent erysipelas concerned 156 patients, of whom 116 had attacks recurring at least 3 times a year. These patients had high levels of streptococcal L Form antigen in their circulating blood. The antigen was significantly lowered after vaccine therapy.[3]

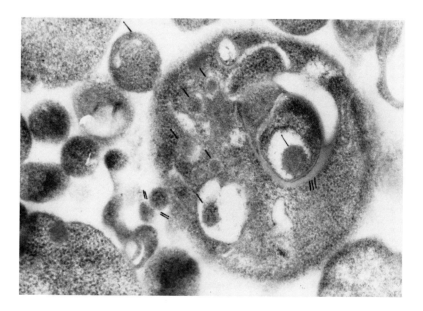

FIGURE 7. Note variation in size of spherical units in culture of L Forms of β-hemolytic streptococci Group A and their formation within a large body. (From Prosorovsky, S. V., Katz, L. N., and Kagan, G. Ya., *AMS USSR H' Meditsina*, 1981. With permission.)

SUMMARY

The strange situation in rheumatic fever of continuous vigorous antibody production with apparent absence of the pathogen seems to be clarified by finding CWD streptococci and their antigens in cardiac tissue, circulating blood, and inflamed joints.

The persistence of streptococcal antigens, thus, stimulating the damaging antibodies, can be explained by the difficulty the body experiences in eliminating the organisms.[9] Why are some variants still found after penicillin therapy when no cardiac sequelae result? One may hypothesize that the variants persisting after antibiotic administration differ from those present without therapy.

As in rheumatic fever, in recurrent erysipelas there is persistence of L Form antigen of hemolytic streptococci.

REFERENCES

1. **Ardamatsky, N. A., Delektorsky, V. V., and Vladimirova, L. N.,** Of microorganisms in the peripheral blood of patients with certain internal diseases, *Klin. Med.* (Moscow), 59(7), 103–106, 1981.
2. **Catanzaro, F. J., Stetson, C. A., Morris, A. J., Chamovitz, R., Rammelkamp, C. H., Jr., Stolzer, B. L., and Perry, W. D.,** The role of the streptococcus in the pathogenesis of rheumatic fever, *Am. J. Med.,* 17, 749–756, 1954.
3. **Cherkasov, U. L. and Gavrilova, G. A.,** Clinical and pathogenetic significance of infection of erysipelas patients, *Klin. Med.,* 64(3), 51–54, 1986.
4. **Cluff, L. E. and Johnson, J. E.,** Post-streptococcal disease, in *Immunological Disease,* Samter, M., Ed., Little, Brown, Boston, 1965.
5. **Collis, W. R. F.,** Bacteriology of rheumatic fever, *Lancet,* 2, 817–820, 1939.
6. **Denny, F. W., Jr., Wannamaker, L. W., Brink, W. R., Rammelkamp. C. H., Jr., and Custer, E. A.,** Prevention of rheumatic fever, treatment of the preceding streptococcal infection, *JAMA,* 143, 151–153, 1950.

7. **Fitzpatrick, T. B., Eisen, A. Z., Wolff, K., Freedberg, I. M., and Austen, K. F., Eds.,** *Dermatology in General Medicine,* 3rd ed., McGraw-Hill, New York, 1987.

8. **Freimer, E. H. and Zabriskie, J. B.,** An immunological relationship between the protoplast membrane of group A streptococci and the sarcolemma, the cell membrane of cardiac muscle, in *Microbial Protoplasts, Spheroplasts, and L Forms,* Guze, L. B., Ed., Williams & Wilkins, Baltimore, 1968, 356–371.

9. **Gilbert, J. and Fox, A.,** Elimination of Group A streptococcal cell walls from mammalian tissues, *Infect. Immun.,* 55, 1526–1528, 1987.

10. **Goldstein, I., Halpern, B., and Robert, L.,** Immunological relationship between streptococcus A polysaccharide and the structural glycoproteins of heart valve, *Nature (London),* 213, 44–47, 1967.

11. **Goncharova, S. A., Gorina, L. G., Vulfovich, Yu. V., Zheverzheeva, I. V., and Labinskaya, A. S.,** Serologic parameters of Group A streptococcus L Forms persistence in rheumatic patients, *Vestn. Akad. Med. Nauk. SSSR,* 6, 34–36, 1989.

12. **Gorina, L. G., Zheverzheeva, I. V., Chumachenko, N. V., Goncharova, S. A., and Labinskaya, A. S.,** Detection of hemolytic streptococcus Group A L Forms in the sera of rheumatic patients after the separation of immune complexes, *Zh. Mikrobiol. Epidemiol. Immunobiol.,* 7, 42–45, 1981.

13. **Gorina, L. G., Kagan, G. Ya., Gavrilova, G. A., Goncharova, S. A., and Cherkasov, V. L.,** Free and bound antigen of Group A *Streptococcus haemolyticus* L Forms and the level of circulating immune complexes in erysipelas patients, *Zh. Mikrobiol. Epidemiol. Immunol.,* 9, 106–109, 1983.

14. **Green, C. A.,** Researches into the aetiology of acute rheumatism; rheumatic carditis; post-mortem investigation of nine consecutive cases, *Ann. Rheum. Dis.,* 1, 86–98, 1939.

15. **Gusman, B. S., Vulfovich, Yu. V., and Kagan, G. Ya.,** Morphogenesis of the heart alteration in experimental infection with streptococcus L Form, *Arch. Pathol.,* 12, 26–32, 1980.

16. **Kagan, G. Ya. and Mikhailova, V. S.,** Biological properties of streptococci reversed from L Forms, isolated from patients suffering from rheumatism and endocarditis, *Antibiotiki,* 9, 791–796, 1963.

17. **Kagan, G., Gusman, B., Vulfovitch, Yu., and Besouglova, T.,** Group A streptococcal L Forms as a trigger of immuno-morphological changes of the heart and immunogenic organs, in, *Spheroplasts, Protoplasts and L Forms of Bacteria, INSERM,* 65, 259–264, 1976.

18. **Kaplan, M. H. and Myeserian, M.,** An immunological cross-reaction between group A streptococcal cells and human tissue, *Lancet,* 1, 706–710, 1962.

19. **Kaplan, M. H. and Svec, K. H.,** Immunologic relation of streptococcal and tissue antigens. III. Presence in human sera of streptococcal antibody cross-reactive with heart tissue. Association with streptococcal infection, rheumatic fever, and glomerulonephritis, *J. Exp. Med.,* 119, 651–666, 1964.

20. **Kawakita, S., Iwamoto, H., and Yamada, J.,** Relationship between streptococcal infection and rheumatic fever, *Jpn. Circ. J.,* 33, 1499–1503, 1969.

21. **Klodnitskaya, N.,** On the question of etiology of scarlet fever, *Zh. Mikrobiol. Epidemiol. Immunobiol.,* 33, 31–35, 1962.

22. **Lannigan, R. and Zaki, S.,** Location of gamma globulin in the endocardium in rheumatic heart disease by the ferritin-labeled antibody technique, *Nature (London),* 217, 173–174, 1968.

23. **Leon, O. and Panos, C.,** Adaptation of an osmotically fragile L Form of *Streptococcus pyogenes* to physiological osmotic conditions and its ability to destroy human heart cells in tissue culture, *Infect. Immun.,* 13, 252–262, 1976.

24. **Markowitz, M. and Kuttner, A. G.,** *Major Problems in Clinical Pediatrics,* Vol. 2, Schaffer, A. J., Ed., W. B. Saunders, Philadelphia, 1965.

25. **Mashkov, A. V.,** A study of the properties of L Forms of bacteria isolated from children, *Zh. Mikrobiol. Epidemiol. Immunobiol.,* 12, 62–65, 1974.

26. **Matsunaga, K., Katoh, K., Takahashi, T., Sakamoto, H., Tani, K., Okuda, K., Tadokoro, I., and Kitamura, H.,** The experimental myocarditis of mice immunized with Group A streptococcal cytoplasmic membrane fraction extracted from L Forms, *Jpn. J. Exp. Med.,* 52(5), 267–270, 1982.

27. **Mortimer, E. A., Jr., Vaisman, B., Vigneau, I. A., Guasch, L. J., Schuster, C. A., Rakita, L., Krause, R. M., Roberts, R., and Rammelkamp, C. H., Jr.,** The effect of penicillin on acute rheumatic fever and valvular heart disease, *N. Engl. J. Med.,* 260, 101–112, 1959.

28. **Prozorovsky, S. V., Katz, L. N., and Kagan, G. Ya.,** L Forms of bacteria (mechanism of formation, structure, role in pathology), *ASM USSR, H.' Meditsina,* 1981.

29. **Rammelkamp, C. H. and Stolzer, B. L.,** The latent period before the onset of acute rheumatic fever, *Yale J. Biol. Med.,* 34, 386–398, 1962.

30. **Scheff, G. J., Marienfeld, C., and Hackett, E.,** Blood cultures in rheumatic fever patients, *J. Infect. Dis.,* 103, 45–51, 1958.

31. **Starnes, R.,** Identification of Streptococcal L Variants by Fluorescent Antibody Technique, M.S. thesis, Wayne State University, Detroit, 1963.

32. **Stollerman, G. H.,** The epidemiology of primary and secondary rheumatic fever, in *The Streptococcus, Rheumatic Fever and Glomerulonephritis,* Uhr, J. W., Ed., Williams & Wilkins,Baltimore, 1964, 311–329.

33. **Thomson, S. and Innes, J.,** Haemolytic streptococci in the cardiac lesions of acute rheumatism, *Br. Med. J.,* 2, 733–736, 1940.

34. **Timakov, B. and Kagan, G. Ya.,** *L Forms Bacteria and Species of Mycoplasma in Pathology,* Moscow, 1973.

35. **Timakov, V. D. and Kagan, G. Ya.,** L Forms of hemolytic streptococcus and their detection in the blood of patients with rheumatic fever and bacterial endocarditis, *J. Reumatisma,* 2, 3–8, 1962.

36. **Vulfovich, Yu. V. and Ambardzhanyan, R. S.,** Experimental study of a chronic relapsing pathological process induced by persistent L Forms of Group A streptococcus, *Vestn. Akad. Med. Nauk. SSSR,* 3, 17–21, 1985.

37. **Wedum, B. G.,** Rheumatic tissues in antibiotic-free tissue culture, *Ann. Rheum. Dis.,* 29, 516–523, 1970.

38. **Wilhelm, G. and Dippell, J.,** Die Beziehung eines neu aufgefundenen Ribonucleoproteids aus Streptokokken sur pathogenese des rheumatischen Fiebers, *Zentralbl. Bakteriol. Parasitenkd. Infektionskr. Hyg. Abt. 1 Orig.,* 214, 383–389, 1970.

39. **Wittler, R. G., Tuckett, J. D., Muccione, V. J., Gangarosa, E. J., and O'Connell, R. C.,** Transitional forms and L Forms from the blood of rheumatic fever patients, *8th Int. Congr. Microbiol. Abstr.* (Montreal), p. 125, 1962.

40. **Zabriskie, J. B., Hsu, K. C., and Seegal, B. C.,** Heart-reactive antibody associated with rheumatic fever, characterization and diagnostic significance, *Clin. Exp. Immunol.,* 7, 147–159, 1970.

Chapter 19

JOINT AND BONE DISEASE

Evidence that CWD forms play a significant role in infections of joint and osseous tissue is steadily accumulating. Incriminated variants include Gonococci, Staphylococci, Clostridia, Salmonella, Corynebacteria, and, perhaps most importantly, Propionibacteria.

OSTEOMYELITIS

RECURRENT STAPHYLOCOCCAL OSTEOMYELITIS

Meticulous details concerning a case of recurrent osteomyelitis were published by Rosner.[31] A patient with osteomyelitis of the femur responded well when treated with methicillin followed by oxacillin. Twenty-six months after his operation, the patient again experienced pain in the area of previous infection and exhibited a low-grade temperature. Antibiotics were withheld until after surgical exploration. A sinus from the surface of the old lesion contained thickened exudate. Cultures of this material were positive only on hypertonic medium, yielding growth which reverted to *Staphylococcus aureus* of the same phage type originally infecting the patient, type 80–81. Direct smears obtained at his second hospitalization showed that exudate from the draining sinus and abscess contained occasional large, poorly staining, pleomorphic bodies, but no recognizable staphylococci.

Data by Hedstrom[17] also support the thesis that recurrent osteomyelitis results from revitalization of latent CWD organisms. In 11 patients with closed operative sites, exacerbations of infection proved to be due to the original staphylococcal phage type. Tetracycline administered to three patients suspected of having L Forms did not prevent recurrence of the osteomyelitis. This agrees with our findings that CWD forms of *S. aureus* are often tetracycline resistant, although a clinical response has been recorded in some cases.

CHRONIC STAPHYLOCOCCAL OSTEOMYELITIS

"Unbalanced growth" of a staphylococcus from the blood of a patient with severe chronic osteomyelitis is shown in Figure 1.[50]

Kagan accrued indirect evidence that L-phase bacteria may be involved in many cases of chronic osteomyelitis by showing that lincomycin is usually therapeutic.[20] This antibiotic was chosen for dual properties: it achieves a good concentration in bone and effectively inhibits staphylococcal L-phase variants.

Gordon et al.[15] studied four cases of chronic osteomyelitis whose tissues and exudates were sterile when cultured in conventional media, but which yielded *S. aureus* in hypertonic medium. In one case, the organism grew initially as a CWD form. A therapeutic response resulted when penicillin was supplemented by or replaced by other antibiotics which included erythromycin, lincomycin, and vancomycin.

STREPTOCOCCAL OSTEOMYELITIS

L Forms of Group B streptococci produce chronic osteomyelitis in rats and mice after a simple intraperitoneal injection of the variants. The variant streptococci also persist in the neuroendocrine system and blood.[38]

LISTERIA OSTEOMYELITIS

Experimentally, atrophy of bone marrow resulted when *Listeria monocytogenes* infection was induced by stable CWD forms.[29] Such bone marrow destruction may explain minimal antibody levels characteristic of the spontaneous disease.

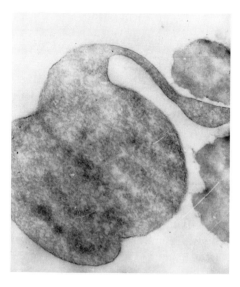

FIGURE 1.   L Form from the blood of severe persisting osteomyelitis. (Original magnification × 64,000.)(Electron micrograph submitted by Ligangirova, N., Ratz, L. N., and Kostantinova, N., 1991.)

SCLEROSING OSTEOMYELITIS

"Primary chronic sclerosing" osteomyelitis is a condition in which one might well seek CWD bacteria. Despite chronicity, no organisms have ever been incriminated. In a case of 11-months duration which responded to surgery and antibiotic, the laboratory examination did not include culture or smear study for CWD forms.[21]

PYOARTHROSIS

CWD forms in cultures of joint exudate have been reported by several groups of investigators, including Hill and Lewis,[18] but little documentation of individual cases has been published. In three cases of pyoarthrosis, variants were found which reverted to classic organisms, as reported by Charache and Kaslick.[6]

NONRHEUMATOID ARTHRITIS

GONOCOCCAL ARTHRITIS

Finding the L phase of the gonococcus in acute arthritis has attracted attention, not only as a "first" report, but also because of the philosophy of the investigators. Holmes et al. conclude that *Neisseria gonorrhea* may commonly inhabit the joints in gonococcal arthritis, more often in variant than parent form.[19] Such arthritis is too commonly regarded as an incomplete form of Reiter's syndrome or as a hypersensitivity reaction to the gonococcus.

Two cases of gonococcal arthritis were identified by LeBar[23] with fluorescent antibody. Reversion of the strains did not occur. It has been reported that even after a week of arthritic involvement, only 50% of gonococcal synovial effusions reveal gonococci by smear or culture for classic organisms.[25]

CLOSTRIDIAL ARTHRITIS

A case clinically typical of rheumatoid arthritis was inexplicably negative in all serological tests. Routine cultures for aerobes and anaerobes were negative, and no organisms were seen in Gram stains. The patient recovered before a diagnosis was made. Synovial fluid, which had been stored at $-70°C$, was cultured on Mycoplasma medium containing yeast extract and 20% unheated agammaglobulin horse serum. The plates were inoculated immediately after solidification by surface streaking and by cutting into the agar, then were placed in jars under anaerobic conditions.

Where inoculation *into* the agar was made, colonies were present after 5 d. The organism was identified as *Clostridium rhamosum,* which has occurred in complications following appendectomy and in the bloodstream in pyrexia, causing some fatalities. It is suggested that the organism was in a wall-deficient stage in the knee fluid, because the growth was initiated in the agar, a site where wall-deficient forms often revert. Furthermore, no growth had resulted when standard methods for isolating anaerobes had been used.

SALMONELLA ARTHRITIS

"Studies of L-forms can be helpful as well as easy to perform" according to Palmer.[26] He supports this thesis by describing a case which initially gave enigmatic findings. A classic *Salmonella enteritidis* grew from synovial fluid containing sufficient ampicillin to kill the salmonella even when the fluid was diluted $32\times$. When cultured on hypertonic medium, the salmonella grew as L Forms, indicating that classic forms were only *in vitro* revertants. Sensitivity testing of variant and parent forms resulted in successful therapy.

Palmer also discussed arthritis in leukemic patients whose arthritis was due to *Aeromonas hydrophila* strains not controlled by antibiotics adequate to inhibit the classic organisms. He noted that septic arthritis could be polymicrobial, with one genus wall deficient and one in the classic state.

Salmonella Saint-paul produced inflammation of a knee in an individual who had suffered severe gastroenteritis. Only CWD stages were seen in smears and primary culture of the synovial fluid. The salmonella was first classified as an enteric organism by fluorescent antibody reaction with 0:14 serum, and the species was identified after reversion.[23]

STAPHYLOCOCCAL ARTHRITIS

CWD staphylococci were the only organisms isolated from three post-therapy cases of infectious arthritis by LeBar.[23] The patients were a 16 year old girl with bilateral pyoarthrosis, a 24-year-old drug abuser whose infectious arthritis followed *Staphylococcus aureus* septicemia, and a 14 year old whose hip had been injured. Each patient had been treated with penicillin or methicillin with only temporary abatement of symptoms. The CWD forms proved sensitive to erythromycin, and two of the patients were successfully treated with this antibiotic. The third patient improved when probenecid was added to the methicillin schedule.

SEPTIC ARTHRITIS OF CHILDREN

In septic arthritis in the young, it is critical to check for CWD forms with the first culturing. Since *S. aureus* is the predominant infector in patients over 4 years of age[13] and *Hemophilus influenzae* the common pathogen between 7 months and 4 years, a diagnosis could frequently be confirmed using corresponding fluorescent antibodies made vs. the classic form, since good cross-reactivity occurs between variant and classic. This requires examining pleomorphic, well-defined microcolonies as well as classic bacteria.

Again referring to acute arthritis of children, Gillespie[13] noted that administration of the correct antibiotic may be necessary within 24 h if residual disability is to be avoided. In his series of 102 children, 23 had permanent malfunction, predominately resulting from delayed diagnosis.

HEMOPHILUS INFLUENZAE IN ARTHRITIS

In some instances the pathogen in acute arthritis has been suggested only by identifying the pathogen in another body site. Morrey et al.[24] described three cases with negative joint culture but with *H. influenzae* isolated from blood of two and spinal fluid of the third. The synovium is a site where pathogens are especially apt to become CWD as indicated by the series of cases recording the invariably high percentage of negative cultures.[1,11,48]

MYCOBACTERIAL ARTHRITIS

Tuberculous arthritis, more often than otherwise, yields only negative cultures by routine methods. In one carefully studied case a biopsy of synovium from the knee grew micro-colonies which were acid fast by intensified staining and had growth stimulation at the edge of isoniazid inhibition zones. The strain resembled *Mycobacterium tuberculosis* isolates from the blood in not reverting.[22]

NOCARDIA ASTEROIDES IN CANINE ARTHRITIS

Buchanan's laboratory described steroid-and-antibiotic-unresponsive fatal polyarthritis in a dog. At most stages of the disease, CWD *Nocardia asteroides* was the only organism cultured. Specific fluorescent antibody located the organism within tissues after death.[5]

STREPTOCOCCAL ARTHRITIS

Type 12 *Streptococcus pyogenes* in the L Form produced inflammation of cartilage, synovial membranes, and surrounding soft tissues in joints of rats and mice given intraperitoneal injections. A change in the ratio of B and T lymphocytes also resulted.[2]

RHEUMATOID ARTHRITIS

The etiology of rheumatoid arthritis (RA) was long one of the most intriguing enigmas in medicine. This was true although presumed causative agents were cultured at least as early as 1893[32] and observed in both direct stains and cultures in 1896[3,33] and 1900.[34]

ARE CLASSIC STREPTOCOCCI INVOLVED?

Because streptococci produce rodent arthritis, they were under continued surveillance as possible etiological agents in man. L Forms of streptococci were inserted into the panorama when Cook et al.[8] found that both L Forms and parent Group A streptococci produce arthritis when injected intraarticularly into rabbits.

LISTERIA-LIKE CWD FORMS IN BLOOD

A Listeria-like CWD form was found in the *blood* of RA patients by Pease, as she studied 76 patients.[27] She noted abundant budding spheroplasts at 3 d in cultures of whole blood, but not serum. The forms preferentially initiated growth in association with erythrocytes. Such variants were also found in small numbers in a few normal persons.

MYCOPLASMA?

Mycoplasmas have been sought by many laboratories investigating RA, with no proof to date[12] of their etiology in the condition in man, despite the amassed wealth of material on participation of these organisms in arthritis of domestic animals and rodents.[40] Contaminant Mycoplasma from animal serum used to enrich tissue culture has been only one of several complications.

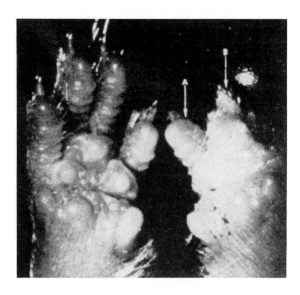

FIGURE 2. Forepaws of third-generation mouse. Right paw had one remaining digit. (From Warren, S. L., Marmor, L., Liebes, D. M., and Hollins, R., *Arch. Intern. Med.*, 124, 629–634, 1969. Copyright American Medical Association. With permission.)

THE VIRUS IN RA

Warren et al. were the first to elucidate the etiology of RA in man. A slurry of synovial cells, not fluid, injected into mice yields a transmissible agent which produces swelling and joint inflammation.[44] The characteristic pathology appeared in 1105 of 2335 mice. It was always possible to adapt it by injection of successive generations to become "inherent" in the animals, thus, spontaneously transmitted *in utero*.[45] The joint disease in a third-generation mouse may be so severe that digits are lost from a foot (Figure 2). Warren and colleagues published numerous studies.[42,43,47]

Subsequent publications by Warren et al. show that rodents can acquire the joint lesions after ingesting synovial tissue from rheumatoid arthritis cases.[46] Their other findings show that the agent causes joint malformations also in cats, turkeys, and rats, and that chicks injected in the embryo stage with synovial slurry show persisting joint malformations after hatching (Figure 3).[41]

In addition to the work from Warren's laboratory, the heat-resistant RNA virus has been demonstrated in mice or tissue culture by Godzeski et al.[14] and at least four other laboratories.[7,35,37] The pathology differs with mouse strain. It seems the situation is not unusual in that not only one virus is involved, rather two.[51]

Crocker and associates found that when immunodeficient mice (C5-deficient) were given RA synovial fluid with cells, the resultant disease was not just arthritis. The mice suffered from deformities of tail and limbs causing limping; some animals exibited runting and ruffled fur (Figure 4). Confirming the finding of Warren et al., the agent could be passed transplacentally in the mouse.[9]

THE POSSIBLE ROLE OF CWD *PROPIONIBACTERIUM ACNES* IN RA

When probing RA synovial fluids for Mycoplasma, the finding was surprising. A Gram-positive pleomorphic organism was consistently seen in cultures and in direct smears (Figure 5). It had CWD characteristics, growing only within soft serum agar and becoming a structureless blur if the slide was fixed by flaming. Continued investigations by Denys,[10]

FIGURE 3.   Two-day chick from embryo injected in amniotic cavity with RA synovial fluid. Maximal contractures in right foot. Left foot has flaccid first toe and rotation of third toe. (From Warren, S. L., Marmor, L., Stamm, M. E., Lewis, H., Horner, H. Liebes, D., and Stanley, W., *Rheumatology,* 6, 361–367, 1975. With permission.)

FIGURE 4.   Mouse injected with rheumatoid arthritis synovial fluid. (From Crocker, J. F. S., Ghose, T., Rozee, W., Woodbury, J., and Stevenson, B., *J. Clin. Pathol.,* 27, 122–124, 1974. With permission.)

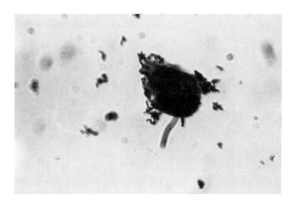

FIGURE 5.   Gram-positive pleomorphic organisms in direct smear of synovial fluid from rheumatic arthritis. Some bacteria are adherent to an exudate cell. (Photo submitted by L. Tunstall.)

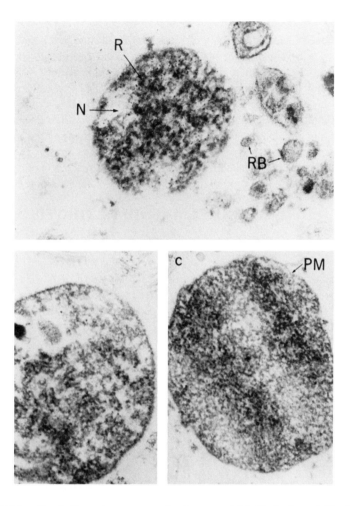

FIGURE 6. Wall-free *Propionibacterium acnes* growing from synovial fluid of a rheumatoid arthritis patient. P = cytoplasmic membrane. N = nuclear material. (From Denys, G. A., Ph.D. dissertation, Wayne State University, Detroit, 1981.)

Frederick Carlock and Lucille Tunstall showed that the organisms were CWD *Propionibacterium acnes*. The organisms stained with acridine orange, fluorescent muramidase (see Frontispiece), and specific fluorescent antibody. Denys followed the reversion from wall free (Figure 6) to classic stage with electron microscopy. When chick embryos were injected with either the classic or CWD stages of *P. acnes*, the chicks hatched with joint deformities resembling the disease caused by direct inoculation of synovial fluid by Warren et al. (Figure 3). Our finding that CWD *P. acnes* heavily populates the fluids was confirmed.[51]

The presence of *P. acnes* in RA is further supported by finding the microbial component muramic acid in RA synovial fluids and not the joint fluid in other conditions.[28] Additionally, Bartholemew and Nelson found *P. acnes* antigen in synovial fluids of RA and not other arthropathies.[4]

That a pathogen exists in the joints of chronic RA is further substantiated by nitroblue tetrazolium chloride (NBT) studies. This compound was vigorously reduced by the synovial fluid leukocytes of 10 of 12 RA patients, giving readings higher than found in the majority of gout and psoriasis patients. As expected NBT reduction was even higher in patients with acute pyogenic arthritis.[16]

RÉSUMÉ RA

It is to be hoped that epidemiology of the RA virus of man will soon be investigated. The virus seems to be consistently present in acute cases, but not demonstrable in later stages. Additionally, the antimicrobial sensitivity of the omnipresent CWD *P. acnes* from human cases should be determined. The sensitivities of CWD variants may differ from sensitivities of the classic stage. Sensitivity to certain antimicrobials may explain the improvement in some RA cases following administration of metronidazole. In other words, if the arthritis is virus initiated but CWD *P. acnes* perpetuated, treating the bacterial variants may be a therapeutic approach.

It is possible that the initial insult from the arthritis virus creates a favorable environment for multiplication of the CWD propionibacterium which is latent in some individuals.[39]

JUVENILE RHEUMATOID ARTHRITIS

Juvenile rheumatoid arthritis (JRA) patients not uncommonly develop chronic uveitis. Wirostko and associates, using electron microscopy, found CWD organisms in the leukocytes of the anterior chamber of the diseased eyes of five cases of JRA. They discussed the rifampin therapy of such cases.[49] The organism found by Wirostko et al. has not as yet been identified. The organism which we found in blood cultures of a JRA case was a highly fastidious bacteroides.

SPONDYLITIS

Tuberculous spondylitis tissue may show prominent granulomas, yet be negative for acid-fast bacilli.[36] In such cases it is routine to confirm the diagnosis only after a month, when cultures for classic tubercle bacilli become positive.

FINDINGS IN VETERINARY MEDICINE

The wealth of information gleaned from culturing synovial fluids in hypertonic medium is exemplified by Roberts and Little.[30] Twenty-three samples of ''sterile'' joint fluids cultured in their veterinary laboratory, employing both Mycoplasma and L Form media, all gave growth. Their findings are listed in Table 1. Isolation of *Mycoplasma hyorhinis* from the hock joints of swine is in accord with the well-established fact that this Mycoplasma causes joint disease in pigs. More unexpected are the CWD isolates; these make a contribution concerning joint disease in all animals. It is assumed that the bacterium existed in the synovial fluids in a wall-deficient stage since routine cultures were negative. The demonstration of wall deficiency in *Erysipelothrix rhusiopathiae* may explain the enigmatic pathology with this organism. It has long been noted that this organism, one of the leading causes for rejecting animals at meat processing plants, cannot be found in the inflamed joints after the early stages of the disease, although the animal's condition continues to worsen. Thus, finding the organism in a wall deficient stage in the joints of sheep may represent the first isolation of this bacillus in its latent stage, solving the mystery of negative cultures. Other isolates may also be the first instances of CWD forms of certain pathogens in animal disease. Isolations of *Pasteurella pseudotuberculosis* can be accepted as causing joint disease without much question. Similarly, *Salmonella dublin* is not unexpected since Salmonella have often been isolated in arthritis of man and domestic animals. However, finding the wall-deficient form of these two pathogens in natural infection is of great interest. The occurrence of *P. pseudotuberculosis* with a coagulase-positive staphylococcus can well be accepted as a dual infection with CWD pathogens.

TABLE 1
Organisms Isolated from "Sterile" Synovial Fluid After
Culture in Hypertonic Media

Animal species	Age (weeks, if known)	Joint (if known)	Bacterial species
Pig	5	Hock	*Mycoplasma hyorhinis*
	12	Hock	*M. hyorhinis*
	9	—	*M. hyorhinis*
	14	—	*Bacillus licheniformis*
	12	Stifle	*B. licheninformis, Staphylococcus epidermidis, Citrobacter freundii*
	3	Hock	*Aerococcus viridans, C. freundii, S. epidermidis*
Sheep	24	—	*Erysipelothrix rhusiopathiae, S. epidermidis, B. pumilus*
	8	—	*Micrococcus Group 6, Streptococcus Group D*
	8	—	*Streptococcus faecalis*
	8	—	*B. licheniformis, S. faecalis*
	8	—	*Micrococcus Group 5, S. epidermidis*
Cattle	8	Hock	*Salmonella dublin*
	—	Stifle	*Streptococcus bovis*
	7	Hock	*S. epidermidis, B. licheniformis*
Chicken	7	Hock	*S. epidermidis*
	—	Hock	*B. licheniformis, S. epidermidis*
	—	Hock	*Pasteurella pseudotuberculosis*
	—	Hock	*P. pseudotuberculosis, Staphylococcus aureus, Alcaligenes faecalis*
	7	Hock	*S. aureus*
	8	Hock	*Corynebacterium xerosis*
	9	Hock	*C. xerosis*
	7	Hock	*C. xerosis*

From Roberts, D. H. and Little, T. W. A., *Br. Vet. J.*, 127, 143–147, 1971.
With permission.

SUMMARY

CWD organisms play an important role in many aspects of rheumatology. Wall-deficient gonococci, Mycobacteria, Clostridia, Salmonella, and Corynebacteria have been found in acute arthritis. It is now known that wall deficient gonococci may explain the "sterile" joint fluids of arthritis accompanying acute gonorrhea. CWD forms of staphylococci have been found occurring alone in recurrent and chronic osteomyelitis. In septic arthritis of children there are excellent reasons for examining for CWD forms with primary smears and cultures.

From rheumatoid arthritis synovial fluid, one can consistently culture a CWD variant which causes deformities in chicks. There is much evidence that chronic infection with the

CWD *Propionibacterium acnes* follows an initial joint insult which is viral initiated. Thus, the chronic arthritis may be perpetuated by this omnipresent secondary invader.

Among the several pathogens found in the CWD stage in arthritic joints of domestic animals, *E. rhusiopathiae* arouses the most interest. Arthritis due to this organism has the mysterious characteristic of worsening after the pathogen in classic form disappears.

REFERENCES

1. **Almquist, E. E.**, The changing epidemiology of septic arthritis in children, *Clin. Orthop. Relat. Res.*, 69, 96–99, 1970.
2. **Ambardzhanyan, R. S., Bartazaryan, N. D., Arzakanyan, E. Kh., Khostikyan, N. G., Vulfovich, Y. V., and Zheverzheeva, I. V.**, Immunological characteristics and histological changes in joints in experimental animals by infected Group A type 12 L Forms of streptococcus, *Zh. Eksp. Klin. Med.*, 28, 109–114, 1988.
3. **Bannatyne, G. A., Wohlmann, A. S., and Blaxall, F. R.**, Rheumatoid arthritis; its clinical history, etiology, and pathology, *Lancet*, 1, 1120–1125, 1896.
4. **Bartholemew, L. E. and Nelson, F. R.**, *Corynebacterium acnes* in rheumatoid arthritis., *Ann. Rheum. Dis.*, 31, 28–33, 1972.
5. **Buchanan, A. M., Beaman, B. L., Pedersen, N. C., Anderson, M., and Scott, J. L.**, *Nocardia asteroides* recovery from a dog with steroid and antibiotic-unresponsive idiopathic polyarthritis, *J. Clin. Microbiol.*, 18, 702–708, 1983.
6. **Charache, P. and Kaslick, D.**, Isolation of protoplasts in human infection, *Clin. Res. 1965*, 293, 1965.
7. **Cohen, B. J., Buckley, M. M., Clewley, J. P., Jones, V. E., Puttick, A. H., and Jacoby, R. K.**, Human parvovirus infection in early rheumatoid and inflammatory arthritis., *Ann. Rheum. Dis.*, 45, 832–838, 1986.
8. **Cook, J., Fincham, W. J., and Lack, C. H.**, Chronic arthritis produced by streptococcal L Forms, *J. Pathol.*, 99, 283–297, 1969.
9. **Crocker, J. F. S., Ghose, T., Rozee, K., Woodbury, J., and Stevenson, B.**, Arthritis, deformities, and runting in C5-deficient mice injected with human rheumatoid arthritis synovium, *J. Clin. Pathol.*, 27, 122–124, 1974.
10. **Denys, G. A.**, Characteristics of Wall Deficient and Classical Forms of *Propionibacterium acnes* from Rheumatoid Arthritis, Ph.D. dissertation, Wayne State University, Detroit, 1981.
11. **Fink, C. W., Dich, V. Q., Howard, J., Jr., and Nelson, J. D.**, Infections of bones and joints in children, *Arthritis Rheum.*, 20, 578–583, 1977.
12. **Fraser, K. B., Shirodaria, P. V., Haire, M., and Middleton, D.**, Mycoplasmas in cell cultures from rheumatoid synovial membranes, *J. Hyg.*, 69, 17–25, 1971.
13. **Gillespie, R.**, Septic arthritis of childhood, *Clin. Orthop. Relat. Res.*, 96, 152–159, 1973.
14. **Godzeski, C. W., Boyd, R., Smith, C. A., and Hammerman, D.**, Viral-like particles in cocultivated rheumatoid synovial cells, *Arthritis Rheum. Abstr.*, 21, 559, 1978.
15. **Gordon, S. L., Greer, R. B., and Craig, C. P.**, Recurrent osteomyelitis: report of four cases culturing L Form variants of staphylococci, *J. Bone Jt. Surg.*, 53A, 1150–1156, 1971.
16. **Gupta, R. C. and Steigerwald, J. C.**, Nitroblue tetrazolium test in the diagnosis of pyogenic arthritis, *Ann. Intern. Med.*, 80, 723–726, 1974.
17. **Hedstrom, S. A.**, The prognosis of chronic staphylococcal osteomyelitis after long-term antibiotic treatment., *Scand. J. Infect. Dis.*, 6, 33–38, 1974.
18. **Hill, E. O. and Lewis, S.**, "L Forms" of bacteria isolated from surgical infections, *Bacteriol. Proc. 1964*, p.48, 1964.
19. **Holmes, K. K., Gutman, L. T., Belding, M. E., and Turck, M.**, Recovery of *Neisseria gonorrhoeae* from "sterile" synovial fluid in gonococcal arthritis, *N. Engl. J. Med.*, 284, 318–320, 1971.
20. **Kagan, B. M.**, Role of L Forms in staphylococcal infection, in *Microbial Protoplasts, Spheroplasts and L Forms*, Guze, L. B. Ed., Williams & Wilkins, Baltimore, 1968, 372–378.
21. **Kopits, S. E and Debuskey, M.**, Primary chronic sclerosing osteomyelitis, *Johns Hopkins Med. J.*, 140, 241–247, 1977.
22. **LeBar, W. D., Mattman, L. H., and Ross, L.**, Isolation of cell wall deficient *Mycobacterium tuberculosis* from a case of chronic arthritis, *Henry Ford Hosp. Med. J.*, 23, 17–20, 1975.
23. **LeBar, W. D.**, Isolation and Identification of Cell Wall Deficient Microorganisms from Infection, M.S. thesis, Wayne State University, Detroit, 1977.

24. **Morrey, B. F., Bianco, A. J., Jr., and Rhodes, K. H.,** Septic arthritis in children, *Orthop. Clin. North Am.,* 6, 923–934, 1975.

25. **O'Brien, J. P., Goldenberg, D. L., and Rice, P. A.,** Disseminated gonococcal infection, a prospective analysis of 49 patients and a review of pathophysiology and immune mechanisms, *Medicine,* 62, 395–406, 1983.

26. **Palmer, D. W.,** Inadequate response to "adequate" treatment of bacterial infection, L Forms and "bactericidal" antibiotic activity, *J. Infect. Dis.,* 139, 725–727, 1979.

27. **Pease, P.,** Morphological appearances of a bacterial L Form growing in association with the erythrocytes of arthritic subjects, *Ann. Rheum. Dis.,* 29, 439–444, 1970.

28. **Pritchard, D. G., Settine, R. L., and Bennett, J. C.,** Sensitive mass spectrometric procedure for the detection of bacterial cell wall components in rheumatoid joints, *Arthritis Rheum.,* 23(5), 608–610, 1980.

29. **Prosorovsky, S., Kotljarova, J., Fedotova, I., and Bakulov, I.,** Pathogenicity of listerial L Forms, in *Spheroplasts, Protoplasts and L Forms of Bacteria,* Roux, J., Ed., *INSERM,* 1976, pp. 265–272.

30. **Roberts, D. H. and Little, T. W. A.,** A note on spheroplasts and Mycoplama in synovial fluid, *Br. Vet. J.,* 127, 143–147, 1971.

31. **Rosner, R.,** Isolation of protoplasts of *Staphylococcus aureus* from a case of recurrent acute osteomyelitis, *Tech. Bull. Reg. Med. Technol.,* 38, 205–210, 1968.

32. **Schüller, M., II,** Untersuchungen über die Aetiologie der sogen. chronischrheumatischen Gelenkentzundungen, *Berl. Klin. Wochenschr.,* 30, 865–868, 1893.

33. **Schüller, M.,** in appendix of Discussion über den Vortag des Herrn Silex, Pathognomonische Kennzeichen der Congenitalen lues, *Berl. Klin. Wochenschr.,* February, 168–175, 1896.

34. **Schüller, M., V,** Polyarthritis chronica villosa und Arthritis deformans, *Berl. Klin. Wochenschr.,* February, 124–128, 1900.

35. **Schumacher, H. R., Jr.,** Synovial membrane and fluid morphologic alterations in early rheumatoid arthritis, microvascular injury and virus-like particles, *Ann. N.Y. Acad. Sci.,* 256, 39–64, 1975.

36. **Scully, R. E.,** Case records of the Massachusetts General Hosp., Case 5–1988, *N. Engl. J. Med.,* 318, 306–312, 1988.

37. **Simpson, R. W., McGinty, L., Simon, L., Smith, C. A., Godzeski, C. W., and Boyd, R. J.,** Association of parvoviruses with rheumatoid arthritis of humans, *Science,* 233, 1425–1428, 1984.

38. **Vartazaryan, N. D., Gorina, L. G., Ambarzhanyan, R. S., Goncharova, S. A., Khostikyan, N. G., Zheverzheeva, I. V., Ter-kasparova, M. R., and Arzakanyan, E. Kh.,** Morphological and immunological characteristics of organ damage in experimental infection with L Form of Group B streptococcus., *Vestn. Akad. Med. Nauk. SSSR,* 7, 22–27, 1986.

39. **Waitzkin, L.,** Latent *Corynebacterium acnes* infection of bone marrow, *N. Engl. J. Med.,* 281, 1404–1405, 1969.

40. **Ward, J. R.,** Host-agent factors in mycoplasmal arthritis of rats, in *Proc. Conf. Relationship of Mycoplasmas to Rheumatoid Arthritis and Related Diseases* (Chicago), Decker, J. L., Ed., U.S. Public Health Serv. Publ. No. 1523, 1966, pp. 255–259.

41. **Warren, S. L., Marmor, L., Boak, R., and Liebes, D. M.,** Transmission of an active agent from rheumatoid arthritis synovial tissue to chicks, *Arch. Intern. Med.,* 128, 619–622, 1971.

42. **Warren, S. L., Marmor, L., Gerken, S. C., and Horner, H. E.,** Correlation of a bioassay with the clinical status of patients with rheumatoid arthritis, *Clin. Orthop.,* 144, 299–304, 1979.

43. **Warren, S. L., Marmor, L., Horner, H. E., Stanley, W. D., Stoeckel, T. A., Gerken, S. C., and Patton, G. A.,** Further studies on the specific infectious agent isolated in rheumatoid arthritis, *J. Rheumatol.,* 6, 135–146, 1979.

44. **Warren, S. L., Marmor, L., Liebes, D. M., and Hollins, R.,** An active agent from human rheumatoid arthritis which is transmissible in mice, *Arch. Intern. Med.,* 124, 629–634, 1969.

45. **Warren, S. L., Marmor, L., Liebes, D. M., and Hollins, R. L.,** Congenital transmission in mice of an active agent from human rheumatoid arthritis, *Nature (London),* 223, 646–64, 1969.

46. **Warren, S. L., Marmor, L., Liebes, D. M., and Rosenblatt, H. M.,** An active agent from rheumatoid arthritis synovial tissue, *Arch. Intern. Med.,* 130, 899–903, 1972.

47. **Warren, S. L., Marmor, L., Stamm, M. E., Lewis, H., Horner, H., Liebes, D., and Stanley, W.,** Thermal inactivation and gradient studies of the active agent in rheumatoid arthritis, *Rheumatology,* 6, 361–367, 1975.

48. **Wiley, J. J. and Fraser, G. A.,** Septic arthritis in childhood, *Can. J. Surg.,* 22, 326–330, 1979.

49. **Wirostko, E., Johnson, L., and Wirostko, W.,** Juvenile rheumatoid arthritis inflammatory eye disease. Parasitization of ocular leukocytes by mollicute-like organisms, *J. Rheumatol.,* 16, 1446–1453, 1989.

50. **Ligangirova,** 1991, personal communication.

51. **Godzeski, C.,** personal communication.

Chapter 20

MYCOBACTERIUM TUBERCULOSIS
AND THE ATYPICALS

"Nearly half the world's population is infected with TB."[24] This chapter reviews aberrant growth of mycobacteria as a natural phenomenon and emphasizes that looking for the variant in disease not only results in earlier diagnosis, it often saves an invasive procedure.

Tuberculosis, the insidious killer that seemed to be fading from view, is again on the rise. Worldwide, there are 30 million cases.[17] Even in developed countries mycobacterial infections have resurged, due to their proclivity to thrive in the immunosuppressed, whether immunosuppression results from cancer therapy, drug abuse, or acquired immune deficiency syndrome (AIDS).[31,42] With multiple antimicrobials available, the major problem is diagnosis. Biopsy of lung tissue is not infrequently resorted to.[62] The diagnosis may be made only at autopsy.[71] It is apparent that in any tissue the tubercle bacillus grows minimally as an acid-fast rod; the predominant growth consists of pleomorphic structures, acid fast only with slightly modified stains (Figure 1).[78] In a long series of cases, employing the sensitive auramine-rhodamine (AO) fluorescent stain, only 50% of the specimens eventually yielding positive cultures showed bacilli in direct smear.[65] Even in AIDS patients' sputum, smears are frequently negative.[18] Diagnosis is also difficult because with time, X-ray findings have changed, as have clinical symptoms.[58] For the *Mycobacterium avium* complex (MAC), sputum smears may be positive only in 16% of cases.[54]

IN VITRO GROWTH OF CWD MYCOBACTERIA

Voluminous literature substantiates that mycobacteria grow more rapidly and with the least fastidiousness as pleomorphic forms, unrecognizable as mycobacteria. Csillag concluded that Mycobacteria are typically biphasic, resembling true fungi.[16] Pla y Armengol[64] found that a large inoculum of the tubercle bacillus grows rapidly on all routine media, appearing as large L-body spheres and also variegated mycelia. Merckx and associates[52] showed that before typical mycobacterial colonies appear on Lowenstein's medium, there is extensive pleomorphic growth, shown by electron microscopy. Xalabarder noted that classic mycobacterial filaments could abandon binary fission and simultaneously divide into 10 to 15 bacilli or develop a variety of other forms (Figure 1 A and B).[77] Lack[44] and Weill-Hallé[75] noted the remarkable syncytial growth which can best be recognized in impression smears.

Sprouting granules observed by Lucksch[47] and Nedelkovitch[59] may stain a dense black. Extensive micrographs by Bassermann[2] and Miguel[53] also document that mycobacteria commonly propagate by modes other than binary fission. Colonies of L-Form tubercle bacilli growing on medium surface and others growing under Millipore filters are shown in Figure 2 A and B.

Brieger[9] at Cambridge followed growth in slide cultures of guinea pig amniotic fluid inoculated with tubercle bacilli. A structureless mass rapidly evolved into tiny nonacid-fast granules. By 72 h, long branching filaments grew from the granules. Brieger observed similar growth when tubercle bacilli were inserted into chicken embryo amnion.[9] Electron micrographs of rods growing from pleomorphic growth of *M. tuberculosis* is depicted in Figure 3.[78]

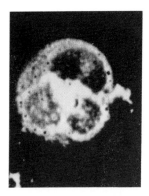
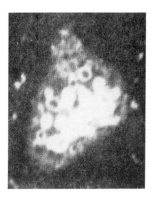

FIGURE 1. Heteromorphic growth of *Mycobacterium tuberculosis* in young cultures. (From Xalabarder, C., *Publ. Inst. Antituberc. Sup.* 7, 1–83, 1970. With permission.)

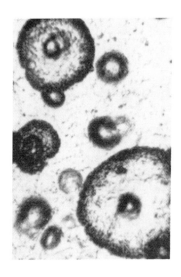
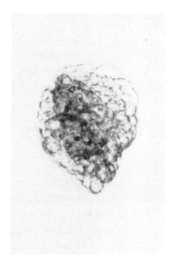

A   B

FIGURE 2. (A) L-Form colony of *M. tuberculosis* growing on medium. (B) L-Form growth beneath millipore filter. (From Xalabarder, C., *Publ. Inst. Antituberc. Suppl.* 7, 1–83, 1970. With permission.)

SPONTANEOUS VARIANT GROWTH IN PLANTS

Much,[57] who investigated typical and atypical forms of the tubercle bacillus for over 27 years, untiringly followed the constant transition from the tiny Gram-positive, nonacid-fast granules, which now bear his name, to acid-fast rods. In one of his most unique studies, he injected tomato and onion plants with large numbers of classic tubercle bacilli. Neither the plants' vigor nor the yield of quality fruit was affected. The mycobacterium multiplied in both plants only in the granular and Actinomyces-like state. It is reassuring that he found a loss of virulence accompanying these morphological changes.

FIGURE 3. Rod with typical shape is emerging from pleomorphic growth. (From Xalabarder, C., *Publ. Inst. Antituberc. Suppl.* 7, 1–83, 1970. With permission.)[78]

WALL-DEFICIENT *M. TUBERCULOSIS* IN BLOOD

Xalabarder, in his laboratory in Barcelona, noted L Forms of *M. tuberculosis* in the blood of tuberculosis patients, and the L Form of BCG in the blood of vaccinated persons.[78]

In Mycobacterioses the wall-deficient variant can usually be seen in 48-h blood cultures. The organism is acid fast with an intensified Kinyoun's stain. Characteristically, in blood culture smears the acid-fast microcolonies are visible with low power magnification (100 ×) and their characteristic morphology revealed at a magnification of 800 × (see Frontispiece). The blood culture is usually positive, providing it is taken during the patient's febrile hours, even when sputum smears and bone marrow biopsy are negative. Acid-fast rods of *M. tuberculosis* were found in the buffy coat of blood of one AIDS patient.[22]

The pleomorphic acid-fast microcolonies develop in any standard blood culture medium in use in the U.S. or Canada. The presumptive diagnosis from the blood culture smear may be substantiated by biochemical reactions as described below, or with fluorescent antibody. Using fluorescent antibody vs. H37Ra to stain CWD colonies in blood cultures of 24 patients, Byrne[81] found good correlation between FA and Kinyoun's staining.

Blood is an excellent source of the organism since approximately half the patients with active disease produce no sputum.[15] Furthermore, there is no confusion caused by the rich normal flora of organisms native to sputum and gastric washings. The blood usually yields a positive culture whether the infection is in lung, meninges, or other organ. Memminger and Whitby found the blood continues to carry the organisms during 3 to 5 d after initiation of specific therapy.[51]

In addition to Kinyoun's stain, the acid-fast characteristics of the variants may be shown by the acid-fast acridine orange stain, by the Victoria Blue stain (see Frontispiece), and the periodic acid method (see Media chapter). Some CWD colonies of *M. tuberculosis* and the sarcoidosis mycobacterium are autofluorescent red in darkfield microscopy (see Frontispiece). The strains fluoresce only in certain cultures, with the conditions fostering fluorescence unknown. If the conditions stimulating fluorescence can be determined, the characteristic will aid in identifying these mycobacteria. There are few accounts of autofluorescent pathogenic genera. However, Ghadially et al. found red fluorescence in Staphylococcal species, in many Enterobacteriaceae, in species of Corynebacterium — *C. diphtheria gravis* and *mitis* but not *intermedius!*[29]

The CWD mycobacteria can be subcultured by transferring large inocula. The morphology varies greatly with pH of the medium (see Frontispiece). The CWD variants, as true for most genera in the wall-deficient state, grow most rapidly at pH 5.0 to 6.0 and, surprisingly, although microaerophilic, grow more rapidly in agitated culture.[39]

A promising approach to detecting mycobacteria in blood concerns isolator tubes. A few microcolonies may be found concentrated in the solvent of unincubated blood and can be stained with specific fluorescent antibody (see frontispiece). Mycobacterial CWD colonies can also be concentrated from cultures by hydrocarbon flotation.[50] An interesting project would compare staining reactivity of colonies concentrated by varied methods.

As recently discovered, slide culture of a drop of patient's blood can give a diagnostic clue which does not even require a microscope for detection. (See the phenol phenomenon in Frontispiece and the chapter on Media, Methods, and Stains.)

ENZYMATIC REACTIONS OF WALL-DEFICIENT TUBERCLE BACILLI

Biochemical reactivity of the variants has been tested by varied approaches. In each case, the patient's diagnosis was confirmed by smear and/or growth of *M. tuberculosis* from sputum. Blood cultures from 26 cases of pulmonary tuberculosis were examined in API 20E micropanels. The substrates attacked by the CWD organisms in blood cultures of the tuberculosis cases were one or more of the following: urea, arginine, lysine, and ornithine (see Frontispiece). Commonly, reactions were obtained in at least three. Also pertinent to diagnosis were lack of reactivity with carbohydrates and failure to produce H_2S, gelatinase, and cytochrome oxidase. Reactions were clearly differentiated from those of CWD forms of other species commonly growing in blood cultures, including the genera Streptococcus, Staphylococcus, Escherichia, Enterobacter, Histoplasma, Proteus, and Pseudomonas. Fresh isolates growing in the original blood culture were used. Growth of at least 48 h was required.

Niacin production can also reinforce the presumptive identification of CWD *M. tuberculosis*. Amprim found the niacin production could be demonstrated by stimulation of niacin-dependent *Lactobacillus plantarum*.[1] Memminger and Whitby[51] used a different approach as they tested 563 blood cultures from tuberculosis patients for niacin content. Seventy-four percent of the cultures had values exceeding control mean values. Twenty control cultures with organisms commonly causing bacteremias had niacin levels near or lower than uninoculated controls. Detection of niacin employed a modified cyanogen bromide/aniline procedure. Production of deaminases by CWD *M. tuberculosis* in API panels correlates with the finding that high levels of adenosine deaminase characterize the spinal fluid in tuberculosis meningitis.[70]

In the examination of blood cultures from over 200 cases of pulmonary tuberculosis, several problems were encountered. Repeated blood cultures of a febrile man had been negative by routine smear and subculture. Examination made for CWD forms revealed organisms suggesting staphylococci at the edge of microcolonies. Jubinski, using fluorescent lysostaphin, confirmed that many colonies of CWD *Staphylococcus aureus* were present.[38] On therapy for staphylococci the patient's temperature decreased slightly, but he expired at 48 h. Autopsy revealed advanced pulmonary tuberculosis. Re-examination of the blood cultures showed CWD mycobacteria also present. Polymicrobialism with classic bacteria is well known, but two genera in CWD stages were not anticipated. Other problems have included detecting growth only at 72 h.

WALL-DEFICIENT VARIANTS IN SPUTUM AND OTHER SPECIMENS

INCIDENCE IN SPUTUM AND SIGNIFICANCE

Although in our experience it has been more profitable to examine blood then sputum, many laboratories find that seeking CWD in sputum greatly increases the sensitivity of tuberculosis diagnosis as shown by the following data.

From a laboratory which examined sputum and other infected specimens, the following results were published which show the improved diagnosis when L Forms are sought:[26]

Classic organism +; L Form −	9.6%
Classic organism +; L Form +	19.2%
Classic organism −; L Form +	43.7%
Classic organism −; L Form −	27.5%

In individuals with limited tuberculosis, L Forms were found in 17% of 700 sputum samples, vs. 2.3% of classic bacilli. Likewise, gastric lavage revealed L Forms more than three times as often as bacillary forms.[41]

It has been noted that the number of L Forms in sputum relate to the severity of the pulmonary infection and to the resistance to therapy.[3] It has also been recorded that if "cured" patients have L Forms of the tubercle bacillus in their sputum, their chance of relapse is 47%, compared with 14% if the L Forms were absent.[20]

Also, in extrapulmonary disease, detection of CWD tubercle bacilli can be an important diagnostic procedure. Of 114 granulomas studied by Dorozhkova and associates, 35% had cultures positive for mycobacterial L Forms but no tubercle bacilli detected by routine methods.[19]

STAINED SMEARS OF EXUDATE

When sputum smears are examined after intensified staining, it may not require microscopy to locate microcolonies, seen as tiny red dots. More often the growth is detectable only at a magnification of 200 × or greater. Rarely, miniature bacilli about half the diameter and length regarded as typical, are seen. However, more often found are microscopic colonies of acid-fast pleomorphic organisms, or acid-fast circles, L-bodies (see Frontispiece). Blood has the obvious advantage of lacking the rich normal flora of sputum which may cause confusion.

AO, so invaluable for rapidly detecting tubercle bacilli in sputum smears, is not easily retained by the CWD Mycobacteria in blood. Atkins found that AO is intensified by pretreating the blood culture smears with periodic acid.[82] However, there is a unique reason for AO staining of *sputa*. Other pleomorphic bacteria as well as CWD mycobacteria may retain the stain when modified by brief decolorizing. Thus, when AO-staining pleomorphic forms are seen, no conclusion is possible. When such stained forms are absent, a presumptive absence of *M. tuberculosis* can be concluded, as tested by Garvin[27] and by Steffen.[72]

CONCENTRATION OF CWD COLONIES

Concentration of tubercle bacilli from spinal fluid or sputum by hydrocarbon flotation, long practiced, continues to be valuable in increasing diagnostic sensitivity.[7] The microcolonies of the wall-deficient bacteria also appear in hydrocarbon concentrates of blood cultures and clinical samples. Such concentrates can be used for fluorescent antibody as well as acid-fast stains.

DISEASE FROM ONLY THE CWD STAGES

Sometimes it appears that CWD stages per se produce tuberculosis. A case in point was a 56-year-old male, with no known immunodeficiency, who expired with miliary tuberculosis. Stable CWD forms were found in all lesions, including those in the brain. It was noted that filterable units of the organism could easily penetrate the blood-brain barrier.[6]

A similar case concerns a 3-year-old child who died of generalized tuberculosis. By culture and guinea pig inoculation only L Forms of *M. tuberculosis* were found in the child's lung, liver, and spleen.[80]

FILTRABLE STAGES OF *M. TUBERCULOSIS*

The filtrable form of *M. tuberculosis,* one of the stages of cell wall deficiency, has been demonstrated in tuberculous individuals in almost every type of secretion and exudate. Thus, this "ultravirus" has been found in pleural fluid,[4,12,14] in ascitic fluid,[33] in the spinal fluid in meningitis,[13,14] in the peripheral blood of women during menstruation,[36] in urine,[23] and in mothers' milk.[28] The list also includes the exudate of therapeutic pneumothorax and the scrotal exudate of hydrocele.[13] This virus stage of the tubercle bacillus was recovered from the blood of individuals suffering from cutaneous tuberculosis and from their dermal lesions.[30] Dzis and associates[21] concluded that accurate diagnosis and therapy of tuberculosis required detection of filtrable as well as bacillary stages of the tubercle bacillus.

Residual infection after treatment is an important problem. Filtrable forms of the tubercle bacillus were almost invariably present in tuberculous tissue examined after 6 months of specific therapy. The variants retained the original pathogenicity of the bacilli and could also cause nonspecific inflammation.[32]

Pitfalls in judging filtrability of mycobacteria include the following: Investigators are aware that swollen lymph nodes in guinea pigs which have received a filtrate of *M. tuberculosis* do not indicate per se that a living filtrable form was in the inoculum.[5] Antigenic material in autoclaved filtrates can cause lymphadenopathy. Likewise, acid-fast granular material in a lymph node injected with a filtrate does not necessarily signify that living forms were in the filtrate.[8,9] Acid-fast particles may pass through a filter and if injected into a lymph node remain there, giving the false impression of multiplication of an acid-fast form. It is also well known that guinea pigs housed in a room with tuberculous animals may become spontaneously infected. It is fortunate that much of the work on the virus stage of the tubercle bacillus was done by men whose primary research interest was tuberculosis and who repeated their work many times with controls.

Molecular methods which recognize mycobacterial antigens or fatty acids in clinical samples may prove functional in detecting CWD mycobacteria in centers with appropriate equipment.[11,25,40,45,79] This can include studies of circulating immune complexes.[63]

Enzyme-linked immunosorbent assay (ELISA) may contribute to diagnosis of mycobacteriosis by quantitating the mycobacterial antigen 85.[76]

Fluorescein-conjugated lectins are promising for identifying *M. fortuitum* and *M. chelonei*.[35] The testing of CWD stages with lectins has not been reported.

Staining and growing the CWD stages of mycobacteria can remain a simple method for diagnosis and confirmation. In addition, the quantity of microbial substance to analyze by molecular methods can be rapidly increased in the CWD stage vs. the bacillary.

MYCOBACTERIUM SCROFULACEUM

M. scrofulaceum and *M. avium* were prominent in lymph node infections of children, comprising approximately 44 and 46% of the mycobacteria cultured. Fifty-four percent of

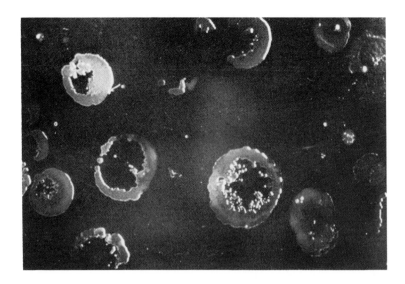

FIGURE 4. L-Form colonies which grew from brain of child who expired from *M. scrofulaceum* infection. (From Korsak, T., *Acta Tuberc. Pneumol. Belg.*, 66, 445–469, 1975.)

the growing organisms could not be identified and were designated MAIS complex to cover avium-intracellulaire-scrofulaceum complex. In approximately 34% of children with lymphadenopathy, considered by virtue of skin tests to be mycobacterial, no organism grew in culture.[37]

A generalized infection with *M. scrofulaceum* which proved antibiotic resistant caused the death of an 11-year-old boy who had no demonstrable immunodeficiency.[43] No viable classic mycobacteria were found in the autopsied tissues. Instead, colonies consisting of nonacid-fast cocci, often mixed with large acid-fast spheres (L-Bodies), grew from bone, brain, lung, liver, and spleen. The morphology of the variants and their colonies differed with isolates from different tissues. Remarkable colonies grew from brain in cultures at 45°C (Figure 4). Lysis of the center of the colonies suggests phage. The small central daughter colonies, which appeared after 8 d, consisted of acid-fast large spheres considered mycoplasma-like. Korsak's publication contains many excellent color micrographs of the variants and their reversion to acid-fast bacilli.

MYCOBACTERIUM AVIUM-INTRACELLULAIRE (MAC COMPLEX)

The common *M. avium-intracellulaire* (MAC) infection in AIDS patients is another development stimulating interest in mycobacteriosis. MAC infections are remarkably difficult to diagnose. In a review of 94 cases, it was noted that only 16% of the infections had acid-fast bacilli in sputum.[54] In another study a positive culture of *M. avium* complex did not appear until 95 d, and early diagnosis was seldom possible even with liver biopsy.[66] Obviously, ancillary diagnostic methods are needed. Detection of acid-fast microcolonies in blood is one such ancillary method.

Bacilli of MAC have been grown from the blood of some AIDS patients[56] and also detected in buffy coats.[60] Blood culture has given not only an earlier diagnosis, but blood is often the only tissue from which the pathogen is isolated.[74] Moffie and associates note that *M. avium* circulates in the blood of patients who do not have AIDS. The case they describe was a patient infected with *M. avium* type 2.[55]

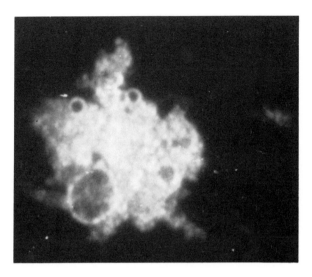

FIGURE 5. CWD colony of *M. intracellulaire* in blood culture. Rim of the large sphere was acid fast.

An early study shows the value of looking for CWD stages in blood.[49] In this case the patient with leukemia was in remission when he developed a pulmonary problem considered to be candidiasis. Six blood cultures contained vacuolated colonies, acid fast by an intensified method (Figure 5). Some colonies reverted to *M. intracellulaire*, identified by The National Jewish Center for Immunology and Respiratory Medicine, as type Watson. The same organism grew from lung tissue examined at autopsy.

Another study concerned three members of a family whose pet dove had died. The blood of all three patients contained acid-fast L-bodies and colonies, and in one individual, bacilli were numerous. In 7 d tiny colonies developed on Lowenstein-Gruft medium and Middlebrook's 7H10 slants and also in impression smears from chocolate agar with all media incubated in 2% CO_2 (see Frontispiece). Most unexpected was the finding that in each case the organisms, including the bacilli, had grown in blood whose only incubation had been during 2-d mail transit. Growth of thick, nonacid-fast mycelia, characteristic of *M. avium*[10] was also abundant. The organisms have not been identified.

MYCOBACTERIUM FORTUITUM

M. fortuitum, which is thought to invade man from its habitat in the soil, has to date been reported in a CWD stage in one patient. The acid-fast pleomorphic organisms comprising microcolonies in blood cultures reverted to slender rods, morphologically resembling mycobacteria. However, transition was not complete as growth in macroscopic colonies on the surface of Lowenstein's medium did not result. Typical growth of *M. fortuitum* developed in the patient's sputum cultures.

MYCOBACTERIUM CHELONEI

M. chelonei suddenly assumed importance when xenographs, porcine heart valves, occasionally were infected with this soil organism. Cell wall deficiency of the bacterium is suggested by the difficulty experienced in finding it by routine methods and the multiple transfers sometimes necessary to establish growth.[46]

FACTORS FOSTERING MYCOBACTERIAL WALL DEFICIENCY

As expected, pertinent antibiotics foster cell wall deficiency in mycobacteria. Streptomycin or cycloserine have been shown to induce such variation in BCG, H37Ra, *M. tuberculosis, M. bovis, M. phlei, M. smegmatis,* and *M. avium.*[27,48,67,68]

Many agents such as lysozyme,[73] macrophages[34] and bacteriophage[73] known to produce CWD stages in bacteria naturally occur in the vertebrate body. Correspondingly, it is interesting that tuberculous lungs contain elevated levels of lysozyme.[61]

"FRAGILITY" OF CWD MYCOBACTERIA

Culture of wall-deficient mycobacteria can be successful without hypertonic medium, differing from the requirement of many genera. Correspondingly, these variants are undamaged when centrifuged at forces reaching 15,000 *g.*[69] CWD variants of many genera have unexpected resistance to prolonged incubation or freeze-storage. The variants often retain their characteristic morphology and staining reactions after the standard sputum decontamination-concentration procedure as shown by a preliminary study done with Herman Kiefer Hospital. However, it was unexpected that wall-deficient *M. tuberculosis, M. intracellulaire, M. fortuitum,* and strains from sarcoidosis could survive 10 to 12 years under refrigeration, as recently noted in our laboratory. Xalabarder early emphasized that L Forms of Mycobacteria were remarkably different from L Forms of other species in resistance to physical and chemical agents.[78]

SUMMARY

From the earliest days of bacteriology, pleomorphic forms and filterable units of *M. tuberculosis* have been studied in cultures, experimental animals, and man.

Wall-deficient mycobacteria, acid fast when Kinyoun's stain is buffered, are usually present in spinal fluid, pleural fluid, or sputum. In the febrile patient mycobacterial infection in any body site is usually complicated by wall-deficient forms of the bacterium in circulating blood. It will be seen in smears of blood cultures after 72 h and may even be seen in concentrates of freshly collected blood. The variants grow well in primary blood cultures using any medium commonly employed in the U.S.

REFERENCES

1. **Amprim, F. L.,** Rapid Identification of *Mycobacterium tuberculosis* by Detection of Characteristic Organic Products, M.S. thesis, Wayne State University, Detroit, 1978.
2. **Bassermann, F. J.,** Die L Form des Tuberkulose-erregers in elektronoptischer Darstellung, *Beitr. Klin. Tuberk.,* 113, 134–135, 1955.
3. **Berezovsky, B. A. and Salobai, R.,** Role of mycobacterial L variants in development and progress of pulmonary tuberculosis relapses, *Probl. Tuberk.,* 4, 32–35, 1988.
4. **Bernstein,** Unbekannte Formenelemente im Sputum von Tuberkulosekranken, *Beitr. Klin. Tuberk.,* 82, 504–505, 1933.
5. **Berry, J. W. and Lowry, H.,** A slide culture method for the early detection and observation of growth of the tubercle bacillus. A preliminary report, *Am. Rev. Tuberc.,* 60, 51–61, 1949.
6. **Biron, M. G. and Soloveva, I. P.,** Acute hematogenic generalization of tuberculosis caused by L Forms of Mycobacteria, *Probl. Tuberk.,* 8, 75–76, 1989.
7. **Biswas, S. K., Singh, P. P., Bhagi, R. P., and Shah, A.,** Detection of tubercle bacilli in sputum smear by floatation method, *Indian J. Chest Dis. All. Sci.,* 29, 8–12, 1987.

8. **Blacklock., J. W. and Williams, J. R.,** The localization of tuberculous infection at the site of injury, *J. Pathol. Bacteriol.,* 74, 119–131, 1957.

9. **Brieger, E. M.,** The host parasite relationship in tuberculous infection, *Tubercle,* 30, 242–253, 1949.

10. **Brieger, E. M. and Glauert, A. M.,** A phase-contrast study of reproduction in mycelial strains of avian tubercle bacilli, *J. Gen. Microbiol.,* 7, 287–294, 1952.

11. **Brooks, J. B., Daneshvar, M. I., Haberberger, R. L., and Mikhail, I. A.,** Rapid diagnosis of tuberculosis meningitis by frequency-pulsed electron-capture gas-liquid chromatography detection of carboxylic acids in cerebrospinal fluid, *J. Clin. Microbiol.,* 28, 989–997, 1990.

12. **Brosbe, E. A., Sugihara, P. T., and Smith, C. R.,** Growth characteristics of *Mycobacterium avium* and group III nonphotochromogenic Mycobacteria in HeLa cells, *J.Bacteriol.,* 84, 1281–1286, 1962.

13. **Calmette, A.,** *Tubercle Bacillus Infection and Tuberculosis in Man and Animals,* Williams & Wilkins, Baltimore, 1923.

14. **Calmette, A. and Valtis, J.,** Virulent filterable elements of the tubercle bacillus, *Ann. Med.,* 19, 553, 1926.

15. **Chawla, R., Pant, K., Jaggi, O. P., Chandrashekhar, S., and Thukral, S. S.,** Fibreoptic bronchoscopy in smear-negative pulmonary tuberculosis, *Eur. Respir. J.,* 1, 804–806, 1988.

16. **Csillag, A.,** Spore formation and "dimorphism" in the mycobacteria, *J. Gen. Microbiol.,* 26, 97–109, 1961.

17. **Daniel, T. M. and Debanne, S. M.,** The serodiagnosis of tuberculosis and other mycobacterial diseases by enzyme-linked immunosorbent assay, *Am. Rev. Respir. Dis.,* 135, 1137–1151, 1987.

18. **Dickerman, S. A., Sherman, A., Balthazar, E. J., and Hazzi, C.,** Ileocecal tuberculosis in a patient with the acquired immune deficiency syndrome, *Am. J. Med.,* 83, 1010–1011, 1987.

19. **Dorozhkova, I. R., Karachunskii, M. A., Kyazimova, L. G., and Bibergal, E. A.,** Comparative effectiveness of different methods used for the microbiological study of excised lung tuberculomas, *Zh. Mikrobiol. Epidemiol. Immunobiol.,* 6, 25–29, 1987.

20. **Dorozhkova, I. R., Karachunsky, M. A., Abdullaeva, E. T., Gamzaeva, N. F., and Kochetkova, E. Ya.,** Detection of tubercle bacilli L Forms as prognostic criterion of tuberculosis relapsing and aggravation in patients with extended residual tuberculous lesions in the lungs, *Probl. Tuberk.,* 3, 14–18, 1989.

21. **Dzis, E. I., Kuzminskaya, A. F., and Samarsky, V. A.,** Characteristic aspects of bacteria elimination in patients with pulmonary tuberculosis, *Vrachebnoe Delo.,* 5, 68–69, 1988.

22. **Eng, R. H. K., Bishburg, E., Smith, S. M., and Mangia, A.,** Diagnosis of *Mycobacterium* bacteremia in patients with acquired immunodeficiency syndrome by direct examination of blood films, *J. Clin. Microbiol.,* 27, 768–769, 1989.

23. **Fontes, A.,** Bemerkungen ueber die tuberculoese infection und ihr virus, *Mem. Inst. Oswaldo Cruz,* 2, 141–146, 1910.

24. **Fox, J. L.,** TB: a grim disease of numbers, *ASM News,* 56, 363–364, 1990.

25. **French, G. L., Chan, C. Y., Cheung, S. W., Teoh, R., Humphries, M. J., and Mahony, G. O.,** Diagnosis of tuberculous meningitis by detection of tuberculostearic acid in cerebrospinal fluid, *Lancet* 117–118, 1987.

26. **Gamzaeva, N. F., Dorozhkova, I. R., and Karachunsky, M. A.,** Increasing the effectiveness of microbiological diagnosis, *Probl. Tuberk.,* 2, 55–57, 1989.

27. **Garvin, D. F.,** Identification of Cell Wall-Deficient Forms from Pulmonary Disease by Chemical and Physical Techniques, Ph.D. dissertation, Wayne State University, Detroit, 1975.

28. **Garvin, D. F.,** quoted by Mattman, L. H., L Forms isolated from infection, in *Microbial Protoplasts, spheroplasts, and L Forms,* Guze, L. B., Ed., Williams & Wilkins, Baltimore, 1968, 472–483.

29. **Ghadially, F. N., Neish, W. J. P., and Dawkins, H. C.,** Mechanisms involved in the production of red fluorescence of human and experimental tumours, *J. Pathol. Bacteriol.,* 85, 77–92, 1963.

30. **Gilkerson, S. W., Moss, M., and Hogue, C.,** Isolation of Mycobacteria by micro- and macro-cultures, *Bacteriol. Proc. 1966,* p.53, 1966.

31. **Golman, K. P.,** AIDS and tuberculosis, *Br. Med. J.,* 295, 511–512, 1987.

32. **Golyshevskaya, V. I., Zemskova, Z. S., and Kovrolev, M. B.,** Characteristics of the filterable forms of *Mycobacterium tuberculosis* and their pathological importance, *Zh. Mikrobiol. Epidemiol. Immunobiol.,* 6, 23–27, 1984.

33. **Gursel, A.,** La valeur du gelose au sang penicilline dans le diagnostic de la tuberculose, *Turk. Hij. Tecr. Biyol. Derg.,* 18, 92–96, 1958.

34. **Hatten, B. A. and Sulkin, S. E.,** Intracellular production of Brucella L Forms. I. Recovery of L Forms from tissue culture cells infected with *Brucella abortus, J. Bacteriol.,* 91, 285–296, 1966.

35. **Jackson, M., Chan, R., Matoba, A. Y., and Robin, J. B.,** The use of fluorescein-conjugated lectins for visualizing atypical mycobacteria, *Arch. Ophthalmol.,* 107, 1206–1209, 1989.

36. **Johnston, F. F. and Sanford, J. P.,** Tuberculous peritonitis, *Ann. Intern. Med.,* 54, 1125–1131, 1961.

37. **Joshi, W., Davidson, P. M., Jones, P. G., Campbell, P. E., and Roberton, D. M.,** Non-tuberculous mycobacterial lymphadenitis in children, *Eur. J. Pediatr.,* 148, 751–754, 1989.

38. **Jubinski, J. C.,** Fluorochrome Labelling of Lysostaphin and Its Application to Identification of Classical and L Phase Staphylococci, M.S. thesis, Wayne State University, Detroit, MI, 1977.

39. **Judge, M. S. and Mattman, L. H.,** Cell wall deficient mycobacteria in tuberculosis, sarcoidosis, and leprosy, in *Cell Wall Deficient Bacteria*, Domingue, G. J., Ed., Addison-Wesley, Reading, MA, 1982, 257–298.

40. **Kadival, G. V., Mazarelo, T. B., and Chaparas, S. D.,** Sensitivity and specificity of enzyme-linked immunosorbent assay in the detection of antigen in tuberculous meningitis cerebrospinal fluids, *J. Clin. Microbiol.,* 23, 901–904, 1986.

41. **Karachunsky, M. A., Dorozhkova, I. R., and Gamzaeva, N. F.,** Microbiological investigation of bronchoalveolar lavage liquid in patients with limited tuberculosis of the lungs, *Probl. Tuberk.,* 3, 17–19, 1988.

42. **Kiehn, T. E., Edwards, F. F., Brannon, P., Tsang, A. Y., Maio, M., et al.,** Infections caused by *Mycobacterium avium* complex in immunocompromised patients, diagnosis by blood culture and fecal examination, antimicrobial susceptibility tests, and morphological and seroagglutination characteristics, *J. Clin. Microbiol.,* 21, 168–173, 1985.

43. **Korsak, T.,** Occurrence of L Forms in a case of generalized mycobacteriosis due to *Mycobacterium scrofulaceum, Acta Tuberc. Pneumol. Belg.,* 66, 445–469, 1975.

44. **Lack, C. H.,** Reproduction of the tubercle bacillus, *Nature (London),* 169, 277, 1951.

45. **Larsson, L., Odham, G., Westerdahl, G., and Olsson, B.,** Diagnosis of pulmonary tuberculosis by selected-ion monitoring: improved analysis to tuberculostearate in sputum using negative-ion mass spectrometry, *J. Clin. Microbiol.,* 25, 893–896, 1987.

46. **Laskowski, L. F., Marr, J. J., Spernoga, J. F., Frank, N. J., Barner, H. B., et al.,** Fastidious Mycobacteria grown from porcine prosthetic-heart-valve cultures, *N. Engl. J. Med.,* 297, 101–102, 1977.

47. **Lucksch, F.,** Körnechenformen und Filtrierbarkeit des tuberkelbacillus, *Beitr. Klin. Tuberk.,* 77, 56–59, 1931.

48. **Mattman, L. H., Tunstall., L. H., Mathews, W. W., and Gordon, D. L.,** L variation in Mycobacteria, *Am. Rev. Respir. Dis.,* 82, 202–211, 1960.

49. **Mattman, L. H.,** L Forms isolated from infections, in *Microbial Protoplasts, Spheroplasts and L Forms,* Guze, L. B., Ed., Williams & Wilkins, Baltimore, 1968, 472–483.

50. **Mattman, L. H.,** Cell wall deficient forms of mycobacteria, *Ann. N.Y. Acad. Sci.,* 174, 852–861, 1970.

51. **Memminger, J. J. and Whitby, J. L.,** Increased niacin levels in blood cultures from known or suspected tuberculosis patients, *Abstr. Ann. Meet. ASM,* 1983, p. 346.

52. **Merckx, J. J., Brown, A. L., Jr., and Karlson, A. G.,** Morphology of two strains of Mycobacteria grown in artificial media as studied with the electron microscope, *Acta Tuberc. Scand.,* 45, 204–220, 1964.

53. **Miguel, J. A.,** Aportaciones al conocimiento de la biologia del bacilo de Koch, *Enferm. Torax,* 4, 159, 1955.

54. **Modilevsky, T., Sattler, F. R., and Barnes, P. F.,** Mycobacterial disease in patients with human immunodeficiency virus infection, *Arch. Intern. Med.,* 149, 2201–2205, 1989.

55. **Moffie, B. G., Krulder, J. W., and de Knijff, J. C.,** Direct visualization of mycobacteria in blood culture, *N. Engl. J. Med.,* 320(1), 61–62, 1989.

56. **Motyl, M. R., Saltzman, B., Levi, M. H., McKitrick, J. C., Friedland, G. H., and Klein, R. S.,** The recovery of *Mycobacterium avium* complex and *Mycobacterium tuberculosis* from blood specimens of AIDS patients using the nonradiometric BACTEC NR 660 medium, *Am. J. Clin. Pathol.,* 94, 84–86, 1990.

57. **Much, H.,** Die Variation des tuberkelbacillus in form und wirkung, *Beitr. Klin. Tuberk.,* 7, 60–71, 1931.

58. **Mühlberger, F., Stengele, G., Schafroth, U., and Walther, A.,** Offene tuberkulosen bei radiologisch nicht nachweisbarem kavernenbefund, *Schweiz. Med. Wschr.,* 117, 1596–1597, 1987.

59. **Nedelkovitch, J.,** Mode du multiplication du bacille de Koch. Morphologie du bacille et des ses colonies. Quelques sources d'erreurs, *Ann. Inst. Pasteur* (Paris), 78, 177–189, 1950.

60. **Nussbaum, J. M., Dealist, C., Lewis, W., and Heseltine, P. N. R.,** Rapid diagnosis by buffy coat smear of disseminated *Mycobacterium avium* complex infection in patients with acquired immunodeficiency syndrome, *J. Clin. Microbiol.,* 28, 631–632, 1990.

61. **Oshima, S., Myrvik, Q. N., and Leake, E.,** The demonstration of lysozyme as a dominant tuberculostatic factor in extracts of granulomatous lungs, *Br. J. Exp. Pathol.,* 42, 138–144, 1961.

62. **Palenque, E., Amor, E., and Bernaldo de Quiros, J. C.,** Comparison of bronchial washing, brushing and biopsy for diagnosis of pulmonary tuberculosis, *Eur. J. Clin. Microbiol.,* 6(2), 191–192, 1987.

63. **Patil, S. A., Sinha, S., and Sengupta, U.,** Detection of Mycobacterial antigens in leprosy serum immune complex, *J. Clin. Microbiol.,* 24, 169–171, 1986.

64. **Pla y Armengol, R.,** Die verschiedenen Formen des tuberkuloseerregers, *Beitr. Klin. Tuberk.,* 77, 47–55, 1931.

65. **Pollock, H. M. and Wieman, E. J.,** Smear results in the diagnosis of mycobacterioses using blue light fluorescence microscopy, *J. Clin. Microbiol.,* 5, 329–331, 1977.

66. **Prego, V., Glatt, A. E., Roy, V., Thelmo, W., Dincsoy, H., and Raufman, J. P.,** Comparative yield of blood culture for fungi and mycobacteria, liver biopsy, and bone marrow biopsy in the diagnosis of fever of undetermined origin in human immunodeficiency virus-infected patients, *Arch. Intern. Med.,* 150, 333–336, 1990.

67. **Rastogi, N. and Venkitasubramanian, T. A.,** Preparation of protoplasts and whole cell ghosts from *Mycobacterium smegmatis, J. Gen. Microbiol.,* 115, 517–521, 1979.

68. **Ratnam, S. and Chandrasekhar, S.,** The pathogenicity of spheroplasts of *Mycobacterium tuberculosis, Am. Rev. Respir. Dis.,* 114, 549–554, 1976.
69. **Ratnam, S. and Chandrasekhar, S.,** The effect of gravitational forces on the viability of spheroplasts of mycobacteria, *Can. J. Microbiol.,* 22, 1397–1399, 1976.
70. **Ribera, E., Martinez-Vazquez, J. M., Ocana, I., Segura, R. M., and Pascual, C.,** Activity of adenosine deaminase in cerebrospinal fluid for the diagnosis and follow-up of tuberculous meningitis in adults, *J. Infect. Dis.,* 155, 603–607, 1987.
71. **Scully, R. E., Mark, E. J., McNeely, W. F., and McNeely, B. U.,** Case records of the Massachusetts General Hospital, Case 35–1988, *N. Engl. J. Med.,* 319, 564–574, 1988.
72. **Steffen, C. M.,** Identification of Cell Wall Deficient Mycobacteria with Auramine Rhodamine Staining and Polyacrylamide Gel Electrophoresis, Ph.D. dissertation, Wayne State University, Detroit, 1985.
73. **Takahashi, S.,** L phase growth of Mycobacteria. I. Cell wall deficient form of Mycobacteria, *Kekkaku,* 54, 63–70, 1979.
74. **Truffot-Pernot, C., Lecoeur, H. F., Maury, L., Dautzenberg, B., and Grosset, J.,** Results of blood cultures for detection of mycobacteria in aids patients, *Tubercle* 70, 187–191, 1989.
75. **Weille-Hallé, B.,** Zur Frage der Virulenzanderung, *Beitr. Klin. Tuberk.,* 77, 39–46, 1931.
76. **Wiker, H. G.,** Quantitative and qualitative studies on the major extracellular antigen of *Mycobacterium tuberculosis* H37Rv and *Mycobacterium bovis* BCG, *Am. Rev. Respir. Dis.,* 141(4 Pt.1), 830–838, 1990.
77. **Xalabarder, C.,** Electron microscopy of tubercle bacilli, *Excerpta Med. Sect. XV Chest Dis.,* 11, 467–473, 1958.
78. **Xalabarder, C.,** Formas L de micobacterias y nefritis cronicas, *Publ. Inst. Antitubercul. Suppl.* 7, 1–83, 1970.
79. **Yanez, M. A., Coppola, M. P., Russo, D. A., Delaha, E., Chaparas, S. D., and Yeager, H., Jr.,** Determination of mycobacterial antigens in sputum by enzyme immunoassay, *J. Clin. Microbiol.,* 23, 822–825, 1986.
80. **Zemskova, Z. S., et al.,** Generalized TB caused by L Forms of TB Mycobacteria in a child, *Probl. Tuberk.,* 2, 64–66, 1985.
81. **Byrne, A.,** unpublished study.
82. **Atkins, A.,** unpublished study.

Chapter 21

SARCOIDOSIS

Sarcoidosis presents a challenge to both clinicians and the microbiologist.[26] Routine methods reveal no organism in smear or culture. This chapter contains information on a mycobacterium blood-borne in over 50 cases of the disease and confirmation that another laboratory has cultured a similar organism from both bronchial washings and blood plasma.[17]

CHARACTERISTICS OF THE DISEASE

Sarcoidosis is described as a disease producing granulomas resembling those of tuberculosis with the difference that the epithelioid-cell tubercles do not become caseated in sarcoid. Additionally, in general, sarcoid differs from tuberculosis in the organs invaded.[23] Unlike tuberculosis, it tends to involve the skin and parotid glands and to cause cystic bone disease and uveitis.[27] It is also peculiar in causing asymptomatic granulomas in striated muscles.[23] It may progress to severe functional impairment of various organs,[8] including the nervous system.[11] Sarcoid is one of the most common causes of sudden, unexpected death in young adults due to invasion of cardiac tissue (Figure 1) and is one cause of blindness which may be preventable. There is some evidence that the disease may be contagious and intrafamilial.[4,7]

KVEIM REACTION

One well-known characteristic of sarcoid is a positive Kveim test. Fifty to eighty percent of patients give a Kveim reaction, a lesion developing after intracutaneous injection of a heat-sterilized suspension of sarcoid spleen or lymph tissue. When biopsied, the nodule histologically resembles sarcoid. It appears to be a delayed reaction to some substance in sarcoid tissue.[12]

ANERGY

Another somewhat unique characteristic of the sarcoid patient is loss of the delayed-type hypersensitivities which characterize normal individuals. Thus, he is usually skin-test negative to tuberculin and to Candida antigen. Often he cannot become sensitized to dinitrochlorobenzene, employed to test for ability to develop delayed-type hypersensitivity. However, the sarcoid patient retains normal antibody responses.[27]

EVIDENCE INVOLVING A MYCOBACTERIUM AS CAUSATIVE AGENT

ANTIBODY

Chapman, in an early study, showed that gel diffusion tests (Ouchterlony) gave substantial support to the theory that a certain mycobacterium is the sarcoid agent. He found, by precipitin bands, that 78% of patients have antibody for atypical mycobacteria but not for tubercle bacilli.[6]

ROLE OF PHAGE?

Mankiewicz and collaborators extensively explored the possible role of mycobacteria plus their phages in sarcoidosis.[19,20] They noted that as a strain of virulent tubercle bacilli became lysogenic it completely altered its pathogenicity and colonial appearance. Especially interesting was a sarcoid patient whose stools contained four mycobacteriophages.[18] The

FIGURE 1.   Sarcoid granulomas on heart wall. (From Roberts, W. C., McAllister, H. A., Jr., and Ferrans, V. J., *Am. J. Med.*, 63, 86–108, 1977. With permission.)

authors found an antibody to mycobacteriophages lacking in patients with sarcoid, but present in individuals with tuberculosis. The theory that a bacteriophage can keep an organism in a CWD stage is supported by Nelson and Pickett's report that, in chronic brucellosis, there is both a circulating atypical form of *Brucella abortus* and a matching phage.[24] Since no phage has been seen in electron micrographs of the sarcoid isolates, perhaps a plasmid may be the pertinent altering factor. No plasmids have been found in *Mycobacterium tuberculosis*.[34]

ACID-FAST ORGANISMS SIGHTED

Acid-fast organisms have been seen in sarcoid tissue by pathologists in several countries.[2] Varied staining methods have revealed acid fast organisms in sarcoid granulomas. In the U.S. they have been especially noted by Cantwell;[5] in Germany the pleomorphic structures have been depicted in detailed electron microscopy by Mascovic[22] and in Spain by Xalabarder.[33] Barth and associates[3] stained acid-fast organisms in the aqueous humor of a sarcoid case complicated by uveitis. Similarly, wall-deficient bacteria in sarcoid uveitis have been photographed within leukocytes by the electron microscopes of Wirostko and associates.[32] Acid-fast bacilli were found by Vaněk in 23 of 23 sarcoid cases. Organisms resembling wall-deficient *M. tuberculosis* were found in 36 of 47 sarcoid patients and in the blood plasma of 10 cases.[30] Chemiluminescent studies by Winsel and colleagues indicate that a microorganism is active in sarcoidosis.[31]

EVIDENCE OF A TRANSMISSIBLE PATHOGEN

The first indications in our laboratory that an organism could, indeed, be found in sarcoid was when scalene lymph nodes from two cases were implanted subcutaneously in guinea pigs. Both nodes were culture negative for classic organisms. One, which contained many pleomorphic acid-fast organisms, caused gross pathology in lungs and liver. The second node, which contained few acid fast pleomorphic organisms, produced definite minimal lesions in a guinea pig.

Research by Mitchell et al. showed that sarcoid tissue injected into mice footpads produced granulomas.[21] Their finding was extended by Taub and associates,[29] who demonstrated that a living agent was involved since granulomas did not result if the sarcoid tissue was frozen and thawed or treated with phenol. This animal model required 15 months for lesion development.

CULTURE OF AN ACID-FAST ORGANISM

Early work by Garvin[10] in our laboratory gave clues to isolation and characterization of the sarcoid agent. From duplicate blood cultures of a sarcoid patient, a CWD form, acid fast by intensified staining, was grown in agitated cultures. Extracts of the growth, in acrylamide-gel electrophoresis, gave a pattern of protein distribution remarkably resembling that of *M. tuberculosis*. The case, followed over a year by Edward Nedwicki, resembled typical sarcoidosis rather than tuberculosis. Garvin's organism was wall deficient, not colonizing on solid medium and lacking classic-type morphology.

Consistent culture of an acid-fast organism from sarcoid patients has only been accomplished in the last decade. Our laboratory has grown a mycobacterium from the blood of 53 of 55 patients. This organism sometimes grows when the patients' blood is incubated without culture medium (see Frontispiece). Growth within red cells is a constant characteristic as also shown in Frontispiece. Extensive study by Judge and Mattman is described below.[13-16] They examined 87 blood cultures from 29 patients. From 28 patients an acid-fast organism grew in 24 to 48 h. Any medium commonly used in the U.S. or Canada was functional. Sometimes growth was superior in aerobic medium, sometimes in hypertonic milieu, and sometimes in anaerobic culture, perhaps resulting from uneven inoculum. The organism is not as dependent on hypertonicity as are many CWD forms.

The isolates from sarcoidosis appear to be identical, with characteristics as follows: Morphologically, the CWD microcolonies resemble those of *M. tuberculosis* and other mycobacteria. They are microaerophilic, growing a few millimeters subsurface in semisolid broth, but are unable to multiply in strictly anaerobic conditions. They are mildly pathogenic for immunocompetent mice, gerbils, and guinea pigs. Infection severity is greatly increased and the time required markedly shortened if one injection of cortisone precedes injection of the CWD organisms. Optimum pathology results when 40 mg of cortisone is given to 4-week-old mice 1 d before inoculation of the bacteria.

The pathologist, Alfred Golden, examining code-labeled slides of animal tissues, was able to distinguish the histological changes resulting from sarcoid organisms from disease caused by CWD *M. tuberculosis*. Susceptibility in the animal species differed in showing ocular disease in over 35% of mice (Figure 2 and Frontispiece) and none in gerbils.

The evidence that sarcoid is caused by a wall deficient Mycobacterium is supported by this Russian report. Filtrable forms indistinguishable from those of the tubercle bacillus were found in bronchial washings of 36 of 47 subjects and in the blood plasma of 10 patients. Some of the tiny forms traversed even a 0.22-μm-pore filter. They were identified by staining with the Ziehl-Nielsen and Morahashi acid-fast stains and by immunofluorescence with antibody made in guinea pigs. They were cultured in liquid medium enriched with 10% plasma. Electron microscope studies were also made.[17]

ANTIBODY VS. THE SARCOID ISOLATES

Five sarcoid isolates stimulated production in rabbits of antibody which reacted to titer with *M. tuberculosis* and *M. fortuitum* but gave little cross-reaction to *M. flavescens*, *M. intracellulaire*, *M. simiae*, *M. szulgai*, and *M. gordonae*.[15,16] Similar cross-reactions occurred between the sarcoid strains and *M. tuberculosis* when the antibody was polyclonal antibody to *M. tuberculosis*.

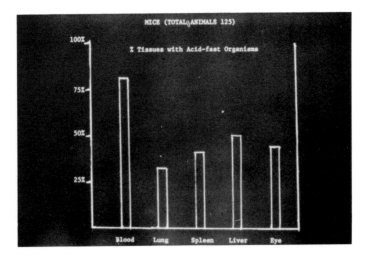

FIGURE 2. Sarcoid variant produces eye disease in a significant percentage of animals. (From Judge, M. S., Ph.D. dissertation, Wayne State University, Detroit, 1979.)

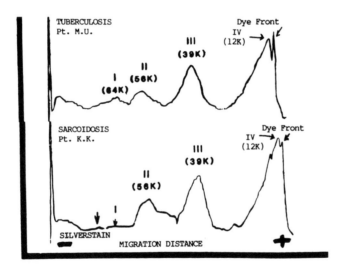

FIGURE 3. Analysis of gel electrophoresis shows similar proteins in CWD organisms from tuberculosis and sarcoidosis. (From Steffen, C. M., Ph.D. dissertation, Wayne State University, Detroit, 1985.)

GEL ELECTROPHORESIS OF PROTEIN CONTENT

Steffen, comparing the cell proteins of *M. tuberculosis* and those of the sarcoid isolates, found no difference,[28] confirming Garvin's work (Figure 3).[10] However, a difference in the extracellular proteins of the two organisms is demonstrable; additional proteins are excreted by the sarcoid strains.[15,16]

ELECTRON MICROSCOPY

Electron micrographs made by Judge showed no detectable differences between CWD *M. tuberculosis* and the CWD sarcoid isolates, except for minute rods in some sarcoid preparations. Electron micrographs of the sarcoid and tuberculosis variants from blood cultures (Figure 4B)[15] remarkably duplicate ones made *in vitro* by antibiotics (Figure 4A).[25]

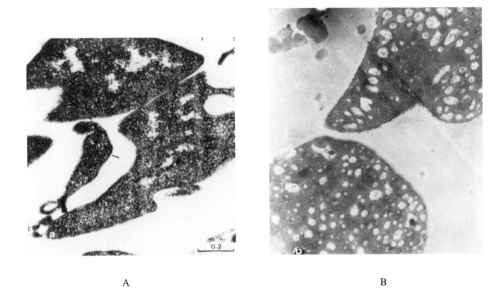

A   B

FIGURE 4. (A) Electron micrograph of CWD *M. tuberculosis* made with antibiotic resembles; (B) CWD organism grown from blood of sarcoidosis patient. (Figure A from Judge, M. S., Ph.D. dissertation, Wayne State University, Detroit, 1979. Figure B from Prozorovsky, S. V., Katz, L. N., and Kagan, G. Ya., *AMS USSR, H'Meditsina,* 1981.)

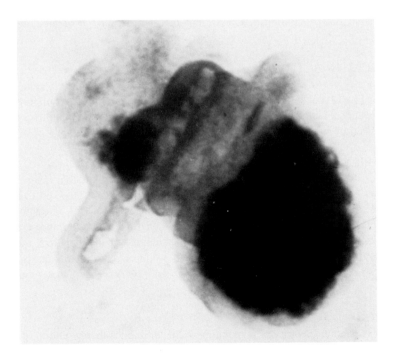

FIGURE 5. Rods grow from sarcoid sphere. (From Judge, M. S., Ph.D. dissertation, Wayne State University, Detroit, 1979.)

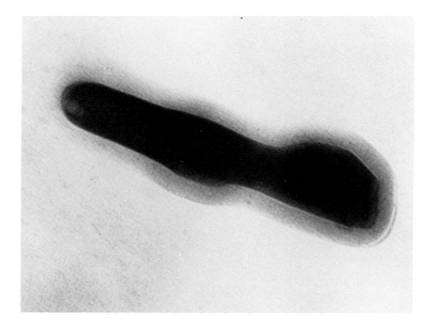

FIGURE 6. Reverted bacillus which did not form macroscopic colonies. (From Judge, M. S., Ph.D. dissertation, Wayne State University, Detroit, 1979.)

Reversion of the sarcoid strains to bacilli occurred (Figures 5 and 6), but this was morphological reversion, not physiological. Classic colonies on the surface of medium never developed.

BIOCHEMICAL REACTIONS

Enzymatically, the sarcoid organisms resemble CWD *M. tuberculosis* in attacking amino acids and urea (see Mycobacteria chapter). Niacin production is evident by supporting growth of a niacin-dependent lactobacillus as shown by Amprin.[1] Tests for indol and H_2S production are negative. Citrate is not utilized, and no acid is produced from sugars.

AUTOFLUORESCENCE OF MICROCOLONIES

The CWD growths of sarcoid and tuberculosis are alike in still another respect. At times a few colonies show fluorescence, varying from bright orange to deep red, viewed with darkfield (see Frontispiece). The pigment usually intensifies under UV. Growth conditions which foster this pigmentation have not been defined. Pigment is not omnipresent, but may have diagnostic significance as CWD forms of other common pathogens lack the pigment, with the exception of *Staphylococcus aureus* which produces a minimal quantity. The bioluminescent pigment varies from that of *Xenorhabdus luminescens* in not requiring dark adaptation.[9]

ANTIMICROBIAL SENSITIVITY

Sarcoidosis is notorious for not responding to the antibiotics therapeutic in tuberculosis. A preliminary search for an antimicrobial which would inhibit the sarcoid strains was made by Evelyn Rosen and Bonnie Simmons. Testing compounds under investigation in leprosy research, they achieved effective growth inhibition of the strains with each of the following: 4,4′-diaminodiphenyl sulfone (DDS) 0.07g/ml, tibione 0.10 g/ml, isonicotinic thioamide 0.08 g/ml, and *p*-isobutyloxyphenyl-*p*′ thiourea (α-pyridyl) 0.03 g/ml. Examination by Judge showed that isoniazid gave greater *in vitro* stimulation of the sarcoid strains than seen with CWD *M. tuberculosis* (Figure 7).[15,16]

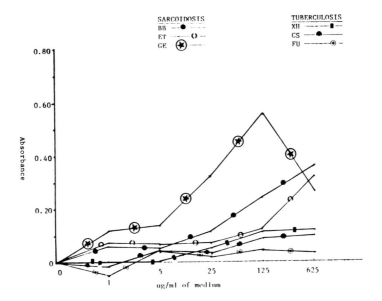

FIGURE 7. Isoniazid (INH)-stimulated growth of isolates from sarcoidosis vs. tuberculosis. Baseline is growth in medium with no inhibitor. Spectrophotometric readings at 525 nm *g*. (From Judge, M. S., Ph.D. dissertation, Wayne State University, Detroit, 1979).

NEW METHODS FOR TUBERCULOSIS AND SARCOID DIAGNOSIS

FINDING THE ORGANISM IN THE BLOOD

1. 48-h blood culture
2. Isolator tube used for blood per se
3. Test phenol phenomenon (see Frontispiece and chapter on Media, Methods and Stains)

IDENTIFYING THE ORGANISM

Acid-fast stains
　　Carbol fuchsin: Kinyoun's intensified
　　Victoria Blue
Biochemical reactions
Cell wall proteins
Extracellular products
Animal pathogenicity: guinea pig, hamster, gerbil, mice

SCREENING METHOD FOR TUBERCULOSIS

Stain sputum with auramine rhodamine and look for pleomorphic as well as classic fluorescing organisms. Stained pleomorphic organisms may not be mycobacteria, but absence of these is a presumptive test to rule out tuberculosis.

SUMMARY

Pleomorphic organisms, acid fast by intensified methods, are repeatedly demonstrated in tissue and smears of sarcoidosis. They can be cultured in 48 h in media suitable for clinical CWD forms. Fortunately, they grow well when blood is inoculated into any blood culture medium commonly used in the U.S. and Canada.

The isolates are especially pathogenic for mice, gerbils, and hamsters when the animals are immunosuppressed, giving pathology similar to but distinguishable from CWD *M. tuberculosis* lesions. The acid-fast pleomorphic organisms are seen in and cultured from granulomas in the animals.

Resemblances of the sarcoid organism to CWD *M. tuberculosis* are found by electron microscopy, biochemical reactions, characteristic protein content, and antigenic reactions. Differences exist not only in histology of the animal lesions, but also in organ tropism, reactivity of cultures to isoniazid, and formation of certain extracellular proteins by the sarcoid strains.

REFERENCES

1. **Amprin, F. L.,** Rapid Identification of *Mycobacterium tuberculosis* by Detection of Characteristic Organic Products, M.S. thesis, Wayne State University, Detroit, 1978.
2. **Aplas, V.,** Eine neue Färberische Nachweismethode für Mykobakterien im Gewebsschnitt, *Arch. Klin. Exp. Dermatol.,* 222, 379–382, 1965.
3. **Barth, C., Judge, M. S., Mattman, L. H., and Hessburg, P.,** Isolation of an acid-fast organism from the aqueous humor in a case of sarcoidosis, *Henry Ford Hosp. Med. J.,* 29 (2), 127–133, 1979.
4. **Biem, J. and Hoffstein, V.,** Aggressive cavitary pulmonary sarcoidosis, *Am. Rev. Respir. Dis.,* 143, 428–430, 1991.
5. **Cantwell, A. R., Jr.,** Histologic observations of variably acid-fast pleomorphic bacteria in systemic sarcoidosis: 3 cases, *Growth* 46, 113–125, 1982.
6. **Chapman, J. S.,** Mycobacterial and mycotic antibodies in sera of patients with sarcoidosis, results of studies using agar double-diffusion technique, *Ann. Intern. Med.,* 55, 918–924, 1961.
7. **Desai, S. G. and Simon, M. R.,** Epidemiology of sarcoidosis, in *Sarcoidosis,* Lieberman, J., Ed., Grune & Stratton, New York, 1985, 32–33.
8. **Fanburg, B. L.,** Sarcoidosis, in *Cecil Textbook of Medicine,* Wyngaarden, J. B. and Smith, L. H., Jr., Eds., W. B. Saunders, Philadelphia, 1988, no. 69.
9. **Farmer, J. J., III, Jorgensen, J. H., Grimont, P. A. D., Akhurst, R. J., Poinar, G. O. Jr., Ageron, E., Pierce, G. V., Smith, J. A., Carter, G. P., Wilson, K. L., and Hickman-Brenner, F. W.,** *Xenorhabdus luminescens* (DNA Hybridization Group 5) from human clinical specimens, *J. Clin. Microbiol.,* 27, 1594–1600, 1989.
10. **Garvin, D. F.,** Identification of CWD Forms from Pulmonary Disease by Chemical and Physical Techniques, Ph.D. dissertation, Wayne State University, Detroit, 1975.
11. **Godwin, J. E. and Sahn, S. A.,** Sarcoidosis presenting as progressive ascending lower extremity weakness and asymptomatic meningitis with hypoglycorrhachia, *Chest,* 97, 1263–1265, 1990.
12. **Johns, C. J.,** Sarcoidosis, in *Principles of Internal Medicine,* 5th ed., Harrison, T. R., Adams, R. D., Bennett, I. L., Jr., Resnik, W. H., Thorn, G. W., and Wintrobe, M. M., Eds., McGraw-Hill, New York, 1962, 1798–1802.
13. **Judge, M. S. and Mattman, L. H.,** Isolation of an acid-fast organism from the blood of sarcoidosis patients, *Bacteriol. Proc. Am. Soc. Microbiol.,* p. 31, 1976.
14. **Judge, M. S. and Mattman, L. H.,** Comparison of animal species as models of sarcoidosis, *Am. Soc. Microbiol. Abstr. Annu. Meet. (Am./Jpn.),* 1979.
15. **Judge, M. S.,** Evidence Implicating a Mycobacterium as the Causative Agent of Sarcoidosis, and Comparison of This Organism with the Blood-Borne Mycobacterium of Tuberculosis, Ph.D. dissertation, Wayne State University, Detroit, 1979.
16. **Judge, M. S. and Mattman, L. H.,** Cell wall deficient Mycobacteria in tuberculosis, sarcoidosis, and leprosy, in *Cell Wall Deficient Bacteria,* Dominigue, G. J., Ed., Addison-Wesley, Reading, MA, 1982, chap 10.
17. **Khomenko, A. G., Golyshevskaya, V. I., and Elshanskaya, M. P.,** Bacteriological study of bronchoalveolar washings and blood plasma in sarcoidosis patients, *Klin. Med. (Moscow),* 65, 80–84, 1987.
18. **Mankiewicz, E.,** Mycobacteriophages isolated from persons with tuberculous and non-tuberculous conditions, *Nature (London),* 191, 1416–1417, 1961.

19. **Mankiewicz, E. and Béland, J.,** The role of mycobacteriophages and of cortisone in experimental tuberculosis and sarcoidosis, *Am. Rev. Respir. Dis.,* 89, 707–720, 1964.
20. **Mankiewicz, E. and Redmond, W. B.,** Lytic phenomena of phage Leo isolated from a sarcoid lesion, *Am. Rev. Respir. Dis.,* 98, 41–46, 1968.
21. **Mitchell, D. N., Rees, R. J. W., and Goswami, K. K. A.,** Transmissible agents from human sarcoid and Crohn's disease tissues, *Lancet,* 2, 761–765, 1976.
22. **Moscovic, E. A.,** Sarcoidosis and Mycobacterial L Forms, histologic studies, in *Cell Wall Deficient Bacteria,* Domingue, G. J., Ed., Addison-Wesley, Reading, MA, 1982, 299–320.
23. **Myers, G. B., Gottlieb, A. M., Mattman, P. E., Eckley, G. M., and Chason, J. L.,** Joint and skeletal muscle manifestations in sarcoidosis, *Am. J. Med.,* 12, 161–169, 1952.
24. **Nelson, E. L. and Pickett, M. J.,** The recovery of L Forms of Brucella and their relation to Brucella phage, *J. Infect. Dis.,* 89, 226–232, 1951.
25. **Prozorovsky, S. V., Katz, L. N., and Kagan, G. Ya.,** L Forms of bacteria (mechanism of formation, structure, role in pathology), *AMS USSR, H'Meditsina,* 1981.
26. **Scully, R. E., Ed.,** Case records of the Massachusetts General Hospital, *N. Engl. J. Med.,* 322, 1728–1738, 1990 and 324, 677–687, 1991.
27. **Sharma, O. P.,** Immunological relationship between sarcoidosis and tuberculosis, *Indian J. Med. Res.,* 58, 1551–1559, 1970.
28. **Steffen, C. M.,** Identification of Cell Wall Deficient Mycobacteria with Auramine Rhodamine Staining and Polyacrylamide Gel Electrophoresis, Ph.D. dissertation, Wayne State University, Detroit, 1985.
29. **Taub, R. N., Sachar, D., Siltzbach, L. E., and Janowitz, H.,** Transmission of ileitis and sarcoid granulomas to mice, *Trans. Assoc. Am. Physicians,* 87, 219–224, 1974.
30. **Vaněk, J.,** Acid-fast bacilli of mycobacterial nature in sarcoidosis, *Beitr. Pathol. Anat.,* 136, 303–315, 1968.
31. **Winsel, K., Christ, R., Eckert, H., Renner, H., Unger, U., and Grollmuss, H.,** Chemiluminescence measurements of alveolar macrophages in patients with sarcoidosis, *Z. Erkrank. Atm. Org.,* 169, 23–32, 1987.
32. **Wirostko, E., Johnson, L., and Wirostko, B.,** Sarcoidosis associated uveitis parasitization of vitreous leukocytes by mollicute-like organisms, *Acta Ophthalmol.,* 67, 415–424, 1989.
33. **Xalabarder, C.,** L Forms of Mycobacteria and chronic nephritis, *Publ. Inst. Antituberc.,* Suppl. 7, 1970.
34. **Zainuddin, Z. F. and Dale, J. W.,** Does *Mycobacterium tuberculosis* have plasmids?, *Tubercle,* 71, 43–49, 1990.

Chapter 22

LEPROSY

"A bespectacled middle-aged beggar sits in a wheelbarrow beside the Ganges in Benares, India. He wears a faded saffron sarong. His legs are swathed in bandages. Leprosy has eaten away his toes and fingers. 'I have not committed any sins in this life. I was an ordinary laborer. Four years ago I was stricken with leprosy and my life changed' ".[24]

Leprosy has not disappeared despite improvement in therapy. Dapsone-resistant strains necessitate multiple-drug treatment.[23] Vaccine development has not reached fruition.[7]

Remarkably, in the U.S. cases occur in individuals who have not traveled outside the northern U. S., and who lack contact with known lepers, thus suggesting carriers. Seeking antibodies to phenolic glycolipid I has been done in an attempt to detect leprosy in healthy people in New Guinea.[2] Similar studies may identify the healthy contact carriers responsible for cases in nonendemic areas of the U.S. This chemically defined glycolipid is believed to be invariably and exclusively associated with *Mycobacterium leprae*.[18]

It is curious, indeed, that the skin lesions of leprosy often reveal myriad acid-fast bacilli, which fail to grow on media suitable for fastidious bacteria and for the tubercle bacillus. Improvements in propagating the agent have been made by many dedicated investigators.

Several lines of evidence indicate that a wall-deficient state is involved in *M. leprae* growth. As early as 1910, Fontes noticed the frequent absence of acid-fast bacilli in many typical leprous lesions.[15,16] Variant growth has been demonstrated in laboratory animals and in cultures in numerous laboratories. Just as 3 years are required for growth of *M. leprae* in the armadillo,[38] its appearance as acid-fast bacilli in culture requires time. This has led to the conception that it cannot be propagated *in vitro*. Many careful studies have resulted in culture of *M. leprae* as described in the following paragraphs.

IN VITRO GROWTH OF *M. LEPRAE*

When the organism of leprosy was cultured by Gürün on his unique medium of wax, almond oil, sucrose, and serum, early growth consisted solely of granules.[17] Much later, acid-fast rods developed in the culture.

An early report of *M. leprae* cultivation was by Twort,[39] distinguished for his discovery of bacteriophage. The secret ingredient of his *M. leprae* medium was ground cells of *M. tuberculosis* added to a medium suitable for the tubercle bacillus. Macroscopic growth occurred after 6 weeks as a colorless film along the inoculating track of the needle.

Ogata's laboratory is 1 of 63 in Japan which have described methods for culture of the leprosy bacillus. Ogata[28] identified the acid-fast rods in his cultures as *M. leprae* by a battery of tests:

1. Injecting an extract of the bacilli into known lepers to initiate a hypersensitivity response.
2. Complement-fixation with antibody from leprosy cases.
3. Granulomas after inoculating the culture intratesticularly in rabbits and subcutaneously in mice.
4. A certain redolence in the cultures which resembled the odor emanating from leprosy patients.

Acid-fast rods did not appear in the cultures until after 2 months. The author felt that during this incubation period an atypical organism propagated. Ogata considered anaerobic cultivation essential, supporting the thesis that initial growth is of a variant form. He remarked

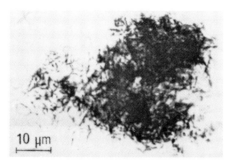

FIGURE 1.   Colony growing from blood plasma of a leprosy case. (From Murohashi, T. and Yoshida, K., *Bull. WHO*, 47, 195–210, 1972. With permission.)

that similarly the atypical stage of the tubercle bacillus frequently required an anaerobic milieu.

Growth of *M. leprae* from patient's nodules was attained in a semisynthetic soft agar medium by Murohashi and Yoshida[27] (Figure 1). Some of the colonies resemble CWD growth of *M. tuberculosis*.

Dhople and associates recently devised a medium which grew *M. leprae* but did not permit subculture.[12] Alexander-Jackson, experienced in following the L cycle in *M. tuberculosis*, also made extensive studies of the variant growth of *M. leprae*.[41]

Desiccated thyroid has been used in culture of *M. leprae*[25] as proposed by Biswas.[5] Hyaluronic acid has been a useful supplement in some media.[5,37] Equally interesting was the observation that the bacteria often grew in the mycelia of an accompanying blastomyces.[25] We have noted that *M. tuberculosis* may grow in the mycelia of Candida. An unanticipated finding was the psychrophilic tendency of *M. leprae* not only growing better at 30 and 10°C than at 37°C, but also multiplying at −20°C.[4,10]

Whereas the growth of *M. leprae* is exceedingly slow as acid-fast rods, growth in a wall deficient state may occur rapidly. Chatterjee[9] found that the organism, in a complex medium at pH 4.0, grows through cyclic changes, all within a few days, bearing little semblance to acid-fast rods. Most significantly, Chatterjee found that Mycobacteria separated from lepromatous tissues consist of the same variegated pleomorphic forms cycling in acid medium. He, therefore, suggested that propagation of *M. leprae* in the host occurs through an L-cycle.

Eight publications on growth of *M. leprae* from the laboratory of Pares were summarized.[29] With improved medium, growth occurred in from 2 to 6 weeks rather than requiring 6 to 12 months as in earlier studies.

When *M. leprae* grows *in vitro*, it exists predominantly within globules, just as it does *in vivo*.[11,35] This suggests that the rods have reverted within the large bodies characteristic of the L-cycle.

A study made in Venezuela in 1973 merits more attention than it received. Thirty-eight cases of tuberculoid and lepromatous leprosy were examined. In 80% of the patients the plasma and/or buffy coat contained a pure culture of a pleomorphic organism which reverted to acid-fast rods (Figure 2). The final organisms resembled *M. leprae* in morphology, in having their acid fastness extractable with pyridine, and in forming nodules in chick chorioallantoic membranes. Macroscopic growth in special media required 2 months; microscopic growth was detectable within 48 h.[3] The large spheres, termed globi by the authors, contained classic bacilli. Such large spheres resembled the L-bodies seen by Tunstall as CWD *M. tuberculosis*, induced by cycloserine, reverted to typical bacilli.

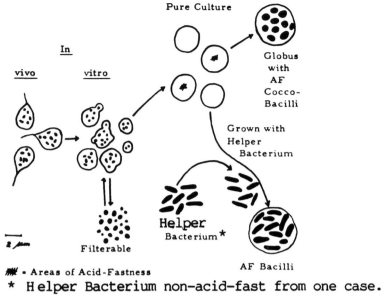

FIGURE 2. Growth from plasma of leprosy cases. Bodies with hyaline "tails" develop into spheres with acid-fast central areas and, finally, if helper bacterium is present, yield acid-fast bacilli. (From Barksdale, L., Convit, J., Kim, K. S., and de Pinardi, M. E., *Biochem. Biophys. Res. Commun.*, 54(1), 290–296, 1973. With permission.)

Growing *M. leprae in vitro* in mouse macrophages has brought great improvement in evaluating therapeutic agents for Hansen's disease.[32] Again, growth is evidenced by increase in a specific phenolic glycolipid. Morphology of the resident bacteria has not been described.

IN VIVO PROPAGATION OF *M. LEPRAE*

GROWTH IN MICE

Evans and Levy reported that during the early growth of *M. leprae* in the mouse footpad, the few organisms detected were within macrophages and bound by a *single* membrane. When organisms were found in the sarcoplasm of striated muscle, they were usually similarly bound by a single membrane. All the organisms appeared to be intact and viable.[14]

Use of the footpad method as devised by Shepard[36] has greatly abetted research on *M. leprae*. While it has often been assumed that the organism remains in the footpad, this is probably not true, as revealed in Reference 34. Rees et al.[34] show that progressive and prolonged infection with *M. leprae* makes no obvious change in the health of the mice, which disseminate *M. leprae* via nasal exudate. Perhaps similar "healthy" human carriers explain the new cases appearing in the U.S. without known source of infection. The specific staining method for *M. leprae* of Campo-Aasen and Convit,[8] described later in this chapter, may facilitate detection of both the rod and aberrant forms of the organism, thereby simplifying recognition of carriers.

LEPROSY IN AN IMMUNOLOGICALLY DEFICIENT ANIMAL

M. leprae multiplies more rapidly in mice pretreated by thymectomy plus total body irradiation. Thymectomy negates the cellular immunity of the mouse, while irradiation prevents antibody formation. The immunologically deficient mouse gives high yields of *M. leprae* growing with normal walls.[33] Pattyn[31] has shown that mice subjected to thymectomy alone do not show increased susceptibility to *M. leprae*.

GROWTH IN THE ARMADILLO

The armadillo (*Dasypus novemcinctus* L.) is an animal susceptible to *M. leprae*, giving a copious yield of bacilli-filled granulomas. Therefore, for the first time it has been possible to harvest large quantities of the bacteria for analyses.[38]

Most surprising, *M. leprae* has been found to cause natural infections in armadillos. The infected mammals were found in Louisiana and Texas.[1,21,26] And equally unexpected is evidence that soil may be a source of the pathogen.[6] One would suspect that it does not grow in soil in the classic stage.

CELL CULTURES IN PERITONEAL CAVITIES

M. leprae mimics some viruses and Rickettsiae in cell cultures, as painstakingly recorded by Jadin et al.[20] Portions of 600 lepromatous lymph nodes seeded in the peritoneal cavities of mice and hamsters yielded growth in the peritoneal macrophages which could be followed for 8 months. Initially the organism grew intranuclearly, only belatedly breaking through the nuclear membrane to grow in the cytoplasm.[20]

A STAIN FOR THE CLASSIC AND VARIANT STAGES OF *M. LEPRAE*

A stain which reacts with the phospholipid of *M. leprae* holds promise for detection of all growth phases of the organism in patients, carriers, inoculated animals, and cultures. The stain described by Campo-Aasen and Convit reveals both typical rods and atypical forms.[8] This stain should be useful in identifying the large numbers of untreated, undiagnosed lepers.

In many cases of leprosy, 10^5 Mycobacteria per milliliter circulates in the blood.[13] These are organisms with typical bacillary morphology. The stain of Campo-Aasen and Convit[8] may show that CWD forms circulate in the blood when the typical bacilli are absent.

ANALYSIS OF IMMUNE COMPLEXES

Circulating immune complexes have been demonstrated in leprosy.[30] The authors concluded that mycobacterial antigens were in the immune complexes.

MONOCLONAL ANTIBODIES

Antibody specific for *M. leprae* has been prepared by Kolk et al.[22] Such monoclonal antibody may not only aid early diagnosis but certify identity of the common, early, rapidly multiplying stages of *M. leprae*. The topic of monoclonal antibodies for Mycobacteria has been reviewed.[19] Using monoclonal antibody, Young found copious amounts of *M. leprae*-specific antigen in tuberculoid leprosy in which acid-fast organisms could not be demonstrated.[40]

SUMMARY

Reputable investigators have cultured *M. leprae* in a wall-deficient stage for decades. Methods have recently improved. Proof that the growth is *M. leprae* has been given by many methods; further confirmation may be forthcoming by identification of glycolipids and by employing monoclonal antibodies. This could lead to early identification of cases and carriers.

In almost 120 years it has not been possible to grow the organisms in macroscopic colonies. Much can be done with the microscopic growth obtainable in 48 h.

Unexpected findings in addition to evidence of carriers in the U.S. include isolating the organism from naturally infected armadillos and from soil, and learning its psychrophilic, acidophilic, and anaerobic preferences.

REFERENCES

1. **Anderson, M.,** Leprosy in an armadillo from Texas, *Lepr. Sci. Mem.,* June, 1978.
2. **Bagshawe, A. F., Garsia, R. J., Baumgart, K., and Astbury, L.,** IgM serum antibodies to phenolic glycolipid-I and clinical leprosy: two years' observation in a community with hyperendemic leprosy, *Int. J. Lepr.,* 58/1, 25–30, 1990.
3. **Barksdale, L., Convit, J., Kim, K. S., and de Pinardi, M. E.,** Spheroidal bodies and globi of human leprosy, *Biochem. Biophys. Res. Commun.,* 54(1), 290–296, 1973.
4. **Bhatia, V. N.,** Effect of temperature, cholesterol and nerve tissue on multiplication of armadillo *M. leprae, Indian J. Lepr.,* 61, 453–457, 1989.
5. **Biswas, S. K.,** Growth of *Mycobacterium leprae* in thyroxine treated culture medium-a preliminary report., *Lepr. India,* 50, 57–63, 1978.
6. **Blake, L. A., West, B. C., Lary, C. H., and Todd, J. R., IV,** Environmental nonhuman sources of leprosy, *Rev. Infect. Dis.,* 9, 562–577, 1987.
7. **Bloom, B. R., Salgame, P., Mehra, V., Kato, H., Modlin, R., Rea, T., Brennan, P., Convit, J., Lugozi, L., Snapper, S., et al.,** Vaccine development. On relating immunology to the Third World: some studies on leprosy, *Immunology,* 2, 87–9; 91–2, 1989.
8. **Campo-Aasen, I. and Convit, J.,** Identification of the noncultivable pathogenic Mycobacteria *M. leprae* and *M. lepraemurium, Int. J. Lepr.,* 36, 166–170, 1968.
9. **Chatterjee, B. R.,** Cytology of *M. leprae, IX Int. Congr. Microbiol.(Moscow),* Abstr. Pap., p. 63, 1966.
10. **Chatterjee, B. R. and Das-Roy, R.,** Growth of *M. leprae* in a redox system. III. Evidence of growth at low temperature (Psychro-phillia) and, further refinement of growth medium, *Indian J. Lepr.,* 61, 458–466, 1989.
11. **Devignat, R.,** Multiplication of Hansen's bacillus in complex symbiosis *in vitro, Nature (London),* 190, 832, 1961.
12. **Dhople, A. M., Green, K. J., and Osborne, L. J.,** Limited in vitro multiplication of *Mycobacterium leprae, Ann. Inst. Pasteur Microbiol.,* 139, 213–223, 1988.
13. **Drutz, D. J., Chan, T. S. N., and Lu, W. H.,** The continuous bacteremia of lepromatous leprosy, *N. Engl. J. Med.,* 287, 159–162, 1972.
14. **Evans, M. J. and Levy, L.,** Ultrastructural changes in cells of the mouse footpad infected with *M. leprae, Infect. Immun.,* 5, 238–247, 1972.
15. **Fontes, A.,** Bemerkungen ueber die tuberculoese Infection und ihr Virus, *Mem. Inst. Oswaldo Cruz,* 2, 141–146, 1910.
16. **Fontes, A.,** Über die Filtrierbarkeit des Tuberkulosevirus vom Standpunkte des Polymorphismus, *Beitr. Klin. Tuberk.,* 77, 1–15, 1931.
17. **Gürün, H.,** A new culture method for the organism of leprosy, tuberculosis, and syphilis, *Ruzgarli Matbaa* (Ankara), pp.1–42, 1957.
18. **Hunter, S. W. and Brennan, P. J.,** A novel phenolic glycolipid from *M. leprae* possibly involved in immunogenicity and pathogenicity, *J. Bacteriol.,* 147, 728–735, 1981.
19. **Ivanyi, J., Morris, J. A., and Keen, M.,** Studies with monoclonal antibodies to mycobacteria, *Monoclonal Antibodies Bacteria,* 1, 59–90, 1985.
20. **Jadin, J., Francois, J., Bisoux, M., Languillon, J., and Moris, R.,** Development intranucleaire de *M. leprae* dans les cellules histiocytaires chez l'animal, *Bull. Acad. Natl. Med. (Paris),* 152, 89–91, 1968.
21. **Kirchheimer, W. F. and Sánchez, R. M.,** Leprosy in the wild, *Lepr. Sci. Mem.,* June 1978.
22. **Kolk, A. H. J., Ho, M. L., Klatser, P. R., Eggelte, T. A., DeJonge, S., and Kuijper, S.,** The use of monoclonal antibodies for the detection of mycobacteria and mycobacterial antigens, *Antonie van Leeuwenhoek,* 51, 429–30, 1985.
23. **Lechat, M. F., Declercq, E. E., Mission, C. B., and Vellut, C. M.,** Selection of MDT strategies through epidemiometric modeling, *Int. J. Lepr.,* 58/2, 296–301, 1990.
24. *Life Mag.,* December 1990.
25. **Massalski, W. K., Shukla, R. R., Lewenstein, W., and Wierzbicki, R.,** Culture of Hansen's bacilli in vitro in a case of lepromatous leprosy, *Mater. Med. Polona,* 10, 35–37, 1978.

26. **Meyers, W. M., Walsh, G. P., Brown, H. L., Rees, R. J. W., and Convit, J.,** Naturally acquired leprosy-like disease in the nine-banded armadillo (*Dasypus novemcinctus*): reactions in leprosy patients to lepromins prepared from naturally infected armadillos, *J. Reticuloendothelial Soc.,* 22/4, 369–75, 1977.

27. **Murohashi, T. and Yoshida, K.,** An attempt to culture *Mycobacterium leprae* in cell-free, semisynthetic, soft agar media, *Bull. WHO,* 47, 195–210, 1972.

28. **Ogata, N.,** Cultivation of human leprosy bacillus (Hansen's bacillus), *Jpn. J. Microbiol.,* 3, 139–143, 1959.

29. **Pares, Y.,** Letter to the Editor, *Int. J. Lepr. Other Microbact. Dis.,* 46, 223–4, 1978.

30. **Patil, S. A., Sinha, S., and Sengupta, U.,** Detection of Mycobacterial antigens in leprosy serum immune complex, *J. Clin. Microbiol.,* 24, 169–171, 1986.

31. **Pattyn, S. R.,** Comparative behavior of a strain of *M. leprae* in 5 different mouse strains and in thymectomized mice, *Zentralbl. Bakteriol. Parasitenkd. Infektionskr. Hyg. Abt. 1 Orig.,* 197, 256–258, 1965.

32. **Ramasesh, N., Krahenbuhl, J. L., and Hastings, R. C.,** In vitro effects of antimicrobial agents on *Mycobacterium leprae* in mouse peritoneal macrophages, *Antimicrob. Agents Chemother.,* 33, 657–662, 1989.

33. **Rees, R. J. W., Waters, M. F. R., Weddel, A. G. M., and Palmer, E.,** Experimental lepromatous leprosy, *Nature (London),* 215, 599–602, 1967.

34. **Rees, R. J. W., Weddell, A. G. M., Palmer, E., and Pearson, J. M. H.,** Human leprosy in normal mice, *Br. Med. J.,* 3, 216–217, 1969.

35. **Roy, A. N.,** Some interesting observations of the cultivation of *M. leprae, Indian Med. J.,* 60, 24–25, 1966.

36. **Shepard, C. C.,** The experimental disease that follows the injection of human leprosy bacilli into foot pads of mice, *J. Exp. Med.,* 112, 445–454, 1960.

37. **Skinsnes, O. K., Matsuo, E., Chang, P. H. C., and Anderson, B.,** In vitro cultivation of leprosy bacilli on hyaluronic acid based medium-a preliminary report, *Int. J. Lepr.,* 43, 193–203, 1975.

38. **Storrs, E. E., Walsh, G. P., and Burchfield, H. P.,** Leprosy in the armadillo, new model for biomedical research, *Science,* 183, 851–2, 1974.

39. **Twort, F. W.,** A method for isolating and growing the lepra bacillus of man, preliminary note, *Proc. R. Soc. London (Biol.),* 83, 156–8, 1911.

40. **Young, S.,** Detection of antigens in tissues, Indo-UK Workshop on Leprosy Research, Central JALMA Institute for Leprosy, Agra, India, February 15–17, 1989.

41. **Alexander-Jackson, E.,** unpublished manuscript.

Chapter 23

CROHN'S DISEASE AND ULCERATIVE COLITIS

A search for the microbes responsible for Crohn's disease (CD) and ulcerative colitis (UC) has continued for approximately 60 years, and methods to find the answers are now on the drawing board.

An excellent description of CD, was given by Mitchell and Parent.[26] The chronic inflammation may involve any part of the gastrointestinal tract from esophagus to anus. The inflammation is often discontinuous, with diseased segments separated by apparently healthy tissue. Ulcers may perforate the wall, resulting in leakage of intestinal content into the abdominal cavity, or even the formation of fistulas connecting intestine and skin, bladder, vagina, or adjacent bowel. Also, noncaseating granulomas are noted histologically with associated regional and mesenteric lymph nodes. These findings suggest that CD is due to infections by mycobacteria or fungi. Surgical removal of diseased bowel is often required (Figure 1).

The etiology of a second important inflammatory bowel disease, UC, is also an enigma. UC lacks the thickening of intestinal wall, narrowing of the intestinal lumen, and the granulomas characteristic of CD.[12,13] Although the colon is the primary site of attack, the disease is systemic in nature and often associated with arthritis, uveitis, venous thromboses, liver disease and varied skin lesions, especially pyoderma gangrenosum. The extraintestinal manifestations are virtually identical with those encountered in CD. In many cases it is difficult to differentiate the two conditions. It is estimated that combined CD and UC currently afflict 2 million persons.[20]

There is little doubt that an infectious agent initiates CD, and much evidence indicates that this is also true for UC. Two laboratories, using homogenates of CD tissue, have produced granulomas in mice footpads,[25,32] whereas control bowel homogenates did not. Similarly, Cave and colleagues[5] showed that preparations from CD ileum cause inflammation and granulomas in rabbit ileum. An infectious etiology of CD is also supported by the improvement which often occurs when patients are given immune globulin intravenously.[18]

Pathogenic microbes transmissible to rabbits reside in both CD and UC tissue, with the resultant disease from CD tissue involving mainly the ileum, whereas the UC tissue causes pathology in rabbit colon.[2] The question is, which of the myriad organisms in the diseased intestines are the essential pathogens in the transmission studies?

The athymic, T-cell deficient (nude) mouse has also been an experimental animal in transmission studies in Crohn's disease. Injections of filtrates of Crohn's intestinal tissue or mesenteric lymph nodes produce hyperplastic lymph nodes in some mice. Fluorescent antibody studies show what appears to be a pertinent antigen in the mouse nodes.[29,36]

IS CWD *PSEUDOMONAS MALTOPHILIA* THE AGENT OF CD?

What appeared to be a most promising lead was made by Parent and Mitchell, who isolated CWD *Pseudomonas maltophilia* from intestinal tissue of 8 CD cases.[28] Even more convincing was finding the pseudomonad in mesenteric lymph nodes in two of the patients. Supporting this evidence was antibody to the revertant Pseudomonas strains in 22 of 25 CD patients, absent in 23 ulcerative colitis patients and 15 control individuals.[30] Despite these findings, a major role for *P. maltophilia* in CD has not emerged. Possibly, the data indicate that pseudomonads proliferate in diseased iliac tissue. In five febrile CD patients, Besserer[37] found one with positive blood cultures, namely for *P. aeruginosa,* typed with fluorescent antibody. The patient's episode subsided without specific therapy. Another investigation did not find *P. maltophilia* antigen in CD iliac tissue,[34] and the authors concluded that the bacterium might reside superficially in the lumen.

FIGURE 1. Regional enteritis (Crohn's disease). Segment of terminal ileum has thickened wall and chronically inflamed mucosa. Note sharp demarcation of diseased segment from normal mucosa on the left. (From Glickman, R. M. and Isselbacher, K. J., *Harrison's Principles of Internal Medicine*, 8th ed., McGraw-Hill, New York 1977. With permission.)

IS THE CWD FORM OF *STREPTOCOCCUS FECALIS* INVOLVED IN THESE INFLAMMATORY BOWEL DISEASES?

How extensively L Forms may be involved in intestinal disease was further shown by Orr and associates.[27] They found that a strain of *Streptococcus fecalis* was innocuous when injected into rabbit ileum, whereas the L-Form variant caused granulomas.

FINDINGS WHEN SEEKING VIRUSES

Searches to detect viruses in CD and UC have been productive without contributing to bowel disease etiology. Commenting that viruses are not known as part of the normal gut flora, or as causing gut infections, Aronson and colleagues[1] and Beeken et al.[2] supplied figures to challenge these concepts. An RNA virus emerged, not only in CD, but also in cancerous and other diseased bowel.

Interesting studies by Gitnick and associates, seeking viruses, produced tissue culture cytopathogenesis (CPE) from CD tissue, which was different from the CPE produced from UC tissue.[9] Surprisingly, further studies by McLaren and Gitnick[22] indicated that the CPE was caused by toxins, not living agents, since the substances resisted UV and were not sedimented. They proved to be toxins distinct from *Clostridium difficile* toxin.

MYCOBACTERIAL INVOLVEMENT? WHICH ONE?

Mycobacterium paratuberculosis in a mask of CWD is the leading candidate for causation in CD. Similar findings, perhaps with other species, apply to UC. Pertinent findings relative to CD are given in Table 1.

In addition to isolations of the organism, antibody in the patient supports a role for *M. paratuberculosis* in CD. CD patients have significantly more antibody to the bacterium than do control individuals.[33]

One might think that the above data plus the fulfillment of Koch's postulates with two strains[6] would firmly establish CD causation. Such has not been the case for several reasons discussed below.

Other Mycobacteria which have been isolated from Crohn's tissue include *M. chelonei*,[10] *M. fortuitum*,[14] *M. avium*,[14] and *M. kansasii*.[4,11] Chiodini has noted that these species are well-known opportunists invading immunocompromised hosts.[6]

TABLE 1
Isolations of *Mycobacterium paratuberculosis* from Crohn's Tissue

Chiodini[6]	4 isolates
Gitnick et al.[11]	2 isolates
Coloe et al.[38]	1 isolate[a]
Haagsma et al.[16]	2 isolates
Thorel et al.[39]	1 isolate

[a] This and other Crohn's patients and Johne's diseased cattle all had antibodies reacting with identical antigens of *M. paratuberculosis*.[6]

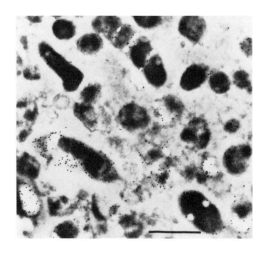

FIGURE 2. Electron micrograph of a bacterial isolate from Crohn's disease stained with anti-*M. kansasii* antibody, followed by protein A-gold. This antibody cross-reacts slightly with Nocardia but not with other genera. The bar represents 1 micron. (Magnification × 27,000.)(Submitted by J. L. Stanford.)

The research is complicated by residence, transitory or otherwise, of varied mycobacterial species in the human intestinal tract.[8,23] One investigator estimates that due to their abundance in the environment, we invariably ingest or inhale millions of Mycobacteria every day.[17] Thus, perhaps it is not surprising that in one study, 40% of biopsies from noninflammatory bowel disease yielded slow growing Mycobacteria.[14] This was more than the isolations from CD or from UC. In contrast, spheroplastic stages of Mycobacteria grew from 12% of Crohn's tissues and 16% of UC tissue and from none of 27 noninflammatory bowel disease tissue. Could this indicate that pathogenesis results only from the spheroplastic stage?

A role for Mycobacteria in UC is supported by Stanford and co-investigators who found pleomorphic variably acid-fast bacteria in 52% of UC patients and only 7% of controls.[31] The researchers also implicated *M. kansasii* in CD (Figure 2). The Kveim test, well recognized as positive in sarcoidosis, is also positive in about half of Crohn's patients. This, too, suggests a mycobacterial etiology for CD.[24–26]

Very promising is identifying the Mycobacterium within the intestinal wall by its specific glycolipid. Such a study, done in 1989, used a monoclonal antibody reactive for a glycolipid present in both *M. avium* and the Mycobacteria from CD. This antibody vs. glycolipid gave evidence that the pertinent Mycobacterium may be found in superficial intestinal tissue of controls, but in CD was present in the submucosa and subserosa.[3]

Similar isolations of Mycobacteria were made from CD and UC by Burnham's group in London.[4] The interpretation of culture isolates will be greatly improved by more precise identification of species by restriction endonuclease analysis.[7,21] Furthermore, improved media giving more rapid growth and reversion of spheroplasts as devised by Graham and associates is expected to clarify the role of pertinent species in these two conditions.[15]

Is it possible CD can be caused by either of two organisms? Erysipelas is well known to have more than one agent, and an important worldwide disease of fish has multiple causative bacteria.[19]

DOES THERAPY SUGGEST MYCOBACTERIAL INVOLVEMENT?

The numerous studies of antimycobacterial therapy directed to Crohn's patients has been summarized.[6] The pivotal antimicrobial agent has been rifabutin, chosen because of its effectiveness in treating the disease in a monkey species in which *M. paratuberculosis* causes naturally occurring regional ileitis. Because of the chronicity of CD, effectiveness has been difficult to evaluate. One of the encouraging studies by Thayer et al.[33] combined streptomycin with rifabutin and indicated that usually 4 to 6 months of therapy are necessary before improvement is definite.

Also encouraging was the improvement in uveitis in a CD patient given rifampin.[35] Again, response to rifampin supports mycobacterial etiology, just as culture studies suggest the bacterium is growing as a spheroplast.

SUMMARY

Viruses and pseudomonads have been supplanted as possible etiological agents in the inflammatory bowel conditions, CD and UC. Especially for CD the focus is on Mycobacteria because a granulomatous intestinal infection in goats, Johne's disease, is caused by *M. paratuberculosis* and because this species has been found in Crohn's disease tissue by five investigative groups in varied countries. Furthermore, revertant *M. paratuberculosis* from Crohn's tissues has produced ileal granulomas when fed to infant goats. A greater amount of pertinent data can be expected as a result of improved medium for growth and reversion of the slowly growing mycobacterial species. Also contributory will be more precise definition of species and identification of isolates by restriction endonuclease analysis.

M. kansasii is also a candidate for initiating CD or UC. *M. kansasii* has been isolated from these tissues by some laboratories. Research has been difficult because of the prolonged time required for growth of these slow-growing species in the classic state. Moreover, reversion from spheroplasts has required extreme patience.

There is a possibility that the conditions result from invasion by mycobacterial spheroplasts rather than classic bacilli. This is suggested by culture studies and by noting that a strain of *S. fecalis* is an intestinal pathogen only when the organism is CWD.

All the studies on etiology do not negate the voluminous data that actual tissue damage results from immune processes which are made in response to microbial invasion.

REFERENCES

1. **Aronson, M. D., Phillips, C. A., Beeken, W. L., and Forsyth, B. R.,** Isolation and characterization of a viral agent from intestinal tissue of patients with Crohn's disease and other intestinal disorders, *Progr. med. Virol.,* 21, 165–176, 1975.
2. **Beeken, W. L., Goswami, K. K. A., and Mitchell, D. N.,** Studies of a viral agent isolated from patients with Crohn's disease and other intestinal disorders, *Gut* 16, 401, 1975.
3. **Blaauwgeers, J. L. G., Das, P. K., Slob, A. W., and Houthoff, H. J.,** Human gut wall reactivity to monoclonal antibodies against *M. avium* glycolipid in relation to Crohn's disease (preliminary results), *Acta Lepr.,* 7 (Suppl. 1), 138–140, 1989.
4. **Burnham, W. R., Lennard-Jones, J. E., Stanford, J. L., and Bird, R. G.,** Mycobacteria as a possible cause of inflammatory bowel disease, *Lancet* 2, 693–696, 1978.
5. **Cave, D. R., Mitchell, D. N., Kane, S. P., and Brooke, B. N.,** Further animal evidence of a transmissible agent in Crohn's disease, *Lancet* 2, 1120–1122, 1973.
6. **Chiodini, R. J.,** Crohn's disease and the mycobacterioses, a review and comparison of two disease entities, *Clin. Microbiol. Rev.,* 2, 90–117, 1989.
7. **Collins, D. M. and DeLisle, G. W.,** Restriction endonuclease analysis of various strains of *Mycobacterium paratuberculosis* isolated from cattle., *Am. J. Vet. Res.,* 47, 2226–2229, 1986.
8. **Edwards, L. B. and Palmer, C. E.,** Isolation of "atypical" Mycobacteria from healthy persons, *Am. Rev. Respir. Dis.,* 80, 747–749, 1959.
9. **Gitnick, G., Rosen, V. J., Arthur, M. H., and Hertweck, S. A.,** Evidence for the isolation of a new virus from ulcerative colitis patients, *Dig. Dis. Sci.,* 24, 609–619, 1979.
10. **Gitnick, G.,** Is Crohn's disease a mycobacterial disease after all?, *Dig. Dis. Sci.,* 29, 1086–1088, 1984.
11. **Gitnick, G., Collins, J., Beaman, B., Brooks, D., Arthur, M., Imaeda, T., and Palieschesky, M.,** Preliminary report on isolation of mycobacteria from patients with Crohn's disease, *Dig. Dis. Sci.,* 34, 925–932, 1989.
12. **Glickman, R. M. and Isselbacher, K. J.,** Diseases of the small intestine, in *Harrison's Principles of Internal Medicine,* 8th ed., Thorn, G. W. et al., Ed., McGraw-Hill, New York, 1977, 1540.
13. **Glickman, R. M.,** Inflammatory bowel disease, in *Harrison's Principles of Internal Medicine,* 12th ed., Wilson, J. D. et al., Ed., McGraw-Hill, New York, 1991, 1268–1281.
14. **Graham, D. Y., Markesich, D. C., and Yoshimura, H. H.,** Mycobacteria and inflammatory bowel disease: results of culture, *Gastroenterology,* 92, 436–42, 1987.
15. **Graham, D. Y., Markesich, D. C., Yoshimura, H. H., and Estes, M. K.,** Microbial etiology of Crohn's disease, Mycobacteria-cause or commensal?, in *Inflammatory Bowel Disease,* Anagnostides, A. A., Hodgson, H. J. F., and Kirsner, J. B., Eds., Van Nostrand Reinhold, New York, 1991, 179–200.
16. **Haagsma, J., Mulder, C. J. J., Eger, A., Bruins, J., Ketel, R. J., and Tytgat, G. N. J.,** Mycobacterium species isolated from patients with Crohn's disease, in *Inflammatory Bowel Disease. Current Status and Future Approach,* MacDermott, R. P., Ed., Elsevier, Amsterdam, 1988, 535–538.
17. **Hermon-Taylor, J., Moss, M., Tizard, M., Malik, Z., and Sanderson, J.,** Molecular biology of Crohn's disease mycobacteria, *Baillière's Clin. Gastroenterol.* 4, 23–42, 1990.
18. **Knoflach, P., Müller, C., and Eibl, M. M.,** Crohn's disease and intravenous immunoglobulin G, *Ann. Intern. Med.,* 112, 385–386, 1990.
19. **Liu, K. C., Mai, L. M., and Chien, C. H.,** Electron microscopic study of the reproductive stages of the gill rot pathogen, *Branchiomyces* sp., of fish, *Fish Pathol.,* 20, 267–272, 1985.
20. **Marlando, J.,** Crohn's disease and ulcerative colitis, *House Calls,* September 1990, p.14.
21. **McFadden, J. J., Butcher, P. D., Thompson, J., Chiodini, R., and Hermon-Taylor, J.,** The use of DNA probes identifying restriction-fragment-length polymorphisms to examine the *Mycobacterium avium* complex., *Mol. Microbiol.,* 1, 283–291, 1987.
22. **McLaren, L. C. and Gitnick, G.,** Ulcerative colitis and Crohn's disease tissue cytotoxins, *Gastroenterology,* 82, 1381–1388, 1982.
23. **Mills, C. C.,** Occurrence of *Mycobacterium* other than *Mycobacterium tuberculosis* in the oral cavity and in sputum, *Appl. Microbiol.,* 24, 307–310, 1972.
24. **Mitchell, D. N., Cannon, P., Dyer, N. H., Hinson, D. F. W., and Willoughby, J. M. T.,** The Kveim test in Crohn's Disease, *Postgrad. Med. J.,* 46, 491–495, 1970.
25. **Mitchell, D. N., Rees, R. J. W., and Goswami, K. K. A.,** Transmissible agents from human sarcoid and Crohn's disease tissues, *Lancet* 2, 761–765, 1976.
26. **Mitchell, P. D. and Parent, K.,** Role of cell wall deficient bacteria in diseases of the gastrointestinal tract, in *Cell Wall Deficient Bacteria,* Domingue, G. J., Ed., Addison-Wesley, MA, 1982.
27. **Orr, M. M., Tamarind, D. L., Cook, J., et al.,** Preliminary studies on the response of rabbit bowel to intramural injection of L Form bacteria (Abstr.), *Br. J. Surg.,* 61, 921, 1974.

28. **Parent, K. and Mitchell, P. D.,** Cell wall defective variants of *Pseudomonas*-like (Group Va) bacteria in Crohn's disease, *Gastroenterology,* 75, 368–372, 1978.

29. **Pena, A. S., Kuiper, I., Walvoort, H. C., Verspaget, H. W., Weterman, I. T., Ruitenberg, E. J., and Das, K. M.,** Seropositivity in Dutch Crohn's disease patients against primed nude mouse lymph nodes, and the difference with lymphocytotoxic antibodies, *Gut* 27, 1426–1433, 1986.

30. **Shafii, A., Sopher, S., Lev, M., and Das, K. M.,** An antibody against revertant forms of CWD bacterial variant in sera from patients with Crohn's disease, *Lancet* 2, 332–333, 1981.

31. **Stanford, J. L., Dourmashkin, R., McIntyre, G., and Visuvanathan, S.,** in *Inflammatory Bowel Disease. Current Status and Future Approach,* MacDermott, R. P., Ed., Elsevier, Amsterdam, 1988, 503–508.

32. **Taub, R. N., Sachar, D. B., Siltzbach, L. E., and Janowitz, H. D.,** Transmission of ileitis and sarcoid granulomas to mice, *Clin. Res.,* 22, 559A, 1974.

33. **Thayer, W. R., Coutu, J. A., Chiodini, R. J., Van Kruiningen, H. J., and Merkal, R. S.,** The possible role of mycobacteria in inflammatory bowel disease. II. Mycobacterial antibodies in Crohn's disease, *Dig. Dis. Sci.,* 29, 1080–1085, 1984.

34. **Whorwell, P. J., Davidson, I. W., Beeken, W. L., and Wright, R.,** Search by immunofluorescence for antigens of Rotavirus, *Pseudomonas maltophilia,* and *mycobacterium kansasii* in Crohn's disease, *Lancet* 697, 1978.

35. **Wirostko, E., Johnson, L., and Wirostko, B.,** Crohn's disease — rifampin treatment of the ocular and gut disease, *Hepatogastroenterology,* 34, 90–93, 1987.

36. **Zuckerman, M. J., Casner, N. A., and Gibbs, M. A.,** Application of an immunofluorescence assay for Crohn's disease using primed nude mouse lymph nodes to a United States/Mexican border population, *Am. J. Med. Sci.,* 296, 260–265, 1988.

37. **Besserer, J.,** unpublished study.

38. **Coloe, P. J., Wilks, C. R., Lightfoot, D., and Tosolini, F. A.,** *Aust. Microbiol.,* 7, 188, 1986.

39. **Thorel, M. F.,** Relationship between *M. avium, M. tuberculosis,* and mycobacteria associated with Crohn's disease, *Ann. Rech. Vet.,* 20, 417, 1989.

Chapter 24

CHARACTERISTICS OF FILTRABLE FORMS

Demonstrating filtrable life in microbial species frustrated some investigators, was accomplished by many pioneer microbiologists, and now can be a classroom exercise.

Tuckett and Moore[68] found a bacterium which predictably demonstrates incredible vitality in a filtrate. When the soil organism, *Cellvibrio gilvus,* is cultivated in broth containing cellobiose, viable units pass through sintered glass filters of varied porosity, all of which retain *Escherichia coli.* A complex life cycle can be followed in the filtrate by dark-field observation. Barely visible particles rapidly grow into normal cells, some of which, in turn, enlarge into amorphic spheres. Within these spheres (large bodies) particles and cells develop which become highly motile, eventually breaking their way through the sphere's flexible wall. The large bodies are fragile and are disrupted by standard staining techniques.

Similar adventitious experiences have greeted investigators attempting to separate phage from host by filtration. Also, in industry, complications arise when a bacterium which synthesizes a desired metabolite fails to be retained by a sterilizing membrane.

In the 1920s an important "school of filtration" was established by Kendall,[29] who found that filtrable units of the *Staphylococcus, Streptococcus,* and *Salmonella typhi* could be coaxed back to their parent forms. He identified the variants of the typhoid bacillus both by fermentation reactions and by agglutinogens. Although William H. Welch regarded Kendall's work as a distinct advance, great skepticism was expressed by most microbiologists. Unfortunately, this was just prior to the demonstration by Klieneberger and by Dienes that filtrable organisms could be grown on solid medium and their sequential reversion steps followed.

PHYSICAL CHARACTERISTICS OF FILTRABLE FORMS

STABILITY AND RESISTANCE OF THE FILTRABLE STAGE

Hadley and co-workers[14] found that the filtrable units of *Shigella shiga* were more stable than the bacilli. The "virus stage", sealed in ampules, could be stored for 2 years at room temperature, as could filtrable units of many other species.[14] It will also be noted that the viable granules in the filtrate (Dick toxin) of scarlatinal streptococci survived storage in phenol.[54]

A NONFILTRABLE L-PHASE ORGANISM

Almost by definition, an L Form has filtrable units, and the intermediate stages between a fixed L Form and its parent usually demonstrate such units in abundance. However, an unstable L-phase Proteus strain of Kellenberger et al.[28] had no filter-passing units. Where a stable L Form of Proteus was filtrable, the unstable penicillin-dependent L-phase strain was not. When the unstable organism was viewed by electron microscopy, the reason for lack of filtrability was obvious. The nonfiltrable organisms usually had a long filamentous outgrowth, whereas the stable filtrable L Forms consisted only of minute spheres.

METHODS TO INCREASE FILTRABILITY

TREATMENT TO RELEASE FILTRABLE UNITS

Many variables influence development and release of filtrable units in a culture. The simple expedient of mechanical shaking increased the yield of filtrable, reverting forms in two strains of *Lactobacillus delbruckii.*[27]

Kellenberger et al.[28] increased filtrability of L-phase Proteus by grinding with a mortar and pestle. They felt they were releasing filtrable particles from L-bodies. Some young cultures of the tubercle bacilli showed filtrable microbes only after sequential washing and shaking with glass beads and sand, refrigeration, then finally prefiltration through paper.[44]

Wyrick and Gooder[75] noted that treating a midlogarithmic L Form culture of streptococci with deoxyribonuclease facilitated the recovery of viable L Forms from a 0.45-μm filter. Without exposure to DNAase, no viable stages passed through the membrane.

INHIBITORS AND INCREASED FILTRABILITY

The species *Candida albicans, C. tropicalis, C. krusei,* and *C. parakrusei* showed abundant filtrable forms when cultured with crystal violet, brilliant green, mycostatin, or amphotericin B. Viable forms were found after passing liquid cultures containing inhibitor through Seitz "sterilizing" pads or after incubation of a paste of the Candida on Saran® membranes with a disk of the inhibitor on the medium. Only rare variant colonies developed in the absence of antimetabolites in Tunstall and Mattman's study.[69]

Remarkably, even established strains of bacterial L Forms may not produce filtrable units unless penicillin is added. A stable L Form of Proteus had no particles which could pass through unglazed porcelain filters until incubated with penicillin in the medium for 7 d, which resulted in formation of numerous filtrable regenerating units.[63]

IN VIVO VS. *IN VITRO* PRODUCTION OF FILTRABLE FORMS

When attempts were made to show a filtrable stage in *Pasteurella pestis,* no life could be demonstrated in culture filtrates. In contrast, homogenates of tissues from infected mice and guinea pigs yielded viable units which reverted to *P. pestis,* sometimes in cultures of the filtrates, sometimes in inoculated animals.[5]

TYPES OF FILTERS

The membranes most commonly employed to retain classic bacteria and permit passage of the wall-deficient stage have a pore size of 0.45 μm. However, many filtrable organisms will pass through 0.20-μm pores. The 0.45-μm membrane cannot be used when sterility of the filtrate is the aim, as contaminating filtrable forms from water and other sources will pass, even if present in relatively small numbers.

Membranes permit passage of small forms better than do filters of diatomaceous earth, porcelain, asbestos, or sintered glass. For example, from a strain of *Proteus vulgaris,* some filtrable forms were recovered after filtration through membrane filters No. 7 and No. 10, but not after passage through Seitz, Schichten, Berkefeld, or glass filters.[60]

However, in countless instances even in the era before membrane filters, filtrable stages of bacteria passed through thick candles or sheets of porous substances (Table 1). Excellent reviews of bacterial filtration were made by Knöll-Jena,[32] Klieneberger-Nobel,[31] and in the book on this topic by Hauduroy,[19] whose portrait is shown in Figure 1.

Only a very few viable bacterial stages traverse filters finer than 0.20 μm when a liquid culture is exposed to suction. In contrast, bacteria propagated on membranes show great ingenuity in maneuvering their mobile fronts through minute crevices.

COLLODIAN MEMBRANES

One method employed sacs of a collodian membrane with porosity not allowing the escape of either diphtheria or tetanus toxin. When such sacs, seeded with a filtrate of the tubercle bacillus, were inserted in guinea pigs' peritoneal cavities, the filtrable stage passed through the membrane and produced glandular tuberculosis in the animal, with many acid fast rods. As is typical of filtrable-form tuberculosis, the pathology was found in lymph

TABLE 1
Demonstration of Filtrable Stages of Microorganisms

Bacillus anthracis: Haag[13] and Heymans[22] produced fatal anthrax in rabbits given intraperitoneally a container of
  B. anthracis coated with 30 layers of collodion.
Brucella abortus: Hatten and Sulkin[15]
Candida albicans, C. krusei, C. parakrusei, C. tropicalis[69]
Cellvibrio gilvus: Tuckett and Moore[68]
Corynebacterium diphtheriae[18] Park 8 strain[14]
Escherichia coli[14,18]
Granulobacter pectinovorum Imsenecki[25]
Lactobacillus acidophilus: Hadley, et al.[14]
L. delbruckii: Kanunnikova[27]
Leptospira icterohaemorrhagica: Inada et al.[26]
Mycobacterium avium[45,73]
Mycobacterium (BCG)[8,42]
Mycobacterium of Johne's Disease (Enteritis of cattle)[40]
M. phlei[45]
Mycobacterium, Runyon's Group II[50]
M. tuberculosis.[2,3,7,10,12,16,46,49,51,52,56,58,64,66,72,73] In blood of man;[1] in blood of infected bovine;[57] in cutaneous lesions
  and in the blood of these patients;[55] in pus from a cold abscess (man)[9]
Pasteurella pestis: Hauduroy and Ghalib[21]
Peptococcus aerogenes, P. anaerobius
Pneumococcus: Bürgers et al.[4]
Proteus: Kellenberger et al.,[28] Mandel and Terranova,[38] Rada,[53] Schellenberg,[60] Silberstein,[63]
Salmonella cholerae suis[14]
S. paratyphi A[14]
S. paratyphi B[14]
S. typhimurium[6,14]
S. typhi[29,36] Filtrable after phage lysis:[11] Koptelova and Bitkova[33]
Shigella[19]
Spirochaeta gallinarum[62]
Staphylococcus:[18] Hoffstadt and Youmans,[23] Marston,[39] Trofimova,[67] Williams[74]
Streptobacillus moniliformis: Klieneberger-Nobel[30,31]
Streptococcus, Group A:[20,54,61] Mortimer,[41] Nicholls,[43] Panos, et al.[47] van Boven et al.[71]
Streptococcus, Group C of equine strangles[70]
S. faecium: Wyrick and Gooder[75]
S. liquefaciens: Young and Armstrong[77]
Vibrio cholerae: Ianetta and Wedgwood[24]
V. metchnikoff[55]

Filtrable forms of *Listeria monocytogenes* demonstrated by Smith and Sword are mentioned in the chapter on
Listeria. Spirochete 277 F from rabbit ticks was successfully filtered through Berkefeld N filters. See data from
Pickens et al. in the chapter discussing spirochetes.

nodes, but not liver or spleen. The work of Xalabarder[78] indicates that kidney damage also
would be expected.

　　Similar collodian membranes were used to demonstrate filtrability of *Bacillus anthracis.*
When sacs containing the bacillus were placed in the peritoneal cavity of rabbits, no animals
survived.[22]

MEMBRANES OF MODIFIED CELLULOSE

　　Cellulose nitrate membranes permitted not only passage of staphylococcal L Forms but
also their in-membrane growth (Figure 2).[65]

　　A most functional type of grow-through filter was first used for CWD variants by Perez-
Miravete and Calderón.[48] They separated the L-phase of *Streptobacillus moniliformis* from
its parent by passage of the fine units through commercial-type cellophane membranes. An

FIGURE 1. Paul Hauduroy. His prolific investigations included studies of microbial variation and filtrability. (From *J. Bacteriol. Virol. Immunol.*, 14, 197, 1935. With permission.)

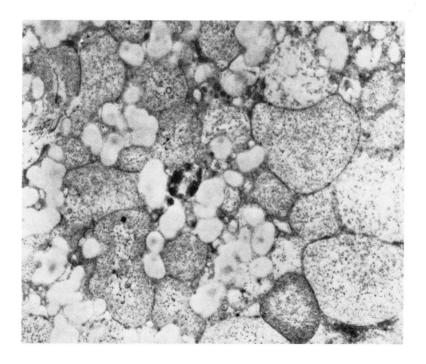

FIGURE 2. Thin section of cultures of staphylococcal L Forms inside 0.60-μm filter. Irregular masses of L-elements are seen. Larger forms are 2 μm in diameter. (From Takahashi, T. and Tadokoro, I., *Jpn. J. Exp. Med.*, 49 (5), 355–360, 1979. With permission.)

excellent membrane is grade 50 Saran® (Dow Chemical Co.). This is the Saran Wrap commonly sold in grocery stores. This simple membrane, after immersion in saline, permits passage of a filtrable form from at least 50% of blood cultures containing a classic bacterium in treated or untreated patients. However, an antibiotic-impregnated disk below the membrane increases the number of wall-deficient forms which develop.

When the growth of a nonfastidious organism above the cellophane membrane is fostered by upward diffusion of nutrient from the medium into the culture on the surface, the filtrable forms become myriad. This growth above the membrane occurs with the Enterobacteriaceae; it is not detectable for Streptococci, Staphylococci, and Candida.

To follow growth from minute granules, it is necessary to have a medium relatively free of dead L-colonies and other debris. Schell[59] removed debris from agar solutions by filtering through commercial Kimwipes® (Kimberly-Clark) before autoclaving.

ISOLATION AND/OR REVERSION OF FILTRABLE FORMS

REVERSION OF FILTRABLE FORMS IN MAN

Having filtrable life revealed by *in vivo* reactions is striking. Unintentionally produced filtrable units of scarlatinal streptococci triggered a remarkable series of events. Among 700 children vaccinated with the Dick toxin by injection of the streptococcal filtrate, typical scarlet fever occurred in 2% of the vaccinees. Prolonged study indicated that the filtrate did contain living microbes, despite the negative sterility check made by culturing the fresh filtrate.[54] When free of variant growth, the toxin was *not* disease producing in rabbits or in mice. However, growth from the filtrate was lethal for the mouse.

REVERSION OF FILTRATES IN LABORATORY ANIMALS

Filtrates of many fastidious organisms have been tested by animal inoculation. Filtrates of beta-hemolytic streptococci of human origin have been injected into mice in four different studies.[20,54,61,70] β-hemolytic streptococci were recovered from 30 to 50% of the animals.

Similarly, equine streptococci from strangles have a filtrable form which is lethal to the mouse within 4 d after injection, yielding typical streptococci in the blood and tissues of the animal.[70] Such filtrable forms exist in the pus of the equine abscesses and in serum broth cultures of the exudate.

The danger of a complicating natural infection is minimal when the bacterium is one not naturally infecting the test animal. This applies to work with *Leptospira icterohaemor-rhagiae* by Inada et al.[26] A filtrate of infected organs or blood, when injected into a new guinea pig, produced the disease and reverting organisms, although spirochetes per se were never seen in the filtrate.

For some microbial species, the filtrable state has grown in infected animals without showing virulence. An example concerns *Salmonella suipestifer*. Organisms in the filtrate reverted and multiplied in swine without producing illness.[37] Conversely, sometimes typical disease results in an animal injected with a filtrate although no classic organisms are found.

ISOLATIONS OF FILTRABLE STATES FROM MAN

Hauduroy, who authored a book on microbial filtrability, described the occurrence of the virus stage of *S. typhi* in the blood of a typhoid case.[17] Although the patient was admitted early in her infection, only the variant was found in blood cultures. Subcultures of the filtrate, initially showing variant growth, reverted to pure cultures of the classic bacillus. Hauduroy postulated that perhaps in typhoid cases the patient commonly acquires and eliminates the organism in the filtrable stage, and only temporarily harbors the classic bacillus.

Lipnicki found CWD stages of the typhoid bacillus in many cases of typhoid after chloramphenicol therapy. He felt that filtrable stages were probably present.[34]

FILTRABLE VIRUSES MAY NOT BE
FILTRABLE EITHER

A glance at Table 1 suggests that all bacteria produce a viable filtrable stage. It seems strange that "filtrable forms of bacteria" were at one time a highly controversial subject. In the days when filtration was not orthodox dogma, Hadley et al. commented that expressing their convictions concerning the common filtrability of bacteria would be professionally foolhardy. "It brings down the strictures of one's friends and enemies alike."[14] It is pertinent here to note that many viruses are reluctantly filtrable. Vaccinia virus often will not pass through a diatomaceous earth or porcelain filter to any detectable degree. Surprisingly, the agent of rabies, described by Pasteur as "too small to be seen", was rarely filtrable with methods used up to 1935. The dimension of rabies virions is approximately 0.1 to 0.15 μm,[76] the size of many viable particles demonstrated in bacterial filtrates.

SUMMARY

Viable filtrable particles have been demonstrated in so many bacterial genera that the characteristic is probably omnipossessed. Filtrability also applies to many fungi. The culture medium components are critical, as is age of the growth, pH, and pretreatment of the culture to release small units. The number of filtrable particles is greatly increased in the presence of antimicrobial agents which stimulate the L-cycle.

REFERENCES

1. **Arloing, F. and Dufourt, A.**, Sur les relations qui existent entre l'ultravirus tuberculeux et les microcultures sur milieu de Loewenstein, *Presse Med.*, 42, 505, 1934.
2. **Beerens, M. J.**, Mise en évidence du virus tuberculeux dans divers produits pathologiques et en particulier dans le sang par la méthode des injections d'extrait acétonique de bacilles de Koch. Caractéres des souches de bacilles tuberculeux ainsi isolés, *Ann. Inst. Pasteur* (Paris), 52, 406–414, 1934.
3. **Bijl, J. P.**, Onderzoekingen over een "filtrabel" tuberculose-virus, *Zentralbl. Bakteriol. Ref.*, 94, 107, 1928.
4. **Bürgers, Th., Lodenkämper, H., and Verfürth, H.**, Studien über Pleomorphie, *Zentralbl. Bakteriol. Parasitenkd. Infektionskr. Hyg. Abt. 1 Orig.*, 138, 58–67, 1936.
5. **Burnet, E.**, Sur la recherche de formes filtrantes des bactéries, *C. R. Soc. Biol.*, 95, 1142–1144, 1926.
6. **Carrére, L., Roux, J., and Mandin, J.**, A propos du cycle L des bactéries: obtention de formes naines, viables et filtrables, en milieu liquide, *C. R. Soc. Biol.* (Montpellier), 148, 2050–2052, 1954.
7. **de Sanctis Monaldi, T.**, Comportement de l'ultra-virus tuberculeux dens l'organisme des cobayes, *Presse Med.*, 37, 326, 1929.
8. **de Sanctis Monaldi, T.**, Les éléments filtrables du BCG, *Presse Med.*, 39, 1052, 1939.
9. **Durand, H. and Vaudremer, A.**, Retour au type classique du bacille tuberculeux filtré apres passage par le péritoine du cobaye, *C. R. Soc. Biol.* (Paris), 90, 916, 1924.
10. **Durand, H. and Charchanski,** Tuberculose experimentale apres inoculation de filtrats tuberculeux, *C. R. Soc. Biol.* (Paris), 93, 499–500, 1925.
11. **Fejgin, B.**, Sur la forme filtrante du bacille d'Eberth, *C. R. Soc. Biol.* (Paris), 92, 1528–1530, 1925.
12. **Floyd, C. and Herrick, M. C.**, An investigation of filterable forms of the tubercle bacillus and of protective substances in the filtrates, *Am. Rev. Tuberc.*, 16, 323–329, 1927.
13. **Haag, F. E.**, Der Milzbrandbazillus, seine Kreislaufformen und Varietaten, *Arch. Hyg.*, 98, 271–318, 1927.
14. **Hadley, P., Delves, E., and Klimek, J.**, The filtrable forms of bacteria. I. A filtrable stage in the life history of the Shiga dysentery bacillus, *J. Infect. Dis.*, 48, 4–159, 1931.
15. **Hatten, B. A. and Sulkin, S. E.**, Viability and filterability of Brucella L Forms, *Bacteriol. Proc. 1971*, p.42, 1971.
16. **Hauduroy, P. and Vaudremer, A.**, Recherches sur les formes filtrables du bacille tuberculeux, *C. R. Soc. Biol.* (Paris), 89, 1276–1278, 1923.

17. **Hauduroy, P.,** Presence des formes filtrantes du bacille d'Eberth dans le sang d'un typhique, *C. R. Soc. Biol.* (Paris), 95, 288–289, 1926.

18. **Hauduroy, P.,** Techniques de culture des formes filtrantes invisibles des microbes visibles, *C. R. Soc. Biol.* (Paris), 95, 1523–1525, 1926.

19. **Hauduroy, P.,** *Les Ultravirus et les Formes Filtrantes des Microbes, Deuxieme Partie, Les Microbes Filtrants Visibles,* Masson et Cie, Paris, 1929.

20. **Hauduroy, P. and Lesbre. P.,** Les formes filtrantes des Streptocoques, *C. R. Soc. Biol.* (Paris), 97, 1394–1395, 1927.

21. **Hauduroy, P. and Ghalib, A.,** Présence du bactériophage antipesteux a Paris, *C. R. Soc. Biol.* (Paris), 100, 1085–1086, 1929.

22. **Heymans, J. Y.,** In vivo comme in vitro les microbes passent a travers la paroi du filtre, *C. R. Acad. Sci.* (Paris), 171, 971–973, 1920.

23. **Hoffstadt, R. E. and Youmans, G. P.,** *Staphylococcus aureus* dissociation and its relation to infection and immunity, *J. Infect. Dis.,* 51, 216–242, 1932.

24. **Iannetta, A. and Wedgwood, R. J.,** Culture of serum-induced spheroplasts from *Vibrio cholerae, J. Bacteriol.,* 93, 1688–1692, 1967.

25. **Imsenecki, A.,** Struktur und entwicklungsgeschichte des *Granulobacter pectinovorum* (Fribes), *Arch. Mikrobiol.,* 5, 451–475, 1934.

26. **Inada, R., Ido, Y., Hoki, R., Kaneko, R., and Ito, H.,** The etiology, mode of infection and specific therapy of Weil's disease (Spirochaetosis Icterohaemorrhagica), *J. Exp. Med.,* 23, 377–402, 1916.

27. **Kanunnikova, Z. A.,** Influence of some external factors on formation of filtrable forms of lactobacilli, *Mikrobiologiia* (AIBS transl.), 27, 172–175, 1958.

28. **Kellenberger, E., Liebermeister, K., and Bonifas, V.,** Studien zur L Form der Bakterien II, *Z. Naturforsch.,* 11b, 206–215, 1956.

29. **Kendall, A. I.,** Filtrable forms of bacteria and their significance, *JAMA,* 99, 67–69, 1932.

30. **Klieneberger-Nobel, E.,** On *Streptobacillus moniliformis* and the filtrability of its L Form, *J. Hyg.,* 47, 393–395, 1949.

31. **Klieneberger-Nobel, E.,** Filterable forms of bacteria, *Bacteriol. Rev.,* 15, 77–103, 1951.

32. **Knöll-Jena, H.,** VI. Über Bakterienfiltration, *Erg. Hyg.,* 24, 266–364, 1941.

33. **Koptelova, E. I. and Bitkova, A. N.,** The antigenic structure of subcultures of Salmonella typhi regenerated from filterable forms, *Zh. Mikrobiol. Epidemiol. Immunobiol.,* 31, 112, 1960.

34. **Lipnicki, B.,** Bacterial L Forms or *Salmonella typhosa* isolated from patients with typhoid fever, *Rec. Adv. Hyg. Med.* (Polish), 12, 113–159, 1958.

35. **Lodenkämper, H. and Kallinich, W.,** Entwicklungsstudien an vibrionen, *Zentralbl. Bakteriol. Parasitenkd. Infektionskr. Hyg. Abt. 1 Orig.,* 142, 376–388, 1938.

36. **Lodenkämper, H.,** Filtrationsversuche mit typhusbazillenkulturen, *Zentralbl. Bakteriol. Parasitenkd. Infektionskr. Hyg. Abt. 1 Orig.,* 146, 155–163, 1940.

37. **Lourens, L. F. D. E.,** Untersuchungen über die filtrierbarkeit der schweinepestbacillen (Bac. suipestifer), *Zentralbl. Bakteriol. Parasitenkd. Infektionskr. Hyg. Abt. 1 Orig.,* 44, 504–512, 1907.

38. **Mandel, P. and Terranova, F.,** Le metabolisme global d'une souche de Proteus ahmed et d'une forme L qui en dérive, *C. R. Acad. Sci.,* 242, 1082–1084, 1956.

39. **Marston, J. H.,** Electron microscopy of staphylococcal L Forms, *Bacteriol. Proc. 1964,* p.71, 1964.

40. **Morin, H. and Valtis, J.,** Sur la filtration du bacille du Johne a travers les bougies Chamberland L$_2$, *C. R. Soc. Biol.* (Paris), 94, 39–40, 1926.

41. **Mortimer, E. A., Jr.,** Production of L Forms of Group A streptococci in mice, *Proc. Soc. Exp. Biol. Med.,* 119, 159–163, 1965.

42. **Negre, L. and Valtis, J.,** Sur les éléments filtrables du bacille bilié de Calmette et Guérin, *Presse Med.,* 29, 1052, 1931.

43. **Nicholls, E. E.,** A study of the organisms recovered from filtrates of cultures of hemolytic streptococci, *J. Infect. Dis.,* 62, 300–306, 1938.

44. **Ninni, C.,** Sur la technique a employer pour obtenir des filtrats contenant des elements filtrables du virus tuberculeux, *Ann. Inst. Pasteur* (Paris), 46, 598–603, 1931.

45. **Ninni, C.,** Les éléments filtrables des bacilles tuberculeux aviaires et des bacilles paratuberculeux, *C. R. Soc. Biol.* (Paris), 107, 615–618, 1931.

46. **Paisseau, G., Valtis, J., and Saenz, A.,** Importance de l'étude des éléments filtrables dans la pathogénie de la tuberculose, *Presse Med.,* 37, 186–187, 1929.

47. **Panos, C., Barkulis, S. S., and Hayashi, J. A.,** Streptococcal L Forms. III. Effects of sonic treatment on viability, *J. Bacteriol.,* 80, 336–343, 1960.

48. **Perez-Miravete, A. and Calderón, E.,** Cellophane paper technic for isolating pleuropneumonialike organisms, *Am. J. Clin. Pathol.,* 26, 685–688, 1956.

49. **Periti, E.,** Le forms filtrabili del bacillo di Koch nelle tubercolosi polmonari cronicissime, *Policlinico (Med.),* 1, 32, 1928.

50. **Pinner, M., IV,** Atypical acid-fast organisms. II. Some observations on filtration experiments, *J. Bacteriol.,* 25, 576–579, 1933.

51. **Popper, M. and Raileanu, C.,** Infection par voie digestive et intracardiaque des cobayes par le filtrat tuberculeux, *C. R. Soc. Biol.* (Paris), 99, 1080, 1928.

52. **Rabinowitsch-Kempner, L.,** Zur Frage der Filtrierbarkeit des Tuberkulosevirus, *Z. Tuberc.,* 52, 18–25, 1928.

53. **Rada, B.,** Filtrierbarkeit der L phase und der globularen form der Bakterien, *Zentralbl. Bakteriol. Parasitenkd. Infektionskr. Hyg. Abt. 1 Orig.,* 176, 85–90, 1959.

54. **Ramsine, S.,** Sur les formes filtrables des Streptocoques et sur la nature de la toxine de Dick, *C. R. Soc. Biol. (Paris),* 94, 1010–1012, 1926.

55. **Ravaut, P., Valtis, J., and van Deinse, F.,** La présence de l'ultra-virus tuberculeux dans le sang d'une malade atteinte de tuberculides cutanées, *Presse Med.,* 38, 912, 1930.

56. **Remlinger, P.,** Peut-il éxister une phase abacillaire virulente à la periode ultime de la tuberculose pulmonaire, *C. R. Soc. Biol.* (Paris), 99, 278–280, 1928.

57. **Rossi, P.,** Présence de l'ultra-virus tuberculeux dans le sang des bovins reagissant a la tuberculine, *C. R. Soc. Biol.* (Paris), 101, 638–639, 1929.

58. **Sanarelli, G. and Alessandrini, A.,** Demonstrazione in vivo e in vitro delle forme filtranti del virus tubercolare, *Ann. Ihiene.,* 40, 489–592, 1930.

59. **Schell, G. H.,** An Artifact Free Medium for the Isolation of L Forms from Clinical Blood Cultures, M.S. thesis, Wayne State University, Detroit, 1968.

60. **Schellenberg, H.,** Untersuchungen ueber die durch Pencillin induziertie Pleomorphie bei *Proteus vulgaris, Zentralbl. Bakteriol. Parasitenkd. Infektionskr. Hyg. Abt. 1 Orig.,* 161, 432–465, 1954.

61. **Sedallian, P. and Gaumond, J.,** Les phase de l'evolution de Streptocoque, *Presse Med.,* 35, 1313–1314, 1927.

62. **Séguin, P.,** *Spirochaeta gallinarum* et formes dites "ultra-virus", *C. R. Soc. Biol.* (Paris), 104, 836–838, 1930.

63. **Silberstein, J. K.,** Observations on the L Forms of Proteus and Salmonella, *Z. Allg. Pathol. Bacteriol.,* pp. 739–755, 1953.

64. **Sweany, H. C.,** The filterability of the tubercle bacillus, *Am. Rev. Tuberc.,* 17, 77–85, 1928.

65. **Takahashi, T. and Tadokoro, I.,** Staphylococcal L colonies grown on and through membrane filters with various pore sizes, *Jpn. J. Exp. Med.,* 49(5), 355–360, 1979.

66. **Toguonnoff, A.,** Sur les éléments filtrables du virus tuberculeux, *C. R. Soc. Biol.* (Paris), 97, 349–351, 1928.

67. **Trofimova, N. D.,** Studies on conditions for the regeneration of filterable forms of staphylococci, *Mikrobiol. Zh. Akad. Sci. Ukraine RSR,* 21, 45–47, 1959.

68. **Tuckett, J. D. and Moore, W. E. C.,** Production of filtrable particles by *Cellvibrio gilvus, J. Bacteriol.,* 77, 227–229, 1959.

69. **Tunstall, L. H. and Mattman, L. H.,** L variants in *Candida* species, *Bacteriol. Proc. 1961,* p. 83, 1961.

70. **Urbain, A.,** Les formes filtrantes du Streptocoque gourmeux, *C. R. Soc. Biol.* (Paris), 97, 1598–1600, 1927.

71. **van Boven, C. P. A., Ensering, H. L., and Hijmans, W.,** Size determination by the filtration method of the reproductive elements of Group A streptococcal L Forms, *J. Gen. Microbiol.,* 52, 403–412, 1968.

72. **Vascellari, G.,** Contributo allo studio della filtrabilita del virus tubercolare, *Policlinico (Med.),* 1, 17, 1928.

73. **Vaudremer, A.,** Étude biologique du bacille tuberculeux. Les formes filtrantes, *Beitr. Klin. Enforsch. Tuberk.,* 77, 16–38, 1931.

74. **Williams, R. E. O.,** L Forms of *Staphylococcus aureus, J. Gen. Microbiol.,* 33, 325–334, 1963.

75. **Wyrick, P. B. and Gooder, H.,** Filterability of streptococcal L Forms, *J. Bacteriol.,* 105, 284–290, 1971.

76. **Yaoi, H., Kanazawa, K., and Sato, K.,** Ultrafiltration experiments on the virus of rabies (virus fixe), *Jpn. J. Exp. Med.,* 14, 73–79, 1936.

77. **Young, L. S. and Armstrong, D.,** Induction, colonial morphology, and growth characteristics of the L Forms of *Streptococcus liquefaciens, J. Infect. Dis.,* 120, 281–291, 1961.

78. **Xalabarder, C.,** L Forms of Mycobacteria and chronic nephritis, *Publ. Inst. Antituberc.,* Suppl. 7, 1970.

Chapter 25

SPIROCHETAE

The recent resurgence of syphilis[43] shows that penicillin has not eliminated the disease. Thus, new parameters are needed to diagnose syphilis and to monitor therapy success or failure. It is now known that unless the disease is diagnosed and treated early, it is usually impossible to eradicate the pathogen.[53,89,90] Recognition of aberrant forms is pertinent here as a way of increasing diagnostic sensitivity in all stages of the infection.

Yobs and associates studied lymph nodes of 45 TPI reactive prisoners who had been treated for syphilis with penicillin in dosages considered adequate.[88] Eleven were found to harbor treponemes. When five patients were retreated, three had persistent infection, as shown by transfer of their lymph nodes to cortisone-treated rabbits. Discouraging findings concerning the longevity of the treponeme in man have also been found by Collart et al.,[12,13] Del Carpio,[15] and Boncinelli et al.[8]

All possible tools should be applied to syphilis diagnosis since no single approach is completely reliable. Despite great improvements in serology, some cases are negative by the most precise methods. Seronegative ocular and neurosyphilis have been reported by Smith and associates.[66] Syphilitic temporal arteritis is found in individuals with no history of syphilis.[67] Rabbit inoculation sometimes, but not always, gives an answer. Tens of thousands of *Treponema pallidum* may not infect every rabbit; high natural resistance in the rabbit may represent cross-reacting antibodies.[88]

As data presented below show, dark-field examination misses some cases. Recognition of the nonspirochetal stage can improve diagnosis at all levels. Very likely, identifying the aberrant stage can make cultures meaningful since the treponeme itself appears in culture rarely.

Fluorescent antibody, an improved tool in recognizing the spirochetal stage,[87] can be expected to assist in identifying variants. Antibody made in response to the variants may prove superior, but for most genera antibody to the classic stage gives good reaction with the CWD variant.

The following paragraphs concern forms of the treponeme which vary from the "classic" textbook picture. Most investigators studying the pathogenesis of the organism agree that the commonly propagating form of the syphilis agent is not the spirochete.

EVIDENCE THAT THE SPIROCHETAL
STAGE IS RARE IN SYPHILIS

Finding the spirochete in syphilis is generally believed to be most likely in the primary chancre, yet even here success does not approach 100%. Finding treponemes in 14 of 17 cases of primary syphilis is considered average success.[20] If the characteristic spirochete is not seen in the initial lesion scraping, there is an anxious waiting period to determine whether serological tests become positive. During this interval, the treponeme continues invasion of neighboring lymph nodes and distant tissues.

Even in early syphilis, the degree of tissue reaction suggests an assault launched by a greater number of invaders than the rare treponemes usually present. When inguinal lymph nodes from six patients were excised before treatment, dark-field examinations of each node for 9 h failed to reveal a single spirochete. Fluorescent antibody (FA) study of node emulsions revealed treponemes in only two samples.[87]

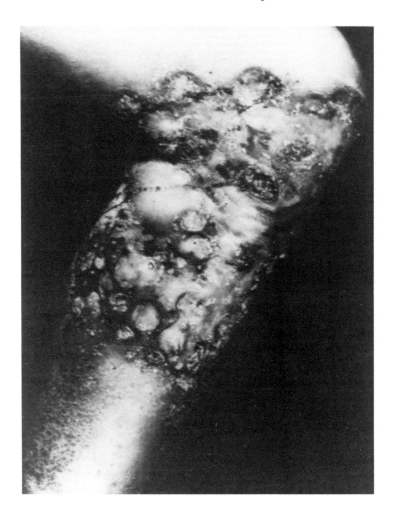

FIGURE 1.   Syphilitic gummas. (From de Graciansky, P. and Boulle, S., *Color Atlas of Dermatology*, Vol. 2, Year Book Medical Publishers, Chicago, 1955.)

When human lymph nodes were studied throughout antisyphilitic treatment, spirochetal forms rapidly disappeared, whereas a granular form of the spirochete persisted. Simon and Mollinedo used a special stain to reveal the granular form but note that certain artifacts cause confusion.[64] FA reactions might be interpreted more easily.

In tertiary syphilis especially, the classic spirochete is difficult to find. This is true for syphilitic aneurysms of the aorta and for gummas, wherever they occur, regardless of severity (Figure 1). Accompanied by numerous "abnormal" forms, a series of shapes can be found leading from small round corpuscles to the typical treponeme.[44] Over 60 years ago it was noted that the routine silver nitrate stain revealed the atypical stages of *T. pallidum* (Figure 2) in luetic aortas.[82]

A similar conclusion concerning syphilitic gummas was made early by Levaditi and Po, who concluded that granules and serrated forms finally evolve to the almost invisible stage of *T. pallidum* and that the transition forms and tiny granules are often the only forms in the brain in paresis.[40] They noted that even Schaudinn and Hoffman[58] had envisioned an evolutive cycle for the agent of syphilis to explain the paucity of forms in the disease.[39] Klieneberger-Nobel stated that Levaditi and Po's concept of granular forms of *T. pallidum* can explain why the latent stages resist chemotherapy.[33]

FIGURE 2. Morphological changes in *T. pallidum* as it was found from superficial to deep layers in a syphilitic aorta. (From Warthin, A. S. and Olsen, R. E., *Am. J. Syphilis*, 14, 433—437, 1930. With permission.)

Wile and Curtis[86] studied the mouse infectivity of *T. pallidum*. Exhaustive dark-field examination through mouse brains showed no spirochetes, yet infection resulted from inoculation of the material into rabbit testicle. Similarly, mice given the tiny granular form of *T. pallidum* develop no lesions but harbor an invisible agent which is virulent when passed back to the rabbit.[41]

Perhaps the most fascinating of all studies of the growth cycle of *T. pallidum* concerns the development of spirochetes from human blood. Roukavischnikoff[57] found that blood in the primary stage of syphilis contains tiny granules which, if properly nurtured, develop into the typical spirochete or more often show stages of growth which he recognized as transitional. It is thought that false negative serological tests for syphilis may be explained because cystic and granule stages of the treponeme have not stimulated antibody reactive with the spirochetal stage.[51]

IN VITRO GROWTH OF NONTREPONEMAL STAGES

SPINAL FLUID AS CULTURE MEDIUM

A very early attempt to culture *T. pallidum* with simple medium demonstrated the atypical stages. Two parts of spinal fluid from an individual with central nervous system syphilis were added to one part of peptone broth. By Day 4 of incubation, many motile ovoid bodies, approximately 2×5 μm were seen which gradually went through a multiplicity of morphologies, only one of which was the typical tightly coiled treponeme.[38]

Gürün grew *T. pallidum* in a beeswax-honey medium. In his experience, every isolate grew first as multitudinous granules.[24]

TISSUE CULTURE

When *T. pallidum* is inoculated into tissue culture, the spirochetes gradually disappear, returning after 20 d only as minute bodies visible in dark-field.[21] This resembles the stabilization of bacterial variants in tissue culture in a viruslike intracellular form (see chapter on Intracellular Growth).

Common L-Form growth of *T. pallidum* in tissue culture and *in vivo* was substantiated by a Russian study.[51] Their work included extensive electron microscopy.

CHICKEN EMBRYO

The thesis that *T. pallidum* has an ''invisible'' form is supported by work in chicken embryos. In several investigations, chorioallantoic membrane from chick embryo inoculated

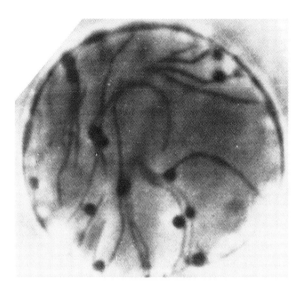

FIGURE 3. *T. pallidum* spirochetes forming in a cyst. (From DeLamater, E. D., Haanes, M., Wiggall, R. H., and Pillsbury, D. M., *J. Invest. Dermatol.*, 16, 231–256, 1951. With permission.)

with *T. pallidum* might be free of spirochetes by dark-field examination yet produce syphilis when inoculated intratesticularly in rabbits.[84,85]

One of the most detailed studies concerned not only cultures, but also propagation in rabbit testes and in chicken embryos. This work, from the Department of Dermatology and Syphilology at the University of Pennsylvania Medical School, by DeLamater et al.[14] described the life cycle of the Nichols, the Kazan, the Reiter, and the Noguchi strains of *T.pallidum.* After 10 d in culture, reproduction by means other than binary fission predominates. Buds, which may separate from the parent spirochete at any stage of their development, form along its length. The buds behave as cysts, from which delicate spiral forms gradually emerge. At times, larger cysts form by the aggregation of several spirochetes. These develop dense granules, which, in turn, become the forerunners of new spirochetes in all the treponemal strains. Their micrographs (see Figure 3) beautifully reveal the development of spirochetes within cysts. The smaller buds from which filamentous spirochetal growth occurs are shown in Figure 4.

ELECTRON MICROSCOPY

If the buds appearing on spirochetae produce miniature spirilla forms, these should be detectable by fine structure sections viewed in the electron microscope. Hampp et al.[26] investigated the Nichols and Noguchi strains of *T. pallidum* as well as strains of *Borrelia vincenti*. Microdrops of liquid cultures were dried on collodion-covered specimen screens. The bud on the spirochetal form in Figure 5 appears similar to the buds noticed by DeLamater et al. Development of spirochetal forms within spheres is shown in Figure 6. The Treponeme and Borrelia gave comparable findings. The authors emphasize that free granules are definitely a phase in the development of spirochetes.[26]

TREPONEMA MICRODENTIUM

Bladen and Hampp[7] have shown the significance of the round bodies of the oral normal flora spirochetes. After extended storage, a culture contains only spheres; yet such a spi-

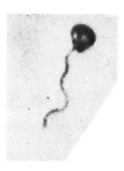

FIGURE 4. Buds on spirochetes from which minute spirochetes gradually develop. (From DeLamater, E. D., Haanes, M., Wiggall, R. H., and Pillsbury, D. M., *J. Invest. Dermatol.*, 16, 231–256, 1951. With permission.)

FIGURE 5. *Borrelia vincenti,* bud on spirochete. (From Hampp, E. C., Scott, D. B., and Wyckoff, R. W. G., *J. Bacteriol.*, 56, 755–769, 1948. With permission.)

rochetefree culture gives rise to spirochetal growth when subcultured.[27] This phenomenon is explained by fine structure exploration of the spheres, revealing that they are packed with tiny microbes in varying stages of development. Some of the intracellular organisms within both *T. microdentium* and *B. vincenti*[7] are devoid of cell walls.

TREPONEMA CALLIGYRUM

T. calligyrum is a nonpathogenic spirochete in human smegma. When cultured in broth fortified with rabbit serum, typical spirochetes appear in young cultures, but at 3 weeks tiny forms abound. Seguin concluded that these were viable, filtrable particles.[61]

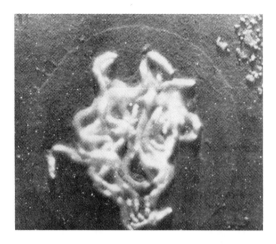

FIGURE 6.   Development of tiny spirochetal forms in cysts or buds of *Borrelia vincenti*. (From Hampp, E. C., Scott, D. B., and Wyckoff, R. W. G., *J. Bacteriol.*, 56, 755–769, 1948. With permission.)

YAWS (*TREPONEMA PERTENUE*)

In tropical countries with heavy rainfall, 70% of the population may suffer from yaws which, if untreated, leads to severe discomfort and disfigurement. The causative agent, *T. pertenue*, is morphologically and antigenically indistinguishable from *T. pallidum*.

Yaws differs from syphilis in being nonvenereal and noncongenital, but resembles syphilis in frequent recurrence after latency and in having stages with aberrant morphology. Both typical spirochetes and stages of the L-cycle occur in the skin and mucous membranes of untreated patients.[40] Gummas, similar to those in syphilis, are not uncommon, and chronic ulcers with a serosanguinous discharge are also encountered. Smears from gummas and from ulcers show no treponemes.[78]

Electron microscopy confirms that *T. pertenue* resembles *T. pallidum* in major points.[50] Like *T. pallidum*, it has cystic as well as spirochetal forms, which can be revealed by negative staining and ultrathin sections.

Ustimenko,[80] studying seven strains of *T. pertenue* isolated from patients' blood, found that penicillin easily induces formation of vesiculated Mycoplasma-like L-colonies in this species.[80] If the medium was enriched with human serum, the penicillin required for L-transformation was sometimes as little as 0.024 U/ml. Neoarsphenamine also induces CWD phase organisms from this treponeme.

SPIROCHAETA MYELOPHTHORA[80] IN MULTIPLE SCLEROSIS

A spiral-shaped organism has been noted in multiple sclerosis (MS) by so many investigators, over so many decades, in so many countries that it can hardly be ignored.

The search for the spirochete concerns three aspects: (1) visualization in brain at autopsy or in spinal fluid, (2) growth of the organism in laboratory animals, (3) culture of the agent. European pathologists led in finding spirochetes by silver staining of brain tissue or spinal fluid sediment.[3,6,23,29,35,45,55,59,70–75] One patient's spinal fluid contained approximately 35 spirochetes per milliliter.[65] Steiner, who named the organism *Spirochaeta myelophthora* (Figure 7), was the most persistent in applying staining. He located the organisms even within the brain's glial cells.[74,75]

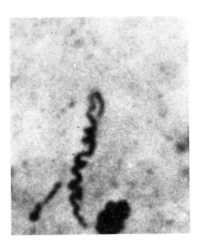

FIGURE 7. *Spirochaeta myelophthora* as photographed from MS autopsy tissue by Gabriel Steiner. (From *J. Neuropathol. Exp. Neurol.*, 13, 221–229, 1954. With permission.)

Regarding animal studies, spirochetes were found when MS spinal fluid was inoculated into guinea pigs,[36,46,60,63,68] monkeys,[1,69] rabbits,[5,23,25,32,76] and hamsters.[91] When young rabbits were injected, the result was paralysis and death.[36] We know of no reported injections in immunosuppressed animals.

Culture studies are given below:

Ichelson[31]	59 of 75 patients	28 controls negative
Newman et al.[49]	5 of 27 patients	13 controls negative
Myerson et al.[48]	3 of 7 patients	21 controls negative

The above investigations were followed by six negative attempts to culture spirochetes, as reviewed by Kurtzke et al.[37] Also damaging were later reports that Ichelson's spirochete was a vibrio, probably a contaminant from water[47,56] not resembling the long spiraling organism of Ichelson's photographs.

Ichelson's organism has also been labeled *T. denticola,* the term under which it has been stored with the American Type Culture Collection. A study found that morphologically it resembled a treponeme more than a vibrio, but that the Ichelson's strain differed in biochemical reactions from a reference *T. denticola* strain.[79] Also pertinent is identifying *T. denticola* antigens associated with blood vessels of MS patients.[16]

Gay and co-workers[91-93] postulate that an oral spirochete is the agent of MS. They state that they have unpublished evidence by Western blot and immunofluorescence that a spirochete is in the central nervous tissue in active MS. They believe the disease is acquired at an age when the barrier between the sphenoidal sinus and subarachnoid space is relatively thin.

How can the discrepancies by explained? For one thing, growing some borrelia-like organisms is an art. It is well known that *Borrelia burgdorferi* is isolated only rarely from human sources. Assuming that a water inhabiting form was in Ichelson's cultures does not explain the positive findings in three series of MS patients vs. negative cultures in other spinal fluids.

In addition to contamination from water which was "sterilized" only by filtration, an important variable in the Ichelson medium is serum. Either pooled human serum or rabbit serum could contain pertinent cross-reacting antibody, which could account for negative series.

A relatively recent study noted that spirochete-like bacteria from autopsy material reacted with sera from over 50 MS cases.[30] The author stressed the pleomorphism of the strains and that antibody indicated the heteromorphic structures were life-cycle phases of one organism. The morphology of the organisms varied with the age of the lesions in the nervous tissue.

Fascinating studies also support a viral etiology for MS. Spongiform degeneration on the central nervous system was produced in sheep and mice inoculated with MS brain tissue. Cultures were negative. Carp et al.[11] and Koldovsky et al.[34] demonstrated an organism by the effect on the neutrophiles of mice. The agent was not cultivable and passed through 50-nm filters but not smaller filters, hence they concluded that the agent was viral. Neither of these characteristics rule out the possibility that the organism is a spirochete prone to forming filterable particles. The relationship of the organism to MS was again shown, this time by its neutralization by sera from patients whose disease was quiescent.

RABBIT SPIROCHETES

T. cuniculi is considered to be the causative agent of rabbit syphilis. Work with *T. pallidum* must be done with rabbits free of this organism. As *T. cuniculi* propagates in the skin and mucous membranes of rabbits, many granules representing the almost invisible stage occur concurrently with the typical spirochete.[40]

Spirochete 277F, from ticks of cottontail rabbits, is apparently infective for no animal except the snowshoe hare. When cultures of the spirochete are inoculated in the yolk sac of chicken embryos, infection spreads throughout the embryo, yielding clear embryonic fluids in which spirochetes are beautifully demonstrated by dark-field examination. However, the first microbial growth consists of groups of basophilic granules. The antigenic relationship between the granules and the spirochetes is readily demonstrated by fluorescent microscopy: the two morphological forms react identically with labeled immune serum. The antibody reveals not only brilliantly stained granules and spirochetes but also spherules elongating into commas.

FREE-LIVING SPIROCHETES

A free-living, pigment-forming spirochete, *S. aurantia,* produces some exceptionally large spherical bodies located terminally or centrally. The "cysts" generally measure from 0.5 to 2 μm but reach 10 μm in diameter when the cells are incubated at 37°C, a temperature at which growth occurs slowly.[9]

There are obvious advantages in studying *T. zuelzerae,* a nonpathogenic treponeme which grows rapidly on a simple medium. Cystic stages are numerous in *T. zuelzerae,*[81] a free-living spirochete isolated from mud and named for Margarete Zuelzer, who described many free-living water spirochetes. As *T. zuelzerae* develops, it produces many large spheroid bodies.

BORRELIA DUTTONI OF AFRICAN RELAPSING FEVER

The agent of relapsing fever in Central Africa is *B. duttoni,* a tick-transmitted spirochete. One of the first observations of variants in spirochetes concerns *B. duttoni.* Dutton and Todd described this organism propagating in its tick host as single granules which become irregular masses of chromatin, in turn developing into short spirochetes.[17]

Inoculation of *B. duttoni* into mice gives encephalitis, but no borrelia are seen when the brain tissue is examined microscopically. However, when brain fragments are placed subcutaneously into new mice, generalized infection rapidly follows, and many Borrelia are demonstrable in sequential samples.

BORRELIA RECURRENTIS OF EUROPEAN RELAPSING FEVER

Patients with European relapsing fever have infectious blood during stages when no spirochetes are visible. Thus, infectivity is exhibited by a nonspirochetal phase of the microbe.[62] Dramatic demonstration of disease production by the granule phase results from examining the flea, which can serve as an intermediate host of *B. recurrentis*. After the flea ingests blood from an infected animal, no borrelia are found in the insect for 8 d, although the entire dissected flea is examined by dark-field microscopy. However, during these 8 d, the flea can infect monkeys. One spirochete in a flea might be overlooked, but, since 122 fleas were scrutinized, the evidence is substantial.[62]

BORRELIA OF FOWL

Relapsing fever spirochetosis also occurs in elephants, camels, antelopes, monkeys, and avian species. The agent of spirochetosis of geese and chicken, *B. anserina*, has been investigated extensively. As in Borrelia of man, infective granules are prominent in pathogenesis, probably explaining how a tick carries an infinite number of infective organisms.[83]

GROWTH PHASES OF A BORRELIA FROM BOVINE RUMEN

A certain Borrelia was found in the rumen of nearly all cattle examined by Bryant.[10] Round bodies developing in cultures appear structureless by phase microscopy, but the electron microscope reveals their content of minute Borrelia. When old cultures containing few typical forms but numerous round bodies are moved to fresh agar medium, many spirochetal colonies develop, indicating many centers where regeneration of spirochetes occurs.

LEPTOSPIRA

Recognition of aberrant stages of Leptospira could facilitate diagnosis in time to guide therapy. Interest in leptospirosis was aroused by infections acquired during Vietnam jungle warfare. The pathogens include nine different species, of which *Leptospira icterohemorrhagiae* and *L. australis* B are responsible for most fatalities.[2]

In view of the difficulties in culturing Leptospira with present methods, the diagnosis is commonly made retrospectively by looking for antibodies in serum. Diagnosis by culture may be possible in the future since Faine has found a method to simplify the medium preparation.[18] Sheep or horse serum, heated at 65°C for 2 h to eliminate the heat-labile IgM natural antibody,[19] can substitute for rabbit serum, which has always been considered essential. It will be useful to remember that Leptospira do not always have spiral morphology with a terminal hook. It is only necessary to transfer a pure culture of Leptospira to any nonideal medium to find growth consisting exclusively of short straight rods and coccoid forms.

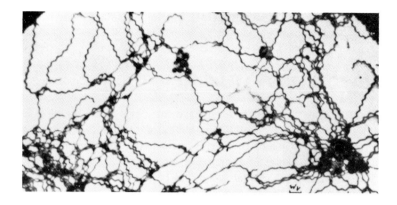

FIGURE 8. Leptospires growing from ''nests'' of cysts. (From Gängel, G. and Themann, H., *Arch. Hyg. Bakteriol.*, 140, 559–568, 1956. With permission.)

LEPTOSPIRA ICTEROHEMORRHAGIAE

Fluorescent antibody studies have shown that the small granular form of *L. icterohemorrhagiae* is common and important *in vivo*. Radu and associates,[52] employing an antiserum which had been carefully adsorbed with homologous tissue, found that certain hepatic or Küpffer cells of infected guinea pigs were filled with specifically staining granules which occupied all the cell except the nucleus. The fluorescent antibody detected almost twice as many infections as dark-field examination or Giemsa staining, and also showed the location of the granular stage of this organism.

Nonbinary fission propagation of *L. icterohemorrhagiae* is beautifully demonstrated in studies by Gängel and Themann.[22] From one active center many Leptospira extend (Figure 8). The authors comment on how greatly their findings with Leptospira resemble the cycles observed in *T. pallidum* by DeLamater and associates.[14] Again, spirochetes form within cysts.

Swain noticed that *L. icterohemorrhagiae* had another characteristic relating it to the cycling of the treponemes.[77] The slender Leptospira with a diameter of only 0.12 μm are sometimes widened by a ''bubble,'' within which a coiled spirochete is seen by fine structure studies. The bubbles appear as early as 5 d.

LEPTOSPIRA POMONA

The continued theme of aberrant stages is illustrated by other Leptospirae. Harrington and Sleight found that *L. pomona* causes death of tissue culture preparations after granules and other atypical structures have developed intracytoplasmically.[28]

AN L FORM LABORATORY LOOKS AT MS SPECIMENS

Here we report studies preliminary, though long term. We find that it is possible to grow a Borrelia-like organism from MS spinal fluids from approximately 90% of MS bloods and from some patients who do not have MS. There is nothing easier than not growing the spirochete. However, we are encouraged that a spirochete can be seen (Figure 9) and grown in serum-free medium with all components autoclaved. It multiplies at temperatures from 22 to 30°C in an agitated culture. It is sometimes motile. It has been stained with acridine orange, with fluorescent muramidase, with Fontana's silver stain (Figure 9), and by using an antibody specific for any class of spirochetes.[54] Identification with antibody used the method of Lukehart and collaborators.[42] The strains are uncooperative, and subcultures do not grow. They multiply in B.S.K. medium[37] designed for cultivation of *B. burgdorferi*, but sparsely. They prefer an alkaline medium and show an oligodynamic reaction to clin-

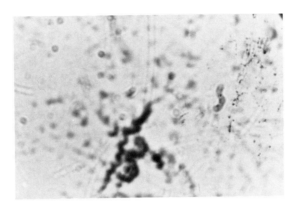

FIGURE 9. Organisms in spinal fluid of MS patient K. I. The fluid was not centrifuged. (Magnification ×
1000.)

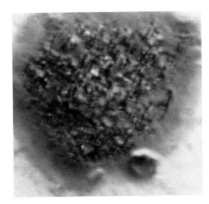

FIGURE 10. Structures which grew in the cultures of the three spinal fluids examined resemble L-bodies with
forms differentiating internally.

damycin. Most interesting are structures seen in unconcentrated samples of spinal fluid (see
Frontispiece). Forms which appear in cultures of spinal fluid also suggest L bodies with
internal growth (Figure 10). These suggest the L bodies of the L cycle and the large spheres
which are found with other species of spirochetes described in this chapter. The significance
of these findings awaits preparing a specific antibody for identification studies. Is there a
normal flora spirochete? Or is this an opportunist? Cooperating in these studies have been
Phillip Hoekstra, Christopher Hussar, Russell Johnson, Carrie Kodner, Hetal Parikh, George
Riviere, and Harold Rossmoore.

SUMMARY OF THE
SPIROCHETAL GROWTH CYCLE

The genera Borrelia, Leptospira, and Treponema are characterized by developing large
cyst-like bodies. The structure and function of the ''cysts'' have been documented with
countless electron micrographs. They resemble the characteristic L-body of the L-cycle in
many respects. Most notably, the classic spirochete may appear in the interior of such cysts.
Secondly, an alternate type of reproduction from these bodies is a sprouting filament which
may become the spirochete. The spirochetal ''cysts'' also resemble L-bodies in that they

form frequently but regenerate erratically when transferred to fresh medium. The spirochetal cysts differ from bacterial L-bodies in usually forming only a few spirochetae rather than the numerous parent forms which may pack a reverting L-body of most species. Secondly, a sprouting cyst usually thrusts out a spirochetal form rather than the infinite varieties of rhizoid growth which can emerge from an L-body of most bacteria.

The formation of tiny refractile granules is also well documented for many species of all genera in the Spirochaetae. Whether these are pathogenic per se remains at this date a controversial point. There is little doubt that even for *T. pallidum* these granules are infective. The multiplication of the granules has been described by careful investigators, and their development into spiral organisms has been described for almost every species.

The identification of aberrant stages of Spirochaetae by staining with fluorochrome-labeled antibody represents significant progress and will greatly facilitate recognition of the granular and cystic stages *in vitro* and *in vivo*. To date, such fluorescent antibody studies have been described for Spirochete 277F of rabbits and for *L. icterohemorrhagiae*. The L Form laboratory at Wayne State University has found that fluorescent antibody to *B. burgdorferi,* supplied by Russell Johnson, reacts with both the spirochete and atypical forms of the organism in culture.

Currently, research is in progress on growing spirochetes from MS cases, based on reports of early experimental work. From blood and spinal fluid of some MS cases it is possible to grow a Borrelia-like organism in serum-free autoclaved medium. Improvement in culture is the next step necessary to obtain a bulk of organisms for antibody production.

Much more research needs to be done to show that the organism is not an opportunist and to answer other questions.

REFERENCES

1. **Adams, D. K.,** Spirochetes in the ventricular fluid of monkeys inoculated from cases of disseminated sclerosis, *Surgo.,* 14, 11, 1948.
2. **Allen, G. L., Weber, D. R., and Russell, P. K.,** The clinical picture of Leptospirosis in American soldiers in Vietnam, *Milit. Med.,* 133, 275–280, 1968.
3. **Austregesilo, A.,** Le schlerose en plaques de form subalque A propos d'un cas, *L'Encephale,* 28, 633, 1933.
4. **Barbour, A. G.,** Isolation and cultivation of Lyme disease spirochetes, *Yale J. Biol. Med.,* 57, 521–525, 1984.
5. **Blacklock, J. W.,** MS agent in rabbits, *J. Pathol. Bacteriol.,* 28, 1, 1925.
6. **Blackman, N.,** Etiologie de la scherose in plaque, These Maurice Levergne, Paris, 1936.
7. **Bladen, H. A. and Hampp, E. G.,** Ultrastructure of *Treponema microdentium* and *Borrelia vincentii, J. Bacteriol.,* 87, 1180–1191, 1964.
8. **Boncinelli, V., Vaccari, R., Pincelli, L., and Lancellotti, M.,** Ricerche sulla Persistenza del *Treponema pallidum* nelle Linfohiandole di Luetici Trattati, *G. Ital. Dermatol.,* 107, 1–14, 1966.
9. **Breznak, J. A. and Canale-Parola, E.,** *Spirochaeta aurantis,* a pigmented, facultatively anaerobic spirochete, *J. Bacteriol.,* 97, 386–395, 1969.
10. **Bryant, M. P.,** The isolation and characteristics of a spirochete from the bovine rumen, *J. Bacteriol.,* 64, 325–335, 1952.
11. **Carp, R. J., Licursi, P. C., Merz, P. A., and Merz, G. S.,** Decreased percentage of polymorphonuclear neutrophils in mouse peripheral blood after inoculation with material from MS patients, *J. Exp. Med.,* 136, 618–629, 1972.
12. **Collart, P., Borel, L. J., and Durel, P.,** Étude de l'action de la pénicilline dans la syphilis tardive, persistance du tréponème pale après traitement, *Ann. Inst. Pasteur* (Paris), 102, 596–615, 1962.
13. **Collart, P., Borel, L. J., and Durel, P.,** Étude de l'action dela pénicilline dans la syphilis tardive, persistance du tréponème pale après traitement, seconde partie, *Ann. Inst. Pasteur (Paris),* 102, 693–704, 1962.
14. **DeLamater, E. D., Haanes, M., Wiggal, R. H., and Pillsbury, D. M.,** Studies on the life cycle of spirochetes. VIII. Summary and comparison of observations on various organisms, *J. Invest. Dermatol.,* 16, 231–256, 1951.

15. **Del Carpio, C.,** Effetti di un trattamento massiccio con penicillina nella sifilde sperimentale del comiglio a diversa distanza dall'infezione. Rapporti fra preserza di anticorpe immobilizzanti oppersistenza dell'infesione, *Riv. Ist. Sieroter Ital.,* 38, 166–173, 1963.
16. **Dick, G. and Gay, D.,** Multiple sclerosis — autoimmune or microbial? A critical review with additional observations, *J. Infect.,* 16, 25–35, 1988.
17. **Dutton, J. E. and Todd, J. L.,** A note on the morphology of *Spirochaeta duttoni, Lancet,* 2, 1523–1525, 1907.
18. **Faine, S.,** Heated sheep or horse serum substitute for rabbit serum in culture media for *Leptospira, Appl. Microbiol.,* 16, 534, 1968.
19. **Faine, S. and Carter, J. N.,** Natural antibody in mammalian serum reacting with an antigen in some leptospires, *J. Bacteriol.,* pp. 280–285, 1968.
20. **Flarer, F.,** On the antitreponemic action of cephaloridine, *Postgrad. Med. J. Suppl.,* 43, 133–134, 1967.
21. **Fodvari, F.,** Conduct of *Spirocheta pallida* in tissue explantations, *Am. J. Syph.,* 16, 145–154, 1932.
22. **Gängel, G. and Themann, H.,** Elektronenmikroskopische Untersuchungen uber die Entwicklungsstadien bei Leptospiren, *Arch. Hyg. Bakteriol.,* 140, 559–568, 1956.
23. **Guirand, P.,** Figures parasitaires dans la scherose en plaques, *L'Encephale,* 26, 349, 1931.
24. **Gürün, H.,** A new culture method for the organisms of leprosy, tuberculosis, and syphilis, *Rüzarli Matbaa* (Ankara), pp. 1–42, 1957.
25. **Gye, F.,** MS agent in rabbits, *Brain,* 14, 213, 1921.
26. **Hampp, E. C., Scott, D. B., and Wyckoff, R. W. G.,** Morphologic characteristics of certain cultured strains of oral spirochetes and *Treponema pallidum* as revealed by the electron microscope, *J. Bacteriol.,* 56, 755–769, 1948.
27. **Hampp, E. C.,** Further studies on the significance of spirochetal granules, *J. Bacteriol.,* 62, 347–349, 1951.
28. **Harrington, D. D. and Sleight, S. D.,** *Leptospira pomona* in tissue culture, preliminary study, *Am. J. Vet. Res.,* 27, 249–256, 1966.
29. **Hassin, G. B. and Diamond, I. B.,** Silver cells and spirochete-like formations in MS and other diseases of the CNS, *Arch. Neurol. Psychiatry,* 41, 471, 1939.
30. **Ibrahim, M. Z. M.,** On the etiology of MS; new observations?, *Abstr. Anat. Rec.,* 214A, 58, 1986.
31. **Ichelson, R. R.,** The cultivation of spirochaetes from spinal fluids of multiple sclerosis cases and negative controls, *Proc. Penn. Acad. Sci.,* 32, 49–54, 1958.
32. **Kaberlah, F.,** MS agent in rabbits, *Dtsch. Med. Work.,* XICII, 102.
33. **Klieneberger-Nobel, E.,** The filterable forms of bacteria, *Bacteriol. Rev.,* 15, 77–103, 1951.
34. **Koldovsky, U., Koldovsky, P., Henle, G., Henle, W., Ackermann, R., and Haase, G.,** Multiple Sclerosis-associated agent, transmission to animals and some properties of the agent, *Infect. Immun.,* 12, 1355–1366, 1975.
35. **Kopeloff, N. and Blackman, N.,** Silver cells (Steiner's Method) in MS compared with their presence in other disease, *Arch. Neurol. Psychiat.,* 34, 1297, 1935.
36. **Kuhn, P. and Steiner, G.,** Uber Die Ursache der MS, *Med. Klin.,* 13, 1007–1009, 1917.
37. **Kurtzke, J. F., Martin, A., Jr., Myerson, R. M., and Lewis, J. I.,** Microbiology in MS, *Neurology* 12, 915–922, 1962.
38. **Leuriaux, C. and Geets, V.,** Culture de *Treponema pallidum* de Schaudinn, *Zentralbl. Bakteriol. Parasitenkd. Infektionskr. Hyg. Abt. 1 Orig.,* 41, 684–688, 1906.
39. **Levaditi, C.,** Gommes syphilitiques et formes anormales du Tréponème, *C. R. Soc. Biol.,* 104, 477–480, 1930.
40. **Levaditi, C. and Po, L. Y.,** Cycle évolutif du *Treponema pallidum,* du *Spirochaeta pertenuis* et du *Spirochaeta cuniculi, C. R. Soc. Biol.,* 104, 736–740, 1930.
41. **Levaditi, C., Lépine, P., and Schoen, R.,** Relation entre le cycle evolutif du *Treponema pallidum* et la genèse des lésions syphilitiques, *C. R. Soc. Biol.,* 104, 72–75, 1930.
42. **Lukehart, S. A., Tam, M. R., Hom, J., Baker-Zander, S. A., Holmes, K, K., and Nowinski, R. C.,** Characterization of monoclonal antibodies to *Treponema pallidum, J. Immun.,* 134, 585–592, 1985.
43. **Lukehart, S. A. and Holmes, K. K.,** Syphilis, in *Harrison's Principles of Internal Medicine,* 12th ed., Wilson, J. D., et al., Eds., McGraw-Hill, New York, 1991, 651–661.
44. **Manouelian, Y.,** Gommes syphilitiques et formes anormales du Tréponème. Ultra-virus syphilitique, *C. R. Soc. Biol.,* 104, 249–251, 1930.
45. **Marburg, D.,** Studies in the pathology and pathogenesis of MS with special reference to Phlebothrombosis and Guiraud's bodies, *J. Neuropathol. Exp. Neurol.,* 1, 3, 1942.
46. **Marinesco, D.,** MS agent in guinea pigs, *Rev. Neurol.,* 35, 481, 1919.
47. **Martin, A., Youngue, E. L., and Kost, P. F.,** The occurrence of *Spirochaeta myelophthora* in cerebrospinal fluid, *Proc. Penn. Acad. Sci.,* 33, 55–93, 1959.
48. **Myerson, R. M., Wolfson, S. M., and Sall, T.,** Preliminary observations on the cultivation and morphology of a microorganism from the cerebrospinal fluid of patients with multiple sclerosis, *Am. J. Multiple Sclerosis,* 236, 677–691, 1958.

49. **Newman, H. W., Purdy, C., Rance, L., and Hill, F. C., Jr.,** The spirochete in multiple sclerosis, *Calif. Med.,* 89, 387–389, 1958.

50. **Ovchinnikov, N. M., Delektorsky, V. V., and Kenigsberg, T. L.,** Experimental yaws and electron microscopy of *Treponema pertenue, Vestn. Dermatol. Venerol.,* 44, 42–49, 1970.

51. **Ovchinnikov, N. M.,** Important problems of serodiagnosis of syphilis, *Vestn. Dermatol. Venerol.,* 8, 22–26, 1981.

52. **Radu, I., Sturdza, N., and Radu, A.,** Etudes des leptospires au moyen des anticorps fluorescents. II. Mise an evidence des leptospires dans l'organisme du cobaye infecte par *L. icterohaemorrhagiae, Arch. Roum. Pathol. Exp. Microbiol.,* 24, 713–726, 1965.

53. **Rice, N. S. C., Dunlop, E. M. C., Jones, B. R., Hare, M. J., King, A. J., Rodin, P., Mushin, A., and Wilkinson, A. E.,** Demonstration of treponeme-like forms in cases of treated and untreated late syphilis and of treated early syphilis, *Br. J. Vener. Dis.,* 46, 1–9, 1970.

54. **Riviere, G. R., Wagoner, M. A., Baker-Zander, S. A., Weisz, K. S., Adams, D. F., Simonson, L., and Lukehart, S. A.,** Identification of spirochetes related to *Treponema pallidum* in necrotizing ulcerative gingivitis and chronic periodontitis, *N. Engl. J. Med.,* 325, 539–543, 1991.

55. **Rogers, H. J.,** The question of silver cells as proof of the spirochetal theory of desseminated sclerosis, *J. Neurol. Psychopathol.,* 13, 50, 1932.

56. **Rosebury, T.,** Vibrio-like bacteria recovered from cultures of *Spirochaeta myelophthora, Proc. Soc. Exp. Biol. Med.,* 105, 134–137, 1960.

57. **Roukavischnikoff, E. J.,** Zur Frage der Entwicklungsstadien des Syphiliserregers, die im Blute des infizierten Menschen und der Versuchstiere zirkulieren, *Zentralbl. Bakteriol. Parasitenkd. Infektionskr. Hyg. Abt. 1 Orig.,* 115, 66–71, 1929.

58. **Schaudinn, F. and Hoffman, P.,** Über *Spirochaeta pallida* bei Syphilis und die Unterschiede dieser Form Gegenüber anderen Arten dieser Gattung, *Berl. Klin. Wochenschr.,* 42, 673–765. 1905.

59. **Scheinken, I. M.,** Uber spirochaten befunde bie MS Sitzunbsber d. Vereins f Psychatrie and Neurol in Wien Jhrb, *Psychiatrie. Neurol.,* 53, 219, 1937.

60. **Schlossman, A.,** MS agent in animals, *Folia Neuropathol. Eston,* 1, 66, 1923.

61. **Seguin, P.,** *Treponema calligyrum* et ultra-virus spirochetique, *C. R. Soc. Biol.,* 104, 247–248, 1930.

62. **Sergent, E. and Foley, H.,** De la periode de latence du spirille chez le pou infecte de fievre recurrente, *C. R. Acad. Sci.,* 159, 119–122, 1914.

63. **Sicard, et al.,** MS spirochetes in animals, *Rev. Neurol.* (Paris), XXLX, 954, 1922.

64. **Simon, C. and Mollinedo, R.,** Diagnostic de la syphilis par la recherche du granule spirochétogène, *Presse Med.,* 48, 513–516, 1940.

65. **Simons, Von H. C. R.,** Is multiple sclerosis a spirochetosis?, *Dtsch. Med., Wochenschr.,* 83, 1196–1200, 1958.

66. **Smith, J. L., Singer, J. A., Moore, M. B., Jr., and Yobs, A. R.,** Sero-negative ocular and neuro-syphilis, *Am. J. Ophthalmol.,* 59, 753–762, 1965.

67. **Smith, J. L., Israel, C. W., and Harner, R. E.,** Syphilitic temporal arteritis, *Arch. Ophthalmol.,* 78, 284–288, 1967.

68. **Steiner, G.,** Guinea pig inoculation with MS tissues, *Arch. Psychiatry Nervenkrankh. Berlin,* 60, 1918.

69. **Steiner, G.,** MS agent inoculation in monkeys, *Z. Neurol. Psychiatry Reger* at Berlin XVLL, 491, 1919.

70. **Steiner, G.,** Untersuchsinger zur Pathogenese der MS, *Z. Neurol. Psychiatry,* 47, 701, 1927.

71. **Steiner, G.,** Zur histopathogenese der MS, *Vehr Dtsch. Nervenarzte, Dresden,* 133, 1939.

72. **Steiner, G.,** Silver staining of MS tissues, *Nervenarzt,* 6, 281, 1932.

73. **Steiner, G.,** Is MS an etiologically uniform infectious disease? *Detroit Med. News Ed. Issue,* 32, 7, 1941.

74. **Steiner, G.,** Acute plaques in MS, their pathogenetic significance and the role of spirochetes as the etiological factor, *J. Neuropathol. Exp. Neurol.,* 11(4), 343, 1952.

75. **Steiner, G.,** Morphology of *Spirochaeta myelophthora* in MS, *J. Neuropathol. Exp. Neurol.,* 13, 221–229, 1954.

76. **Stephanopoulo, F.,** Spirochetes in CSF of MS, *Bull. Med. Pans.,* 30, 595, 1922.

77. **Swain, R. H. A.,** Electron microscopic studies of the morphology of pathogenic spirochetes, *J. Pathol. Bacteriol.,* 69, 117–128, 1955.

78. **Taneja, B. L.,** Yaws: clinical manifestations and criteria for diagnosis, *Indian J. Med. Res.,* 56, 100–113, 1968.

79. **Trevathan, C. A., Smibert, R. M., and George, H. A.,** Lipid catabolism of cultivated treponemes, *Can. J. Microbiol.,* 28, 672–678, 1982.

80. **Ustimenko, L. M.,** L Forms of *Treponema pertenue, Vestn. Akad. Med. Nauk SSSR,* 20, 46–50, 1965.

81. **Veldkamp, H.,** Isolation and characteristics of *Treponema zuelzerae* nov. spec., an anaerobic, free-living spirochete, *Antonie van Leeuwenhoek,* 26, 103–125, 1960.

82. **Warthin, A. S. and Olsen, R. E.,** The granular transformation of *Spirocheta pallida* in aortic focal lesions, *Am. J. Syph.,* 14, 433–437, 1930.

83. **Wenyon, C. M.,** *Protozoology, A Manual for Medical Men, Veterinarians, and Zoologists,* William Wood, New York, 1926.

84. **Wile, U. J. and Snow, J. S.,** The chick embryo as a culture medium for *Spirocheta pallida, J. Invest. Dermatol.,* 4, 103–109, 1941.

85. **Wile, U. J. and Johnson, S. A. M.,** Further study of the chick embryo as a culture medium for the *Spirochaeta pallida, Am. J. Syph.,* 28, 187–191, 1944.

86. **Wile, U. J. and Curtis, A. C.,** in discussion of DeLamater, E. D. et al., Studies on the life cycle of spirochetes. VIII. Summary and comparison of observations on various organisms, *J. Invest. Dermatol.,* 16, 231–256, 1951.

87. **Yobs, A. R., Brown, L., and Hunter, E. F.,** Fluorescent antibody technique in early syphilis, *Arch. Pathol.,* 77, 220–224, 1964.

88. **Yobs, A. R., Rockwell, D. H., and Clark, J. W., Jr.,** Treponemal survival in humans after penicillin therapy, a preliminary report, *Br. J. Vener. Dis.,* 40, 248–253, 1964.

89. **Yobs, A. R., Olansky, S., Rockwell, D. H., and Clark, J. W., Jr.,** Do treponemes survive adequate treatment of late syphilis?, *Arch. Dermatol.,* 91, 379–389, 1965.

90. **Yobs, A. R., Clark, J. W., Jr., Mothershed, S. E., Bullard, J. C., and Artley, C. W.,** Further observations on the persistence of *Treponema pallidum* after treatment in rabbits and humans, *Br. J. Vener. Dis.,* 44, 116–130, 1968.

91. **Gay, D. and Dick, G.,** Is Multiple Sclerosis caused by an oral spirochaete?, *Lancet,* July 12, 1986, 75–77.

92. **Gay, D. and Dick, G.,** Spirochaetes, Lyme disease and Multiple Sclerosis, *Lancet,* September 20, 1986, 685.

93. **Gay, D. and Esiri, M.,** Blood-brain barrier damage in acute multiple sclerosis plaques. An ammunocytological study, *Brain,* 114 (Pt 1B), 557–72, 1991.

Chapter 26

TOXIN FORMATION

THE CLOSTRIDIA

TETANUS

When competent investigators disagree, an enigmatic situation exists. The old adage must apply: "When experienced researchers have different findings, the same experiment was not done."

Publications on toxigenicity of the *Clostridium tetani* L-phase were contradictory, but the discrepancy has been clarified. A first report[6] indicated that the L-variant lacks pathogenicity, but subsequently Rubio-Huertos and Vasquez[22] recorded that the wall-deficient stage makes as much or more toxin than the parent. This report from Spain dealt with variants produced by large quantities of glycine in the culture medium. Was there an essential difference in toxigenicity between glycine-induced variants and those induced by penicillin?

The induction method was not the explanation. In Uruguay, Scheibel and Assandri[24,25] induced four strains of the tetanus bacillus into the L-phase with penicillin. Three L-phase strains were toxigenic, producing the pathognomonic spastic paralysis in mice (Figure 1). The identity of the toxin was confirmed by neutralization with tetanus antitoxin. Scheibel and Assandri explained the discrepancy between their work and the previous results with penicillin-induced L-phase *C. tetani*. Their nontoxigenic strain was a stock culture, probably the same culture used in the early negative study. This strain produces some nontoxic clones. This loss of toxin-forming ability, they note, can also occur with cultures in the classic stage.

The growth of *C. tetani* in an L-phase probably does not complicate the diagnosis of this syndrome, which is based on unique symptoms (Figure 2). However, Scheibel and Assandri conclude that there is a false feeling of security from administration of penicillin unsupported by *C. tetani* antitoxin. The possibility exists that not tetanus bacilli per se but rather their CWD variants may survive antibiotic therapy and, with or without reversion, produce clinical tetanus. Larned observed that when mice received injections of *C. tetani* intramuscularly, the bacilli rapidly became pleomorphic bodies, and only atypical forms could be detected at 24 and 48 h.[37]

BOTULISM

C. botulinum needs neither cell wall nor spore to manufacture its toxin. Toxigenic L-Form lines of *C. botulinum* types A and E, stable for more than 150 transfers and lacking both wall and spore,were investigated. The type A L-Form culture produced 500,000 LD_{50} of toxin per milliliter; the type E L Form titered 40 LD_{50} before tryptic digestion and 3200 LD_{50} per milliliter after trypsinization.[2] The authors, Brown et al.[2] produced propagating L Forms of Clostridia by combined lysozyme and penicillin. If lysozyme, which hydrolyzes the beta-1,4 linkage of the cell wall mucopeptide,[17] is used alone, replicating L Forms do not result, as shown by Duda and Slack[7] and Kawata et al.[15]

It is clear that the L Forms of *C. botulinum* have the potential for contributing to disease of man and lower forms. Botulism applies not only to food poisoning of man, but also to gas gangrene,[5,9] the morbidity of wild fowl,[12,26] and fish in natural waters.[30]

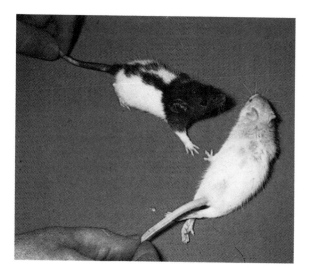

FIGURE 1.   Mouse on the right received an injection of *Clostridium tetani* into tail vein to identify the bacterium. His leg twitches uncontrollably. Mouse on the left received the same culture but was protected by preinjection of 750 U of *C. tetani* antitoxin. (From Stratford, B. C., Ed., *An Atlas of Medical Microbiology, 1977.*)

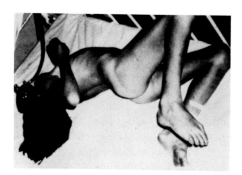

FIGURE 2.   Boy with clinical tetanus showing opisthotonos. (From *Anaerobic Bacteria,* Willis, A. T. and Balows, A., Eds., Charles C. Thomas, Springfield, IL. With permission.)

GAS GANGRENE
Clostridium perfringens

Toxigenicity of *C. perfringens* L-stages was shown by Bittner and Voinesco.[1] When inoculated intramuscularly into guinea pigs, the cultures produced the typical swelling with gas and other characteristics typical of infection with the classic organism. The toxin produced in culture varied from 4 to 64 minimum lethal doses per milliliter, which was less than made by the parent strains. However, it is evident that L Forms of *C. perfringens* produced by other methods are not toxigenic. *C. perfringens,* types A, B, C, D, E, and F, induced into the L-stage by penicillin by Kawatomari, were all avirulent, although some still produced a little hemolysin.[16]

C. perfringens has been found in man in the CWD form, termed transitional by the authors Hill and Lewis,[10] implying that it did revert *in vitro* to the classic spore former. Thus, variants of this organism may play a role in pathogenesis under natural situations, whether by permitting latency or by toxin production per se. In addition, when the role of

CWD variants of *C. perfringens* in natural situations is investigated, the important part which this Clostridium plays in food poisoning of man and domestic animals must be considered.

Clostridium sordelli and Other Toxigenic Clostridia

CWD variants of *C. sordelli* were found in successive blood cultures of a hospitalized patient. The reverted organism made a toxin lethal for guinea pigs, but the patient responded to combined antibiotic therapy without antitoxin.[20] This indicates that the variant can be pathogenic without producing toxin.

The toxigenicity of the CWD variants of all the gas-gangrene Clostridia has not as yet been explored. Bittner advises that his laboratory has produced toxigenic L Forms of *C. septicum*.

THE CORYNEBACTERIA

THE DERMONECROTOXIN OF *CORYNEBACTERIUM HEMOLYTICUM*

C. pyogenes hominis, better known in the U.S. as *C. hemolyticum*, causes sporadic single cases and outbreaks of dermal infections in tropical and temperate climates. The irritating, necrotizing product of the bacterium is found in culture filtrates of both parent and CWD variants, as shown by Patočka and Kalvodova.[21] The L Forms produce phospholipase-D, a potent epidermal irritant, characteristic of the original bacterium. L-Form phospholipase is neutralized by antitoxin produced against phospholipase from the parent organism.

THE DIPHTHERIA BACILLUS

Toxin production is retained by CWD *C. diphtheriae*.[14] The extent to which this complicates diphtheria diagnosis and detection of carriers remains to be ascertained. Retention of toxin production indicates that the phage carrying the genetic information is present in the variant. Figure 3 shows the morphology of a CWD variant produced *in vitro*.

TOXIC PRODUCTS FROM VIBRIO

Neuraminidase is produced by CWD *Vibrio cholerae*. This was shown by early work of Madoff et al.[18]

STAPHYLOCOCCAL ENTEROTOXINS

For determining the role of a wall in formation of bacterial toxins, the food-poisoning Staphylococci constitute a profitable area. Enterotoxin B is not made by stable L-colonies. Czop and Bergdoll[3] state that their finding probably supports Friedman's evidence[8] that Enterotoxin B is actually synthesized on the surface of the coccal wall. The same phenomenon may apply to Enterotoxin C because it, too, was missing in cultures of stable L Forms of Staphylococci whose parents make the C toxin. Furthermore, Staphylococci temporarily growing as L-phase variants in the presence of penicillin, on reversion were found incapable of making C toxin.

Unfortunately for the gourmet who relishes aged foods, the common staphylococcal food toxin, type A, is made by L Forms whenever the parent coccus can produce this protein. Bergdoll[36] hypothesizes that Enterotoxin E may be made by L Forms of Staphylococci because of its similarity to the A toxin.

Enterotoxin production by staphylococcal CWD stages is a vast subject which has had only exploratory investigation. It remains to be learned. e.g., whether staphylococcal variants in natural phenomena, such as bovine mastitis, can be involved in food poisoning.

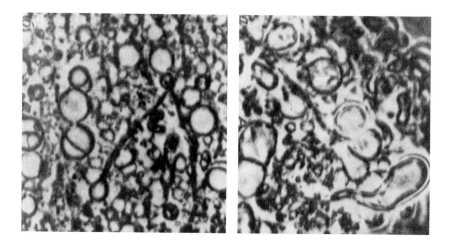

FIGURE 3. Stages in formation of *C. diptheriae* L Forms by penicillin. (From Timakov, B. and Kagan, G. Ya., *L Form Bacteria and Species of Mycoplasma in Pathology*, Moscow, 1973.)

ENDOTOXINS

ENDOTOXICITY IN MICE MADE HYPERSENSITIVE BY BCG

Endotoxin was sought in L Forms of *Salmonella paratyphi* B by Dasinger and Suter.[4] Toxicity was measured in mice made hyperactive to endotoxin by intravenous injection of BCG, given 10 d prior to administration of the L Form or parent organism. Lethality for the sensitive mice was tested by intravenous injection. The LD_{50} of the variant ranged from approximately 7.8 to 11.7 μg. It required from 5 to 29 times as much of the variant as the bacterial form to show toxicity.

SHWARTZMANN REACTION WITH MICROBIAL L FORMS

The basic studies of Sanarelli[23] and Shwartzmann[28] showed that endotoxins of Gram-negative bacteria may sensitize skin so severely that the same antigen intravenously causes hemorrhage in the sensitized area. The relative ability of parent organism and L-variant to sensitize dermis was investigated by injecting classic bacteria and stable L Forms of *Proteus mirabilis* into rabbits. The resultant sensitivity was tested by intravenous injection of either *P. mirabilis* L Forms or *Escherichia coli* endotoxin. Rather surprisingly, the more intense reaction occurred in skin sensitized with L Forms.[13]

L-variants need not be viable in order to cause sensitization. This observation, coupled with the demonstration that the L Forms disappear from the skin after 4 h, in contrast to classic bacteria which remained for at least 72 h, suggests that L Forms rapidly release their endotoxin and that this is the mechanism for the pathogenesis.[13] This conclusion that bacterial L Forms sensitize skin because the endotoxin is released into the tissue is reinforced by finding that the endotoxin extractable from L Forms is about one third that from an equal weight of normal bacteria.[32] Therefore, CWD forms may cause endotoxic damage in the dermis and elsewhere because in a normotensive environment they lyse, rapidly releasing their entire complement of toxin.

LETHALITY IN CHICK EMBRYO

The chick embryo is killed by some endotoxin-containing CWD forms. The yolk sac is an especially favorable locus for propagation of Proteus L Forms, tending to cause embryonic death by 2 to 13 d. In contrast, inoculation into the allantoic cavity or into the chorioallantoic membrane gives irregular results.[27]

ENDOTOXINS OF HEMOPHILUS AND BORDETELLA

H. influenzae, like the Enterobacteriaceae, demonstrates increased toxicity in a CWD stage. Wittler found that glycine-induced spheroplasts injected intraperitoneally in mice give a more rapidly fatal toxemia than parent bacteria having an equal nitrogen content.[34]

Bordetella pertussis spheroplasts, also glycine-induced, were examined by Mason. Substantial amounts of endotoxin were found associated with the variants.[19]

NEUROTOXINS

A fraction of *P. mirabilis* containing high neurotoxicity has been studied by Izdebska.[11] This fraction, which constitutes 1.48% of the dry weight of the parent *P. mirabilis,* is found as 2.55% of the L Form. However, the neurotoxin from the L Form has changes in carbohydrate components and is less potent than the parental neurotoxin. The stable L Form employed in this study might not resemble any occurring physiologically *in vivo* as the strain had been passaged as a variant for 3 years. Other toxic substances retained in the CWD variants are discussed in the chapter on identification.

SUMMARY

Most toxigenic Clostridial species continue to make their specific toxin when cell wall deficient. In *Clostridium tetani* the toxin is increased in some instances; in other species the titer is unchanged, or decreased. The dermonecrotoxin of *Corynebacterium hemolyticum* is similarly found intact in filtrates of this organism grown in the CWD state.

Unfortunately for the epicure, Enterotoxin A, the common staphylococcal food "poison" is made by L Forms of the cocci as well as by the parent cocci. It is probable that Enterotoxin E is also made by staphylococcal L Forms. In contrast, Enterotoxins B and C have not been found in cultures of stable staphylococcal L Forms.

Endotoxins are usually decreased in CWD forms, but because the variants lyse in certain tissues, damage resulting from explosive release of endotoxin may be remarkable.

Chick embryos are killed by propagation of CWD forms of Proteus. Spheroplasts of *H. influenzae* produce a rapidly fatal toxemia in mice. Endotoxin is retained in variable degrees in spheroplasts of *H. influenzae* and *B. pertussis.*

REFERENCES

1. **Bittner, J. and Voinesco, V.,** Formes "L" toxigènes du *Clostridium perfringens, 9th Int. Congr. Microbiol. Abstr.* (Moscow), p. 353, 1966.
2. **Brown, G. W., Jr., King, G., and Sugiyama, H.,** Penicillin-lysozyme conversion of *Clostridium botulinum* Types A and E into protoplasts and their stabilization as L Form cultures, *J. Bacteriol.,* 104, 1325–1331, 1970.
3. **Czop, J. K. and Bergdoll, M. S.,** Synthesis of enterotoxin by L Forms of *Staphylococcus aureus, Infect. Immun.,* 1, 169–173, 1970.
4. **Dasinger, B. L. and Suter, E.,** Endotoxic activity of L Forms derived from *Salmonella paratyphi* B, *Proc. Soc. Exp. Biol. Med.,* 111, 399–400, 1962.
5. **Davis, J. B., Mattman, L. H., and Wiley, M.,** *Clostridium botulinum* in a fatal wound infection, *JAMA,* 146, 646–648, 1951.
6. **Dienes, L.,** Isolation of L type cultures from Clostridia, *Proc. Soc. Exp. Biol. Med.,* 75, 412–415, 1950.
7. **Duda, J. J. and Slack, J. M.,** Toxin production in *Clostridium botulinum* as demonstrated by electron microscopy, *J. Bacteriol.,* 97, 900–904, 1969.

8. **Friedman, M. E.**, Inhibition of staphylococcal enterotoxin B formation by cell wall blocking agents and other compounds, *J. Bacteriol.*, 95, 1051–1055, 1968.

9. **Hampson, C. R.**, A case of probable botulism due to wound infection, *J. Bacteriol.*, 61, 647, 1951.

10. **Hill, E. O. and Lewis, S.**, "L Forms" of bacteria isolated from surgical infections, *Bacteriol. Proc. 1964*, p. 48, 1964.

11. **Izdebska, K.**, Thermolabile proteins (neurotoxins) from *Proteus mirabilis* and its stable L Forms, *Acta Microbiol. Pol.*, 14, 41–54, 1965.

12. **Jensen, W. I. and Gritman, R. B.**, An adjuvant effect between *Clostridium botulinum* types C and E toxins in the mallard duck *(Anasplatyrhynchos)*, *Proc. Botulism Conf.*, Inst. Assoc. Microbiol. Soc. (Moscow), 1967, 407–413.

13. **Kalmanson, G. M., Kubota, M. Y., and Guze, L. B.**, Production of the Shwartzman reaction with microbial L Forms, *J.Bacteriol.*, 96, 646–651, 1968.

14. **Kanei, C., Uchida, T., and Yoneda, M.**, Isolation of the L phase variant from toxigenic *Corynebacterium diphtheriae* C7 (B), *Infect. Immun.* 20, 167–172, 1978.

15. **Kawata, T., Takumi, K., Sato, S., and Yamashita, H.**, Autolytic formation of spheroplasts and autolysis of cell walls in *Clostridium botulinum* type A, *Jpn. J. Microbiol.*, 12, 445–455, 1968.

16. **Kawatomari, T.**, Studies on the L Forms of *Clostridium perfringens*. I. Relationship of colony morphology and reversibility, *J. Bacteriol.*, 76, 227–232, 1958.

17. **Lehninger, A. L.**, *Biochemistry*, Worth, New York, 1970, 234.

18. **Madoff, M., Annenberg, S. M., and Weinstein, L.**, Production of neuraminidase by L Forms of *Vibrio cholerae*, *Proc. Soc. Exp. Biol. Med.*, 107, 776–777, 1961.

19. **Mason, M. A.**, The spheroplasts of *Bordetella pertussis*, *Can. J. Microbiol.*, 12, 539–545, 1966.

20. **Mattman, L. H., Dowell, V. R., and Neblett, T. R.**, Intraerythrocytic forms in a case with *Clostridium sordelli* bacteremia, *Bacteriol. Proc. 1971*, p. 69, 1971.

21. **Patočka, F. and Kalvodova, D.**, L Formy *Corynebacterium pyogenes hominis*, *Cs. Epidemiol. Mikrobiol. Immunol.*, 18, 5–6, 1969.

22. **Rubio-Huertos, M. and Vasquez, C. G.**, Morphology and pathogenicity of L Forms of *Clostridium tetani* induced by glycine, *Ann. N.Y. Acad. Sci.*, 79, 626–634, 1960.

23. **Sanarelli, G.**, De la pathogénie du choléra. IX. La choléra expérimentale, *Ann. Inst. Pasteur (Paris)*, 38, 11–72, 1924.

24. **Scheibel, I. and Assandri, J.**, Isolation of toxigenic L phase variants from *C. tetani*, *Acta Pathol. Microbiol. Scand.*, 46, 333–338, 1959.

25. **Scheibel, I. and Assandri, J.**, In vitro investigation into the sensitivity of different strains of *C. tetani* to antibiotics, *Acta Pathol. Microbiol. Scand.*, 47, 435–444, 1959.

26. **Sciple, G. W.**, Avian botulism, information on earlier research, *U.S. Fish Wildl. Serv. Spec. Sci. Rep. Wildl.*, p. 23, 1953.

27. **Shakovsky, K. P. and Levashov, V. S.**, Pathogenicity of L Form of *Proteus vulgaris* for chick embryos, *Zh. Mikrobiol. Epidemiol. Immunobiol.*, 43, 29–33, 1966.

28. **Shwartzmann, G.**, *Phenomenon of Local Tissue Reactivity*, Paul B. Hoeber, New York, 1937.

29. **Stratford, B. C., Ed.**, *An Atlas of Medical Microbiology*, 1977.

30. Symposium on problems of botulism in Japan, *Jpn. J. Med. Sci. Biol.*, 16, 303–312, 1963.

31. **Timakov, B. and Kagan, G. Ya.**, *L Form Bacteria and Species of Mycoplasma in Pathology*, Moscow, 1973.

32. **Weibull, C., Bickel, W. D., Haskins, W. T., Milner, K. C., and Ribi, E.**, Chemical, biological and structural properties of stable Proteus L Forms and their parent bacteria, *J. Bacteriol.*, 93, 1143–1159, 1967.

33. **Willis, A. T. and Balows, A., Eds.**, *Anaerobic Bacteria*, Charles C Thomas, Springfield, IL, 1974.

34. **Wittler, R. G.**, L Forms, protoplasts, spheroplasts, A survey of *in vitro* and *in vivo* studies, in *Microbial Protoplasts, Spheroplasts, and L Forms*, Guze, L. B., Ed., Williams & Wilkins, Baltimore, 1968, 200–211.

35. **Bittner, J.**, personal communication.

36. **Bergdoll, M. S.**, personal communication.

37. **Larned, P.**, unpublished study.

Chapter 27

FUNGI

This chapter describes the unique characteristics of wall-deficient fungi, the reasons why they have not been commonly recognized in infections, and their occurrence *in vitro*, either spontaneous or induced.

HOW IS WALL DEFICIENCY OF FUNGI RECOGNIZED?

Wall-deficient bacteria are called fungoidal as they produce yeast-like budding spheres or simulate molds with elongated branching threads. How, then, does one solve the dilemma of recognizing a wall-deficient *fungus*? One can start with the vital activity in a fungal filtrate of *Candida albicans* where the tiny 0.15-μm particles cannot possibly possess the wide hard walls of the parent. Colonies developing are usually comprised of twisted Gram-negative skeins (Figure 1) so delicate that their course is interrupted by submicroscopic gaps. These fine threads of growth have never been described as part of the classic growth of fungi. The minuscule hyphae are accompanied by budding ovoids and triangles which tend to dissolve when Gram staining is attempted.[35] If staining succeeds, the forms are Gram-negative until reversion is complete.

HUGE L-BODIES

The L-bodies (giant spheres) which develop as CWD stages of fungi may be even larger than those from bacteria. They are distinct from chlamydospores or structures described as classically fungal. The L-body of Candida demonstrates a unique characteristic: its wall can give rise to yeast cells (Figure 2). Such reversion of yeasts from L-bodies of both *C. albicans and C. tropicalis* was followed in time-lapse studies by Tunstall and Mattman.[35] This is in accord with the findings of Nečas[20] that the yeast cell wall is extremely important. Thus, without resort to electron microscopy or chemical analysis, physical characteristics distinguish fungal CWD stages from their classic cycles.

SPONTANEOUS WALL-DEFICIENT VARIANTS *IN VITRO*

NATURAL PROTOPLASTS OF *HISTOPLASMA CAPSULATUM*

The two types of *H. capsulatum* differ in their inclination to form wall deficient bodies. The yeast phase of the brown mycelial type spontaneously forms protoplasts which are stable in 2 *M* MgSO$_4$. In contrast, the albino stage as a yeast requires the addition of 2-deoxy-D-glucose to inhibit formation of cell wall glycan before protoplasts are released. Berliner and Reca noted that the protoplast escapes at the bud scar, which is frequently limited only by a membrane.[1]

SPONTANEOUSLY WALL-DEFICIENT SACCHAROMYCES AND CANDIDA

As early as 1956 a Hungarian study demonstrated filtrable, viable, regenerating units of *Saccharomyces cerevisiae*.[8] Candida likewise spontaneously produces wall deficient colonies as shown by L. H. Tunstall[38] and J. L. Wilson.[39] *C. albicans, C. tropicalis, C. krusei,* and *C. parakrusei* behave similarly in synthetic-type medium, where approximately 50% of colonies are wall deficient. Initial growth is inapparent and subsurface in the agar. However, by 3 d, microscopic colonies appear on the surface consisting of intermediate-type colonies which readily revert to the classic stage. Pour plates of Candida in *Sabouraud's* agar result

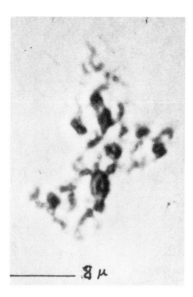

FIGURE 1. Early growth of a transitional colony in brilliant green has developed from initial thin filaments. (Magnification × 1000.)

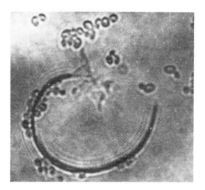

FIGURE 2. Growth of classic cells from incomplete wall of a giant L-body in filtrate culture of *C. albicans.*

in 30 to 50% wall-deficient colonies, even without osmotic stabilizer. Many media will not support CWD growth of Candida. The addition of human blood seems to render media unsuitable.

GROWTH OF NUTRIENT-DEPLETED CANDIDA CELLS

"Starved" cells of Candida initiate growth exclusively as wall-deficient forms which are almost unrecognizable as viable material. Starved cells of Candida, produced by propagation in a 1.0% dextrose solution in a shaker water bath for 48 h, deplete their nitrogenous stored compounds.[35] These cells inoculated into pour plates of the semisynthetic medium of Medill and O'Kane, yield colonies exclusively wall deficient. Initially, they are fine filaments deep in the agar, progressing to the intermediate type of colony, reverting as pseudomycelia grow from the large cells at the edge of the intermediate colony. By 2 weeks the growth is entirely classic in morphology.

FILTRATION TO DEMONSTRATE SPONTANEOUSLY WALL-DEFICIENT CANDIDA

Candida species consistently demonstrate viable units filtrable through Seitz pads of sterilizing porosity, sintered glass filters, or cellulose filters (Gelman) of 0.2 μm-pore diameter. However, a superior yield of filtrable microorganisms is obtained employing single thickness Saran® membrane which is dipped in saline just before use. The Saran filtrates also produce the best reversion rate from filtered organisms, but even here reversion is rare. In contrast to reversion from filtration, CWD growth produced by antimicrobial inhibitors has a high reversion rate if such growth is left in the original medium.

PLEOMORPHISM AFTER INJECTION INTO PLANTS

Penicillium commune injected into *Inpatiens holstii* plants develops pleomorphic giant cells packed with minute rods, unlike anything known in classic colonies of this mold. Also, pasty gymnoplasts frequently form. Meinecke considers this a natural development of CWD forms without microbial inhibitors.[17] Such bizarre growth in plants is reminiscent of the early finding of Much, who inoculated *Mycobacterium tuberculosis* into tomato and onion plants.

FUNGAL VARIANTS IN INFECTIONS

CWD *C. albicans* has been pathogenic when injected into laboratory animals.[10] Candida CWD variants have been noted in endocarditis[26] and in mycohemia (see Frontispiece).[3,33] A clue to identification of *C. albicans* is given by polyacrylamide gel (Figure 3).[19]

Blastomyces dermatitidis in the wall deficient stage has been recognized in the blood of an infected patient (see Frontispiece).[29] A yeast-like organism in the L-stage was found in gallbladder disease.[7] A CWD Cryptococcus which reverted after long refrigeration was in the spinal fluid in meningitis. Its original morphology was that of a diphtheroid.[14] CWD *H. capsulatum* strains were identified by fluorescent antibody in the blood of an endocarditis case[15] and in the spinal fluid of a child after operation to correct hydrocephalus (unpublished study). An unpublished study of pleural fluid found growth of a yeast-like organism (Figure 3A) only in soft or ''soupy'' media supplemented with separately autoclaved yeast extract. Reports from reputable mycology reference laboratories were ''No Growth'', even though the characteristics of the organism were submitted with the cultures. Unfortunately, fluorescent antibody studies were not done. Similarly, a fungus from endocarditis (Figure 3B), would not grow on standard media to permit its identification by laboratories specializing in mycology.

Cell wall-deprived forms of dermatophytes may not be mere curiosities of the laboratory. Cooper[40] noted that dermatophyte-infected skin, cultured on medium selective for fungi, yields colonies predominantly of minute CWD forms. However, fluorescent antibody is needed to determine which of the aberrant colonies might be the pathogenic fungus. Approximately one fifth of the CWD cultures from uveitis suggested fungi in their incompletely reverted stage (see Frontispiece).[6]

More often than not, in fungal infections cultures are negative, and diagnosis is made by seeing organisms with typical morphology in biopsy or autopsy tissue. Recent examples concern aspergillus.[27,28] Also, in a reported series of 34 cases of aspergillus endocarditis, ante-mortem diagnosis was made in only 26%.[12]

The difficulty in diagnosing even the common Candida mycemia is shown by looking for Candida enolase rather than attempting to grow the organism.[37] Diagnosis of fungal infections is enormously complicated by pathogenic potential of fungi commonly considered

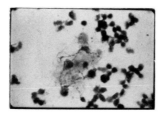

A

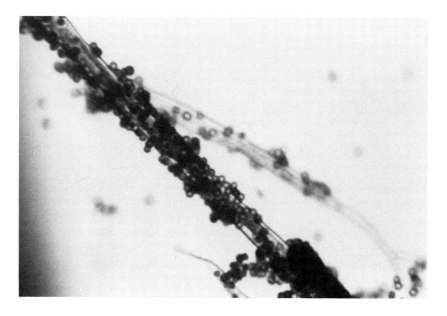

B

FIGURE 3. (A) Yeast-like fungus from pleural fluid grew only in media suitable for CWD forms. (B) Organism cultured from blood of a subacute bacterial endocarditis case could not be grown on standard media. (Photograph submitted by Ira Gantz.)

strictly saprophytes. Rinaldi states, "Given the right immunocompromised host, virtually any fungus can kill a human being." An example is mushroom endocarditis.[24] Studies in our laboratory indicate that although classic growth from infections is uncommon and delayed, growth of the CWD stage occurs rapidly from pleural fluid, blood, or spinal fluid. In broth supplemented with a trace of agar and with separately autoclaved yeast extract, CWD growth may be expected in 48 h, even when the genus is Histoplasma.

MIMICKING OF TISSUE CELLS BY FUNGAL WALL-FREE VARIANTS

Are fungi in smears overlooked because they are confused with blood cells? Meinecke has been impressed with the many ways wall-deficient fungal cells imitate blood cells of vertebrates.[16] The aberrant wall-deficient growth of Ascomycetes can imitate erythrocytes, lymphocytes, plasma cells, platelets, even the "signet ring cell" of carcinoma. Meinecke speculated that at times CWD variants of fungi may appear *in vivo* and be overlooked, being confused with tissue cells of the host.

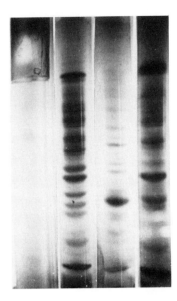

FIGURE 4. Polyacrylamide gel patterns from extracts of medium sediment. Reading left to right, uninoculated medium: classic *C. albicans* in blood culture; CWD *C. albicans* in skip blood culture; classic *C. albicans* in blood culture. (From Motwani, N., Ph.D. dissertation, Wayne State University, Detroit, 1976.)

FLUORESCENT REAGENTS AND POLYACRYLAMIDE GEL TO IDENTIFY FUNGAL VARIANTS

Fluorescent muramidase labels CWD fungi as well as bacterial variants. As expected, both the fluorescein and rhodamine conjugates are functional. Thus, this enzyme helps to differentiate pleomorphic microbial growth from distorted tissue cells noted by Meinecke.[16] For identification a more specific label is needed, such as fluorescent antibody.

Fluorescent antibody aids in identification of highly atypical Candida growth. This was shown with labeled antisera kindly supplied by Leo Kauffman of the Centers for Disease Control and with antisera for classic and wall-deficient Candida, employed in the indirect method of immunofluorescence. The three sera behaved similarly, with one exception: the antiserum made in response to the wall-deficient variants did not stain the surprisingly heavy capsular coat enveloping the classic Candida. The capsule is absent from the CWD stages. Although most stages stained well, the very fine mycelia of the CWD forms were not stained by any of the sera.

Polyacrylamide gel studies show that the total proteins of CWD *C. albicans* resemble those of the classic yeast. Electrophorograms of CWD *C. albicans* from blood cultures are shown in Figure 4.[19]

CWD FUNGI FORMED BY ANTIBIOTICS, DYES, OR ENZYMES

RAMICIDIN, MYCOSTATIN, AMPHOTERICIN B, CRYSTAL VIOLET, AND BRILLIANT GREEN

There are many agents which create wall-deficient fungi from their classic parents. For example, the antibiotic ramicidin induces wall-deficient forms of Ascomycetes as reported by Riddell and Stewart.[23]

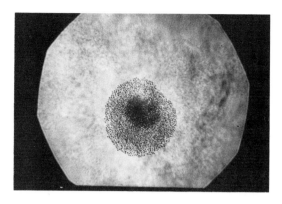

FIGURE 5.   Mature transitional colony which has absorbed brilliant green from the medium. The round units are fragile. This colony type usually reverts to the parent. (Magnification × 880.)

Many inhibitors increase wall deficiency in *C. albicans, C. krusei, C. parakrusei,* and *C. tropicalis.* Differences between species concern the concentration of inhibitor required and not the resultant morphology or reversibility. Penicillin seems totally incapable of L-phase induction in Candida. In contrast, 1000 U of *Mycostatin* in the center of an agar plate yields a multiplicity of variants composed of fine anastomosing filaments or irregular amorphous masses, with a central body corresponding to the syncytium of bacterial wall deficiency or to the expanding irregular growth of a spheroplast.

Amphotericin B is comparable to mycostatin in inducing filtrable wall-deficient Candida in solid or liquid medium. This is of interest with reference to Rosner's case of endocarditis, where only wall-deficient forms were found on a heart valve after amphotericin therapy.[26]

Crystal violet and brilliant green likewise are potent stimulators of cell wall deficiency in Candida, giving colonies with identical morphology to those induced by amphotericin B and mycostatin. Strominger[30] found that crystal violet interferes with cell wall formation of Staphylococci, with resultant intracellular accumulation of uridine nucleotides. Presumably a similar process may operate in Candida.

CWD variants differ from their parents in thriving in antifungal agents, both dyes and antibiotics. *C. albicans* wall-deficient forms propagate in concentrations of brilliant green which color the variant colony (Figure 5).[35]

FILTRATION OF CANDIDA CULTURES GROWN WITH INHIBITORS

The yield of filtrable forms is enormously increased when Candida is cultured with the antimicrobials described above. Filtrates are desirable because they are free of the artifacts contributed to the induction medium by dyes and antibiotics. Findings reported here concern growth of the filtrates in Medill-O'Kane medium rendered almost detritus-free by filter-sterilizing the ingredients (see Figure 1). The micrographs are of sections of agar from pour plate cultures.

Growth improves when the forms are concentrated by centrifugation, results being more satisfactory in slide cultures than in pour plates. Aging the filtrate, useful in fostering reversion of Staphylococci,[34] is ineffectual here perhaps because reversion factors are removed by filtration as found by Keleti et al. in Saccharomyces filtrates.[8]

ENZYMES OF THE SNAIL *HELIX POMATIA*

L Forms of *C. albicans*[5] and of *C. utilis*[31] form after exposure to gut enzymes of the snail *H. pomatia.*[10]

Protoplasts of *S. cerevisiae* are made by snail enzymes,[21] but if enzyme exposure is brief, "prosperoplasts" develop. These are osmotically sensitive cells with normal morphology but shown by electron microscopy to have loosened walls.[4,32]

The many species of dermatophytes found by Koch and Lange to respond similarly to *H. pomatia* enzymes are the following: *Trichophyton mentagrophytes, T. rubrum, T. ferrugineum, T. verrucosum, T. schonleinii, T. violaceum, Microsporum gypseum, M. cookei, M. vanbreuseghemii, Keratinomyces ajelloi, Ctenomyces serratus*, and *Arthroderma curreyi*.[11] Regeneration of the protoplasts to typical mycelia occurs when transplants are made onto appropriate media.

Factors operative in formation of such dermatophyte variants were defined by Rosenthal et al.[25] Cysteine is almost an absolute requirement for formation of "protoplasts" from some species of the genus Trichophyta.

β-Glucuronidase from *H. pomatia* makes spheroplasts from *Aspergillus parasiticus*. Aflatoxin is made not only by classic mycelia and spheroplasts, but also by lysates of the mycelia and of the spheroplasts.[36]

The precise morphology of protoplasts, spheroplasts, and yeast cells of *B. dermatitidis* was elucidated by electron microscopy in the laboratory of Miegeville et al. Again, the inducing agent was contributed by the snail.[18]

STREPTOENZYME FROM STREPTOMYCES

An enzyme from a streptomyces produces spheroplasts from *Drechslera dematioidea*. The action resembles that of snail gut enzymes. A similar effect results merely from exposure to distilled water, again with reversion possible.[13]

ENZYME FROM MATING GAMETES

A wall-deficient mutant was produced from *Chlamydomonas reinhardtii* with a wall-lysing enzyme made by mating gametes of the wild-type organism. The mutant released most of its carbonic anhydrase into the growth medium, confirming that the formation of the enzyme is above the protoplasmic membrane.[9]

REVERSION OF PROTOPLASTS

See Figure 6 for reversion of the yeast *Schizosaccharomyces pombe* after protoplasts were produced by enzyme action.[22] Similarly, reversion of *Histoplasma capsulatum* protoplasts has been followed with electron microscopy by Berliner et al.[2]

ANTIBIOTIC-DEPENDENT GYMNOPLASTS

Gymnoplasts, i.e., fungal protoplasts which never had a cell wall, develop from Aspergillus, Penicillium, and other Ascomycetes.[17] Remarkably, these can exhibit the phenomenon of antibiotic dependency, appearing only in the presence of certain antibiotics in media so impoverished that little classic growth develops. Aspergillus species, on a low-nutrient medium of iron-sugar gelatin, grow sparsely as simple mycelia. In contrast, growth is heavy around disks impregnated with certain antibiotics. Microscopic observation of this antibiotic-dependent growth frequently shows lateral swellings along the hyphae which prove to be gymnoplasts (Figure 7). Thus, stimulation by antibiotics on nutritionally poor medium is one of the best methods to demonstrate gymnoplasts consistently from mold-like organisms. Stimulation results from aureomycin, kanamycin, penicillin, and erythromycin but not from chloramphenicol, furadantin, penetracyne, or sulfonamides. The concentration of antibiotics in the disks is moderate, resembling concentrations attainable *in vivo*, e.g., aureomycin 30 μg and penicillin 5 U.

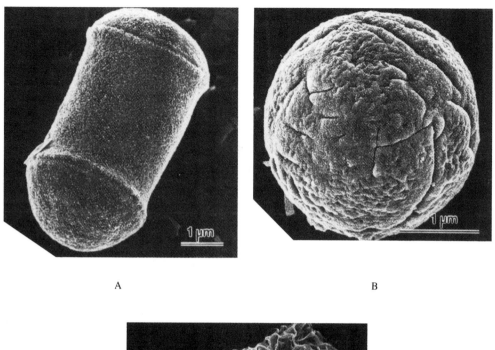

A   B

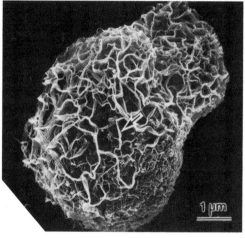

C

FIGURE 6.   Regeneration of protoplasts of *Schizosaccharomyces pombe*. (A) Whole cell, (B) protoplasts made by exposure to NovoZym 234, (C) regenerating yeast cell at 5 h. (From Osumi, M., Yamada, N., Kobori, H., Taki, A., Naito, N., Baba, M., and Nagatani, T., *J. Electron Microsc.*, 38, 457–468, 1989. With permission.)

This phenomenon of CWD fungal growth stimulated by commonly employed antibiotics may be associated with the fungal infections often complicating therapy. It has always been thought that the mycological problems resulted solely from inhibition of normal flora bacteria.

SUMMARY

In a variety of situations, saprophytic and parasitic fungi grow as wall-deficient organisms, sometimes as a natural event, sometimes induced by antimicrobial agents.

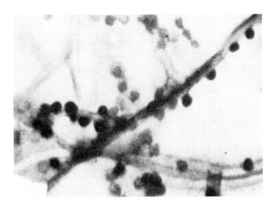

FIGURE 7. Aspergillus stimulated by erythromycin, aureomycin, kanamycin, or penicillin yields wall-free forms (gymnoplasts). (From Meinecke, G., *Antibiose,* Wilhelm Goldmann Verlag, Munich, 1971.)

Finding variants which include *C. albicans,*[10] *B. dermatitidis,* and *H. capsulatum* in the nonclassic state in human infections indicates their pathogenic potential. It is noted that in the majority of fungal infections cultures are negative although typical fungal morphology is seen in biopsy or autopsy tissue. Experience shows that pertinent growth does occur in media suitable for CWD variants. Suitable media include veal infusion broth made semiliquid with purified agar or gelatin and supplemented with a final concentration of 1.0% yeast extract, separately autoclaved and added as the medium is used. Media is aged at least 1 d before inoculating.

Fungi growing feebly may be stimulated to heavy growth by certain antibiotics and simultaneously produce wall-free forms. This is antibiotic-dependent growth of the fungus, rather than variant induction. This may be one reason why fungi invade antibiotic-treated patients.

Molds invading plant cells may find it convenient to propagate as CWD bodies. Fluorescent antibody and other stains are needed to differentiate fungal variants in plants from host material, just as specific staining is needed for clinical isolates.

REFERENCES

1. **Berliner, M. D. and Reca, M. E.,** Release of protoplasts in the yeast phase of *Histoplasma capsulatum* without added enzyme, *Science,* 167, 1255–1257, 1970.
2. **Berliner, M. D., Carbonell, L. M., and Biundo, N., Jr.,** Regeneration of protoplasts of *Histoplasma capsulatum*: a study by light microscopy, *Mycologia,* 64, 708–721, 1972.
3. **Charache, P.,** Atypical bacterial forms in human disease, in *Microbial Protoplasts, Spheroplasts, and L Forms,* Guze, L. B., Ed., Williams & Wilkins, Baltimore, 1968, 484–494.
4. **Darling, S., Theilade, J., and Birch-Andersen, A.,** Kinetic and morphological observations on *Saccharomyces cerevisiae* during spheroplast formation, *J. Bacteriol.,* 98, 797–810, 1969.
5. **Georgopapadakou, N. H. and Smith, S. A.,** Chitin synthase in *Candida albicans,* comparison of digitonin-permeabilized cells and spheroplast membranes, *J. Bacteriol.,* 162, 826–829, 1985.
6. **Hessburg, P., Mattman, L. H., Barth, C., and Dutcheshen, L.,** Aqueous microbiology: the possible role of CWD bacteria in uveitis, *Henry Ford Hosp. Med. J.,* 17, 177–193, 1969.
7. **Kazancheva, A. M.,** On the formation of L Forms of microbes in the organism, *Klin. Med.* (Moscow), 40(3), 32, 1962.
8. **Keleti, T., Lendavi, Z., Takac, L., and Szabolcsi, G.,** Untersuchungen über die lebensfahigen erweisskorper (filtrierbare form) von *Saccharomyces cerevisiae.* II. Methodische problem bei der untersuchung der lebensfahigen hefeeiweisskorper, *Acta Physiol. Acad. Sci. Hung.,* 9, 407–414, 1956.

9. **Kimpel, D. L., Togasaki, R. K., and Miyachi, S.,** Carbonic anhydrase in *Chlamydomonas reinhardtii.* I. Localization, *Plant Cell Physiol.,* 24(2), 255–259, 1983.

10. **Kobayashi, G., Friedman, L., and Kofroth, J.,** Some cytological and pathogenic properties of spheroplasts of *Candida albicans, J. Bacteriol.,* 88, 795–801, 1964.

11. **Koch, H. A. and Lange, G.,** Untersuchungen an Protoplasten von Dermatophyten, *Mykosen,* 10, 33–36, 1967.

12. **Lang, D. M., Leisen, J. C. C., Elliott, J. P., Lewis, J. W., Wendt, D. J., and Quinn, E. L.,** Echocardiographically silent *Aspergillus* mural endocarditis, *West. J. Med.,* 149, 344–338, 1988.

13. **Leone, R.,** Spheroplasts of *Drechslera dematioidea, Allionia (Turin),* 25, 37–50, 1982.

14. **Louria, D. B., Kaminski, T., Grieco, M., and Singer, J.,** Aberrant forms of bacteria and fungi found in blood or cerebrospinal fluid, *Arch. Intern. Med.,* 124, 39–48, 1969.

15. **Mattman, L. H. and Judge, M. S.,** Septicemia and some associated infections, demonstration of CWD bacteria, in *Cell Wall Deficient Bacteria,* Domingue, G. J., Ed., Addison-Wesley, Reading, MA, 1982, 427–451.

16. **Meinecke, G.,** Milieuanpassungen bei Schimmelpilzen, *Zentralbl. Bakteriol. Parasitenkd. Infektionskr. Hyg. Abt. 1 Orig.,* 180, 273–278, 1960.

17. **Meinecke, G.,** *Antibiose,* Wilhelm Goldmann Verlag, Munich, 1971.

18. **Miegeville, M., Bouillard, C., Marjolet, M., and Vermeil, C.,** New contribution of scanning electron microscopy for the morphological study of *Blastomyces dermatitidis* yeasts spheroplasts and protoplasts, *C. R. Seances Acad. Sci. Ser. III Sci. Vie.,* 292(9), 657–664, 1981.

19. **Motwani, N.,** Identification of Cell Wall Deficient Bacteria by Immunofluorescence and Polyacrylamide Gel Electrophoresis, Ph.D. dissertation, Wayne State University, Detroit, 1976.

20. **Nečas, O.,** Giant yeast cells, *Folia Biol.,* 3, 101–107, 1957.

21. **Oertel, W. and Goulian, M.,** Deoxyribonucleic acid synthesis in permeabilized spheroplasts of *Saccharomyces cerevisiae, J. Bacteriol.,* 132, 233–246, 1977.

22. **Osumi, M., Yamada, N., Kobori, H., Taki, A., Naito, N., Baba, M., and Nagatani, T.,** Cell wall formation in regenerating protoplasts of *Schizosaccharomyces pombe,* study by high resolution, low voltage scanning electron microscopy, *J. Electron Microsc.,* 38, 457–468, 1989.

23. **Riddell, R. and Stewart, G.,** *Fungous Diseases and Their Treatment,* Butterworths, London, 1958.

24. **Rinaldi, M.,** Fungi as human pathogens: sublime and bizarre, *ASM News,* 57, 7, 1991.

25. **Rosenthal, S. A., Fine, H. L., and Baer, R. L.,** Cell wall deficient forms of dermatophytes produced in vitro, *J. Invest. Dermatol.,* 49, 449–455, 1967.

26. **Rosner, R.,** Isolation of Candida protoplasts from a case of Candida endocarditis, *J. Bacteriol.,* 91, 1320–1326, 1966.

27. **Scully, R. E.,** Case records of the Massachusetts General Hosp., Case 7–1988, *N. Engl. J. Med.,* 318, 427–440, 1988.

28. **Scully, R. E.,** Case records of the Massachusetts General Hosp., Case 45–1990, *N. Engl. J. Med.,* 323, 1329–1336, 1990.

29. **Senatore, G.,** Presumptive Identification of Classical and Cell Wall Deficient Microorganisms by Gas-Liquid Chromatography, M.S. thesis, Wayne State University, Detroit, 1983.

30. **Strominger, J. L.,** Enzymatic reactions in bacterial cell wall synthesis sensitive to penicillins, cephalosporins, and other antibacterial agents, in *Antibiotics I: Mechanism of Action,* Gottlieb, D. and Shaw, P. D., Eds., Springer-Verlag, Berlin, 1967, 705–713.

31. **Svihla, G., Schlink, F., and Dainko, J. L.,** Spheroplasts of the yeast *Candida utilis, J. Bacteriol.,* 82, 808–814, 1961.

32. **Svobodá, A., Farkas, V., and Bauer, S.,** Response of yeast protoplasts to 2-deoxyglucose, *Antonie van Leeuwenhoek,* 35, (Suppl.) B11–B12, 1969.

33. **Swieczkowski, D. M., Mattman, L. H., Truant, J. P., and Wilner, F. M.,** Cell wall deficient forms of *Candida albicans* in mycohemia, *Lab. Med.,* 1, 41–42, 1970.

34. **Trofimova, N. D.,** Studies of the conditions for regeneration of filterable forms of staphylococci, *Microbiol. J. Acad. Sci. Ukraine RSR,* 21, 45, 1959.

35. **Tunstall, L. H. and Mattman, L. H.,** L variation in Candida species, *Bacteriol. Proc. 1961,* p. 83, 1961.

36. **Tyagi, J. S., Tyagi, A. K., and Venkitasubramanian, T. A.,** Preparation and properties of spheroplasts from *Aspergillus parasiticus* with special reference to the *de novo* synthesis of aflatoxins, *J. Appl. Bacteriol.,* 50, 481–491, 1981.

37. **Walsh, T. J., Hathorn, J. W., Sobel, J. D., Marz, W. G., Sanchez, V., Maret, M., Buckley, H. R., Pfaller, M. A., Schaufele, R., Sliva, C., Navarro, E., Lecciones, J., Chandrasekar, P., Lee, J., and Pizzo, P. A.,** Detection of circulating candida enolase by immunoassay in patients with cancer and invasive candidiasis, *N. Engl. J. Med.,* 324, 1026–1031, 1991.

38. **Tunstall, L. H.,** unpublished study.

39. **Wilson, J. L.,** unpublished study.

40. **Cooper, T.,** unpublished study.

Chapter 28

SENSITIVITY TO ANTIMICROBIAL AGENTS

ANTIBIOTIC SENSITIVITIES

In general, antibiotics inhibiting cell wall structuring are innocuous to wall defective bacteria. Correspondingly, antibiotics which interrupt protein synthesis are inhibitive to the variant if the parent is sensitive. However, exceptions are numerous. Antibiotics considered metabolic inhibitors, which are exquisitely toxic to the parent bacterium, may, for inexplicable reasons, be tolerated by the variant. This is often true of the tetracyclines.[4,6,26]

In therapy of infections where CWD forms are found, it may be necessary to treat both the wall deficient stage and the classic organism[31] since suppression of one stage may foster the propagation of the other. It is impossible to know at the time the patient is being treated for a CWD organism in the blood whether the classic form of the pathogen is also within his tissues. This may explain why it is often true that therapy must be bivalent, directed against both the parent and variant, as in cases of pneumococcal meningitis which fail to respond to either chloramphenicol or penicillin alone but which do recover on the combined therapy.

At other times it has been found that penicillin alone may eliminate a meningitis with CWD forms.[19] Has the variant retained enough wall to be β-lactam sensitive? Such exceptional instances are included in an accompanying table.

SENSITIVITY OF CWD VARIANTS
FROM INFECTIONS

Variants from infections tend to grow poorly, and pertinent data are scarce. However, recent improvements in culture of variants, especially those from the urinary tract, have permitted antibiogram readings in several laboratories. Examples of *in vitro* sensitivity tests on clinical samples are given in the chapters of specific diseases. Standard antibiotic sensitivity testing will be applicable to all CWD variants when media are found which yield luxurious growth.

Figure 1 shows an antibiotic-impregnated disk on medium inoculated with a wall-deficient isolate from a blood culture. The logical interpretation here might be that a single antibiotic-resistant colony developed in the diffusion zone around the antibiotic. However, interpretation is complicated by growth in original cultures as microcolonies. Thus, what appears to be new growth may be merely inoculum. More accuracy results when plates are inoculated as identically as possible and duplicate plates refrigerated. Ideally, a ring of stimulated growth marks the periphery of the no growth circle, demonstrating the oligodynamic effect well known for microbial inhibitors. Chopra[51] in our laboratory obtained rather heavy debris-free growth of CWD forms from blood cultures by streaking the inoculum on a 1.2-μm-pore Metrecel membrane placed on the culture medium. It is likely that this method can be applied to testing antibiotic sensitivity by the disk method. Surface inoculation is more likely to succeed if the medium is serum agar or chocolate agar enriched with separately autoclaved yeast extract. Minute colonies may be seen with a hand lens around disks in zones where classic organisms are completely inhibited. One, thus, can make readings for the classic stage vs. the variants.

Sensitivity tests may be done in tubes with 2 ml of broth, a trace of agar, 1% separately autoclaved yeast extract and, if the CWD strain requires, swine serum. One disk containing

FIGURE 1.   With the exception of one colony, there is no growth of the wall-deficient forms around the antibiotic disk.

the antimicrobial is added to a tube. If the antibiotic is active in an acid pH a reaction of 5.5 is ideal, as it gives accelerated growth in an incubated shaker.

Considering the longer generation time of wall-deficient microbes, there is need for methods which give accelerated antibiotic sensitivity readings. McClatchy early published a rapid method to determine the antibiotic sensitivity of classic bacteria by following the incorporation of ^{14}C tryptophan into cellular protein.[27] Motwani[30] found that CWD bacteria and fungi revealed their presence in routine blood cultures, using the BACTEC method. Among 210 blood cultures the CWD organisms were usually detectable by a radiometric reading higher than baseline but lower than levels given by classic organisms.[30]

Thermoduric CWD contaminants in animal sera (not killed complement is heat inactivated) may be almost completely eliminated by two filtrations through 0.22-μm membranes, with intermittent storage of the serum at room temperature for 24 h.[50]

Erythromycin is one of the better antibiotics for therapy of CWD variants as suggested by Table 1. A modification, 6-*O*-methylerythromycin (Te-031), is even more potent in restricting growth of L Forms of *Staphylococcus aureus, Streptococcus pyogenes,* and *Escherichia coli.* Furthermore, after oral administration, the blood level of Te-031 is maintained higher than when erythromycin is given.[29]

The sensitivity of L Forms in culture does not always correspond with response within tissue cells. Girardi human heart cells and human fibroblast cells protect Group A streptococcal L Forms, necessitating about 50 times as much erythromycin, lincomycin, or tetracycline as if the bacteria were extracellular.[2]

SENSITIVITY OF *IN VITRO*-PRODUCED VARIANTS

Antibiotic sensitivity of CWD forms reported in the literature has been done predominantly with isolates which were induced into the variant state *in vitro* with antibiotics. The strains were selected because their dependable growth permitted antibiotic susceptibility testing by standard methods routinely applied to classic bacteria. Antibiotic sensitivity readings of such *in vitro*-induced variants have been compiled in a table in the first edition of this book. Anyone who has worked with clinical CWD forms knows that great differences exist between variants produced *in vivo* and *in vitro.*

A detailed study was made of three stable L Forms of *Pseudomonas aeruginosa.* They were highly resistant to carbenicillin, piperacillin, gentamicin, streptomycin, polymyxin B, and colistin. Practically speaking they were resistant to all antibiotics widely used to treat

TABLE 1
Clinical Response of CWD Forms to Antibiotic Administration

Organism and/or disorder	Clinically improved by	Resistant to
Crohn's disease[7]	Rifabutin and streptomycin	
Cryptococcus neoformans[23]	Amphotericin B	
Streptococcus pneumoniae		
Meningitis[26]	Penicillin + chloramphenicol	
Escherichia coli		
Urinary tract infection[24]	Nitrofurantoin	
Interstitial lung disease[48]	Rifampin	
(idiopathic in man, mouse)		
Listeria monocytogenes	Chloramphenicol and	
Encephalitis[6]	tetracycline	
Mima polymorpha[28]		
Urinary tract infection	High doses of ampicillin	
with bacteremia		
Proteus		
Cystitis[9]		Tetracycline
Pyelonephritis[9]	Erythromycin	Chloramphenicol
Pyelonephritis[10]		Chloramphenicol, sulfonamides, tetracycline
Urinary tract infection[14]	Ampicillin, followed by erythromycin	
Pseudomonas aeruginosa		
Canine endocarditis[3]		Erythromycin, tetracycline
Meningitis[9]	Ampicillin + keflin	
Staphylococcus aureus		
Abscesses[5]	Chloramphenicol, but patient relapsed	
Acute arthritis[21]	*In vitro* sensitivity to keflin, erythromycin, novobiocin, tetracycline	
Cystic fibrosis[18]	Erythromycin	Methicillin, polymyxin B
Urinary tract infection[9]		Keflin
Staphylococcus epidermidis[5]	Methicillin	Penicillin
Streptococcus		
Cystitis[9]		Kanamycin
Urinary tract infection[9]		Mandelamine
Septicemia[26]	Chloramphenicol	Penicillin, streptomycin, tetracycline
Septicemia[41]	Chloramphenicol + spiramycin	
Streptococcus sanguis		
L Forms dog, man[8]	Erythromycin	
Subacute bacterial endocarditis	Erythromycin + keflin	Massive doses penicillin, vancomycin
Streptococcus viridans		
Endocarditis	Erythromycin, keflin	Penicillin, tetracycline
Streptococcus and *Staphylococcus*		
Idiopathic hematuria (Domingue, G. J., unpubl.)	Macrodantin	
Unidentified organisms		
8 cases of meningitis[20]	Megadoses of penicillin	
Endocarditis	Chloramphenicol + tetracycline	Penicillin

TABLE 1 (continued)
Clinical Response of CWD Forms to Antibiotic Administration

Organism and/or disorder	Clinically improved by	Resistant to
Whipples disease[43]	Tetracycline for 2 years, ampicillin + chloramphenicol	
Whipples disease[5]	Chloramphenicol, erythromycin	Tetracycline
Whipples disease[1]	Ceftriaxone	
Whipples disease[11]	Tetracycline, 2 months to 3 years	
	Tetracycline, penicillin, and streptomycin, 2 weeks	
	Penicillin, 3 months	
	Penicillin and streptomycin, 3 weeks	

See Chapter on Urinary Tract Infections for more data.

Pseudomonas infections. In contrast, the variants were more susceptible than the parent bacilli to tetracyclines, chloramphenicol, and erythromycin.[49]

CWD FORMS INHIBITED
BY CELL WALL INHIBITORS

It was unexpected that a cell wall inhibiting antibiotic could, in turn, be bacteriostatic for wall-deficient variants. However, many examples of this are now known.

BACITRACIN

Bacitracin, one of the antibiotics concerned, probably damages bacterial cytoplasmic membranes as well as interrupts wall synthesis.[32,38] Thus, no wall may be needed for its action.

FINDINGS WITH PENICILLIN

As Table 1 shows, penicillin can be effective in therapy of some wall-deficient microbes, perhaps those retaining considerable wall. Remarkable is the inhibition of *Mycoplasma neurolyticum* by penicillin, cell wall free though it must be. Surr[39] has noted that ribonucleic acid formation in a bacterial membrane fraction is inhibited by penicillin, showing that membrane damage can result from this antibiotic.

FINDINGS WITH D-CYCLOSERINE

Grula and Grula found that D-cycloserine, commonly considered an antibiotic acting by cell wall inhibition, severely damages the cell membrane.[12] By several investigative approaches, this was substantiated, including the failure of protoplasts of *Streptococcus faecalis* to take up D-alanine in the presence of the antibiotic, whereas there is no interference in uptake by whole cells. Furthermore, the antibiotic does not inhibit growth or division, at least of Erwinia, if the milieu is made hypertonic with NaCl or certain other stabilizers. The investigators believe this strongly suggests that the main effect of D-cycloserine is not vs. the cell wall.

READING ANTIBIOTIC SENSITIVITY TESTS

With some strains antibiotic sensitivity tests may be complicated by growth of CWD variants.[17,33] Discrepancies in antibiotic sensitivity testing caused by CWD growth have been investigated by Johnson and associates.[17] Their report was not concerned with whether the variants existed *in vivo*. Rather, they emphasized that the surface haze developing in MIC testing of antimicrobials consists of CWD growth. High concentrations of an antibiotic are lethal and give no growth on the blood agar surface. Lower concentrations may not kill the microorganism but foster the growth of CWD variants. Development of such variants can, of course, explain antibiotic failures. Four species of Fusobacterium grew exclusively as CWD variants in low concentrations of ampicillin, cefoxitin, moxalactam, piperacillin, and imipenem.

Similarly, antibiotic testing in tubes can be confusing. A cloudy tube may show no growth on solid medium when subcultured to confirm viability. The variants can be seen in an alcohol-fixed smear stained with acridine orange. Growth in such tubes does, of course, indicate antibiotic resistance of the variant.

CWD FORMS RESISTANT TO AGENTS LETHAL
FOR CLASSIC ORGANISMS

Notoriously, the wall-deficient forms cause persistent infections because of resistance to commonly employed antibiotics. This resistance applies not only to penicillin, but can involve resistance or tolerance to bacitracin[46] or almost any other agent.

CWD FORMS INHIBITED BY AGENTS TOLERATED
BY THE CLASSIC ORGANISM

It was at first assumed that wall-deficient variants would have the same antibiograms as their parents, with the exception that inhibitors of wall synthesis could not limit the propagation of an organism already depleted of wall. While generally true,[46,47] this premise has many exceptions, a conspicuous one being erythromycin, which suppresses many CWD variants in species where the parent organism is erythromycin resistant.[37]

One explanation of this erythromycin susceptibility exclusively in the CWD variant is uncomplicated. *Proteus mirabilis,* which has an insensitive parent form, is markedly inhibited in its wall-deficient stage by erythromycin and the other macrolide antibiotics: magnamycin, leukomycin, spiramycin, oleandomycin, and angolamycin. Taubeneck concluded that with the commonly used concentrations, the natural resistance of *P. mirabilis* to macrolides is due to the inability of these antibiotics to penetrate the walls of the normal rods and not to insensitivity of the metabolism of the Proteus.[42]

CWD streptococci may lose antibiotic resistance by another mechanism. These variants can lose a plasmid simultaneously with loss of resistance to erythromycin, chloramphenicol, lincomycin, tetracycline, and pristinamycin.[35] An excellent article reviewing mechanisms of bacterial resistance[16] notes that not only plasmids but transposons may give genetic resistance to antibiotics. Transposons are DNA units which jump from one chromosomal or plasmid site to another.

Kagan found that Staphylococci, also, usually are more inhibited by erythromycin when in the L phase than as classic cocci. It is common for an L-phase Staphylococcus to be inhibited by half the concentration of erythromycin needed for the parent.[18]

Lincomycin is another antibiotic which may be more potent vs. CWD stages of some Staphylococci than against the parent. Here the amount of antibiotic needed for the variant might be less than one tenth the inhibitive level for the parent.[18]

An antibiotic susceptibility which may have a complex basis was noted by Lederberg and St. Clair. A mutant of *E. coli* which tolerated 1.0 mg/ml of streptomycin as a bacillus was inhibited by 20 µg/ml of streptomycin when it grew as a penicillin-induced spheroplast.[22]

REVERTANTS MAY DIFFER FROM PARENTS IN ANTIBIOTIC SENSITIVITY

It is the general experience of investigators that revertants often differ from the original classic bacterium. This may include antibiotic sensitivities.[36] Resistance in the parent can be absent in the revertant as shown by Schönfeld.[36] Staphylococcal strains lost their original resistance to chloramphenicol, tetracycline, and erythromycin as they reverted from penicillin-induced L Forms. They were passaged in the L Form from 2 to 12 times.

In contrast, inexplicably, resistance may be gained in the revertants in contrast to the sensitivity of the parent. An example is given below.[45]

	Parent streptococcus	Revertant streptococcus
Penicillin	0.01	8.
Keflex	0.03	32.
Erythromycin	0.01	1.
Lincomycin	0.06	32.
Tetracycline	0.25	2.

ERADICATION OF CWD FORMS WHEN TISSUE CULTURES ARE CONTAMINATED

A fluorescent dye was found to serve a dual role. First, it reveals the presence of contamination with mycoplasma or CWD organisms. Second, with several transfers of the cells in dye-containing medium the contaminants are permanently eliminated (Figure 2).[15]

SUMMARY

With the exception of avoiding the penicillins there are few rules to guide therapy of CWD infections. Erythromycin, chloramphenicol, and, occasionally, the tetracyclines have been inhibitive. Newer antimicrobials look promising. With cooperative isolates, sensitivity testing can be done on the surface of serum agar. Usually MICs must be read in broth containing serum and a trace of agar. Sometimes therapy must be directed against both parent microbe and variant.

Antimicrobial resistance can appear or disappear as a classic stage microbe loses part of its wall or as it reverts to the parent stage from cell wall deficiency. Thus, many mechanisms which may include movement of plasmids or transposons are involved, not all of which have been elucidated.

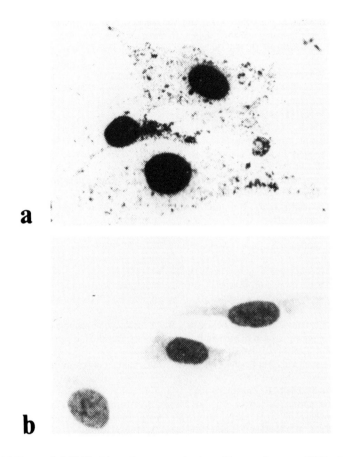

a

b

FIGURE 2. (a) Human diploid fibroblasts show contamination with mycoplasma or CWD microorganisms. (b) The tissue culture has lost its contaminants after two passages with the fluorescent dye Hoechst 33258. (From Hellkuhl, B. and Grzeschik, K. H., *Cytogenet. Cell Genet.*, 36, 584–585, 1983. With permission.)

REFERENCES

1. **Adler, C. H. and Galetta, S. L.,** Oculo-facial-skeletal myorhythmia in Whipple Disease: treatment with ceftriaxone, *Ann. Intern. Med.,* 112, 467–469, 1990.
2. **Azhukaitene, R. P. and Eshmaintaite, N. A.,** Study of the antibiotic sensitivity of Group A Streptococcal L Forms in an artificial nutrient medium and in human cell cultures, *Antibiot. Khimioter.,* 34(10), 741–745, 1989.
3. **Bone, W. J.,** L Form of *Pseudomonas aeruginosa* the etiologic agent of bacterial endocarditis in a dog, *Vet. Med. Small Anim. Clin.,* 65, 224–226, 1970.
4. **Brem, A. M. and Reidt, W.,** A cell wall deficient form of the Pneumococcus in a case of pneumonia, *Chest,* 61, 200–202, 1972.
5. **Charache, P.,** Atypical bacterial forms in human disease, in *Microbial Protoplasts, Spheroplasts, and L Forms,* Guze, L. B., Ed., Williams & Wilkins, Baltimore, 1968, 484–494.
6. **Charache, P.,** Cell wall defective bacterial variants in human disease, *Ann. N.Y. Acad. Sci.,* 174, 903–911, 1970.
7. **Chiodini, R. J.,** Crohn's disease and the Mycobacteriosis: a review and comparison of two disease entities, *Clin. Microbiol. Rev.,* 2, 90–117, 1989.
8. **Chmel, H.,** Graft infection and bacteremia with a tolerant L Form of *Streptococcus sanguis* in a patient receiving hemodialysis, *J. Clin. Microbiol.,* 24, 294–295, 1986.

9. **Conner, J. F., Coleman, S. E., Davis, J. L., and McGaughey, F. S.,** Bacterial L Forms from urinary-tract infections in a veterans hospital population, *J. Am. Geriatr. Soc.,* 16, 893–900, 1968.

10. **Domingue, G. J. and Schlegel, J. U.,** The possible role of microbial L Forms in pyelonephritis, *J. Urol.,* 104, 790–798, 1970.

11. **Fleming, J. L., Wiesner, R. H., and Shorter, R. G.,** Whipple's Disease: clinical, biochemical, and histopathologic features and assessment of treatment in 29 patients, *Mayo Clin. Proc.,* 63, 539–551, 1988.

12. **Grula, M. M. and Grula, E. A.,** Action of cycloserine on a species of Erwinia with reference to cell division, *Can. J. Microbiol.,* 11, 453–461, 1965.

13. **Gutman, L. T., Turck, M., Petersdorf, R. G., and Wedgwood, R. J.,** Significance of bacterial variants in urine of patients with chronic bacteriuria, *J. Clin. Invest.,* 44, 1945–1952, 1965.

14. **Gutman, L. T., Shaller, J., and Wedgwood, R. J.,** Bacterial L Forms in relapsing urinary-tract infections, *Lancet,* 1, 464–466, 1967.

15. **Hellkuhl, B. and Grzeschik, K. H.,** Elimination of mycoplasma contamination from mammalian cell cultures by the bibenzimidazole derivative Hoechst 33258, *Cytogenet. Cell Genet.,* 36, 584–585, 1983.

16. **Jacoby, G. A. and Archer, G. L.,** New mechanisms of bacterial resistance to antimicrobial agents, *N. Engl. J. Med.,* 324, 601–612, 1991.

17. **Johnson, C. C., Wexler, H. M., Becker, S., Garcia, M., and Finegold, S. M.,** Cell wall defective variants of *Fusobacterium, Antimicrob. Agents Chemother.,* 33, 369–372, 1989.

18. **Kagan, G. Ya.,** Role of L Forms in staphylococcal infection, in *Microbial Protoplasts, Spheroplasts, and L Forms,* Guze, L. B., Ed., Williams & Wilkins, Baltimore, 1968, 372–378.

19. **Kagan, G. Ya.,** Some aspects of investigations of the pathogenic potentialities of L Forms of bacteria, in *Microbial Protoplasts, Spheroplasts, and L Forms,* Guze, L. B., Ed., Williams & Wilkins, Baltimore, 1968, 422–443.

20. **Kagan, G. Ya., Koptelova, E. I., and Pokrovsky, B. M.,** Isolation of pleuropneumonia-like organisms, L Forms and heteromorphous growth of bacteria from the cerebrospinal fluid of patients with septic meningitis, *J. Hyg. Epidemiol.,* 9, 310–317, 1965.

21. **Lebar, W. D.,** Isolation and Identification of Cell Wall Deficient Microorganisms from Infection, M.S. thesis, Wayne State University, Detroit, 1977.

22. **Lederberg, J. and St. Clair, J.,** Protoplasts and L type growth of *E. coli, J. Bacteriol.,* 75, 143–158, 1958.

23. **Louria, D. B., Kaminski, T., Grieco, M., and Singer, J.,** Aberrant forms of bacteria and fungi found in blood or cerebrospinal fluid, *Arch. Intern. Med.,* 124, 39–48, 1969.

24. **Mardh, P. A., Fritz, H., Kohler, L., and Schersten, B.,** Hypoglucosuria and L Forms of *E. coli* in the urine, *Acta Med.Scand.,* 186, 549–552, 1969.

25. **Masri, A. F. and Grieco, M. H.,** Bacteroides endocarditis, report of a case, *Am. J. Med. Sci.,* 263, 357–367, 1972.

26. **Mattman, L. H. and Mattman, P. E.,** L Forms of *Streptococcus faecalis* in Septicemia, *Arch. Intern. Med.,* 115, 315–321, 1965.

27. **McClatchy, J. K.,** Rapid method of microbial susceptibility testing, *Infect. Immun.,* 1, 421–422, 1970.

28. **Merline, J. R. and Mattman, L. H.,** Cell wall deficient forms of Actinomyces complicating two cases of leukemia, *Mich. Acad.,* 3, 113–121, 1971.

29. **Morimoto, S., Nagate, T., Sugita, K., Ono, T., Numata, K., et al.,** Chemical modification of erythromycins. III. In vitro and in vivo antibacterial activities of new semisynthetic 6-*O*-methylerythromycins A, Te-031 (Clarithromycin) and Te-032, *J. Antibiot.,* 43(3), 295–305, 1990.

30. **Motwani, N.,** Identification of Cell Wall Deficient Bacteria by Immunofluorescence and Polyacrylamide Gel Electrophoresis, Ph.D. dissertation, Wayne State University, Detroit, 1976.

31. **Nativelle, R. and Deparis, M.,** Formes évolutives des bactéries dans les hémocultures, *Presse Med.,* 68, 571–574, 1960.

32. **Rieber, M., Imaeda, T., and Cesari, I. M.,** Bacitracin action on membranes of Mycobacteria, *J. Gen. Microbiol.,* 55, 155–159, 1969.

33. **Roberts, D. E., Ingold, A., Want, S. V., and May, J. R.,** Osmotically stable L Forms of *Haemophilus influenzae* and their significance in testing sensitivity to penicillins, *J. Clin. Pathol.,* 27, 560–564, 1974.

34. **Rosner, R.,** Isolation of Candida protoplasts from a case of Candida endocarditis, *J. Bacteriol.,* 91, 1320–1326, 1966.

35. **Schmitt-Slomska, J., Caravano, R., and El-Solh, N.,** Loss of plasmid-mediated resistance after conversion of a Group B Streptococcus strain to a stable cell wall deficient variant, *Ann. Microbiol. (Inst. Pateur),* 130A, 23–27, 1979.

36. **Schönfeld, J. K.,** "L" forms of Staphylococci; their reversibility; changes in the sensitivity pattern after several intermediary passages in the "L" phase, *Antonie van Leeuwenhoek,* 25, 325–331, 1959.

37. **Schuhmann, E. and Taubeneck, U.,** Stabile L Formen verschiedener *E. coli*-Stamme, *Z. Allg. Mikrobiol.,* 9, 297–313, 1969.

38. **Snoke, J. E. and Cornell, N.,** Protoplast lysis and inhibition of growth of *Bacillus licheniformis* by bacitracin, *J. Bacteriol.,* 89, 415–420, 1965.
39. **Surr, J. C.,** Ribonucleic acid in a "membrane" fraction of *E. coli* and its relation to cell wall synthesis, *J. Bacteriol.,* 84, 1061–1070, 1962.
40. **Swieczkowski, D. M., Mattman, L. H., Truant, J. P., and Wilner, F. M.,** Cell wall deficient forms of *Candida albicans* in mycohemia, *Lab. Med.,* 1, 41–42, 1970.
41. **Tanret, P. and Solignac, H.,** Sur deux cas de maladie d'Osler a hemo-ovocultures positives, *Soc. Med. Hop. Paris Bull. Mem.,* 113, 247–251, 1963.
42. **Taubeneck, U.,** Susceptibility of *Proteus mirabilis* and its stable L Forms to erythromycin and other macrolides, *Nature (London),* 196, 195–196, 1962.
43. **Von Herbay, A. and Otto, H. F.,** Whipple's disease: a report of 22 patients, *Klin. Wochenschr.,* 66(12), 533–539, 1988.
44. **Voureka, A.,** Bacterial variants in patients treated with chloramphenicol, *Lancet,* 1, 27–28, 1951.
45. **Vulfovich, Yu. V., Krasilnikova, O. Ya., Gamova, N. A., and Konstantinova, N. D.,** The resistance to antibiotics of reverted cultures of streptococcus of Group A and not classifiable streptococcus, obtained from the blood of people ill with rheumatic fever, *Antibiot. Med. Biotechnol.,* 8, 627–632, 1986.
46. **Ward, J. R., Madoff, S., and Dienes, L.,** In vitro sensitivity of some bacteria, their L Forms and pleuro-pneumonia-like organisms to antibiotics, *Proc. Soc. Exp. Biol. Med.,* 97, 132–135, 1958.
47. **Waterbury, W. E., Boydstun, J., Castellani, A. G., Freedman, R., and Gavin, J. J.,** Effect of nitrofurans on cell wall synthesis and L phase growth of *Staphylococcus aureus, Antimicrob. Agents Chemother.,* pp. 339–344, 1965.
48. **Wirostko, E., Johnson, A., and Wirostko, W. J.,** Mouse interstitial lung disease and pleuritis induction by human Mollicute-like organisms, *Br. J. Exp. Pathol.,* 69, 891–902, 1988.
49. **Yamamoto, A. and Homma, J. Y.,** L Form of *Pseudomonas aeruginosa.* II. Antibiotic sensitivity of L Forms and their parent forms, *Jpn. J. Exp. Med.,* 48, 355–362, 1978.
50. **Swieczkowski, D. M.,** personal communication.
51. **Chopra, C.,** unpublished study.

Chapter 29

MISCELLANEOUS DISEASES

LUNG AND BRONCHIAL DISEASE

THE HISTORIC STUDY WITH *STREPTOBACILLUS MONILIFORMIS*

The first report in the U.S. of a pathogenic L Form was by Dienes and Edsall.[22] The middle ear of a rat, showing typical disease "twisting", was opened and the small amount of pus planted on a medium containing equal volumes of centrifuged boiled blood agar and horse serum. By 3 d six typical *S. moniliformis* colonies developed, but the remainder of the plate was densely covered with tiny colonies corresponding to Kleineberger's L-organism. The L-organism preserved its peculiar character and did not revert to the bacterial form. The investigators injected white mice subcutaneously and intraperitoneally with small pieces of agar containing the dense, pure growth of tiny L-colonies. The mice died in 1 to 10 d. Animals dying by the third day showed no pathological change except a slight pleural exudate. Mice dying later showed abundant clear pleural exudate and sometimes a thick subcutaneous edema over the entire ventral surface of the body. From the pleural, peritoneal, and subcutaneous exudates, the L-organisms grew abundantly in pure culture, but no classic bacteria were demonstrable either by direct dark-field microscopic examination or by various staining methods. The heart blood was sterile in all mice except one, an animal which died within 24 h, yielding a few L-colonies. It is interesting to note, commented the investigators, that the disease produced by the L Form variant differed in some respects from that ascribed to the Streptobacillus itself. They also noted that the pathogenicity of the "L-organism" for mice was not uniform. When first tested, a strain killed mice in 4 d. After two animal passages, death occurred after 5, 6, and 10 d. Some fully grown white mice showed no symptoms following injection of L-organisms. The mice which died in 1 to 3 d, on the other hand, were half-grown animals; thus, it seemed that both the passage history of the L Forms' strain and the age of the test animal influence the pathogenicity.

It is interesting that under other circumstances L Forms of *S. moniliformis* were not mouse pathogenic. It might be pertinent that Dienes and Edsall in the work describing pathogenicity used L Forms from infections where their predominance indicated that they, per se, were causing the disease in the animal.

PLEURISY AND INTERSTITIAL LUNG DISEASE BY CWD FORMS FROM UVEITIS

Pleurisy developed in 17 of 100 mice given CWD forms from human uveitis. Interstitial lung disease developed in 21 of the mice. Wirostko and co-investigators noted that in man interstitial lung disease is usually of unknown etiology and speculated that rifampin might be beneficial since infected mice showed some response to this antibiotic.[68] The variants did not revert and their identity was not determined.

PNEUMONIA COMPLICATING A KIDNEY GRAFT

A woman became febrile after immunosuppressive therapy following a kidney transplant. Chest X-rays revealed pneumonia. Cultures of sputum, blood, and urine were uninformative. She was treated empirically with large doses of penicillin, without response. Following autopsy a piece of consolidated lung was sent to the laboratory for culture of CWD organisms. This specimen, examined by Wickers[75] in our laboratory, yielded a Gram positive pleomorphic organism in hypertonic medium containing serum and separately autoclaved yeast extract. With successive transplants, the morphology and capsule suggested pneumococci (see Frontispiece). The strain gave inconsistent surface growth as alpha-hemolytic, very mucoid colonies. It was bile soluble, and capsular swelling occurred with Type III pneumococcal antibody. This case appears to indicate pathogenesis for CWD organisms in the transplant recipient.

L Forms of pneumococci have been found in varied manifestations of pneumococcal infection in children, including pneumonia.[42] Mere growth of pneumococci in human cell lines can result in loss of classic organisms and multiplication of the variants.[32]

Ermakova and colleagues noted that patients with acute leukemia tend to have L Form infections. Correspondingly, they found that mice immunodepressed by a variety of methods became susceptible to pneumococcal L Forms.[25]

PNEUMOCOCCI ROUTINELY PRESENT IN THE BLOOD OF PNEUMONIA CASES

The pneumococcus, albeit CWD, seems to be routinely present in circulating blood in pneumococcal pneumonia. We find that even when pneumococcal pneumonia cases lack classic forms in sputum, the CWD stages are usually present in pretherapy blood cultures and can be identified by agglutination with antiserum. The pool or type of antiserum reacting correlates with the type of pneumococcus isolated from sputum when the pathogen is found there. The isolates studied in our laboratory usually have not reverted. CWD pneumococci can be presumptively identified by polyacrylamide gel. Certain bands are accentuated in the variants, but reference electrophorograms can be helpful.[48]

Such CWD pneumococci can persist after therapy. In a patient who expired with pneumococcal pneumonia, pneumococcal CWD forms were found in blood cultures. The patient had received tetracycline before hospitalization. The bacteria in the blood were Gram-negative pleomorphic forms which gradually reverted to encapsulated pneumococci pathogenic for mice. Classical pneumococci were isolated from the patient's sputum only intermittently in this study by Brem and Reidt.[15]

HEMOPHILUS PARAINFLUENZAE

"Protoplasts" of *H. parainfluenzae* were tested for pathogenicity by McKay et al.[44] The wall-deficient stage was produced by aging cultures, then the resultant "protoplasts" propagated in a culture of swine kidney cells. One-week-old piglets from specific pathogen-free sows were inoculated intratracheally. When sacrificed after 20 d, all inoculated pigs had gross pneumonic lesions with infected lobes plum red, atelectic, and consolidated. It is interesting that in seven animals where only variants were found by smear and culture, the infiltrating leukocytes were lymphocytes and histiocytes. The two injected animals showing a polymorphonuclear exudation had rods in exudate smears, and classic bacilli were cultured from one.

Pasteurella multocida and *P. hemolytica* as well as *H. parainfluenzae* were implicated as probably causing naturally occurring enzootic-like pneumonia in pigs when the bacilli are in a wall-deficient form. McKay et al. noticed that often the classic stage of these organisms was recovered only after the tissue had been stored in the frozen state.[44] Such necessary freezing is similar to the cold storage of brain often required before attaining growth of *Listeria monocytogenes*. This "cold enrichment modification" merits remembering.

THE L-PHASE OF *BORDETELLA PERTUSSIS*

Wittler[70] produced lung disease in mice by intranasal inoculation of L-phase variants of *H. pertussis*. In almost every instance, the variant reverted when lung tissue was cultured.

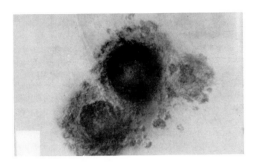

FIGURE 1. L colonies of microaerophilic streptococci from pleural fluid. (Magnification × 170.)(From Kenny, J. F., *South. Med. J.,* 71, 180–190, 1978. With permission.)

However, impression smears from mice killed at 1 to 2 d showed only an increase in L-phase organisms and no revertants. Wittler concluded, "The L-form appears to be pathogenic."[70]

CHRONIC BRONCHITIS

In chronic bronchitis antigens of Group A streptococci circulate in the blood of some patients. Antigens of both the classic and the CWD stages are present,[31] suggesting a persistent low-grade infection.

Spheroplasts of *H. influenzae* grow from the sputum of patients with persistent bronchial infection. To obtain such growth an essential ingredient of the medium is *N*-acetyl-*O*-glucosamine. The presence of some wall is shown by electron microscopy. Subculture of the variants brings reversion.[54]

PLEURISY AND CHRONIC EMPYEMA

The L Forms in a pleural fluid sample studied by Kenny[39] reverted to microaerophilic streptococci, (Figure 1). In 97 patients with chronic empyema, bacillary *Mycobacterium tuberculosis* was demonstrable in only 34% of the patients, whereas CWD stages of the pathogen were isolated from 80%.[14]

OCULAR DISEASE

Anterior uveitis, inflammation of the iris and ciliary body, is not uncommon.[72] This condition, characterized by pain and blurred vision, can progress to blindness. Bacteria or fungi have been isolated rarely: *Clostridium perfringens*,[66] *Neisseria gonorrheae*,[57] *Staphylococcus aureus*,[11] Leptospira,[62] Actinomyces,[33] Coccidioides,[53] Pneumococcus,[47] Listeria,[30] Streptococcus,[3] and Treponema.[58] The opinion of Woods[72] that the condition is probably an autoimmune phenomenon or of Coles and Nathaniel[20] that bacteria are absent is generally accepted. Reviewing 340 uveitis cases, Spencer[60] found that in only two cases was there evidence of endogenous infection.

In a preliminary study seeking either Mycoplasma or CWD stages, Hessburg et al. found CWD forms in aqueous humor of 11 of 19 eyes with unexplained uveitis. The variants were carried in successive culture and studied by acridine orange staining.[34] Active infection was suggested by the complement found within leukocytes stained with fluorescent anti-complement antibody, implying immune complexes with microbes. Often the growth suggested fungi. These findings are illustrated in the Frontispiece.

The above work was extended by Barth[6] and others[7,8,34] studying 63 uveitis patients.

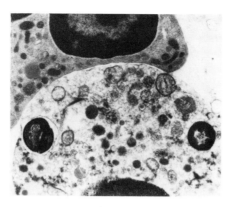

FIGURE 2.   Leukocyte within aqueous humor contains CWD organisms. Uranyl acetate-lead citrate stain. (Magnification × 32,812.)(From Wirostko, E., Johnson, L. A., and Wirostko, W. J., *Virchows Arch. A. Pathol. Anat.*, 413, 349–355, 1988. With permission.)

The majority of the variants exhibited some cell wall, as revealed by staining with rhodamine-labeled muramidase.

Fourteen of the Barth uveitis isolates were injected intraperitoneally into mice who received a single injection of cortisone. When sacrificed at 1 month, diseased organs varied with the strain. Involved in some instances were kidney, liver, lung, and testes. The inoculated CWD strain was recovered from involved tissue, which usually included the spleen. Attempts to obtain reversion of the organisms were not successful. The morphology of several strains suggested fungi.

The identity of the variants isolated can be further explored with protein bands in electrophoresis, Ouchterlony, and fluorescent antibody tests. Fungi remain suspect as well as bacteria since they also react with fluorescent muramidase.

The uveitis studies have been furthered by Wirostko and associates with intriguing and enlightening findings. Electron microscopy detected CWD microbes in 2 to 10% of the leukocytes invading diseased aqueous humor (Figure 2).[69] Leukocytes from 4 patients were injected into 100 mice sacrificed after 12 months. Controls consisted of 200 uninoculated animals. Prominent in the test mice were both cardiac and uveal vasculitis. Some animals had disease in other areas, including lung, pleura, or gut. Some mice developed exophthalmic chronic orbital inflammatory disease. Slow virus diseases have long been known, but there is much less awareness of slow bacterial disease.

Bacterial involvement and even perhaps clues to etiology are indicated by skin tests on patients with anterior uveitis. As high as 64% of these patients may show sensitization to streptococcal or staphylococcal protein.[4]

Jacobs and Golden[35] culturing eyes in their ophthalmology practice found CWD forms in approximately (1) 30% of corneal ulcers, (2) 14% of conjunctivitis, and (3) 40% of dry eye syndrome. The aqueous humor of seven patients who had routine cataract removal grew CWD forms in three cases. Some of the isolates reverted to genera which included Staphylococcus, Neisseria, and Pseudomonas.

In a long-term study of CWD forms of Nocardia, Beaman and Scates[9] found ocular disease resulting from long-persisting CWD variants of *Nocardia caviae;* no classic Nocardia were detected in the opacification in eyes of mice.

CHRONIC SINUSITIS

In a study of chronic sinusitis, Frederick and Braude[28] found combined aerobic and anaerobic cultures inadequate for diagnosis of this common almost intractable condition

TABLE 1
Results of Specimens Obtained by
Lynch and Caldwell-Luc Procedures
in Chronic Sinusitis

Result	No. of specimens
No growth	21
Aerobic growth only	19
Anaerobic growth only	26
Both aerobes and anaerobes	17
Total	83

(Table 1). Culture for CWD organisms probably would have detected the noncultured bacteria seen in Gram stain and others undetected in a heat-fixed Gram stain. The lysozyme in nasal mucous could contribute to maintenance of the variants.

Another study, in Finland, had similar findings in acute maxillary sinusitis. Although aerobic and anaerobic cultures were done immediately after removing the exudate, 24% of the infections yielded no growth.[36]

TONSILLITIS

L Forms are frequently found in cases of recurrent exudative tonsillitis. The species involved include the *Streptococcus mitis* group.[61]

WHIPPLE'S DISEASE

Whipple's disease, fortunately rare, has challenging aspects. Once the diagnosis is made, therapy may be successful but may need to be uninterrupted for 2 years.

In the pretetracycline era, Whipple's disease was uniformly fatal. The remaining greatest problem is diagnosis, which may be delayed as long as 10 years, the average being $3^1/_2$ years.[65] Multiple organs are involved, as shown by the diverse aspects of the condition.[27] Weight loss is prominent, since rather than absorbing lipids the patient passes fats in his stool. Other symptoms include intestinal cramps, arthritis, fever, anemia, and neurological aberrations. Fatal myocarditis may occur with intestinal involvement.[16,59,64,67]

Once the diagnosis is suspected it is usually confirmed by intestinal biopsy. Of greatest diagnostic significance is glycoprotein in the macrophages which infiltrate the villi of the jejunum, revealed in PAS stains of specimens obtained by biopsy or otherwise. The macrophages also contain pleomorphic bacilli as noted by Yardley and Hendrix[74] and by Chears and Ashworth.[18] The pleomorphic bodies are shown in phagocytic glial cells from the hypothalamus in a case described by Feurle et al. (Figure 3).[26]

Attempts to identify the bacteria in the intestinal macrophages are impractical due to contamination with intestinal flora.[27] Isolation of a blood-borne organism related to the disease was first made by Charache,[17] as she found CWD Group D streptococci in 9 blood cultures and a lymph node. Specific therapy was successful in treating this tetracycline-resistant case. Currently, therapy often continues to be "shotgun" with tetracycline administered first. Ceftriaxone, administered intravenously and followed by doxycycline, has been one successful regimen.[2]

However, it is apparent that diagnosis can be facilitated and treatment made more pertinent if growth of the CWD pathogen is made and its antibiotic sensitivities determined. Finding the pathogen in blood in a CWD stage is a good possibility as mentioned above.[17]

The consistent morphology of the organism in macrophages suggests that a single species is involved.

However, certain cases of Whipple's syndrome may be due to an organism other than streptococci. A careful study in Amsterdam concerned a case in which *H. influenzae* type E was isolated from multiple biopsies of the small intestine.[63] The electron micrography of these investigators suggested that the bacterium was wall deficient, not in tissue, but in culture.

Search for the causative agent of Whipple's disease has been complicated by several factors. Some isolations have been multiple organisms. The most extensive applications of techniques to confirm an actual connection between an isolate and the disease have been at McMaster's Department of Medicine in Ontario[19] and at Johns Hopkins Medical School.[40] In the first study, the α-hemolytic streptococcus isolate could be grown intracellularly in tissue culture within human colon cells, producing the PAS staining material characteristic of the syndrome. IgG in the patient's serum reacted with organisms within the patients lymph nodes and with the isolated streptococci. Hypertonic medium was necessary for primary isolation of the streptococcus.

The Johns Hopkins study examined the bacterial antigen in the foamy macrophages of 3 cases of Whipple's disease. The authors found that the bacterial antigen from the 3 cases reacted with 23 varied bacterial antisera with a consistent pattern. The investigators proved their point that a single species may be involved in Whipple's disease.

BEHÇET'S SYNDROME AND CANKER SORES (APHTHOUS ULCERS)

Behçet's syndrome, a world wide affliction, was noticed in patients by the dynamic practitioner, Hulusi Behçet, in Istanbul and had been studied earlier in Greece.[1,10] The ordinary canker sore is clinically and pathologically indistinguishable from the mouth ulceration of Behçet's disease. However, the unfortunate victim of Behçet's syndrome experiences simultaneous oral, genital, and/or ocular ulcers. Fifty percent of the cases also suffer joint changes.[46] Lesions may appear in the skin (Figure 4).[51] Other tissues may be attacked, as shown by the reported fatalities from diffuse encephalitis[71] or ulcerative bowel disease.[13] Its etiology has been enigmatic. Inclusion bodies resembling viruses or chlamydial organisms have been seen in synovial fluids.[12,45] Other agents considered as possible in the etiology have been Mycoplasma and L Forms.[43]

A most pertinent finding has been in Tokyo by Namba and associates.[49] Twenty-six patients with Behçet's disease had a high titer of *Streptococcus pyogenes* antigen in circulating blood preceding and during an episode. The antigen was absent in 30 control patients whose uveitis was due to tuberculosis, sarcoidosis, Vogt-Koyanagi-Harada's disease, toxoplasmosis, or trauma. Structures resembling CWD streptococci were observed in the leukocytes of the buffy coat of Behçet's patients' blood (Figure 5). Namba also found that the Behçet's patients had abnormally low antibody titers to streptococci.[50]

That Behçet's episodes may be triggered by immune reactions is suggested by the following studies. One investigation shows that the complement titer drops precipitously just before an attack (Figure 6).[56]

Ancillary studies on the role of streptococcal antigen in Behçet's syndrome have been profitable. Four of 16 Behçet's disease patients skin tested with antigens of streptococci experienced severe reactions in eyes or intestines or extensive oral aphthae and facial ulcers. Two of the reacting patients received injections of both *S. pyogenes* and *S. faecalis*. Two had been given only the *S. pyogenes* extract. Skin tests with other killed bacteria did not cause flares of Behçet's symptomatology. Mizushima et al. also found that dental manip-

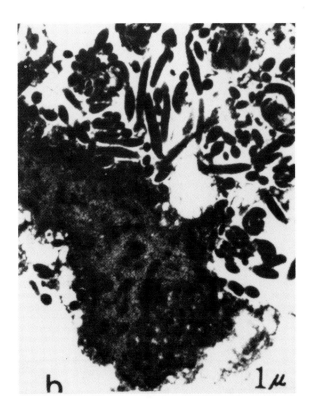

FIGURE 3. Hypothalamus glial cell in Whipple's disease (fatal case). (From Feurle, G. E., Volk, B. and Waldherr, R., *N. Engl. J. Med.*, 300, 907–908, 1979. With permission.)

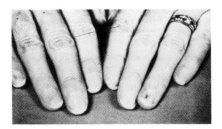

FIGURE 4. Behçet's syndrome patient has hemorrhagic infarctions in left ring and right little finger. (From O'Duffy, J. D., Carney, J. A., and Deodhar, S., *Ann. Intern. Med.*, 75, 561–570, 1971. With permission.)

ulations, well known to flood the bloodstream with α-hemolytic streptococci, can also cause flares of Behçet's disease symptoms.[46]

Not only is circulating streptococcal antigen high in Behçet's syndrome, but antibody to the CWD stage is low. That this is not a general immunodeficiency is shown by normal titers to staphylococcal CWD forms.[52]

It is thought that Behçet's patients have chronic foci of infection which feed streptococcal antigen into the blood.[37] Mizushima et al. suggest that antimicrobial therapy in Behçet's disease needs reevaluation.[46] Several physicians have administered dapsone to their Behçet's patients.[21]

There is evidence that the actual damage in the syndrome results from immune reactions.

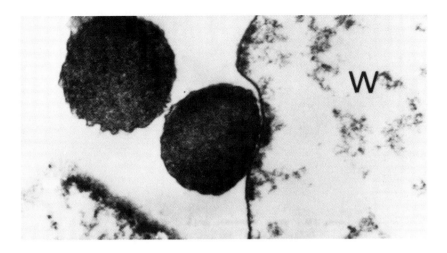

FIGURE 5. Streptococcal L Form-like structures in white blood cell (W) layer of Behçet's patient. Two structures are in contact with each other, and one adheres to white blood cell. Stratified and uneven protoplasmic membranes are seen on the L Form. (From Namba, K., Ueno, T., and Okita, M., *Jpn. J. Ophthalmol.*, 30, 385–401, 1986. With permission.)

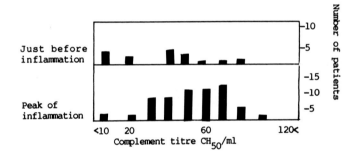

FIGURE 6. Complement titers in sera from patients with Behçet's disease in attacks of ocular inflammation. The height of the bars shows the number of patients with a complement titer indicated. (From Shimada, K., Kogure, M., Kawashima, T. and Nishioka, K., *Med. Biol.*, 52, 234–239, 1974. With permission.)

IgM and Group D antigen are found in blood vessel walls. Cytological analysis has revealed participation by activated T cells, macrophages, and natural killer cells.[37] A predisposed tissue type is HLA-B5 [23] or more specifically HLA-B51.[24]

Immunological reactions to streptococci are also implicated by studies of Yanase. His leukocyte migration test showed migration inhibition when patients' leukocytes were exposed to streptococcal antigens.[73]

SPECIFIC THERAPY IN BEHÇET'S DISEASE

An aspect favoring the theory that infection actually underlies the autoimmunity in both Behçet's disease and aphthous ulcers is the improvement that sometimes follows administration of the sulfone dapsone.[21,55]

CANKER SORES (APHTHOUS ULCERS)

Similar etiology may explain recurrent canker sores. Most individuals occasionally suffer with this condition, which are single or multiple necrotizing ulcers of the oral mucosa.

FIGURE 7. Immunofluorescent staining shows reaction with antistreptococcus Group D at the infiltration site of an aphthous ulcer. (Magnification × 240.)(From Kaneko, F., Takahashi, Y., Muramatsu, Y., and Miura, Y., *Br. J. Dermatol.*, 113, 303–312, 1985. With permission.)

Research indicates that in the quiescent state the individual harbors a wall deficient form of *S. sanguis* in his oral cavity. Supporting this theory, Barile et al.[5] found that during a quiescent period, scar tissue of a previous lesion yielded only L-colonies on biopsy and culture. A fluorescent antibody study indicates that streptococcal antigen may be concentrated in an aphthous ulcer (Figure 7).[38]

BRUCELLOSIS

Brucellosis has long been noted as an infection in which the bacterium is difficult to isolate. Conceivably this is related to a finding by Freeman and Rumack that spheroplasts can be more virulent than the bacilli.[29] Spheroplasts made by several inducing agents were lethal for monocytes; classic bacilli were innocuous (Figure 8).

NEUROENDOCRINE SYSTEM DAMAGE

Few reports concern CWD Group B streptococci. The results when these are given to a laboratory animal differ markedly from consequences of injecting CWD Group A streptococci. White rats and mice given a single intraperitoneal injection of stable L Forms of Streptococcus Group B had progressive degeneration in the hypothalamus, pituitary gland and adrenals. After 6 to 12 months there was often tissue regeneration with only partial recovery of organ function.[41]

KAPOSI'S SARCOMA

A fungus which may or may not be related to the disease has been isolated from the blood and, in some cases, also from the involved tissue in 6 cases of Kaposi's sarcoma. The organism appears to be biphasic. Growing at pH 5.5 the organisms resemble CWD mycobacteria, but at neutral and alkaline pH the colonies are fungoidal. The growth is always acid fast when stained with the intensified Kinyoun's method. Electron microscopy has not

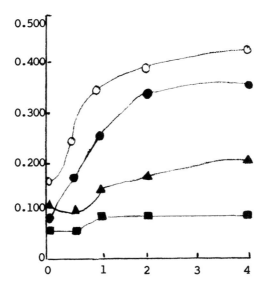

FIGURE 8.   Lethality of Brucella spheroplasts vs. normal Brucella. Release of trichloroacetic acid-soluble cellular constituents from monocytes infected with normal and spheroplast cells of smooth Brucella suis 32P. Symbols: ○, penicillin-induced spheroplasts; ●, glycine-induced spheroplasts; ▲, normal Brucells; ■, uninfected monocyte control supernatant fluid. (From Freeman, B. A. and Rumack, B. H., *J. Bacteriol.*, 88, 1310–1315, 1964. With permission.)

as yet been done. The organism acts wall deficient only in not growing on the surface of any medium standard for fungi or bacteria in aerobic or anaerobic atmosphere. Reference mycology laboratories in the U.S., South America, and Great Britain have been unable to identify the organism. E. Beneke and L. L. Rogers found that after intraperitoneal injection mice harbor the organism for 2 months in liver and spleen. However, the animals show no sign of disease. In the laboratory of John Lohr attempts to propagate the fungus in epithelioid-type tissue culture cells gave negative results. The organism grows at temperatures of 10 to 40°C, less slowly at 30 to 40°C. It is not serophilic. It is seen in small numbers after 7 d of incubation in semiliquid medium (see Frontispiece). Microscopic colonies appear after several weeks in soft agar pour plates containing 10% sucrose. Finger puncture blood on a slide incubated at 35°C, for 4 d in auramine rhodamine dye (Difco) plus NaH_2CO_3 shows growth of characteristic spores. Broth media needs 1% separately autoclaved yeast extract and agar or silica gel. This organism was not found in blood cultures of 27 AIDS patients who did not have signs of Kaposi's sarcoma.

SUMMARY

Pulmonary diseases associated with CWD microbes include pneumonia, bronchitis, interstitial pneumonitis, pleurisy, and chronic empyema.

Uveitis, a condition long considered hypersensitivity, has been shown by culture, electron microscopy, and animal inoculations, to harbor CWD microbes of diverse genera. Other ophthalmic problems appear to be in some instances also caused by wall-deficient bacteria or fungi.

Whipple's disease, once considered invariably fatal, now responds to therapy with one or multiple antibiotics. Organisms culturable only in media suitable for L Forms are thought to be responsible.

Behçet's syndrome has forms suggesting CWD streptococci in the patient's blood. Immune reactions to CWD streptococci are also atypical in this condition.

The very common canker sore (aphthous ulcer) may involve similar latency with streptococci.

The hypothalamus, pituitary gland, and adrenals can be damaged by CWD stages of Group B streptococci.

REFERENCES

1. **Adamantiades, B.**, A case of recurrent hypopyon iritis, *Bull. Athens Med. Soc.*, 586, 1930.
2. **Adler, C. H. and Galetta, S. L.**, Oculo-facial-skeletal myorhythmia in Whipple Disease: treatment with ceftriaxone, *Ann. Intern. Med.*, 112, 467–469, 1990.
3. **Amsler, M.**, New clinical aspects of the vegetative eye, *Trans. Ophthalmol. Soc. U.K.*, 68, 45–74, 1948.
4. **Asaoka, T., Kinugawa, K., and Komuro, S.**, Skin test by azostaphylococcal and azostreptococcal protein in suppurative disease and endogenous uveitis esp. Behçet's disease, *Acta Soc. Ophthalmol. Jpn.*, 66, 96–109, 1962.
5. **Barile, M. F., Francis, T. C., and Graykowski, E. A.**, *Streptococcus sanguis* in the pathogenesis of recurrent aphthous stomatitis, in *Microbial Protoplasts, Spheroplasts, and L Forms*, Guze, L. B., Ed., Williams & Wilkins, Baltimore, 1968, 444–456.
6. **Barth, C. L.**, Studies on the etiology of uveitis, Microbiological, immunological, hematological, Ph.D. dissertation, Wayne State University, Detroit, 1974.
7. **Barth, C. L., Judge, M. S., Mattman, L. H., and Hessburg, P. C.**, Isolation of an acid-fast organism from the aqueous humor in a case of sarcoidosis, *Henry Ford Hosp. J.*, 27, 127–133, 1979.
8. **Barth, C. L. and Hessburg, P. C.**, Role of cell wall deficient bacteria in uveitis, in *Cell Wall Deficient Bacteria*, Domingue, G. J., Ed., Addison-Wesley, Reading, MA, 1982, 465–488.
9. **Beaman, B. L. and Scates, S. M.**, Role of L Forms of *Nocardia caviae* in the development of chronic mycetomas in normal and immunodeficient murine models, *Infect. Immun.*, 33, 893–907, 1981.
10. **Behçet, H.**, Über rezidivierende aphthöse, durch ein Virus verursachte geschwüre am Mund, am Auge und an den Genitalien, *Dermatol. Wochenschr.*, 105, 1152–1157, 1937.
11. **Berens, C.**, in discussion in Woods, A. C., Nongranulomatous uveitis in man and experimental animals, *Surv. Ophthalmol.*, 4, 327–369, 1959.
12. **Bloch-Michel, H., Benoist, M., Siboulet, A., et al.**, Aphthose et rheumatisme inflammatoire. Forme atypique du syndrome de Behçet?, *Rev. Rhum. Mal. Osteoartic.*, 32, 408–414, 1965.
13. **Bøe, J., Dalgaard, J. B., and Scott, D.**, Mucocutaneous-ocular syndrome with intestinal involvement; a clinical and pathological study of four fatal cases, *Am. J. Med.*, 25, 857–867, 1958.
14. **Bogush, L. K., Karachunskii, M. A., Dorozhkova, I. R., Zhangireev, A. A., and Abramov, E. L.**, L Forms of *Mycobacterium tuberculosis* in tuberculosis patients with chronic Pyothorax, *Probl. Tuberk.*, 12, 19–23, 1981.
15. **Brem, A. M. and Reidt, W.**, A cell-wall deficient form of the pneumococcus in a case of pneumonia, *Chest*, 61, 200–202, 1972.
16. **Bruckstein, A. H.**, Whipple's disease — effective therapy, elusive etiology, *Hospital Pract.*, 24(5A), 16–70, 1989.
17. **Charache, P.**, Atypical bacterial forms in human disease, in *Microbial Protoplasts, Spheroplasts, and L Forms*, Guze, L. B., Ed., Williams & Wilkins, Baltimore, 1968, 484–494.
18. **Chears, W. C., Jr. and Ashworth, C. T.**, Electron microscopic study of the intestinal mucosa in Whipple's disease. Demonstration of encapsulated bacilliform bodies in the lesion, *Gastroenterology*, 41, 129–138, 1961.
19. **Clancy, R. L., Tomkins, W. A. F., Muckle, T. J., Richardson, H., and Rawls, W. E.**, Isolation and characterization of an aetiological agent in Whipple's disease, *Br. Med. J.*, 3, 568–570, 1975.
20. **Coles, R. S. and Nathaniel, A.**, Role of streptococcus in pathogenesis of anterior uveitis, *AMA Arch. Ophthalmol.*, 61, 45–49, 1959.
21. **Convit, J., Goihman-Yahr, M., and Rondon-Lugo, A. J.**, Effectiveness of dapsone in Behçet's disease, *Br. J. Dermatol.*, 111, 629–630, 1984.
22. **Dienes, L. and Edsall, G.**, Observations on the L-organism of Klieneberger, *Proc. Soc. Exp. Biol. Med.*, 36, 740–746, 1937.
23. **Dündar, S. V., Gençalp, U., and Şimşek, H.**, Familial cases of Behçet's disease, *Br. J. Dermatol.*, 113, 319–321, 1985.

24. **Ehrlich, G. E.,** Behçet's disease: current concepts, *Comprehens. Therapy,* 15(1), 27–30, 1989.

25. **Ermakova, G. L., Vlasova, A. G., Gusenova, F. M., and Martynova, V. A.,** Experimental infection of mice with L Forms of bacteria against the background of immunodepressants, *Zh. Mikrobiol. Epidemiol.,* 51, 91–94, 1974.

26. **Feurle, G. E., Volk, B., and Waldherr, R.,** Cerebral Whipple's disease with negative jejunal histology, *N. Engl. J. Med.,* 300, 907–908, 1979.

27. **Fleming, J. L., Wiesner, R. H., and Shorter, R. G.,** Whipple's disease, clinical, biochemical, and histopathologic features and assessment of treatment in 29 patients, *Mayo Clin. Proc.,* 63, 539–551, 1988.

28. **Frederick, J. and Braude, A. I.,** Anaerobic infection of the paranasal sinuses, *N. Engl. J. Med.,* 290, 135–137, 1974.

29. **Freeman, B. A. and Rumack, B. H.,** Cytopathogenic effect of *Brucella* spheroplasts on monocytes in tissue culture, *J. Bacteriol.,* 88, 1310–1315, 1964.

30. **Goodner, E. K.,** in discussion in Kimura, S. J., Microbiologic aspects of experimental uveitis, historic approach, in *Immunopathology of Uveitis,* Maumenee, A. E. and Silverstein, A. M., Eds., Williams & Wilkins, Baltimore, 1964, 28.

31. **Gorina, L. G., Romanova, R. Yu., and Goncharova, S. A.,** Microbial association in chronic respiratory diseases, *Zh. Mikrobiol. Epidemiol. Immunobiol.,* 4, 18–21, 1986.

32. **Gosteva, V. V., Klitsunova, N. V., Gotvyanskaya, T. P., Semenenko, T. A., and Vasileva, V. I.,** Interaction of *Streptococcus pneumoniae* with continuous human cell lines, *Zh. Mikrobiol. Epidemiol. Immunobiol.,* 12, 32–36, 1981.

33. **Harley, R. D. and Wedding, E. S.,** Syndrome of uveitis, meningoencephalitis, alopecia, poliosis, and dysacousia. Report of a case due to Actinomyces, *Am. J. Ophthalmol.,* 29, 524–535, 1946.

34. **Hessburg, P. C., Mattman, L. H., Barth, C., and Dutcheshen, L. T.,** Aqueous microbiology, the possible role of cell wall deficient bacteria in uveitis, *Henry Ford Hosp. Med. J.,* 17, 177–194, 1969.

35. **Jacobs, Y. and Golden, B.,** Aberrant bacterial forms from various ocular sites, *Arch. Ophthalmol.,* 94, 933–936, 1976.

36. **Jousimies-Somer, H. R., Savolainen, S., and Ylikoski, J. S.,** Bacteriological findings of acute maxillary sinusitis in young adults, *J. Clin. Microbiol.,* 26, 1919–1925, 1988.

37. **Kaneko, F., Kaneda, T., Ohnishi, O., Kishiyama, K., Takashima, I., Fukuda, H., Kado, Y., and Endo, M.,** Behçet's syndrome and infection allergy. I. Detection of chronic infectious foci and immune responses to bacterial vaccines *in vivo* and *in vitro, Jpn. J. Allergol.,* 27, 440, 1978.

38. **Kaneko, F., Takahashi, Y., Muramatsu, Y., and Miura, Y.,** Immunological studies on aphthous ulcer and erythema nodosum-like eruptions in Behçet's disease, *Br. J. Dermatol.,* 113, 303–312, 1985.

39. **Kenny, J. F.,** Role of cell wall defective microbial variants in human infections, *South. Med. J.,* 71, 180–190, 1978.

40. **Keren, D. F., Weisburger, W. R., Yardley, J. H., Salyer, W. R., Arthur, R. R. and Charache, P.,** Whipple's disease, demonstration by immunofluorescence of similar bacterial antigens in macrophages from three cases, *J. Hopkins Med. J.,* 139, 51–59, 1976.

41. **Khostikyan, N. G., Vartazaryan, N. D., and Gorina, L. G.,** Morphological changes in the neuroendocrine system after inoculation of animals with L Form of Streptococcus B, *Arkh. Patol.,* 49(5), 64–70, 1987.

42. **Kvetnaya, A. S., Chistyakova, Yu. V., Abramova, N. V., and Volkova, M. O.,** Biological characterization of *Streptococcus pneumoniae* strains isolated from children with pneumococcal diseases, *Zh. Mikrobiol. Epidemiol. Immunobiol.,* 4, 25–30, 1989.

43. **Lehner, T.,** Behçet's syndrome and autoimmunity, *Br. Med. J.,* 1, 465–467, 1967.

44. **McKay, K. A., Abelseth, M. K., and Vandreumel, A. A.,** Production of an enzootic-like pneumonia in pigs with "protoplasts" of *Haemophilus parainfluenzae, Nature (London),* 212, 359–360, 1966.

45. **Mitrovic, D., Kahn, M. F., Ryckewaert, A., and Seze, S. de,** Étude en microscopie électronique des inclusions présentes dans les cellules de type macrophagique du liquide synovial au cours d'un cas de syndrome de Behçet, *Sem. Hop.* Paris, 44, 2527–2535, 1968.

46. **Mizushima, Y., Matsuda, T., Hoshi, K., and Ohno, S.,** Induction of Behçet's disease symptoms after dental treatment and streptococcal antigen skin test, *J. Rheumatol.,* 15(6), 1029–1030, 1988.

47. **Mortada, A.,** Subacute and chronic inflammatory iris nodules due to pyogenic bacteria, *Br. J. Ophthalmol.,* 46, 669–673, 1962.

48. **Motwani, N.,** Identification of Cell Wall Deficient Bacteria by immunofluorescence and Polyacrylamide Gel Electrophoresis, Ph.D. dissertation, Wayne State University, Detroit, 1976.

49. **Namba, K., Ueno, T., and Okita, M.,** Behçet's disease and streptococcal infection, *Jpn. J. Ophthalmol.,* 30, 385–401, 1986.

50. **Namba, K.,** Titer of anti-streptococcal antibody in Behçet's disease, *Acta Soc. Ophthalmol. Jpn.,* 92(2), 269–273, 1988.

51. **O'Duffy, J. D., Carney, J. A., and Deodhar, S.,** Behçet's Disease, report of 10 cases, 3 with new manifestations, *Ann. Intern. Med.,* 75, 561–570, 1971.

52. **Ogawa, T., Namba, K., Nishiyama, Y., and Sawada, K.,** The antibody titer to streptococcal and staphylococcal L Form in Behçet's disease, *Acta Soc. Ophthalmol. Jpn.* 94, 408–412, 1990.

53. **Pettit, T. H., Learn, R. N., and Foos, R. Y.,** Intraocular coccidioidomycosis, *Arch. Ophthalmol.,* 77, 655–661, 1967.

54. **Roberts, D., Rutman, A., Higgs, E., and Cole, P.,** Isolation of spheroplasts of *Haemophilus influenzae* in primary culture on supplemented selective medium from the sputum in persistent bronchial infection, a reservoir for reinfection, *Thorax,* 39, 700, 1984.

55. **Sharquie, K. E.,** Suppression of Behçet's disease with dapsone, *Br. J. Dermatol.,* 110, 493–494, 1984.

56. **Shimada, K., Kogure, M., Kawashima, T., and Nishioka, K.,** Reduction of complement in Behçet's disease and drug allergy, *Med. Biol.,* 52, 234–239, 1974.

57. **Sidler-Huguenin,** Ueber metastatische Augenentzundungen namentlich bei Gonorrhoe, *Arch. Augenh.,* 69, 346–378, 1911.

58. **Smith, J. L., Israel, C. W., and Harner, R. E.,** Syphilitic temporal arteritis, *Arch Ophthalmol.,* 78, 284–288, 1967.

59. **Southern, J. F., Moscicki, R. A., Magro, C., Dickerson, G. R., Fallon, J. T., and Bloch, K. J.,** Lymphedema, lymphocytic myocarditis, and sarcoidlike granulomatosis. Manifestations of Whipple's disease, *JAMA,* 261(10), 1467–1470, 1989.

60. **Spencer, W. H.,** in Aronson, S. B., Ed., *Sloan Symposium on Clinical Entities in Uveitis with Systemic Manifestations,* C. V. Mosby, St. Louis, 1968.

61. **Sprinkle, P. M.,** Current status of mycoplasmatales and bacterial variants in chronic otolaryngic disease, *Laryngoscope,* 82, 737–747, 1972.

62. **Sturman, R. M., Laval, J., and Weil, V. J.,** Leptospiral uveitis, *Arch. Ophthalmol.,* 61, 633–640, 1959.

63. **Tytgat, G. N., Hoogendijk, J. L., Agenant, D., and Schellekens, P. Th.,** Etiopathogenetic studies in a patient with Whipple's disease, *Digestion,* 15, 309–321, 1977.

64. **Vanderschueren, D., Dequeker, J., and Geboes, K.,** Whipple's disease in a patient with longstanding seronegative polyarthritis., *Scand. J. Rheumatol.,* 17, 423–426, 1988.

65. **Von Herbay, A. and Otto, H. F.,** Whipple's disease: a report of 22 patients, *Klin. Wochenschr.,* 66(12), 533–539, 1988.

66. **Walsh, T. J.,** Clostridial ocular infections. Case report of gas gangrene panophthalmitis, *Br. J. Ophthalmol.,* 49, 472–477, 1965.

67. **Wilcox, G. M., Tronic, B. S., Schecter, D. J., Arron, M. J., Righi, D. F., and Weiner, N. J.,** Periodic acid-schiff-negative granulomatous lymphadenopathy in patient with Whipple's disease., *Am. J. Med.,* 83, 165–170, 1987.

68. **Wirostko, E., Johnson, L. A., and Wirostko, W. J.,** Mouse interstitial lung disease and pleuritis induction by human Mollicute-like organisms, *Br. J. Exp. Pathol.,* 69, 891–902, 1988.

69. **Wirostko, E., Johnson, L. A., and Wirostko, W. J.,** Mouse exophthalmic chronic orbital inflammatory disease, *Virchows Arch. A. Pathol. Anat.,* 413, 349–355, 1988.

70. **Wittler, R. G.,** The L Form of *Haemophilis pertussis* in the mouse, *J. Gen. Microbiol.,* 6, 311–317, 1952.

71. **Wolf, S. M., Schotland, D. L., and Phillips, L. L.,** Involvement of nervous system in Behçet's syndrome, *Arch. Neurol.,* 12, 315–325, 1965.

72. **Woods, A. C.,** *Endogenous Inflammations of the Uveal Tract,* Williams & Wilkins, Baltimore, 1961.

73. **Yanase, K.,** Immunological investigation of Behçet's disease, *Skin Res.,* 20, 23–48, 1978.

74. **Yardley, J. H. and Hendrix, T. R.,** Combined electron and light microscopy of Whipple's disease, *Bull. Johns Hopkins Hosp.,* 109, 80–98, 1961.

75. **Wickers, L.,** unpublished study.

Chapter 30

PHAGE

Unique reactions distinguish the interplay of phage with wall-deficient microbes. Spheroplasts may lose their receptors for intact phage, but they become fantastically efficient in accepting and duplicating the naked phage nucleic acid strand. These interesting innovations are discussed below.

In other respects, phage attack is similar for classic bacteria and variants. Thus, CWD microorganisms may be parasitized by the typical phage with a tail, by the rounded phage which lacks an appendage, or by the thread-like phage virions.

INTERACTIONS OF CWD FORMS
WITH COMPLETE PHAGE

RECEPTORS FOR PHAGE

It was early learned that receptors for bacteriophages can be on incomplete walls since the spectacular large L-bodies of *Proteus vulgaris* adsorb phage.[7,53] Sometimes, in transition to the spheroplastic state, there appears to be quantitative retention of receptors, as shown by two *Escherichia coli* strains which adsorbed T2, T5, and λ_{hc} phages at the same rate as the parent. This was true in spheroplasts made with phage enzyme, lysozyme, glycine, or by diaminopimelic acid deprivation.[27]

However, walls of spheroplasts do not always retain their phage receptors. Spheroplasts susceptibility can vary depending on the agents responsible for wall-components loss. Spheroplasts of *E. coli* developed with penicillin and glycine were sensitive to ϕ X 174 phage, whereas serum-produced spheroplasts were immune.[20]

The composition of receptors is specific for each type of phage. The adsorption sites for staphylococcal phages appear to contain *O*-acetyl groups.[52] Mycobacterial phages, in contrast, seem to attach to polysaccharide units. Differences in mycolic acids account for phage resistance or susceptibility as shown by Imaeda and San Blas,[31] and loss of mycolic acid may be assumed when phage receptors disappear from a mycobacterium wall.

METABOLISM IN SPHEROPLASTS AFTER PHAGE INVASION

When phage injects DNA into a spheroplast, metabolism in the CWD cell can cease as rapidly as in the parent bacterium. β-Galactosidase synthesis stops immediately after T_2 phage invades *E. coli* B.[16]

Phage Susceptibility and L-Phase Colony Type

In discussing L-phase organisms, it is helpful to differentiate between the large "fried egg" and the smaller colony, usually only slightly if at all vesiculated. In a study with *P. mirabilis* the "fried egg", mycoplasma-like colonies were found to lyse completely when bacteriophage was dropped on the culture. The smaller granulated colonies were unaffected by the same phage.[55,56]

Revertibility Related to Phage Adsorption

A definite but not invariable association has been found between revertible variants and the presence of wall adequate for phage attachment. Two unstable "L Forms" of *E. coli* were able to host their respective phages.[48] Two stable CWD variants of *P. mirabilis* were resistant to temperate and virulent phages which attacked the parent Proteus.[57]

An exception to receptor presence being necessary for reversibility concerns an isolate of *Propionibacterium acnes* studied by Zierdt and Wertlake.[63] This variant, from fatal subacute bacterial endocarditis had lost its phage receptors in all stages except the completely reverted classic form.[63]

Phage Type and Susceptibility to L-Phase Induction by Antibiotics

Staphylococci of different phage types have varied responses to methicillin in production of L-phase colonies. Group III phage type is the most easily converted to L colonies as shown by Hamburger and Carleton[24] and by Fatma and Abla.[18]

MECHANISMS OF PHAGE RELEASE

A remarkable characteristic of some classic bacteria also occurs in their wall-deficient variants. This is the release of a large yield of phage without a detectable rupture of the bacterial wall. This occurs with phage X ϕ 174, a tail-bearing phage with large capsomeres containing single-stranded DNA, as noted by Bradley et al.[11]

SENSITIVITY TO PHAGE ACQUIRED BY SPHEROPLASTING

The loss of phage receptors in wall-depleted microbes is expected. Conversely, a surprising phenomenon was noticed by Hara and co-workers. When a phage-resistant strain of *E. coli* was induced to spheroplasts with EDTA-lysozyme or polymyxin E, it acquired exquisite sensitivity to the coliphage T_4.[25] This suggests that certain phage receptors may be inaccessible in the classic bacteria.

Penetration of phage RNA into spheroplasts can be complete in 60 s with the spheroplast membrane participating in the transport.[43]

COMPARATIVE YIELD OF FREE PHAGE FROM PARENT AND CWD STAGES

The development of phage within a spheroplast may be quantitatively normal, restricted, or increased. The rate of phage production may be normal for an *E. coli* T-phage system.[16] In many instances there has been no evidence that the phage yield is diminished in the CWD variant.[20,35] Less phage production is exemplified by CWD *Salmonella typhimurium,* which adsorbs phage P22 C at a normal rate and in normal numbers, yet the phage yield is only one fourth that from the parent strain.[42] When certain wall-deficient variants produce phage, 12 to 23 times as many phage units may be ejected as from the parent organism.[35] Thus, large amounts of phage excreted might be active in natural situations.

GENERAL CHARACTERISTICS OF PHAGE RELATED TO ABILITY TO ATTACK CWD VARIANTS

Not unexpectedly, CWD forms of *Proteus mirabilis* were completely immune to a phage which required flagella for attachment.[6]

It is interesting that specific antiserum to phage can enter spheroplasts and neutralize the virus. This was shown for bacteriophage T4 and *E. coli* B.[58]

LYSOGENY IN CWD FORMS

EXISTENCE AND DURATION OF LYSOGENY IN CWD FORMS

As might be anticipated, the prophage of lysogeny often continues to be replicated in CWD forms. Lysogeny in stable L Forms of *Staphylococcus aureus* was first demonstrated by Takemasa and nine co-investigators.[54] Electron microscopic evidence of phage production in CWD forms has been supplied by Schmid[46] for *S. aureus* (Figure 1), by Imaeda[32] for *Mycobacterium phlei,* and by Gumpert and associates[22] for *Bacillus subtilis* and *Streptomyces*

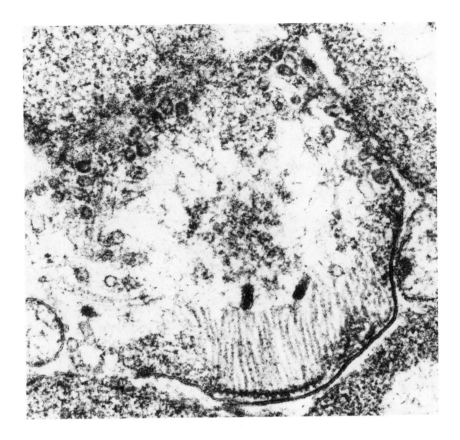

FIGURE 1. Phages in L-Form cell with cross-striated phage tails originating adjacent to the cytoplasmic membrane. L-Form cells treated with 1 μg mitomycin per milliliter for 120 min. (From Schmid, E. N., *Ann. Inst. Pasteur Microbiol.*, 137B(3), 283–289, 1986. With permission.)

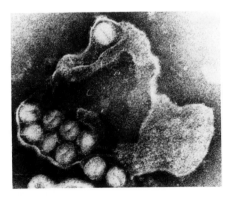

FIGURE 2. Phage occurring intracellularly in an L-Form cell of *Streptomyces hygroscopicus*. (From Gumpert, J., Zimmermann, I., and Taubeneck, U., *J. Basic Microbiol.*, 26(1), 15–25, 1986. With permission.)

hygroscopicus (Figure 2). In some instances the phage was naturally occurring within an L-Form colony.

Stable L Forms and newly made variants of *Salmonella typhosa* and *S. typhimurium* produced by glycine or penicillin all maintained the lysogenic state observed in the parent

TABLE 1
Susceptibility to Phage Induction

Organism	Agent tested for phage induction	Reaction compared with parent bacterium
Salmonella typhimurium[42]	Temperature elevation	No phage produced
	Mitomycin C	No phage produced
Salmonella, CWD forms of nine strains[21]	UV light	Only one strain produced high titer of phages
Staphylococcus aureus cla[47]	Mitomycin C	Same amount Mitomycin needed; L Form and parent L Form produced some phage with polyheads
S. aureus EMT[54]	UV light	One type of phage made by L Form; two different phages produced by parent
S. aureus 209 P[54]	UV light	One identical phage made by parent and L Form
Escherichia coli[2]	UV light	Phage induced in parent only
	Temperature elevation	Same phage induction in spheroplasts and parent

organism.[21] Lysogenic strains of wall-deficient *E. coli*[49] and *Pseudomonas aeruginosa*[4] have also been studied extensively. In general, however, there is a tendency for lysogeny to be lost more often than in parent strains.

PRODUCTION OF VIRULENT PHAGE FROM LYSOGENIC CWD FORMS

Lysogenic strains of classic bacteria continuously release a small amount of phage from a few bacteria which spontaneously produce the phage rather than replicate only the phage nucleic acid. To increase phage yield, one may induce the production of lytic phage particles by treating the lysogenic bacteria with such agents as ultraviolet (UV) light, hydrogen peroxide,[44] mitomycin C, or temperature elevation. However, when a lysogenic bacterium loses all or part of its cell wall, it is less readily stimulated to produce phage (Table 1).

Why removal of cell wall or wall components may bring this resistance to forces naturally inducing phage production may be explained by the following data. After irradiation a lysogenic organism produces a factor which triggers phage synthesis.[8] This factor is evidenced by its transfer during conjugation to a nonirradiated bacterium, which causes the recipient to produce phage. It would appear that this substance is produced in CWD forms only with great difficulty and that the factor triggering the shift from lysogenicity to virion production is usually made by wall-associated enzymes.

Borek and Ryan[8] have shown that lysogenic protoplasts induced to make phage with UV exposure may retain this sensitivity only 15 min after formulation of the protoplasts by a converting agent. Thus, it might be reasoned that the entity in the cell which responds to UV light is a preformed substance which remains in functional amounts only briefly after the cell wall is lost.

LYSOGENY REPRESSES OR FOSTERS SPHEROPLASTIC TENDENCIES

The DNA of a phage (as it is integrated in a lysogenic bacterium) can mysteriously repress the tendency of the bacterium to become spheroplastic. A nonlysogenized *E. coli* may need only 10,000 U of penicillin per milliliter for spheroplasting, where the phage-carrying duplicate requires 30,000 U.[49]

Latent phage may restrict stability of a CWD form. Lysogenized *E. coli* K12 treated with penicillin produces unstable CWD variant colonies. In contrast, when phage-free, the *E. coli* converts to an easily stabilized variant.[49]

IN VIVO ASSOCIATIONS OF PHAGE AND HOST MICROBE

The *in vivo* associations of phage and bacterial hosts suggest that phage may hold a bacterium in a wall-depleted state, as in latent brucellosis[41] and in sarcoid, according to some evidence.[36]

SUSCEPTIBILITY OF PROTOPLASTS TO PHAGE

The generalization has been made that phage receptors are on the cytoplasmic membranes of some Gram-positive bacteria, but are lacking on Gram-negative organisms. Correspondingly, certain *B. subtilis* phages lysed protoplasts only, and some actinophages caused protoplast lysis.[22] Gumpert and associates[22] reported that phages lytic for *B. subtilis* protoplasts are small DNA double-stranded viruses.

Jacobson and Landman have examined *Bacillus subtilis* phages in detail.[33,34] One bacillus strain which has 700 receptor sites on the wall was found to have approximately one third as many on its protoplast membranes. Strains of *B. megaterium* have been found to lose phage sensitivity when protoplasts.[12,15,60]

No protoplasts of *E. coli* are known to adsorb phage, although the phage receptors on the intact bacterial wall may be very close to the membrane. Bayer published a provocative finding that in *E. coli*, the sites for phage attachment are in the areas where the membrane protrudes up into the wall, near the surface. For each *E. coli* bacillus there are from 200 to 400 such localized receptor areas.[3]

SPHEROPLASTS AS HOSTS FOR
PHAGE NUCLEIC ACIDS

In 1960, when it was learned that DNA from bacteriophage can infect bacterial spheroplasts, molecular geneticists eagerly initiated the harvest of this new approach. The spheroplast system has several advantages. In studies of nucleic acid function, the bacterial cell wall of the host is usually nonpertinent, constituting nuisance contamination. Spheroplast walls contribute a minimal quantity of such unwanted material. Other benefits of the phage-spheroplast system include the alacrity with which phage-related proteins are formed and the variety of genetic information which can be introduced.

Spheroplasts are also superior to whole bacterial cells as recipients of phage nucleic acid. For example, single-stranded DNA from phage ϕ X174 infects spheroplasts but only rarely penetrates classic bacterial cells. Spheroplasts made from some entirely immune bacterium may serve as acceptors of phage DNA.[23] Guthrie and Sinsheimer[23] concluded that this infectivity resembles that of tobacco mosaic virus-RNA in that no specific attachment mechanism is involved.

Despite the comparative ease with which phage DNA infects spheroplasts, it does not indiscriminately enter all strains and species. Even when a parent cell can accept a complete phage, the bacterial spheroplast may inexplicably reject the strands of pure DNA. Conceivably, treatment used to make spheroplasts may sometimes damage cytoplasmic membranes.

Why wall-deficient microbes do not require helper phage[14] is intriguing in view of evidence that helper phage assists by adsorbing intracellular DNAase which tends to destroy the transforming DNA.

Another enigma concerns findings of Wahl and associates:[59] DNA of phage φ X174, extracted from *Shigella paradysenteriae,* infects spheroplasts of this strain but is strangely noninfective for spheroplasts of *E. coli* strains susceptible to the intact phage. Hofschneider is one of the few investigators who found that DNA from phage φ X174 was infectious for both a parent *E. coli* line and its spheroplasts.[28]

DOUBLE-STRANDED DNA

Meyer and co-workers[38] found that the double-stranded DNA in carefully controlled conditions can, indeed, infect spheroplasts, although excluded from classic bacterial cells. The pores in the intact cell wall apparently are too small to allow passage of the double-stranded DNA into the cytoplasm and admit single-stranded DNA only with great difficulty. Classic bacteria, therefore, exhibit a lower level of competence than do spheroplasts.

Various normal components of media influence phage DNA penetration into spheroplasts. Casamino acids, gelatin, NH_4Cl, glycerol, and $CaCl_2$ diminish infectivity, and NaCl, KCl, KH_2PO_4, and $MgSO_4$ destroy it completely.[38]

Not all double-stranded bacteriophages can serve as donors of nucleic acid. Meyer et al. found that DNA from coliphage T2 could not infect spheroplasts unless a protein remnant was left attached to the nucleic acid.[38] Nor can all spheroplasts be recipients. Dityatkin succeeded in infecting lysozyme spheroplasts but not penicillin spheroplasts or those obtained by freeze-thawing in the presence of lysozyme.[17]

SPHEROPLASTS INTERPRET DNA AND REPAIR BREAKS IN NUCLEIC ACIDS

Can a partly denuded bacterium receiving phage DNA interpret the message of the penetrating nucleic acid? Results indicate that the phage genetic information is translated efficiently and that analyzing the products can be done with more ease in the absence of cell wall components.

The capability of a host protoplast is confirmed when damaged DNA is the infecting molecule. Breaks in DNA strands of coliphage are readily repaired by the spheroplast, just as by the intact cell.[30] As many as three heat-induced strand breaks per phage genome are repaired by the spheroplast.[29]

DNA REPLICATION IN SPHEROPLAST

The spheroplast system has contributed information on the configuration of the DNA replication within the host. Young and Sinsheimer concluded that multiplication of Δ phage DNA occurs only as a twisted circle up to the time of phage maturation.[62] When phage is maturing, linear phage DNA, containing cohesive ends, is finally synthesized and encapsulated with protein. The circular form of DNA from Δ phage can infect spheroplasts but not complete bacterial cells. Harvesting circular DNA can be done when the host has not yet complicated the picture with production of protein for phage heads and tails. Thus, free DNA molecules can be obtained, and their infectivity for spheroplasts permits them to function as carriers of genetic information.

PHAGE RNA REACTION WITH A PSYCHOTROPIC DRUG

Chlorpromazine (CPZ), a psychotropic drug, has an undesirable side effect of causing photosensitization of skin, injuring cellular membranes. Connected with this is the photobinding of CPZ to RNA and DNA. The damage of phage RNA by CPZ under UV illumination has been employed to learn what happens to the RNA.[37] The phage loses the ability to infect *E. coli* spheroplasts.

FORMATION OF CWD CELLS BY PHAGE

PHAGE PENETRATES CLASSIC ORGANISM

Under appropriate conditions, a classic bacterium may become wall impoverished on hosting a bacteriophage. Thus, spheroplasts may result when *E. coli* is infected with the tailless phage α3. Formation of phage particles is continuous in infected cells, which release phage through a small hole formed by the rupture of a bulge on the surface of the organism.[11]

Spheroplasts of BCG are produced by infection with a specific bacteriophage. Such mycobacterial spheroplasts produced by Gaudier et al., although round, retained their acid-fastness and were Gram-positive.[19] Spheroplasts were made from strains of the tubercle bacillus by exposure to an adapted mycobacteriophage.[39]

Whether a bacterial strain forms spheroplasts or is completely disintegrated is a function of the invading phage rather than of the bacterium, as shown in studies on *P. aeruginosa*. When one strain was infected by an RNA phage, the result was spheroplast formation, whereas a contractile phage (PB-1) fragmented the cell wall without any possibility of intermediate steps to permit survival as spheroplasts.[9,10]

CWD FORMATION AFTER CONTACT WITH INACTIVATED PHAGE

Phage particles inactivated by osmotic shock[26] can convert an organism to the spheroplast state,[13] whereas active phage lyses the bacteria. It was concluded that only active phage affects the protoplasmic membrane.

PHAGE-ASSOCIATED LYSINS

Phage-associated lysin (PAL) is the enzyme made by intracellular phage which permits its escape into the environment.

One of the first who investigated a phage-produced lysin was Sertić,[51] examining a feces-isolated *E. coli* phage. He found that the periphery of a lysis plaque was free of phage, containing only the PAL lysin. Other studied PALs include the potent bacterial wall lysin from *B. megaterium*.[40]

A lysin produced by phage of one strain may have its maximum activity vs. the wall substrate of another strain. *Streptococcus faecalis* var. *zymogenes,* parasitized by a phage of sewage origin, released a lysin which made protoplasts from *S. faecalis* var. *liquefaciens*.[5]

A PAL of Klebsiella has as its substrate the Klebsiella capsule rather than wall. Interestingly, this enzyme has a therapeutic benefit for mice infected with the same Klebsiella strain.[1]

PAL from Group C streptococci has been used to make protoplast membranes of type 12 Group A streptococci, famous for involvement with acute glomerulonephritis.[61] The authors concluded the membranes could facilitate defining the role of streptococcal membrane composition in the kidney damage.

GENETICS AND PROTOPLASTS

Protoplasts, currently tools as useful in genetic investigations as drosophila, have many obvious advantages over the fruit fly. Protoplasts are infected with nucleic acids to enlighten problems in medicine, agriculture, and basic molecular biology. One example concerns industry and a species important in amino acid production. Phage DNA can effectively be introduced into protoplasts of *Brevibacterium lactofermentum* as an initial step in analyzing the bacterium's genome.[45]

SUMMARY

Adsorption of phage, lysogeny, and production of lytic phage can all be found in any degree of cell wall deficiency. However, Gram-negative bacteria, when protoplasts, appear to lose their phage receptors as do many strains of Gram-positive bacteria.

A complete bacterial wall is sometimes necessary for efficient phage induction, as induction of phage by UV does not always occur with spheroplasts. However, temperature elevation and mitomycin C are effective in inducing phage production in the absence of complete cell wall.

RNA phages have been tools in studying the mechanism of RNA inactivation by chlorpromazine. The *in vivo* reactions of this psychotropic drug are not well understood.

PALs have a place in preparing bacterial membranes and cell wall components for assessing their relation to disease.

Protoplasts are used for myriad studies using phage as carriers of genetic information.

REFERENCES

1. **Adams, M. H. and Park, B. H.,** An enzyme produced by a phage-host cell system. II. The properties of the polysaccharide depolymerase, *Virology,* 2, 719–736, 1956.
2. **Bailone, A., and Levine, A.,** Lack of lysogenic induction in "diaminopimelic acid spheroplasts", *J. Bacteriol.,* 136(1), 441–443, 1978.
3. **Bayer, M. E.,** Adsorption of bacteriophages to adhesions between wall and membrane of *Escherichia coli, J. Virol.,* 2, 346–356, 1968.
4. **Bertolani, R., Elberg, S. S., and Ralston, D.,** Variations in properties of L Forms of *Pseudomonas aeruginosa, Infect. Immun.,* 11, 180–192, 1975.
5. **Bleiweis, A. S. and Zimmerman, L. S.,** Formation of two types of osmotically fragile bodies from *Streptococcus faecalis var. liquefaciens, Can. J. Microbiol.,* 7, 363–373, 1961.
6. **Bloss, L.,** Further studies on phage reaction with spheroplasts and L-forms of *Proteus mirabilis, Arch. Mikrobiol.,* 49, 81–85, 1964.
7. **Böhme, H. and Taubeneck, U.,** Die wirkung von bakteriophagen auf normalformen und "large bodies" von *Proteus mirabilis, Naturwissenschaften,* 45, 296, 1958.
8. **Borek, E. and Ryan, A.,** The transfer of a biologically active irradiation product from cell to cell, *Biochim. Biophys. Acta,* 41, 57–73, 1960.
9. **Bradley, D. E.,** The structure and infective process of a *Pseudomonas aeruginosa* bacteriophage containing ribonucleic acid, *J. Gen. Microbiol.,* 45, 83–96, 1966.
10. **Bradley, D. E. and Robertson, D.,** The structure and infective process of a contractile *Pseudomonas aeruginosa* bacteriophage, *J. Gen. Virol.,* 2, 247–254, 1968.
11. **Bradley, D. E., Dewar, C. A., and Robertson, D.,** Structural changes in *Escherichia coli* infected with a φ X 174 type bacteriophage, *J. Gen. Virol.,* 5, 113–121, 1969.
12. **Brenner, S. and Stent, G. S.,** Bacteriophage growth in protoplasts of *Bacillus megaterium, Biochim. Biophys. Acta,* 17, 473–475, 1955.
13. **Carey, W. F., Spilman, W., and Baron, L. S.,** Protoplast formation by mass adsorption of inactive bacteriophage, *J. Bacteriol.,* 74, 543–544, 1957.
14. **Chen, K. C.,** Further studies on the deoxyribonucleic acid helping effect during transformation, *Mol. Gen. Genet.,* 112, 323–340, 1971.
15. **D'Adda, I. and Cavallero, F.,** Rapporti tra fago BM e forme L del *B. megaterium, Ig. Mod.,* 50, 57–62, 1957.

16. **Dénes, G. and Polgár, L.,** T2 phage infection of ''protoplasts'' obtained by penicillin treatment of *Escherichia coli* B, *Nature (London),* 183, 696–697, 1959.

17. **Dityatkin, S. Y.,** Biologically active DNA from bacteriophage, *Zh. Mikrobiol. Epidemiol. Immunobiol.,* 42, 6–10, 1965.

18. **Fatma, A. M. and Abla, M. E. M.,** L Form colony production by methicillin and its relation to different staphylococcal bacteriophage types, *J. Egyptian Pub. Health Assoc.,* 51(4), 209–213, 1976.

19. **Gaudier, B., Roos, P., and Ballester, L.,** Spheroplastes de B.C.G. induits par un bacteriophage specifique, *Ann. Inst. Pasteur Lille,* 20, 75–83, 1969.

20. **Gemsa, D. and Davis, S. D.,** Failure of bacteriophage φ X 174 to multiply in serum spheroplasts, *Nature (London),* 215, 176–177, 1967.

21. **Goryachkina, N. S. and Levashev, V. S.,** Lysogenicity of bacterial variants with a modified cellular wall reproduced on the spheroplasts and L-form model, *Vestn. Akad. Med. Nauk. SSSR,* 20, 37–39, 1965.

22. **Gumpert, J., Zimmermann, I., and Taubeneck, U.,** Phage adsorption and productive lysis in stable protoplast type L-forms of *Bacillus subtilis* and *Streptomyces hygroscopicus, J. Basic Microbiol.,* 26(1), 15–25, 1986.

23. **Guthrie, G. D. and Sinsheimer, R. L.,** Infection of protoplasts of *Escherichia coli* by subviral particles of bacteriophage φ X 174, *J. Mol. Biol.,* 2, 297–305, 1960.

24. **Hamburger, M. and Carleton, J.,** Staphylococcal spheroplasts and L colonies. II. Conditions conducive to reversion of spheroplasts to vegetative staphylococci, *J. Infect. Dis.,* 116, 544–550, 1966.

25. **Hara, K., Nakata, A., and Kawamata, J.,** The infection of polymixin-treated *E. coli* by urea-processed T4 phage (TL), *Jpn. J. Bacteriol.,* 24, 588–589, 1969.

26. **Herriott, R. M.,** Nucleic-acid-free T2 virus ''ghosts'' with specific biological action, *J. Bacteriol.,* 61, 752–754, 1951.

27. **Hofschneider, P. H.,** T$_i$ and λ phage adsorption on protoplast-like bodies of *Escherichia coli, Nature (London),* 186, 568–569, 1960.

28. **Hofschneider, P. H.,** Über ein infektiöses desoxyribonucleinsäure-agens aus dem phagen φ X 174, *Z. Naturforsch, (B),* 15, 441–444, 1960.

29. **Hotz, G. and Mauser, R.,** Infectious DNA from coliphage T1. I. Some properties of the spheroplast assay system, *Mol. Gen. Genet.,* 104, 178–194, 1969.

30. **Hotz, G., Mauser, R., and Walser, R.,** Infectious DNA from coliphage T1. III. The occurrence of single-strand breaks in stored, thermally-treated, and U. V. irradiated molecules, *Int. J. Radiat. Biol.,* 19, 519–536, 1971.

31. **Imaeda, T. and San Blas, F.,** Adsorption of mycobacteriophage on cell wall components, *J. Gen. Virol.,* 5, 493–498, 1969.

32. **Imaeda, T.,** Ultrastructure of L-phase variants isolated from a culture of *Mycobacterium phlei, J. Med. Microbiol.,* 8(3), 389–402, 1975.

33. **Jacobson, E. D. and Landman, O. E.,** Interaction of protoplasts, L Forms, and bacilli of *B. subtilis* with 12 strains of bacteriophage, *J. Bacteriol.,* 124(1), 445–8, 1975.

34. **Jacobson, E. D. and Landman, O. E.,** Adsorption of bacteriophages phi 29 and 22a to protoplasts of *B. subtilis* 168, *J. Virology,* 21(3), 1223–7, 1977.

35. **Landman, O. E., Burchard, W. K., and Angelety, L. H.,** Lysogeny and bacteriophage adsorption in stable and reverting L-forms of *Salmonella paratyphi* B and *Escherichia coli, Bacteriol. Proc. 1962,* p.58, 1962.

36. **Mankiewicz, E. and van Walbeek, M.,** Mycobacteriophages: their role in tuberculosis and sarcoidosis, *Arch. Environ. Health,* 5, 122–128, 1962.

37. **Matsuo, I., Ohkido, M., Fujita, H., Kawano, K., and Suzuki, K.,** In vitro study of chlorpromazine photosensitization: photoinactivation of bacteriophage RNA in the presence of chlorpromazine, *J. Dermatol.,* 10(2), 125–128, 1983.

38. **Meyer, F., Mackal, R. P., Tao, M., and Evans, E. A.,** Infectious deoxyribonucleic acid from λ bacteriophage, *J. Biol. Chem.,* 236, 1141–1143, 1961.

39. **Millman, I.,** Formation of protoplasts from mycobacteria by mycobacteriophage, *Soc. Exp. Biol. Med. Proc.,* 99, 216–219, 1958.

40. **Murphy, J. S. and Righthand, V. F.,** A phage-associated enzyme of *Bacillus megaterium* which destroys the bacterial cell wall, *Virology,* 4, 563–581, 1957.

41. **Nelson, E. L. and Pickett, J. J.,** The recovery of L Forms of brucella and their relation to brucella phage, *J. Infect. Dis.,* 89, 226–232, 1951.

42. **Ray, R. K. and Burma, D. P.,** Multiplication of bacteriophage P22 in penicillin-induced spheroplasts of *Salmonella typhimurium, J. Virol.,* 5, 45–50, 1970.

43. **Rymar', S. E. and Kordjum, V. A.,** Penetration of phage MS 2 RNA into spheroplasts of *Escherichia coli, Mikrobiol. Zh.,* 38(5), 552–556, 1976.

44. **Salton, M. R. J. and McQuillen, K.,** Bacterial protoplasts. II. Bacteriophage multiplication in protoplasts of sensitive and lysogenic strains of *Bacillus megaterium, Biochim. Biophys. Acta,* 17, 465–472, 1955.

45. **Sánchez, F., Peñalva, M. S., Patiño, C., and Rubio, V.,** An efficient method for the introduction of viral DNA into *Brevibacterium lactofermentum* protoplasts, *J. Gen. Microbiol.,* 132, 1767–1770, 1986.

46. **Schmid, E. N.,** Bacteriophages in L Form of *Staphylococcus aureus, J. Bacteriol.,* 164, 397–400, 1985.

47. **Schmid, E. N.,** Temperate bacteriophages and polyheads produced in the L Form of *S. aureus* after mitomycin induction, *Ann. Inst. Pasteur Microbiol.,* 137B(3), 283–289, 1986.

48. **Schuhmann, E. and Taubeneck, U.,** Stable L-formen verschiedener *Escherichia coli*-Stämme, *Z. Allg. Mikrobiol.,* 9, 297–313, 1969.

49. **Schuhmann, E. and Taubeneck, U.,** L-Formen von *Escherichia coli* K 12 (λ). I. Induktion und Vermehrungsweise, *Z. Allg. Mikrobiol.,* 11, 205–219, 1971.

50. **Sekiguchi, M., Taketo, A., and Takagi, Y.,** An infective deoxyribonucleic acid from bacteriophage φ X 174, *Biochim. Biophys. Acta,* 45, 199–200, 1960.

51. **Sertić, V.,** Untersuchungen über einen lysinzonen bildenden bakteriophagen. I. Der aufbau der bakteriophagenkolonien, *Zentralbl. Bakteriol. Parasitenkd. Infektionskr. Hyg. Abt. 1 Orig.,* 110, 125–139, 1929.

52. **Shaw, D. R. D. and Chatterjee, A. N.,** O-Acetyl groups as a component of the bacteriophage receptor on *Staphylococcus aureus* cell walls, *J. Bacteriol.,* 108, 584–585, 1971.

53. **Stempen, H.,** Demonstration of a cell wall in the large bodies of *Proteus vulgaris, J. Bacteriol.,* 70, 177–181, 1955.

54. **Takemasa, N., Hirachi, Y., Kato, Y., Kurono, M., Toda, Y., Kotani, S., et al.,** Demonstration of lysogeny in stable L Forms of *Staphylococcus aureus, Biken Journal,* 27(4), 177–81, 1984.

55. **Taubeneck, U., Böhme, H., and Schuhmann, E.,** Untersuchungen über die L-phase von *Proteus mirabilis* mit hilfe von bakteriophagen, *Biol. Zbl.,* 77, 663–673, 1958.

56. **Taubeneck, U. and Böhme, H.,** Der einfluss von Bakteriophagen auf die L-phase von *Proteus mirabilis, Z. Naturforsch. (B),* 13, 471–472, 1958.

57. **Taubeneck, U.,** Die phagenresistenz der stabilen L-form von *Proteus mirabilis, Z. Naturforsch. (B),* 16, 849–850, 1961.

58. **Tsutsaeva, A. A., Mikukinskii, I. E., Vysekantsev, I. P, and Degtiarev, A. V.,** Intracellular neutralization of bacteriophage T4 by antiphagic serum, *Zh. Mikrobiol. Epidemiol. Immunobiol.,* 10, 74–5, 1984.

59. **Wahl, R., Huppert, J., and Emerique-Blum, L.,** Production de phages par des "protoplastes" bacteriens infectes par des preparations d'acide desoxyribonucleique, *C. R. Acad. Sci. (Paris),* 250, 4227–4229, 1960.

60. **Weibull, C.,** The isolation of protoplasts from *Bacillus megaterium* by controlled treatment with lysozyme, *J. Bacteriol.,* 66, 688–695, 1953.

61. **Wheeler, J., Holland, J., Terry, J. M., and Blainey, J. D.,** Production of group C streptococcus phage-associated lysin and the preparation of *S. pyogenes* protoplast membranes, *J. Gen. Microbiol.,* 120(1), 27–33, 1980.

62. **Young, E. T. and Sinsheimer, R. L.,** Vegetative bacteriophage DNA. I. Infectivity in a spheroplast assay, *J. Mol. Biol.,* 30, 147–164, 1967.

63. **Zierdt, C. H. and Wertlake, P. J.,** Transitional forms of *Corynebacterium acnes* in disease, *J. Bacteriol.,* 97, 799–805, 1969.

Chapter 31

BACTERIOCINS

Bacteriocins may be thought of as armament for civil war: they kill members of their same species. Sometimes related species are also susceptible. Bacteriocins are commonly defined as nonsedimentable complex proteins approximately 60 kD in size, encoded on plasmids.[11] A single molecule can make a lethal hit on a bacterium.[12]

The relative lethality of bacteriocins for CWD forms vs. lethality for classics has been studied. The production of CWD variants by bacteriocins has been described as discussed below. The secretion of bacteriocins by CWD bacteria has also been published.

Bacteriocins are being examined anew since it was discovered that they kill certain malignant cells, while innocuous for normal tissue.[10,13] Some examples concern human cell lines. Several colicins inhibit DNA synthesis in rat glioma cells.[18] In some instances the reactions of pertinent bacteriocins with wall-deficient bacteria are being analyzed to learn the characteristics of cell membranes responsible for bacteriocin susceptibility vs. resistance.

The early expectations that bacterial infections can be treated or prevented with bacteriocins continue to stimulate investigations. An example is the promising result when a bacteriocin from *Streptococcus sobrinius* prevented dental plaque in rats.[8] Similarly, a mutacin from *S. mutans* was bacteriocidal for most indicator strains of *S. mutans* and *S. salivarius*.[6] Colicins have also been suggested for therapy in outbreaks of bacillary dysentery.[2] Staphylococcin in an ointment was applied to 81 patients with pyoderma. When the infection was mainly staphylococcal, encouraging results were obtained.[24]

Vaccines for enteropathogenic *Escherichia coli* and *Vibrio cholerae* made by treatment of the organisms with bacteriocins have been protective.[3] *E. coli* infections in mice have responded to treatment with colicine V.[21] Negative findings include failure of pyocines to limit pseudomonas infections.[23]

RELATIVE SUSCEPTIBILITY OF CWD FORMS TO BACTERIOCINS

An *E. coli* strain can have colicin resistance in both outer and cytoplasmic membranes, can have sensitivity in both these areas, or can have sensitivity in only one of the membranes. Furthermore, the response of a strain may differ for different colicins.[17] The reactions were analyzed by comparing bacillary with protoplast form. Data on 5 genera are given in Table 1.

The majority of colicins are protein and hence digested by trypsin. If bacterial cells immersed in a bacteriocin suspension are then exposed to trypsin within a 5-min interval, it is possible to reach and destroy the bacteriocin, thus canceling its effect. Most interesting, the time of trypsin rescue of wall-free L Forms exposed to colicin is decreased to zero. There is no canceling of the colicin effect, even if the L Forms are treated with trypsin immediately after contact with colicin.[16]

CELL WALL DEFICIENCY PRODUCED BY BACTERIOCINS

Certain bacteriocins, in certain concentrations, do not kill their prey, but rather cause deletions in the bacterial wall. Examples of this are shown in Figure 1 and Table 2.

TABLE 1
Bacteriocin Sensitivity of CWD Forms Compared to Parent's Sensitivity

Bacillus megaterium
  Parent and protoplasts have similar sensitivities to megacins[22]
Escherichia coli
  Protoplasts are sensitive to E2, Ia, K to which the parent is resistant[17]
Neisseria meningitidis (85 strains)
  L Forms are only rarely susceptible to a bacteriocin lethal to most strains in classic state[1]
Proteus mirabilis D52
  L Form is sensitive to the same and additional colicins as parent[16]
Streptococcus cremoris
  Protoplasts are resistant to Lactostreptocin 5, to which parent is sensitive[25]

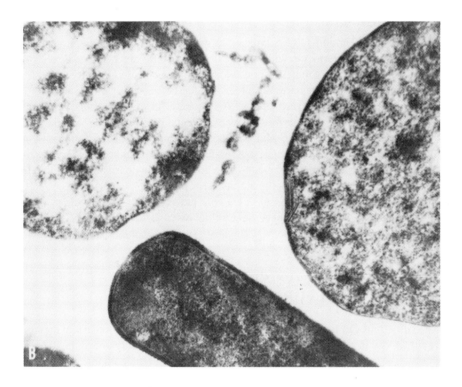

FIGURE 1. Viable spheroplasts made by exposure of *Clostridium perfringens* to a bacteriocin. (Magnification × 42,000.) (From Mahoney, D. E., Butler, M. E., and Lewis, R. G., *Can. J. Microbiol.*, 17, 1435–1442, 1971. With permission.)

TABLE 2
Bacteriocins Known to Cause Cell Wall Deficiency

Clostridium perfringens
  One strain is altered to spheroplast form by 4 bacteriocins (Figure 1)[9]
Yersinia pestis
  Produces a bacteriocin which makes spheroplasts from certain other strains[15]
  Makes a pesticin causing spheroplast formation from *Y. pestis*, *Y. pseudotuberculosis*, and *Y. enterocolitica*[5]

TABLE 3
Bacteriocin Production by CWD Forms

CWD Organism

Streptococcus FF22
    Makes Streptococcin A-FF22, a well-characterized molecule[7]
Yersinia pseudotuberculosis
    Spheroplasts make a pseudotuberculocin not produced by parent[4]

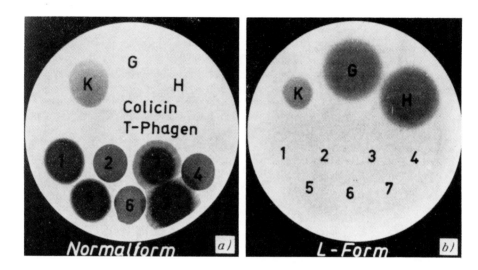

FIGURE 2. (a) Normal *E. coli* cells lysed by all tested phage strains and by one of the colicins. (b) In the L Form, phage susceptibility is lost, but receptors for colicins, inaccessible in the parent strains, are now exposed. (From Schuhmann, E. and Taubeneck, U., *Z. Allg. Mikrobiol.*, 9, 297–313, 1969. With permission.)

BACTERIOCIN SYNTHESIS BY CWD FORMS

A complete cell wall is not needed for bacteriocin synthesis as shown by their production by CWD stages (Table 3).

ARE PHAGE AND COLICIN RECEPTORS IDENTICAL?

The question of whether phages and colicins share the same receptors is graphically answered in Figure 2. Phage receptors of seven types are lost from an *E. coli* strain in the spheroplast state, but new colicin receptors are exposed.

In this chapter the term "colicin" has been limited to protein molecules rather than including the phage tails sometimes termed colicins. It is expected that common wall-component receptors may be found for phages and the phage-tail colicins.[19,20]

SUMMARY

Bacteriocins are moderately sized molecules, commonly protein, produced by many strains of bacteria. In contrast to antibiotics, they attack specific receptors on the cell membrane of a target bacterium and commonly cause immediate death by degradation of

RNA and/or DNA, rather than interfere with some aspect of protein or cell wall synthesis. "Hits" by bacteriocins on the target may be lethal or in some instances result in viable spheroplasts.

When a bacterium becomes an L Form it may lose certain phage receptors and reveal new bacteriocin receptors.

The prophylactic and therapeutic use of bacteriocins is considered promising by investigators who have performed preliminary animal studies. An unexpected finding is that malignant cells are damaged by bacteriocins, whereas normal cells are immune.

REFERENCES

1. **Barbuti, S., Montagna, M. T., and Quarto, M.,** Researches on the production of inhibitory substances by strains of *"N. meningitidis"*, *Annali Sclavo*, 21(4), 435–447, 1979.
2. **Burés, J., Horák, V., and Duben, J.,** Importance of colicinogeny for the course of acute bacillary dysentery., *Zentralbl. Bakteriol. Parasitenkd. Infektionskr. Hyg. Abt. 1 Orig., Reihe A, Med. Mikrobiol. Parasitol.*, 245(4), 469–475, 1979.
3. **Evans, D. J., Jr., Evans, D. G., Opekun, A. R., and Graham, D. Y.,** Immunoprotective oral whole cell vaccine for enterotoxigenic *E. coli* diarrhea prepared by in situ destruction of chromosomal and plasmid with colicin E2, *FEMS Microbiol. Immunol.*, 1(1), 9–18, 1988.
4. **Gramotina, L. I.,** Spheroplast bacteriocin from strains of the causative agent of pseudotuberculosis and its identification by the antibacterial activity spectrum and the morphology of the inhibition zones, *Antibiotiki*, 22(8), 704–708, 1977.
5. **Hall, P. J. and Brubaker, R. R.,** Pesticin-dependent generation of osmotically stable spheroplast-like structures, *J. Bacteriol.*, 136(2), 786–789, 1978.
6. **Hamada, S., Imanishi, H., and Ooshima, T.,** Isolation and mode of action of a cell-free bacteriocin (mutacin) from serotype g *Streptococcus mutans* MT3791, *Zentralbl. Bakteriol. Parasitenkd. Infektionskr. Hyg. A*, 261, 287–298, 1986.
7. **Hryniewicz, W. and Tagg, J. R.,** Bacteriocin production by group A streptococcal L-forms, *Antimicrob. Agents Chemother.*, 10(6), 912–914, 1976.
8. **Kitamura, K., Masuda, N., Kato, K., Sobue, S., and Hamada, S.,** Effect of a bacteriocin-producing strain of *Streptococcus sobrinus* on infection and establishment of *Streptococcus mutans* on tooth surfaces in rats, *Oral Microbiol. Immunol.*, 4(2), 65–70, 1989.
9. **Mahony, D. E. and Li, A.,** Comparative study of ten bacteriocins of *Clostridium perfringens*, *Antimicrob. Agents Chemother.*, 14(6), 886–892, 1978.
10. **Musclow, C. E, Farkas-Himsley, H., Weitzman, S. S., and Herridge, M.,** Acute lymphoblastic leukemia of childhood monitored by bacteriocin and flowcytometry, *Eur. J. Cancer Clin. Oncol.*, 23(4), 411–418, 1987.
11. **Neville, D. M., Jr. and Hudson, T. H.,** Transmembrane transport of diphtheria toxin, related toxins, and colicins, *Annu. Rev. Biochem.*, 55, 195–224, 1986.
12. **Nomura, M. and Witten, C.,** Interaction of colicins with bacterial cells. III. Colicin tolerant mutations in *E. coli*, *J. Bacteriol.*, 94, 1093, 1967.
13. **Saito, H. and Watanabe, T.,** Effect of a bacteriocin produced by Mycobacterium smegmatis on growth of cultured tumor and normal cells, *Cancer Res.*, 39(12), 5114–5117, 1979.
14. **Schuhmann, E. and Taubeneck, U.,** Stabile L-formen verschiedener *Escherichia coli*-Stämme, *Z. Allg. Mikrobiol.*, 9, 297–313, 1969.
15. **Sikkema, D. J. and Brubaker, R. R.,** Resistance to pesticin, storage of iron, and invasion of HeLa cells by Yersiniae, *Infect. Immun.*, 55(3), 572–578, 1987.
16. **Smarda, J. and Taubeneck, U.,** Situation of colicin receptors in surface layers of bacterial cells, *J. Gen. Microbiol.*, 52, 161–172, 1968.
17. **Smarda, J. and Schuhmann, E.,** Studies of colicin action on wall-less stable L-forms of *Escherichia coli*. III. A colicin-tolerant mutant in a wall-less stable L-form, *J. Basic Microbiol.*, 25(7), 457–460, 1985.
18. **Smarda, J, Smarda, J., Jr., and Vrbicka, Z.,** Colicins E7 and E8 degrade DNA in sensitive bacteria, *Folia Microbiol.*, 35, 348–352, 1990.

19. **Smit, J. A., Hugo, N., and de Klerk, H. C.,** A receptor for a *Proteus vulgaris* bacteriocin, *J. Gen. Virol.,* 5, 33–37, 1969.
20. **Smit, J. A., Stocken, L. A., and de Klerk, H. C.,** A receptor for *Proteus morganii* bacteriocin 336, *J. Gen. Microbiol.,* 65, 249–251, 1971.
21. **Smith, H. W. and Huggins, M. B.,** Treatment of experimental *Escherichia coli* infection in mice with colicine V, *J. Med. Microbiol.,* 10(4), 479–482, 1977.
22. **Von Tersch, M. A. and Carlton, B. C.,** Megacinogenic plasmids of *Bacillus megaterium, J. Bacteriol.,* 155(2), 872–877, 1983.
23. **Williams, R. J.,** Treatment of *Pseudomonas aeruginosa* infections with pyocines, *J. Med. Microbiol.,* 9(2), 153–161, 1976.
24. **Wolowiec, Z., Kindykiewicz, L., and Olszewski, W.,** Therapeutic value of Staphylococcin A in pyoderma, *Przeg. Derm.,* 65(4), 433–438, 1978.
25. **Zajdel, J. K., Ceglowski, P., and Dobrazanski, W. T.,** Mechanism of action of lactostrepcin 5, a bacteriocin produced by *Streptococcus cremoris* 202, *Appl. Environ. Microbiol.,* 49(4), 969–974, 1985.

Chapter 32

ENTOMOLOGY

STERILITY FROM CWD STREPTOCOCCI

A new role for wall-deficient bacteria was discovered by Ehrman et al. (Figure 1).[7] CWD streptococci, residing peacefully in Drosophila, produce sterility in male offspring when subspecies of Drosophila are cross-mated. The streptococcal variants inhabit the testes, causing no disease in their original host unless large numbers of the CWD bacteria are injected into the insect, in which case likewise sterility or loss of viability in offspring may result as though from a cross-match. Originally the symbionts were thought to be mycoplasma, shown untrue by electron microscopy (EM) and culture studies.

The Mediterranean meal moth, *Ephestia kuehniella,* also harbors a wall deficient microbe. The Ephestia symbiont has been shown to be harmless to Drosophila, but larvae of the moth are killed by the CWD streptococci from Drosophila.[8]

A LEGACY IN INSECTS OF
WALL-DEFICIENT MICROBES

HOUSING THE LEGACY IN COLORFUL MYCETOMES

In some insects and lower animals, microbial flora is not incidental, but carefully provided for by special organs which house the guests. Even if an insect is freed of its symbiotic bacteria by artificial means, its descendants may continue to carry such organs designed for bacterial propagation for at least 25 successive generations.[12]

MYCETOMES AND THEIR CONTENT

The brightly colored organs which hold microbial symbionts are termed mycetomes, literally meaning "fungus bodies".[12] Their color is due to pigmented grains of unknown function which pack the cells lining the host mycetomes.

Because of the great pleomorphism of the microbes which inhabit the mycetomes, their content was originally interpreted in a variety of erroneous ways: as supplementary kidneys, accessory ovaries, or stored food.[4] It is generally recognized that extreme pleomorphism, characterizing cell wall deficiency, is exhibited by nearly all symbiotic bacteria living in plants and lower animals.[10] Perhaps in such protected environments less microbial wall is needed.

SYMBIONTS ESSENTIAL FOR HOST LIFE

Frequently, a symbiont is absolutely necessary for the continued survival of the host. It is well established that microbial symbionts are essential for the basic metabolism to the kissing bug (Triatominen)[9] and to the human body louse, *Pediculus humanus.*[1-3]

VARIATION AND CYCLING IN THE SYMBIONTS

The best known example of wall deficient cycling in these symbionts concerns a bacterium inhabiting the locust (Cicada), as shown in Figure 2.[16] Three characteristics stand out: (1) the bacterium develops amazingly large syncytial masses; (2) the cycle of bacterial development varies, depending on the sex of its host; (3) the development of the bacterium is regulated by maturation of the host, as the host passes through embryo, first, and second instar larval stages. This tendency of the host growth stage to regulate bacterial morphology is frequently encountered, e.g., in the cattle tick, *Boophilus annulatus,* which serves as the

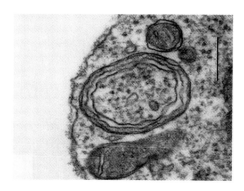

FIGURE 1. CWD streptococcal form in female gonadal tissue of *Drosophila paulistorum.* Two of the outer membranes are host-contributed. The symbiont is flanked by two mitochondria. (From Ehrman, L., Factor, J. R., Somerson, N., and Manzo, P., *Am. Nat.,* 134, 890–896, 1989. With permission.)

vector of Texas fever.[6] Steinhaus noted that morphology of the symbiont was often determined not only by age and developmental stage of the insect, but also by its state of health.[18]

Microscopic studies show infinite examples of similarity between morphology of insect symbionts and the L-cycle. A commonly occurring form in the cycles of insect symbionts mimics the large body of the "L-cycle" of vertebrate-parasitizing and free-living bacteria. For example, L-bodies form in the symbiont of the cutworm, *Euxoa segetum.*[17]

In the carpenter ant, *Camponotus,*[4] symbionts sequentially appear as small rods, thick sausages, tennis rackets, and amorphic spheres requiring a DNA stain to confirm their cellular nature. Often, as a symbiotic bacterium prepares to enter the new generation of the insect, it undergoes changes correlated with its invasive ability. In the leafhopper, *Cixius nervosius,* the gradual adaptation of the symbiont to the infective stage is related to the appearance of conspicuous vacuoles in the host cells.[4] Ingredients within the vacuoles appear to be the active substances responsible for change in infectivity and morphology of the symbionts.

Symbionts of the horse fly, *Hippobosca capensis,* may consist of round, oval, pear, and disk-shaped microbial fragments held together by thin strands. These strange forms circulate not only intracellularly, but also into the fluid secretion of the milk glands.[4] The proteinaceous "milk" nourishes the larvae of the fly in their early developmental stage in the fly's genital organs.

SYMBIONTS IN THE BLOOD

Paillot found an interesting bacterium in larvae blood of the sandfly, *Neurotoma nemoralis.* The isolates grew as long rods in broth but when subcultured to solid medium produced huge L-bodies with a compact nucleus-like mass.[17]

FINE STRUCTURE STUDIES

Only a few laboratories have explored the fine structure of the insect symbionts. EM hints that insect symbionts may have some unique structures. After examining intracellular symbionts of the cockroach, *Blatta orientalis,* Meyer and Frank[15] concluded that the living intracellular bacteria differed structurally from any other bacterium described to date, being unique in the arrangement of nuclear material and possessing fine tubes which, in some cases, led to the exterior by penetrating the cell wall.

CULTURE OF THE BACTERIAL SYMBIONTS

For many years attempts to cultivate the symbionts of roaches met with no success,

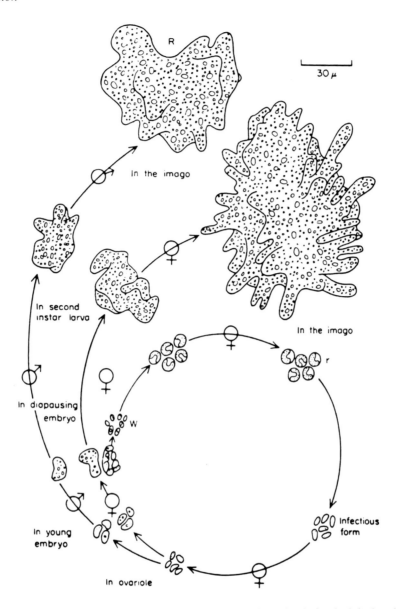

FIGURE 2. Sex of locust host influences the cycling of its bacterial parasite. In female, infectious forms may develop. In locust of both sexes, giant forms appear. (From Müller, H. J., *Forsch. Fortschr.*, 18, 183–187, 1942. With permission.)

until Keller[11] enriched medium with roach fat. The symbionts, which Keller called *Rhizobium uricophilum,* grew *in vivo* in a fat body.

SYMBIONTS MAY MAKE N_2 AVAILABLE FOR RECYCLING BY THE HOST

Keller concluded that *R. uricophilum* in cockroaches, by utilizing uric acid, contributes nitrogen to nourish the roach.[11] Other investigators have made similar deductions. Thus, the symbiotic bacteria in insects may perform the same function for insects as soil bacteria accomplish in the nitrogen cycle for plants. In both circumstances, polymorphism of the symbionts is common. Aphids are heavily populated with such bacteria, which go through

FIGURE 3. The blood-sucking eel, *Hirudo medicinalis,* profits from its bloodmeal only after Pseudomonas symbionts housed in the outpouchings of the intestine digest the erythrocytes. (After Kükenthal, W., *Handbuch der Zoologie,* Vol. 2, Walter de Gruyter, Berlin, 1934. With permission.)

the morphological cycle so well known for *Azotobacter,* beginning with a vesicle which develops a multitude of small rods, which, in turn, expand into giant forms.

EFFECTS OF ANTIBIOTICS ON SYMBIONTS

The pubic louse, *Phthirus pubis,* has a symbiont which has been named *Pseudococcus citri.*[13] When penicillin is given to lice, the Pseudococcus first forms large bodies and wide mycelia and finally is almost eliminated by lysis. The insects expire 10 d after the death of their symbionts. Paper chromatography of the lice as they occur normally and symbiont-free suggest that the symbionts utilize certain free amino acids which, not being incorporated into proteins, are toxic to their hosts.[13]

Steinhaus et al. noted that symbionts of the stinkbug, *Murgantic histrionica,* which are normally convoluted and pleomorphic, become even more bizarre, developing many vacuolated forms in insects receiving streptomycin.[19] The streptomycin was administered ingeniously by injection into the vascular system of the host plant, kale.

THE MICROBIAL CHEF
FOR THE LEECH

Figure 3 delineates the structures of the common leech, *Hirudo medicinalis,* well-known as the blood-sucking eel. It is not generally known that the "therapeutic" characteristics of the leech are entirely bacterial dependent. Its symbiont-carrying pouches, the mycetomes, bilaterally line the digestive tract, filled with bacteria.[14] The symbiont eliminates any other bacterial flora by producing a wide-spectrum antibiotic. Remarkably, the bacteria process each blood meal for the leech, as they possess the proteolytic enzyme necessary for erythrocyte digestion which the host lacks.[5] When the intestine of Hirudo is made germ-free with chloramphenicol, digestion of blood stops.

SUMMARY

Two new facts emerged from Drosophila studies. Sterility may be caused by CWD streptococci growing in testicular tissue. Genetic susceptibility to CWD pathogenesis has been shown.

The microbial flora of insects is more than just "normal" in many instances: symbiont bacteria are often essential for host life. Antibiotic-produced death of *Pseudococcus citri,* the symbiont of the pubic louse, is followed shortly after by host death, possibly because the bacteria utilize excess amino acids lethal to the louse. In general, symbionts seem to serve their insect host much as soil bacteria do plants in varied aspects of the nitrogen cycle. Symbiont polymorphism during the process is common.

The insect host houses its symbionts in spacious mycetomes, organs specialized for this purpose. Bacterial development within mycetomes, which is regulated by host insect sex and age, exhibits cycling, with syncytial, large body, and polymorphous stages. Insects at various stages of their growth produce substances which direct bacterial morphology and infectivity.

The medicinal leech, *Hirudo medicinalis,* depends on symbiont bacteria housed in mycetomes along its digestive tract to process its blood meals. The symbiont pseudomonad, which eliminates all other bacterial flora of the leech with a wide spectrum antibiotic, possesses the enzyme necessary for erythrocyte digestion lacking in the host leech. The bacterial symbionts of insects and of the leech are found to be wall deficient.

REFERENCES

1. **Aschner, M.,** Experimentelle Untersuchungen über die Symbiose der Kleiderlaus, *Naturwissenschaften,* 20, 501–505, 1932.
2. **Aschner, M. and Ries, E.,** Das Verhalten der Kleiderlaus bei Ausschaltung ihrer Symbionten, *Z. Morphol. Oekol. Tiere,* 26, 529–590, 1933.
3. **Aschner, M.,** Studies on the symbiosis of the body louse. 1. Elimination of the symbionts by centrifugalization of the eggs, *Parasitology,* 26, 21–314, 1934.
4. **Buchner, P.,** *Endosymbiosis of Animals with Plant Microorganisms,* Engl. transl., Interscience, New York, 1965.
5. **Büsing, K. H., Doll, W., and Freytag, K.,** Die Bakterienflora der medizinischen Blutegel, *Arch. Mikrobiol.,* 19, 52–86, 1953.
6. **Cowdry, E. V.,** Group of microorganisms transmitted hereditarily in ticks and apparently unassociated with disease, *J. Exp. Med.,* 41, 817–830, 1925.
7. **Ehrman, L., Factor, J. R., Somerson, N., and Manzo, P.,** The *Drosophila paulistorum* endosymbiont in an alternative species, *Am. Nat.,* 134, 890–896, 1989.
8. **Gottlieb, F. J., Simmons, G. M., Ehrman, L., Inocencio, B., Kocka, J., and Somerson, N.,** Characteristics of the *Drosophila paulistorum* male sterility agent in a secondary host, *Ephestia kuehniella, Appl. Environ. Microbiol.,* 42, 838, 1981.
9. **Gumpert, J. and Schwartz, W.,** Untersuchungen über die Symbiose von Tieren mit Pilzen und Bakterien. X. Die Symbiose der Triatominen I. Aufzucht symbiontenhaltiger und symbiontenfreier Triatonminen und Eigenschaften der bei Triatominen vorkommenden Mikroorganismen, *Z. Allg. Mikrobiol.,* 2, 209–225, 1962.
10. **Hornbostel, H.,** Ueber die bakteriologischen Eigenschaften des Darmsymbionten beim medizinischen Blutegel (*Hirudo officinalis*), nebst Bemerkungen zur Symbiosefrage, *Zentralbl. Bakteriol. Parasitenkd. Infektionskr. Hyg. Abt. 1 Orig.,* 148, 36–47, 1941.
11. **Keller, Von H.,** Die Kultur der intrazellularen Symbionten von *Periplaneta orientalis, Z. Naturforsch.,* 5b, 269–273, 1950.
12. **Koch, A.,** Symbiosestudien. II. Experimentelle Untersuchungen an *Oryzaephilus surinamensis* L. (Cucujidae, Coleopt.), *Z. Morphol. Oekol. Tiere,* 32, 137–180, 1936.
13. **Kohler, M. and Schwartz, W.,** Untersuchungen uber die Symbiose von Tieren mit Pilzen und Bakterien. IX. Über die Beziehungen zwischen Symbionten und Wirtsorganismus bei *Pseudococcus citri, Ps. maritimus* und *Orthezia insignis, Z. Allg. Mikrobiol.,* 2, 190–208, 1962.

14. **Lehmensick, R. von,** Ueber einen neuen bakteriellen Symbionten im Darm von *Hirudo officinalis* L., *Zentralbl. Bakteriol. Parasitenkd. Infektionskr. Hyg. Abt. 1 Orig.,* 147, 317–321, 1941.

15. **Meyer, G. F. and Frank, W.,** Elektronenmikroskopische Studien zur intracellularen Symbiose verschiedener Insekten, *Z. Zellforsch.,* 47, 29–42, 1957.

16. **Müller, H. J.,** Formende Einflusse des tierischen Wirtskorpers auf symbiontische Bakterien, *Forsch. Fortschr.,* 18, 183–187, 1942.

17. **Paillot, A.,** Sur deux bacteries parasites des larves de *Neurotoma nemoralis, C. R. Acad. Sci.,* 178, 246–249, 1924.

18. **Steinhaus, E. A.,** Observations on the symbiotes of certain *Coccidae, Hilgardia,* 24, 185–206, 1955.

19. **Steinhaus, E. A., Batey, M. M., and Boerke, C. L.,** Bacterial symbiotes from the caeca of certain Heteroptera, *Hilgardia,* 24, 495–518, 1956.

Chapter 33

ECOLOGY

INTRODUCTION

In the earliest notations on bacterial variants, there was great interest in morphological response to environment. Temperature shifts and moisture restriction changed a bacterium into a still propagating but unrecognizable form. It was demonstrated that virulence and configuration were synchronized; sometimes pleomorphism signaled loss of virulence, sometimes an increase. Therefore, the experimentalists sought an explanation of seasonal disease and of epidemics in microbial response to annual modifications in soil and waters. Whether microbial variations and mutations occur predominantly in animal hosts or in air, soil, or water remains to be determined. The dependable seasonal variations in enteric and respiratory infections remain enigmatic to this day. The effects of sunlight, temperature, and nutrition on the bacterial wall, and, hence, on its morphology, are studied largely in laboratory-created environments, as described in the following paragraphs.

ATYPICAL pH

Changes in pH occur *in vitro*, in soil, and *in vivo*. A most responsive species is *Hemophilus influenzae;* strains from both spinal fluid and the respiratory tract are invariably markedly pleomorphic at pH 5.5 to 6.0 in medium containing 1% heated blood. At the other pH extreme, 8.5 to 8.8, similar but less severe structural alterations develop.[35]

Escherichia coli grown at pH 2.5 becomes almost unrecognizable, consisting of fine branching forms or large amorphic masses.[32] Similar changes occur in *Mycobacterium leprae* multiplying at low pH (see Leprosy chapter.)

TEMPERATURE

Refrigeration or even mild temperature changes have long been known to cause L-type structuring in some species.[16] A *Klebsiella pneumoniae* strain propagates as oversized spheres and branching strands at 20°C. However, *anaerobically* morphology is normal at all incubation temperatures in an O_2-free atmosphere.[4] Brzin noted a reverse situation in a fresh isolate of *Bacterium anitratum* which had classic morphology only at low temperature.[7]

Vibrio piscium, a spirillum pathogenic for carp, becomes nonmotile at 37°C and expands into pear-shaped organisms.[14] Most unexpectedly, many of the large forms become Gram-positive, resembling Staphylococci.

MURALYTIC MOLECULES

Ecological explorations may well reveal that bacteria become wall deficient from exposure to biochemical secretions of other organisms in their environment. Many microbial extracellular enzymes act on other species or other strains of the same species. An edible mushroom, *Clavatia utriformis*, can cause wall deficiency in *Alcaligenes faecalis*.[9] An extracellular lysozyme-like enzyme frequently produced by Staphylococci is, incidentally, correlated with pathogenicity and was identified by Wädstrom and Hisatsune as hexosaminidase.[42]

The bovine rumen is a sea where many microbes are turned into spheroplasts or protoplasts by enzymes secreted by the normal flora Streptomyces, Pseudomonas, Bacillus, and Arthrobacter.[2]

In a natural situation not only are L Forms made by products of other organisms, but

L Forms can produce bacteriocins with a spectrum of activity. Kalmanson and associates were among the first to investigate this, discovering bacteriocin production by *Streptococcus faecalis* L Forms.[26] Also see the Bacteriocins chapter.

CHALAROPSIS B ENZYME (AN *N*-ACETYL HEXOSAMINIDASE)

In turn, the walls of Staphylococci are solubilized by enzymes from the fungus Chalaropsis, releasing units of the wall of *Staphylococcus aureus* for examination. It has been difficult to free Staphylococci of their walls by surface lysis since they are largely insensitive to lysozyme.[24]

ENZYMES FROM PSEUDOMONAS

An enzyme from *Pseudomonas aeruginosa* and *P. fluorescens* also lyses staphylococcal walls. This enzyme appears in trypticase soy broth, but not in a synthetic medium, and makes staphylococcal spheroplasts which are good sources of highly polymerized DNA.[10] The authors, Burke and Pattee,[10] found a similar enzyme in Flavobacterium and Cytophaga, perhaps correlating with the DNA guanine plus cytosine content of these genera and indicating a relationship of certain species of the *Pseudomonadaceae*. A similar study investigated an enzyme from *P. aeruginosa* which produces L-colonies from *S. aureus*.[19]

A staphylolytic enzyme from Flavobacterium has also been described by Kato et al.[27] The pseudomonads and Flavobacteria, abundant in lakes, may well remove walls from cohabitating Gram-positive cocci.

PHOSPHOMANNANASE, AN EMZYME FROM *B. CIRCULANS*

This enzyme, termed the PR-factor since it produces protoplasts from yeasts, depolymerizes phosphomannans in the wall of *Hansenula* species and degrades the phosphate-rich walls of *Saccharomyces*.[31] The segments of yeast wall released into the medium measure were large, indicating that very few bonds need be split to release the wall.

ACTINOMYCETE ENZYMES

Wall-lysing enzymes from Actinomycetes are of two classes.[22] Micromonospora AS has enzymes practically devoid of lipase activity, yet active on fungal conidia. In contrast, the enzyme from Streptomyces RA digests lipid rapidly. Thus, the Actinomycetales in nature must attack fungal walls efficiently.

L FORMS SUPPORTING GROWTH OF A CILIATE

Both the L Form and classic stage of *Agrobacterium tumefaciens* proved excellent substrates for propagation of *Tetrahymena pyriformis*. Thus, variants play a role in establishing the flora of natural waters.[1]

ALCOHOLS

Ethyl alcohol in a concentration of approximately 7% converts *E. coli* into L-phase colonies which greatly resemble those formed by penicillin stimulus. The conversion can take place on the surface of medium when the alcohol diffuses from disks, or in liquid medium.[6]

LITHIUM CHLORIDE

An effect of LiCl on bacterial morphology was noted as early as 1900 by Gamaleia.[21] Some strains of *E. coli* require no additive except 2.5% LiCl to produce spheroplasts which continue growth as L-type colonies.[34]

Eisler found among a great many bacterial species cultured with LiCl that only the spore formers fail to produce soft-walled half moon, trypanosome-like forms or spheres with horns.[18] A reverse of the expected concerns a strain of *E. coli* which propagates as enormous nonmotile globules *unless* the medium contains LiCl.[25]

TELLURITE

Tellurite in soil must play a role in the ever-shifting stages of flora from rigid-walled units to ramifying or budding plastic-walled microbes. An example concerns *Pasteurella pseudotuberculosis,* which on tellurite medium may grow as barely visible colonies of many enlarged round and triangular cells among the normal coccobacilli.[8]

SALT CONTENT AND IDENTIFICATION OF SPECIES

Wall-deficient pleomorphism caused by high concentrations of salts gives clues to identification of pathogens. This principle is especially applied in Eastern countries when dealing with organisms of public health importance.

In a very early report, Hankin and Leumann[23] remarked that the plague bacillus was difficult to identify. They recommended the following procedure: the new isolate is first cultured on standard medium, then on agar containing approximately 3% NaCl. If the culture is the plague microbe, every bacillus will be swollen into pear-shaped or round forms with extruding filaments or bodies resembling Torulae.[23] The ideal concentration of NaCl varies with the basic medium.

It was noted in the laboratory of Kitasato that large spindle forms are of diagnostic significance in recognizing both the plague and the dysentery bacillus when cultured in medium containing 4% salt.[29]

INDUCTION OR FOSTERING BY UNIDENTIFIED
PRODUCTS OF MICROBIAL GROWTH

One of Dienes' first observations of L-phase growth concerned two strains of Proteus approaching each other on agar. Not only could they not quite merge, but enormous spheres developed at the borders.[16] This inhibition by an antigenically different strain, known as the Dienes phenomenon, is useful for typing Proteus isolates. Falkow found that L-type colonies of the vesiculated type are induced in nutrient agar pour plates if they contain a high concentration of filtrates of 10-d broth cultures of nonidentical strains. Unexpectedly, the filtrate can induce a few L-phase colonies from the same strain.[20]

Another organism which produces L Forms when exposed to its own metabolic products is a Clostridium studied by Dienes.[17] The extensive liquefaction of subcutaneous tissue in guinea pigs suggests the organism was probably *Clostridium histolyticum.*

Release from rigid monomorphism as a result of microbial antagonism was recorded 60 years ago when *H. influenzae* strains grew as ramifying fungi in the presence of streptococci (see Figure 1).[41]

The growth of *H. influenzae* L Forms is enhanced by filtrates of cultures of normal respiratory tract bacteria. The fostering strains include alpha-hemolytic streptococci, non-hemolytic strep, *Micrococcus roseus,* and coagulase-negative staphylococci. The authors comment on the significance of this fostering of L Form residence of the pathogen.[37]

Aeromonas hydrophila excretes an enzyme which changes *S. aureus* to replicating protoplasts.[12]

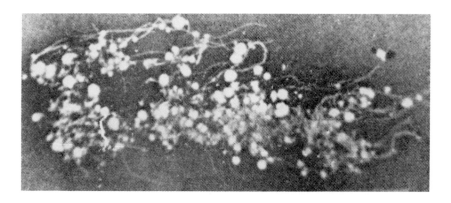

FIGURE 1. *Hemophilus influenzae* grown with β-hemolytic streptococci. (From Wade, H. W. and Manalang, C., *J. Exp. Med.*, 31, 95–103, 1920. By copyright permission of the Rockefeller University Press.)

ANTIBIOTICS PRODUCED IN SOIL

No antibiotic has ever been identified as naturally occurring in soil. However, the work of numerous investigators, described by Brian,[5] shows that of 13 fungi and 2 bacterial species added to sterile soil, only one failed to produce its antibiotic.

EXPOSURE TO UV

Sunlight quantitatively varies so much with season that the flora of water and air is bombarded irregularly with UV rays, a potent agent affecting microbial walls. In the laboratory there is a gamut of results, depending not only on microbial species, but also on the exposure procedure. Studies by Kelner,[28] Barner and Cohen,[3] and Cooper[13] indicate that some bacilli have decreased ability to form L-type colonies on penicillin agar after preexposure to UV light. In other circumstances, UV light in itself induces CWD cells.[11]

Repeated radiation can produce subtle changes not demonstrable after a single exposure. *Agrobacterium tumefaciens*, twice irradiated with a UV dose which kills 90% of the cells, yields spherical CWD forms.

When *Corynebacterium diphtheriae* is exposed to UV rays, the results are similar to those with *A. tumefaciens*. The maximum number of CWD organisms are produced by intermittent exposure to the light, alternating with periods of incubation on blood agar.[15]

BDELLOVIBRIOS

The bdellovibrios, amazing minute bacteria which parasitize bacteria of normal size,[38] are called ''leech-vibrios'' because they dart against their prey with tremendous force, which, by itself, appears sufficient to rupture the host wall. Shortly after invasion occurs, some change is evoked which causes the host wall to lose its rigidity, although only one relatively small area is violated by the entrance of the vibrio.[36] This spheroplasting of the host consistently precedes multiplication of the bdellovibrios. Bdellovibrios are numerous in soil and water and are commonly hosted by saprophytic or amphobiotic genera such as Pseudomonas or Escherichia (Figure 2).

When penicillin is added to cultures of such parasitized bacteria, the bdellovibrios also respond by losing their comma shape and rounding into spheres.[39,40] In the situations studied

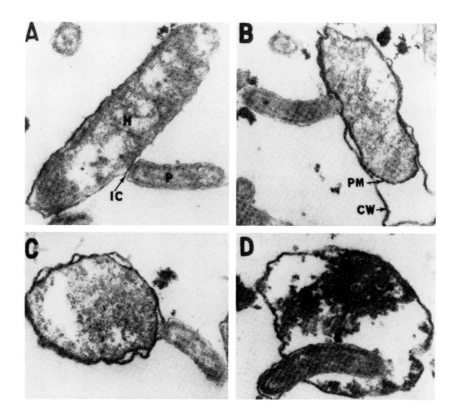

FIGURE 2. Penetration of *Pseudomonas fluorescens* by *Bdellovibrio bacteriovorus*. Host wall loses rigidity, and spheroplast results soon after initial contact of the parasite. (From Scherff, R. H., DeVay, J. E., and Caroll, T. W., *Phytopathology*, 56, 627–632, 1966. With permission.)

to date, the vibrios have lost viability after exposure to penicillin. Whether the parasite may survive in this cell wall depleted stage is not known.

ALCALIGENES L FORMS AND HEPATITIS VIRUS

L Forms of *Alcaligenes faecalis* were found in the blood of carriers of Hepatitis virus B. The variants were identified after reversion. Correlated and equally remarkable, the feces of patients with active hepatitis B infection had L Forms of *A. faecalis* predominating in feces, almost replacing the normal flora (see Figure 3).[33]

FURUNCULOSIS IN FISH

The salmon pathogen, *Aeromonas salmoncida*, survives in the aquatic environment in a nonculturable state.[30] The investigators, McIntosh and Austin,[30] postulate that survival in the L Form may explain the mysteries of culture-negative infective water and culture-negative fish carriers. Naturally occurring L Forms were found in the kidney and spleen of some diseased salmon (Figure 4). The authors concluded that the best medium formulated to date is still woefully inadequate. They found brain heart infusion superior to fish peptone, sucrose superior to NaCl or polyvinyl pyrrolidone as an osmotic stabilizer, and agar better than silica as a jelling agent.

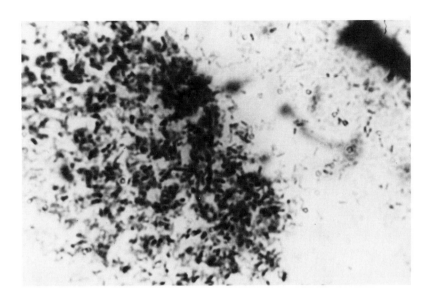

FIGURE 3.   L Forms ("empty" spheres) and reverting bacilli in *Alcaligenes faecalis* from blood of hepatitis B carrier. (From Pal, S. R., Sethi, R., Lyton, W. P., Paul, K., Usha, R., et al., *Bull. P.G.I.*, 14, 11–17, 1980. With permission.)

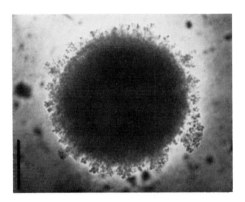

FIGURE 4.   L-Form colony of *Aeromonas salmonicida* from kidney of Atlantic salmon diseased with furunculosis. Note diffuse edge of the colony. Bar = 0.11 mm (Dienes stain). (From McIntosh, D. and Austin, B., *J. Appl. Bacteriol Symp. Suppl.*, 70, 1S–7S, 1991. With permission.)

SUMMARY

Do epidemics and seasonal infectious diseases result from the impact of environmental changes on the pathogen? The important question of whether the host or the pathogen is modified by the weather remains to be answered. Many studies have shown that wintering over can change the virulence of an organism. In some instances this has been associated with propagation in an atypical, highly pleomorphic stage.

Whether experiencing a cycle of cell wall deficiency more often increases or decreases virulence is hard to state since loss of virulence has also been noted. Since it is difficult to follow the fate of specific pathogenic Enterobacteriaceae discharged into natural waters, data are predominantly from laboratory studies. Cyclic changes in bacteria may be marked following changes in temperature, salinity, pH, nutrition, and sharing a milieu with other microbes excreting muralytic enzymes. However, as methods of cultivating CWD forms improve, seeking wall-deficient microbes in natural waters and in air promises to be a fruitful area of research.

REFERENCES

1. **Alonso, P., Gil, I., and Rodriguez, D.,** Nutritional efficiency of different bacterial species on the growth of *Tetrahymena pyriformis, Protistologica* 17(2), 199–202, 1981.
2. **Bagnyuk, V. M., Kulikova, I. Ya., and Berestecky, O. A.,** The role of lytic bacteria in the rumen of animals, *Izv. Akad. Nauk SSSR Ser. Biol.,* 4, 552–561, 1985.
3. **Barner, H. D. and Cohen, S. S.,** The relation of growth to the lethal damage induced by UV irradiation in *Escherichia coli, J. Bacteriol.,* 71, 149–157, 1956.
4. **Benulic, T., Brzin, B., and Sinkovec, J.,** Unusual forms of *Klebsiella pneumoniae* found at unfavorable incubation temperatures, *Life Sci.,* 3, 595–598, 1964.
5. **Brian, P. W.,** The ecological significance of antibiotic production, *7th Symp. Soc. Gen. Microbiol.* (London), pp. 168–188, 1957.
6. **Bringmann, G.,** Zur pleomorphie von *E. coli* unter Einfluss geringer mangen äthylalkohol, *Zentralbl. Bakteriol. Parasitenkd. Infektionskr. Hyg. Abt. 1 Orig.,* 107, 497–501, 1952.
7. **Brzin, B.,** The influence of temperature on the size and shape of *Bacterium anitratum, Acta Pathol. Microbiol. Scand.,* 57, 188–198, 1963.
8. **Brzin, B.,** Morphological change of *Pasteurella pseudotuberculosis* on tellurite media, *Zentralbl. Bakteriol. Parasitenkd. Infektionskr. Hyg. Abt. 1 Orig.,* 211, 212–215, 1969.
9. **Brzin, B. and Dezman, B.,** The effect of the fungus *Clavatia utriformis* on the morphological picture and physiological behaviour of bacteria, *Zentralbl. Bakteriol. Mikrobiol. Hyg.,* 133, 738–744, 1978.
10. **Burke, M. E. and Pattee, P. A.,** Purification and characterization of a staphylolytic enzyme from *Pseudomonas aeruginosa, J. Bacteriol.,* 93, 860–865, 1967.
11. **Challice, C. E. and Gorrill, R. H.,** Some observations on the morphological changes in *E. coli* accompanying induction by ultraviolet light, *Biochim. Biophys. Acta,* 14, 482–487, 1954.
12. **Coles, N. W. and Gross, R.,** Preparation of metabolically active *Staphylococcus aureus* protoplasts by use of the *Aeromonas hydrophila* lytic enzyme, *J. Bacteriol.,* 115, 746–751,1973.
13. **Cooper, P. D.,** Site of action of radiopenicillin, *Bacteriol. Rev.,* 20, 28–48, 1956.
14. **David, H.,** Beiträge zur Morphologie der Bakterien, *Zentralbl. Bakteriol. Parasitenkd. Infektionskr. Hyg. Abt. 1 Orig.,* 70, 1–29, 1927.
15. **Davis, J. C. and Mudd, S.,** Cytological effects of ultraviolet radiation and azaserine on *Corynebacterium diphtheriae, J. Gen. Microbiol.,* 14, 527–532, 1956.
16. **Dienes, L.,** The development of Proteus cultures in the presence of penicillin, *J. Bacteriol.,* 57, 529–546, 1949.
17. **Dienes, L.,** Isolation of L type cultures from Clostridia, *Proc. Soc. Exp. Biol. Med.,* 75, 412–415, 1950.
18. **Eisler, M.,** Ueber wirkungen von salzen auf bakterien, *Zentralbl. Bakteriol. Parasitenkd. Infektionskr. Hyg. Abt. 1 Orig.,* 51, 546–564, 1909.
19. **Falcon, M. A., Mansito, T. B., Carnicero, A., and Gutierrez-Navarro, A. M.,** L form-like colonies of *Staphylococcus aureus* induced by an extracellular lytic enzyme from *Pseudomonas aeruginosa, J. Clin. Mecrobiol.,* 27, 1650–1654, 1989.
20. **Falkow, S.,** L Forms of Proteus induced by filtrates of antagonistic strains, *J. Bacteriol.,* 73, 443–444, 1957.
21. **Gamaleia, N.,** Heteromorphismus der Bakterien unter dem Einfluss von Lithiumsalzen, in *Elemente der allgemeinem Bakertiologie,* A. Hirschwald, Berlin, 1900, 204–205.
22. **Garcia Acha, I., Aguirre, M. J. R., Lopez-Belmonte, F., and Villanueva, J. R.,** Use of lipase and strepzyme for the isolation of 'protoplasts' from fungal mycelium, *Nature (London),* 209, 95–96, 1966.
23. **Hankin, E. H. and Leumann, B. H. F.,** A method of rapidly identifying the microbe of bubonic plague, *Zentralbl. Bakteriol. Parasitenkd. Infektionskr. Hyg. Abt. 1 Orig.,* 22, 438–440, 1897.
24. **Hash, J. H.,** Purification and properties of staphylolytic enzymes from *Chalaropsis* sp., *Arch. Biochem.,* 102, 379–388, 1963.

25. **Hussong, R. V.**, A large dissociative form of *Escherichia coli* and its relation to the so called Pettenkofer bodies, *J. Bacteriol.*, 25, 537–544, 1933.

26. **Kalmanson, G. M., Hubert, E. G., and Guze, L. B.**, Effect of bacteriocin from *Streptococcus faecalis* on microbial L Forms, *J. Infect. Dis.*, 121, 311–315, 1970.

27. **Kato, K., Matsubara, T., Mori, Y., and Kotani, S.**, "Protoplast" formation in *Staphylococcus aureus* using the lytic enzyme produced by a Flavobacterium, *Biken J.*, 3, 201–203, 1960.

28. **Kelner, A.**, Growth, respiration, and nucleic acid synthesis in ultraviolet-irradiated and in photoreactivated *Escherichia coli*, *J. Bacteriol.*, 65, 252–262, 1953.

29. **Kitasato, S.**, Die Widerstandfahigkeit der Cholerabacterien gagen das Eintrocknen und gagen Hitze, *Z. Hyg.*, 5, 134–140, 1889.

30. **McIntosh, D. and Austin, B.**, The role of cell wall deficient bacteria (L Forms; spheroplasts) in fish diseases, *J. Appl. Bacteriol. Symp. Suppl.*, 70, 1S–7S, 1991.

31. **McLellan, W. J., Jr. and Lampen, J. O.**, Phosphomannanase (PR-factor), an enzyme required for the formation of yeast protoplasts, *J. Bacteriol.*, 95, 967–974, 1968.

32. **Meyer-Rohn, J. and Tudyka, K.**, Gestaltsänderungen von *Escherichia coli* unter der Einwirkung von Antibiotica, Sulfonamiden, chemischen und physikalischen Noxen, *Arzneim. Forsch.*, 7, 151–157, 1957.

33. **Pal, S. R., Sethi, R., Lyton, W. P., Paul, K., Usha, R., et al.**, *Alcaligenes faecalis* in the faecal samples of Hepatitis B antigen positive subjects. II, *Bull. P. G. I.*, 14, 11–17, 1980.

34. **Pitzurra, M. and Szybalski, W.**, Formation and multiplication of spheroplasts of *Escherichia coli* in the presence of lithium chloride, *J. Bacteriol.*, 77, 614–620, 1959.

35. **Reed, G. and Orr, J. H.**, The influence of H ion concentration upon structure. I. *H. influenzae*, *J. Bacteriol.*, 8, 103–112, 1922.

36. **Scherff, R. H., DeVay, J. E., and Caroll, T. W.**, Ultrastructure of host-parasite relationships involving reproduction of *Bdellovibrio bacteriovorus* in host bacteria, *Phytopathology*, 56, 627–632, 1966.

37. **Shishido, H.**, Growth-enhancing effects of culture filtrates of sputum isolates on the L Forms of *Haemophilus influenzae*,*Tohoku J. Exp. Med.*, 149, 271–282, 1986.

38. **Stolp, H. and Starr, M. P.**, *Bdellovibrio bacteriovorus* gen. et sp. n., a predatory, ectoparasitic, and bacteriolytic microorganism, *Antonie van Leeuwenhoek*, 29, 217–248, 1963.

39. **Varon, M. and Shilo, M.**, Interaction of *Bdellovibrio bacteriovorus* and host bacteria. I. Kinetic studies of attachment and invasion of *E. coli* B by *Bdellovibrio bacteriovorus* 109, *J. Bacteriol.*, 95, 744–753, 1968.

40. **Varon, M. and Shilo, M.**, Interaction of *Bdellovibrio bacteriovorus* and host bacteria. II. Intracellular growth and development of *Bdellovibrio bacteriovorus* in liquid cultures, *J. Bacteriol.*, 99, 136–141, 1969.

41. **Wade, H. W. and Manalang, C.**, Fungous developmental growth forms of *Bacillus influenzae*, *J. Exp. Med.*, 31, 95–103, 1920.

42. **Wädstrom, T. and Hisatsune, K.**, A bacteriolytic hexosaminidase from *Staphylococcus aureus*, *J. Gen. Microbiol.*, 57, xxiv–xxv, 1969.

Chapter 34

SOIL MICROORGANISMS

BEEF STEAK AND PEANUT BUTTER

Without microbial nitrogen fixation there would be no plants, no cattle, and eventually no life, as all nitrogen would be released into the atmosphere. This is well known, as is the major role of the Rhizobia (root-dwelling) bacteria in recapturing atmospheric nitrogen. An acre of alfalfa may fix 400 lb of nitrogen in a single season.[32] It is estimated that root nodule bacteria of leguminous plants restore 5 to 6 million tons of nitrogen to U.S. soil yearly.[34] The subsequent utilization of nitrogen by herbivorous, then carnivorous, animals makes initial assimilation by soil bacteria a vital step in basic nutrition of higher life forms.

Some of the early studies implicating CWD bacteria in nitrogen fixation were made in Pretoria by Staphorst and Strijdom.[33] Sectioning the root nodule of the peanut plant, *Arachis hypogaea,* they found spheroplasts rather than the rod-like Rhizobium expected. These spheres were identified as bacteroids of Rhizobium by using fluorescent antibody. Other studies showed that at a later stage the spheroplast walls were lost, leaving protoplasts.[20]

In root nodules where Rhizobia assimilate nitrogen, the normally small bacterium becomes enlarged and often strangely distorted, with branches or clubbed ends. The expansion of the cells, now termed "bacteroids", correlates with reduction in the rigid layer of the bacterial wall.[22] Fine structure studies by MacKenzie et al. showed that the resultant bacterial cell wall is only half as thick as the wall of the normal, nonbacteroidal Rhizobia.[25]

Interestingly, freeze-etch studies reveal that the wall of the bacteroid becomes not only thinner; it also is much more pliable.[25] Rhizobium bacteroids are known to be relatively susceptible to lysozyme, trypsin, sodium deoxycholate,[8] and chloramphenicol.[15] As commonly observed in spheroplast development, cell wall changes are concomitant with alterations in the plasma membrane and other cellular components. As the bacteroids develop, there is loss of fibrillar material from the nucleoid region, and redistribution of the interior ribosome-like particles.[16] All these modifications typify spheroplast fine structure.

In accord with its new metabolic role, the bacteroid exhibits changes in the enzymes associated with its membrane. In *Rhizobium japonicum,* cytochromes A, A3, and O, present in the membrane of the rods, are absent in the bacteroids. However, the bacteroids gain certain soluble, autooxidizable, CO-reactive, and biologically reducible pigments.[1–3]

Biochemical analyses confirm the impression that the components missing in the bacteroid wall often represent depletions in the rigid peptidoglycan layer. As shown in Table 1, both the total carbohydrate and the amino sugars may be diminished in the bacteroid. Muramic acid, although not quantitated, may be present in considerably less amounts in the bacteroid wall, and diaminopimelic acid shows a reduction of almost 50%. There may be decreases in the quantity of most amino acids.

The above and additional data suggest that Rhizobium bacteroids usually are cells with severely reduced mucopeptide content.[35] Brussel and co-workers measured less lipopolysaccharide in the bacteroids of *R. leguminosarum* than in their bacterial stage.[9]

Some rhizobia infecting root nodules in peanuts may be atypical among the root nodule symbionts in having the bacteroid lose one wall and synthesize a second wall.[5]

When rhizobia fix N_2 while growing free in the soil, they resemble symbionts in forming the swollen pleomorphic bacteroids. The asymbiotic bacteroids of the peanut *Bradyrhizobium* lose several polypeptides.[30] The lima bean *Rhizobium* sp. 127E15 fixed N_2 when grown asymbiotically and pleomorphic bacteroids were demonstrated (Figure 1 A and B).[31]

TABLE 1
Depletions in Bacteroid Wall

	Small rods	Bacteroids
Total carbohydrate	8.6	6.4
Amino sugars	5.3	2.6

Note: Values expressed as a percentage of cell
wall on a dry wt basis.

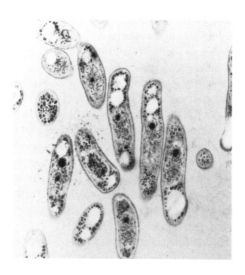
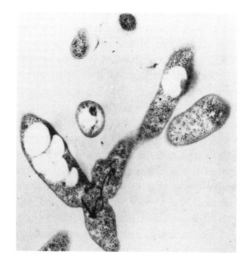

FIGURE 1. Rhizobium in the lima bean rhizosphere shows normal form (A) and larger N_2-fixing bacteroids (B). (From Ramaswamy, P. and Bal, A. K., *Curr. Microbiol.*, 15, 223–228, 1987. With permission.)

Price and co-investigators, studying the nodules of the tropical tree, Parasponia, in Australia, found that the morphology of the nitrogen-fixing Rhizobium bacteroids was similar to that of the Rhizobium bacteroids growing in legume nodules.[29]

There has been a great deal of interest in the benefits which would accrue if N_2-fixing genes were incorporated into plants not normally nitrogen fixing. A step in this direction was taken in Hungary by Nghia et al.[27] who were able to insert Rhizobium cells into algae by mixing them with algal protoplasts in the presence of polyethylene glycol. However, although the Rhizobium appeared to take permanent residence in the algae, no nitrogen-fixation was demonstrated.

L FORMS AND THE TOBACCO INDUSTRY

The association of a photosynthetic bacterium with tobacco cells has been investigated with electron microscopic (EM) studies. The photosynthetic *Anabaena variabilis* apparently was transformed to the L-stage when growing within tobacco protoplasts.

PATHOLOGY OF PLANTS CAUSED BY CWD BACTERIA

TUMORS IN BEAN AND CARROT PLANTS

Tumors in the stems of bean plants resulted where stable L Forms of *Agrobacterium tumefaciens* were inoculated. Similar disease can be produced in carrot tissue. The authors highlight their finding that sterile nucleic acids from the stable L Forms of Agrobacterium can alone induce the tumors.[6]

ASTER YELLOWS DISEASE

In periwinkle plants suffering from aster yellows disease (Figure 2), a CWD bacterium, isolated in hypertonic medium, reverted to a Gram-negative rod with blue fluorescing colonies. The bacterium stained with Dienes' stain and grew on solid medium containing thallous acetate. This pathogen was previously classified as a mycoplasma-like organism (MLO), since its microscopic forms had been seen but not cultured.[28]

THE MYRIAD OF SOIL BACTERIA COUNTED BY PLATING AND ELECTRON MICROSCOPY

POUR PLATES COMPARED WITH DIRECT MICROSCOPY

The bacteria of the earth have long been thought to utilize the wall-thinned stage to permit the packaging of fantastic numbers per gram of soil. Counts as high as 8×10^{12} have been recorded by Jagnow[19] in soil samples from meadows and pastures, suggesting that the bacteria would not have space for classic walls. The area around plant roots, the rhizosphere, is particularly heavily populated. The highest counts are obtained by plating the soil samples in agar, revealing many more organisms than found in direct smears, suggesting that many forms which grow into colonies are too small to be detected by direct microscopy.

FINDINGS BY ELECTRON MICROSCOPY

EM of soil confirmed the astronomical number of organisms recorded by pour plate counts. The soil bacteria are predominantly midgets, having diameters less than 0.3 μm.[4] A few of the dwarf cells have a diameter only 0.08 μm, well below the resolution of light microscopy. Thus, it is possible that many organisms of the flora are wall defective. Fragility is also suggested by some forms seen in the electron micrographs which have a vague, stellate outline. The actual incidence of CWD forms in soil remains to be evaluated.

THE L-CYCLE OF AZOTOBACTER AND ARTHROBACTER

Azotobacter, a common bacterium of the soil, fixes N_2 as it grows in a free-living state. To date, no relationship has been shown between its nitrogen-fixing activity and morphology. However, this genus has long been known to use the L-cycle as a common propagation route. Jones followed the spontaneous production of budding forms and syncytia which later yielded propagating granules.[21] Two species of Arthrobacter, *A. pascens* and *A. terregus,* were also found by Chaplin to exhibit cycling phenomena when cultured *in vitro.*[10]

GONIDIA CONTRASTED WITH THE CWD CYCLE

Many bacteria of soil and natural waters have a gonidia-forming cycle in which they develop atypical walls and propagate by modes other than binary fission. By contrast, the gonidia-forming bacteria make the L-cycle appear a very natural phenomenon. Löhnis' time-honored definition of *gonidia* still applies. "Gonidia are organs of asexual reproduction formed by the contraction of the plasmatic cell content, which leave the parent cell either by breaking the cell wall, or which become liberated when the cell dissolves."[24] Improvements in microscopic resolution show that the gonidia are complex microorganisms (Figure 3).[7] The gonidia are actively motile, flagellated little creatures which live a busy independent

FIGURE 2. Periwinkle plant was injected with a CWD bacterium from aster yellows disease. (From Ploaie, P. G., *Rev. Roum. Biol. Ser. Biol. Veg.*, 28(2), 109–114, 1983. With permission.)

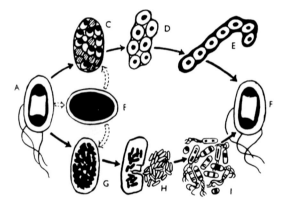

FIGURE 3. Life cycle of Azotobacter. A classical bacterial cell is capable of producing two different types of gonidia. (From Bisset, K. A., *The Cytology and Life-History of Bacteria,* 3rd ed., E & S. Livingstone, Edinburgh and London, 1970, 113. With permission of the author and Cambridge University Press.)

life until they finally enlarge to become counterparts of the parent. Gonidia are formed in cyst-like bodies often termed "mother cells". The gonidial cycle differs from the L-cycle in several major respects. The minute gonidia are uniform in size, structure, and flagellation, and after some multiplication by binary fission, they develop into a form resembling their parent. There is little uncertainty in their progression to large mother cells and no tendency to stop off at any level of development. Unrealistic as these creatures are, they are commonly observed in the Azotobacter and Rhizobia. Thus, in addition to depletion of wall as it fixes N_2 in root nodules, Rhizobium must be a cycling organism in the soil, as it is in culture. Changes characterizing the complexities of the gonidial cycle have been carefully followed by Bisset (Figure 3).[7]

CORES AND CRYSTALS IN STREPTOMYCES

Two species of the soil-dwelling genus Streptomyces are unusual in displaying cytoplasmic rods, both coarse and crystalline inclusions. The large rods occur in approximately

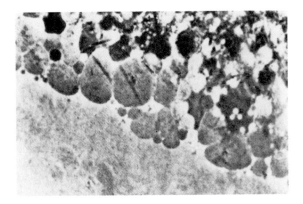

FIGURE 4. Coarse rods bisect peripheral cells in colony of *Streptomyces hygroscopicus*. Phase contrast 629:1. (From Gumpert, J., *Z. Allg. Mikrobiol.*, 23(10), 625–633, 1983. With permission.)

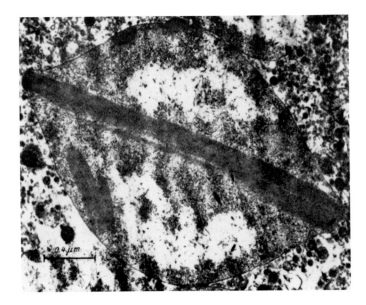

FIGURE 5. Cytoplasmic core is longer than the cell diameter. *Streptomyces hygroscopicus* grown for 10 d. (From Gumpert, J., *Z. Allg. Mikrobiol.*, 23(10), 625–633, 1983. With permission.)

2% of the peripheral cells of a colony (Figure 4) and also in broth cultures. At least one terminus of each rod is attached to the cytoplasmic membrane, but the rods are never associated with nucleoid material. These rods, termed "cores", vary in diameter from 1 to 4 μm in length and from 0.05 to 0.20 μm in width. EM studies show homogeneity. Long cores may distort the cytoplasmic membrane (Figure 5). The cores are not fixation artifacts since they are seen in living cells viewed by phase-contrast microscopy. Gumpert concluded that these bodies are labile repositories of cellular protein.[17] Similar cores have been noticed in L-Form cells of Group D streptococci[12,13,26] and *Pseudomonas aeruginosa*.[18]

Smaller rods with different structure are also found in the nucleoid regions of both *Streptomyces hygroscopicus* and *S. griseus*. Again, these are considered to be stored protein.

Paracrystalline structures are the third type of inclusions seen in the L Forms of the

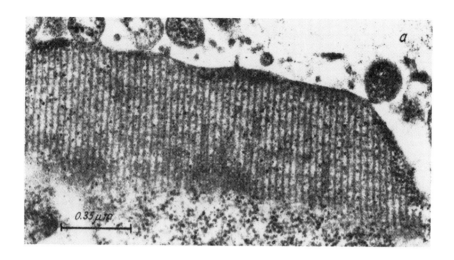

FIGURE 6.   Paracrystalline inclusion bodies in L-Form cell of *Streptomyces hygroscopicus*. (From Gumpert, J., *Z. Allg. Mikrobiol.*, 23(10), 625–633, 1983. With permission.)

Streptomyces (Figure 6). These may consist of lipoprotein. Similar crystalline formations have been noted in L Forms of Proteus,[23] Streptococcus,[14] and *Escherichia coli*.[11]

SUMMARY

The food chain of plant and animal life has a first step largely dependent on the CWD stage of Rhizobia. Usually only after the microbial wall is minimized to become thin and flexible is nitrogen fixation initiated.

CWD microbes also constitute a large part of the soil flora. The unrealistic numbers of bacteria not only in roots, but also found packed into 1 g of fertile earth is partly explained by organisms which have thin elastic walls. The L-cycle has been demonstrated repeatedly in culture of the common soil flora, and, thus, natural propagation of these microbes may include CWD stages.

Stable L Forms of Agrobacterium or their nucleic acids induce tumors where inoculated into the bean, *Phaseolus vulgaris* L, or into carrot tissue. Aster yellows disease is caused by L Forms which revert with difficulty. This pathogen was previously described as an MLO.

REFERENCES

1. **Appleby, C. A.,** A soluble haemoprotein P 456 from nitrogen-fixing Rhizobium bacteroids, *Biochim. Biophys. Acta,* 147, 399–402, 1967.
2. **Appleby, C. A.,** Electron transport systems of *Rhizobium japonicum.* I. Haemoprotein P 456, other co-reactive pigments, cytochromes and oxidases in bacteroids from N_2-fixing root nodules, *Biochim. Biophys. Acta,* 172, 71–87, 1969.
3. **Appleby, C. A.,** Electron transport systems of *Rhizobium japonicum.* II. Rhizobium haemoglobin, cytochromes and oxidases in free-living (cultured) cells, *Biochim. Biophys. Acta,* 172, 88–105, 1969.
4. **Bae, H. C., Cota-Robles, E. H., and Casida, L. E., Jr.,** Microflora of soil as viewed by transmission electron microscopy, *Appl. Microbiol.,* 23, 637–648, 1972.

5. **Bal, A. K., Sen, D., and Weaver, R. W.,** Cell wall (outer membrane) of bacteroids in nitrogen-fixing peanut nodules, *Curr. Microbiol.,* 12, 353–356, 1985.

6. **Beltra, R. and De La Rosa, C.,** Action of stable L Forms of *Agrobacterium tumefaciens* on vegetal tissue cultures, in *Spheroplasts, Protoplasts and L Forms of Bacteria,* Ed., J. Roux, *INSERM,* 65, 273–278, 1976.

7. **Bisset, K. A.,** *The Cytology and Life-History of Bacteria,* 3rd ed., E. & S. Livingstone, Edinburgh, 1970.

8. **Brussel, A. A. N. Van,** Ph.D. thesis, University of Leiden, The Netherlands, 1973.

9. **Brussel, A. A. N. Van, Planque, K., and Quispel, A., The wall of** *Rhizobium leguminosarum* in bacteroid and free-living forms, *J. Gen. Microbiol.,* 101, 51–56, 1977.

10. **Chaplin, C. E.,** Life cycles in *Arthrobacter pascens* and *Arthrobacter terregens, Can. J. Microbiol.,* 3, 103–106, 1957.

11. **Cho, K. Y. and Doy, C. H.,** Ultrastructure of spermine-treated *E. coli,* including a polar organelle concerned with envelope synthesis, *Aust. J. Biol. Sci.,* 25, 543–551, 1972.

12. **Cohen, M., McCandless, R. G., Kalmanson, G. M., and Guze, L. B.,** Core-like structures in transitional and protoplast forms of *Streptococcus faecalis,* in *Microbial Protoplasts, Spheroplasts and L Forms,* Guze, L. B., Ed., Williams & Wilkins, Baltimore, 1968, pp. 94–109.

13. **Coleman, S. E. and Bleiweis, A. S.,** Ultrastructural, physiological, and cytochemical characterization of cores in group D streptococci, *J. Bacteriol.,* 129, 445–456, 1977.

14. **Corfield, P. S. and Smith, D. G.,** Ultrastructural changes during propagation of a group D streptococcal L Form, *Arch. Mikrobiol.,* 75, 1–9, 1970.

15. **Coventry, D. R. and Dilworth, M. J.,** Inhibition of protein synthesis by D-threo-chloramphenicol in the laboratory and nodule forms of *Rhizobium lupini, J. Gen. Microbiol.,* 90, 69–75, 1975.

16. **Dart, P. J. and Mercer, F. V.,** Development of the bacteroid in the root nodule of barrel medic (*Medicago tribuloides Desr.*) and subterraneum clover (*Trifolium subterraneum L.*), *Arch. Mikrobiol.,* 46, 382–401, 1963.

17. **Gumpert, J.,** Ultrastructural characterization of core structures and paracrystalline inclusion bodies in L Form cells of streptomycetes, *Z. Allg. Mikrobiol.,* 23(10), 625–633, 1983.

18. **Hubert, E. G., Potter, C. S., Hensley, T. J., Cohen, M., Kalmanson, G. M., and Guze, L. B.,** L Forms of *Pseudomonas aeruginosa, Infect. Immun.,* 4, 60–72, 1971.

19. **Jagnow, G.,** Numbers and types of bacteria from the rhizosphere of pasture plants, possible occurrence of L-forms, *Nature (London),* 191, 1220–1221, 1961.

20. **Jansen van Rensburg, H., Hahn, J. S., and Strijdom, B. W.,** Morphological development of *Rhizobium* bacteroids in nodules of *Arachis hypogaea* L, *Phytophylactica,* 5, 119–122, 1973.

21. **Jones, D. H.,** Further studies on the growth cycle of Azotobacter, *J. Bacteriol.,* 5, 325–333, 1920.

22. **Jordan, D. C. and Grinyer, I.,** Electron microscopy of the bacteroids and root nodules of *Lupinus luteum, Can. J. Microbiol.,* 11, 721–725, 1965.

23. **Kanda, Y., Eda, T., Mori, Ch., and Kimura, S.,** Phage-like particles in an unstable L Form of *Proteus mirabilis, J. Electron Microsc.,* 26, 215–217, 1977.

24. **Löhnis, F.,** Studies upon the life cycles of the bacteria, *Mem. Natl. Acad. Sci.,* 16, 1–335, 1921.

25. **MacKenzie, C. R., Vail., W. J., and Jordan, D. C.,** Ultrastructure of free-living and nitrogen-fixing forms of *Rhizobium meliloti* as revealed by freeze-etching, *J. Bacteriol.,* 113, 387-393, 1973.

26. **McCandless, R., Cohen, M., Kalmanson, G. M., and Guze, L. B.,** Cores, microbial organelles possibly specific to group D streptococci, *J. Bacteriol.,* 96, 1400–1412, 1968.

27. **Nghia, N. H., Gyurjan, I., Stefanovits, P., Paless, G., and Turtoczky, I.,** Uptake of Azotobacters by somatic fusion of cell wall mutants of *Chlamydomonas reinhardii, Biochem. Physiol. Pflanzen,* 181, 347–357, 1986.

28. **Ploaie, P. G.,** Structures resembling cell wall deficient forms of bacteria associated with aster yellows disease and isolated in axenic culture, *Rev. Roum. Biol. Ser. Biol. Veg.,* 28(2), 109–114, 1983.

29. **Price, G. D., Mohapatra, S. S., and Gresshoff, P. M.,** Structure of nodules formed by rhizobium strain anu289 in the nonlegume parasponia and the legume siratro (Macroptilium atropurpureum), *Bot. Gaz.,* 145, 444–451, 1984.

30. **Ramaswamy, P. and Bal, A. K.,** Media-induced changes in the asymbiotic nitrogen-fixing bacteroids of *Bradyrhizobium* sp. 32H1, *Curr. Microbiol.,* 14, 181–185, 1986.

31. **Ramaswamy, P. and Bal, A. K.,** Asymbiotic nitrogen fixation by *Rhizobium* sp. 127E15 in culture and in the lima bean rhizosphere, *Curr. Microbiol.,* 15, 223–228, 1987.

32. **Stanier, R. Y., Doudoroff, M., and Adelberg, E. A.,** *The Microbial World,* 3rd ed., Prentice-Hall, Englewood Cliffs, NJ, 1970.

33. **Staphorst, J. L. and Strijdom, B. W.,** Some observations on the bacteroids in nodules of *Arachis* spp. and the isolation of *Rhizobia* from these nodules, *Phytophylactica,* 4, 87–92, 1972.

34. **Virtanen, A. I.,** The biology and chemistry of nitrogen fixation by legume bacteria, *Camb. Phil. Soc. Biol. Rev.,* 22, 239–269, 1947.

35. **MacKenzie, C. R. and Jordan, D. C.,** personal communication.

Chapter 35

MICROBES AND MALIGNANCIES

Many viruses are recognized as causing tumor growth in vertebrates, from the agent of plebian warts[8,87,95] to the spectacular big-liver disease of chickens (lymphoid leukosis),[18] which is transmitted not only by saliva and feces, but also transovarian to the egg. A virus-produced adenoma of the kidney in many leopard frogs (*Rana pipiens*) is similar to the Wilm's tumor of children.[62] Blood dyscrasias of fowl,[75] rodents, and horses are caused by well-characterized viruses.[30,44,81,97] Two viral-associated tumors of man of popular interest are carcinoma of the cervix[25,57,98] and Burkitt's lymphoma,[17] whose agent resembles that of common infectious mononucleosis.

Bacteria and fungi are less well known as causes of tumors, but the list of such agents is constantly expanded. A product of Aspergillus, aflatoxin, is but one example of carcinogenic excretions of fungi.[96] Neoplastic growth, termed "cryptococcoma", consistently follows infection by a strain of the yeast *Cryptococcus neoformans*.[78] Thus, Protista as a class are involved in malignancies.

Many points remain to be clarified. What is the interplay between carcinogenic agents and oncogenes? What are the factors activating oncogenic viruses? Do carcinogenic exogenous agents merely upset a balance within virus-infected cells? The questions are difficult to answer because virus particles are scarce, even in known virus-induced neoplasms. The Bittner virus,[6] which causes mouse mammary tumors, is a convenient model for discussion. The infected mouse develops the carcinoma only at a certain age, presumable as the virus is activated by estrogen. Production of virulent bacteriophage is another favorite model for reference. The phage is quiescent, as DNA only, until some environmental change allows it to mastermind the machinery of the cell.[10,11] Do tumor viruses stay latent until activated by provocative stimuli from other microbes?

ETIOLOGY OF HODGKIN'S DISEASE

Hodgkin's disease is intriguing as a malignancy with evidence of infectious causation and perhaps even contagiousness.[39,40,63,73,89] There is statistical evidence that those who acquire the disease were infected around or before their ninth year.[70]

Just as tonsillectomy removes a protective barrier for poliomyelitis, tonsillectomized individuals increase their susceptibility to Hodgkin's disease threefold.[90] Data showing Hodgkin's disease in unrelated propinque individuals concern two surgeons accidentally inoculated with Hodgkin's tissue, a technician who handled urine and feces from a case,[48] and the syndrome in five married couples.[26,42,67] In a case described by Berliner and Distenfeld, the illness of the husband was evident about 18 months after that of his wife.[4] In addition, there is a report of four cases in a high school class.[89] The occurrence of dual cases in families is thought not to be genetic.[26,80]

An associated virus is actively being sought in Hodgkins disease.[66,71] Stewart et al., examining a cell line from a Hodgkin's lymph node[87] by electron microscopy, discerned two types of virus particles, one being herpes-virus type. Marasca and associates reviewed the evidence for a herpes virus in Hodgkin's and using the polymerase chain reaction found the herpes virus type 6 genome in one case.[66] Epstein-Barr virus genome has been found in 20% of Hodgkin's involved cells.[93] The Reed-Sternberg cells, pathognomic of the disease, contain inclusions which could be microbial. Extensive studies of bodies in Reed-Sternberg cells, which resemble viral products, have been reviewed by Bostick.[12] Thus, work by several investigators suggests a viral etiology.

The other possibility for a causative agent of Hodgkin's is a bacterium as elusive as a virus and with filtrable intracellular stages. Investigators who have been satisfied with their positive findings include individuals with impressive reputations. These studies are described in the following paragraphs.

One of the first Americans to become interested in the bacteriology of Hodgkin's disease was Bunting,[16] a University of Wisconsin pathologist. Convinced that there was a microbial factor, he, with colleagues, spent four fruitless years looking for a microorganism. However, following changes in culture procedure, a certain bacterium was isolated from the glands of almost every untreated case of Hodgkin's disease. The organism was so pleomorphic that in working with the first isolates much time was spent in purification passages. In contrast to others, who have usually been unable to demonstrate animal pathogenicity, Bunting adapted a strain to cause disease in monkeys, producing the blood leukocyte picture which he regarded as typifying Hodgkin's disease in man. This included the spectacular Reed-Sternberg cells.[52] Attempts to ameliorate the disease in man by vaccine or specific serum were inconclusive. Bunting believed he had confirmed the earlier work of de Negri and Mieremet[72] and of Fraenkel[35] and Fraenkel and Much.[34]

Another who painstakingly concentrated on Hodgkin's disease was Mazet,[68] who isolated 26 bacterial strains from Hodgkin's blood cultures. His strains were extremely pleomorphic aerobes. He stated it should be easy to repeat his work if only one were patient with slowly and faintly growing organisms. The bacteria passed from granule stages to forms imitating Actinomyces and yeasts. He commented, perhaps with humor, that cultures were apt to be discarded as they passed through the stage which he termed "pseudo-subtilis". Mazet's strains can be characterized by their unique effect in guinea pigs: intratesticular inoculation caused both orchitis and encephalitis.

It is provocative that the nitroblue tetrazolium (NBT) test, which in most instances signifies infection, was positive in all cases of Hodgkin's disease examined by Chang and associates.[23] They comment that the marked increase of NBT reduction "can be interpreted as an active phagocytic and bactericidal response to an unknown stimulus".

CONCLUSIONS REGARDING HODGKIN'S DISEASE

Many investigators have probed the bacterial-etiology theory for Hodgkin's as listed in Table 1. One definite by-product of this research emerges. Diseased lymph glands, especially if in the abdominal cavity, feed diphtheroids and other pleomorphic organisms into the bloodstream.[33] Blood cultures may give a diagnostic clue in such conditions. The possible role of any corynebacterium in Hodgkin's causation remains to be confirmed by identifying isolates with all taxonomic devices currently available and preserving the strains for reference. The fragility, fastidiousness, and pleomorphism of the bacteria isolated have frequently suggested that they are wall deficient when they parasitize Hodgkin's disease patients.

WALL-DEFICIENT BACTERIA IN MALIGNANCIES OTHER THAN HODGKIN'S

DILLER'S MURINE ORGANISM

A most critical point in assessing the significance of a bacterium in malignancies is whether the bacterial parasitism precedes or follows the neoplastic growth. This question can only be investigated in an animal model, where maturation of numerous individuals is followed. The studies of Diller et al.[28,29] involving 1500 mice indicated that a tumor-bearing

TABLE 1
Classical Bacteria Isolated from Hodgkin's Disease
(Many strains had characteristics of wall deficiency)

Organism and Investigator	Comments
Nonacid-fast rod, Fraenkel and Much,[34] Fraenkel[35]	
Corynebacterium, de Negri and Mieremet[72]	With tissue produced lymphadenopathy in guinea pigs
Corynebacterium in lymph gland slices, Bunting[16]	Hodgkin's-like disease in monkeys
Miscellaneous diphtheroids from macerated glands, Bloomfield[9]	No antibody in patients and no hypersensitivity in skin tests
From glands of Hodgkin's and other conditions, an anaerobic diphtheroid, Torrey[88]	Concluded nonpathogenic for animals, a parasite of diseased tissue.
Mycobacterium inferred from skin sensitivity in inoculated animals, L'Esperance[59]	Chickens inoculated with Hodgkin's nodes developed tubercular lesions; series small
Variably acid-fast diphtheroid, pleomorphic, Mazet[68]	Orchitis and encephalitis when strains inoculated intratesticularly into guinea pigs
Corynebacterium in blood cultures, Fleisher[33]	Concluded were assorted and signified diseased lymph nodes with varied pathology
Acid-fast rods, Alexander-Jackson[2]	
Unidentified pleomorphic bacteria, Carpenter et al.[22]	Lethal to 5-d-old chick embryos
Acid-fast organism, Seibert[83]	
Propionibacterium acnes-like organism, Cantwell and Kelso[20,21]	
Variably acid-fast coccoids, Cantwell[19]	Similar organisms found in 4 cases

strain of mouse remained free of the neoplasm unless it carried a certain bacterium. Similarly, giving the provocative bacterium to mice of a strain which normally develops only a few tumors significantly increased their malignancy incidence. Diller's bacterium is a slowly growing fastidious variegating bacillus which resembles both Corynebacteria and Mycobacteria.

NUZUM'S COCCUS

Another organism which grows largely in the wall-deficient stage produces transplantable carcinoma in albino mice, metastasizing mammary carcinoma in canines, and primary epithelioma in man. This is a coccus found in 38 of 41 early human breast cancers by Nuzum.[74] When first isolated, the growth consists of minute granules 0.1 to 0.3 μm in diameter which develop only in anaerobic culture. Filtrability is a dominant characteristic of the organism which finally stabilizes as a Staphylococcus, giving hemolytic pinpoint colonies on human blood agar. The wall-deficient variants of Nuzum,[74] Diller and Medes,[27] Clark,[24] Brehmer,[14] Seibert et al.,[82,83] Livingston and Alexander-Jackson,[60] and Livingston and Livingston[61] have been tumorigenic in laboratory animals, as described in Table 2. Malignancies in laboratory and domestic animals clearly can be triggered by wall-deficient microbes. Examples are also listed in the table.

CLASSIC ORGANISMS WHICH PRODUCE EXPERIMENTAL MALIGNANCIES

One of the most lucid examples of an oncogenic bacterium is the Mycobacterium which initiates tumors in the newt. Another organism, *C. neoformans,* which produces malignant tumors in mice, became carcinogenic only after prolonged subculture in artificial media.[78] If the nucleic acids of the original and mutant yeast are analyzed, the answer to tumorigenesis

TABLE 2
CWD Forms from Malignancies of Man or Animals

Organism, source, and investigator	Comment
Staphylococcus evolving from anaerobic granular stage, Nuzum,[74] 38 human breast cancers and from metastases	Tumorigenic in mice, canines, man
Typical staphylococcus; pleomorphic, Gram + rod. Resists 80°C 10 min., Stearn et al.[86] 18 human tumors	
Granules, cocci, rods; Glover,[41] found organism in 85% of 3000 cases	Antiserum made to his organism was given an extensive trial, and some ''hopeless'' cancer cases had long-term remissions
Complete L-cycle plus sporulating rod, Brehmer[14]	Carcinoma in inoculated mice
CWD forms with filtrable stages, Clark[24]	Pathogenic in the little animal work done. Worked with Glover's medium; findings coincide
CWD forms from 1000 human cancers, Gregory[43]	Speculated a plasmid may be involved. Vaccination sometimes preventative for experimental animals
CWD rods with acid-fast stages, Seibert et al.[83]	Tumors in ICR/albino mice
Organism with filtrable cycle; acid-fast stages, Livingston and Alexander-Jackson[60] and Livingston and Livingston[61]	Tumorigenic in experimental animals
L Form was one of 4 organisms in leukemic bone marrow, Kagan et al.[54]	
CWD forms from 4 malignancies, Cantwell[19]	Isolated from dermal metastases
Acid-fast pleomorphic organism from miscellaneous neoplasms in carcinogen-fed rats, Diller and Medes[27]	
Spheroplast from bovine lympho-sarcoma, McKay, et al.[69]	Cultured from two animals; demonstrated in five of six by fluorescent antibody

by this organism might be revealed. These and other tumorigenic microorganisms are described in Table 3. An association has been noted between infection with *Helicobacter pylori* and gastric adenocarcinoma.[76] The authors conclude ''Infection with *H. pylori* is associated with an increased risk of gastric adenocarcinoma and may be a cofactor in the pathogenesis of this malignant condition.''

BACTERIA AND FUNGI OF UNKNOWN PATHOGENICITY FOUND IN NEOPLASIA

The significance of organisms in naturally occurring neoplasia is weighted by the number of cases showing a particular organism. The 3000 cases of Glover stand out in this respect.[41] Many of the strains isolated exhibit cyclic growth, as shown in Table 2. For any isolate it is exceedingly difficult to judge whether the organism represents merely normal flora invasion of dead tissue, or is pathogenic per se.

HOW MAY A BACTERIUM BE CARCINOGENIC?

ALTERING THE HOST'S ANTIBODY RESPONSE?

Does a carcinogenic bacterium or fungus prevent a normal antibody response by the

TABLE 3
Classic Organisms Which Produce Experimental Malignancies

Bacterium or fungus, investigator	Experimental host
Aspergillus flavus, Boyland[13]	Rat (liver cancer)
Mycobacterium thamnopheos, Inoue and Singer[51]	Newt
A. flavus, Inoue and Singer[51]	Murine carcinoma
Epidermophyton, Inoue and Singer[51]	Murine carcinoma
Microsporum, Inoue and Singer[51]	Murine carcinoma
Scopulariopsis, Inoue and Singer[51]	Murine carcinoma
Streptomyces, Inoue and Singer[51]	Murine carcinoma
Trichophyton, Inoue and Singer[51]	Murine carcinoma
Candida parasilosis, Inoue and Singer[51]	Murine carcinoma
Candida (species unknown), Inoue and Singer[51]	Murine carcinoma
C. albicans, Blank et al.[7]	Murine sarcoma, leukemia,
C. parasilosis, Blank et al.[7]	carcinoma (varied organs)
Scopulariopsis, Blank et al.[7]	

host? The Friend leukemia virus causes leukemia only after the mice receive antigens, which may tie up complement or other immune substances.[56]

ANERGY

Some cancer viruses such as the Gross leukemia agent leave antibody formation intact but prevent cellular defense.[36] Does a bacterium, fungus, or fungal product work through this approach? This subject is reviewed in an excellent article by Ebbesen.[32]

DOES A CARCINOGEN-FOSTERING BACTERIUM ACT AS A HELPER VIRUS?

There are now many examples of viruses which mature only with the aid of an envelope coating supplied by another virus.[49] Can bacteria behave like helper viruses in contributing proteins and polysaccharides to coat the oncogenic DNA?

If Mycoplasma are necessary for the leukemogenic action of some viruses, presumably a similar contribution can be made by the protoplast stage of bacteria. Leukemia can be the result of dual infection with the Rauscher murine leukemia virus and *Mycoplasma laidlawii.* Separately, neither agent induced the disease.[53]

In contrast, another study indicates that some viruses may be oncogenic of themselves. Mice kept germ free except for their inherent leukemia virus develop malignancies at the same age and as frequently as their conventional counterparts known to be parasitized by a Mycoplasma.[38]

DOES A CANCER BACTERIUM WORK BY *IN VIVO* SYNTHESIS OF A CARCINOGENIC COMPOUND?

This phenomenon may, indeed, occur at times, as indicated by the carcinogenesis of the glucoside cycasin only in conventional rats. Germ-free rats cannot convert the cycasin to the aglycone, which is the actual carcinogen.[58]

DO CARCINOGEN-STIMULATING BACTERIA CARRY A VIRUS OF MALIGNANCY?

This possibility has been little explored. A fact suggesting this is the increased tumorigenesis of *Agrobacterium tumefaciens* after exposure to UV or mitomycin C, factors known to increase the formation of mature phage particles.[45-47] However, no carcinogenic phage is known, although phages have been found in malignant growths and in bacteria isolated from tumors.[65]

There are many examples now known of bacterial wall-deficient microbes hosting viruses. Spheroplasts may also at times hold virus particles firmly adsorbed to their surfaces, as shown for *Aerobacter aerogenes* and Influenza A virions.[15] More realistic than hosting entire virions is the possibility that a bacterium can carry just the deadly nucleic acid, whether DNA or RNA, to act by reverse transcriptase.[84,85] *Bacillus subtilis* is a bacterium which replicates the tumorigenic polyoma virus.[3]

The possibilities that may explain the reported series of oncogenic bacteria are discussed in a review by Macomber.[64] Included is the thesis that cancer-associated bacteria could carry oncogenes. He has reviewed the extensive studies of Gregory who found virus-like CWD bacteria in 1000 malignancy biopsies and none in 100 benign tumors.[43]

DO ONCOGENIC BACTERIA FLOOD THE HOST WITH HORMONE-LIKE SUBSTANCES?

A connection between sex hormones and malignancies has long been recognized. Some bacteria have been found to produce estradiol and estrone. *Staphylococcus haemolyticus and Streptococcus bovis* carry a substance resembling human choriogonadotropin (hCG).[31]

INHERENT DIFFICULTIES IN THE RESEARCH

Why has investigation through a 60-year span failed to satisfy the scientific world that bacteria trigger the common malignancies? Much of the media has been exotic, e.g., Glover's concoction of sunflower seeds, Iceland moss, and Irish moss.[41] Nuzum employed an unusual ratio of 3 parts of ascitic fluid to 1 part nutrient agar.[74] Joseph Merline, in our laboratory, cultured over 200 bloods from lymphoma patients and in no instance found the bacterium of Glover, Nuzum, or of more modern descriptions. Merline was not seeking a tumor-instigating bacterium, and such procedures as washing the red cells to eliminate antibody or aging the blood at room temperature to void complement were not followed. His study rather shows that careful standard technique to isolate aerobes, anaerobes, and CWD forms does not grow a carcinogen-fostering bacterium.

On the other hand, there is no justification to conclude that the bacteria isolated from cancer patients are contaminants. Most of the positive studies have been long term and controlled with tissue or blood from noncancer patients. One critical publication noted that positive results were not realized if cultures were made under ultraviolet (UV).[55] It is now well established that exposing medium to UV destroys its ability to support growth of fastidious organisms, although still suitable for established, hardy bacteria.

Tumor production in experimental animals usually necessitates a long-term study and is hard to reproduce due to genetic differences in susceptibility. Even biochemical reactions of bacteria are often not reproducible, due to varied peptones in basal media.

Last, interpretation of findings in lymph node cultures is not easy. Lymph nodes of man, like those of animals, have a "normal" flora.[1]

PRACTICAL APPLICATION OF BACTERIA-IN-CANCER STUDIES

VACCINES FOR PROPHYLAXIS

The aim of those who seek a bacterium as the provocative stimulus for cancer is to prepare a vaccine. Vaccines can immunize against some oncogenic agents, as found for the common and serious Marek's disease of poultry, which has an associated herpes-type virus, as shown by Purchase and associates[79] and others.[5]

ANTISERUM FOR THERAPY

That antiserum to a bacterium can be therapeutic in cancer is still uncertain. If the bacterium causes cancer in the same way that *A. tumefaciens* causes plant tumors, i.e., by altering host cells and then disappearing, measures to eliminate a bacterium after a malignancy is established will not be pertinent. However, the results with Glover's antiserum constitute a record worth remembering.[94]

DIAGNOSIS OF CANCER BY SEROLOGICAL APPROACH

A microbe from the blood of cancer patients is the antigen in a serological test employed by Villequez. In his experience, the microbe agglutinates erythrocytes and latex particles. Sera from cancer patients exhibit blocking antibody which interferes with both the hemagglutination and the clumping of latex. In testing over 200 bloods from patients with malignancies of diverse nature, he found approximately 95% exhibited the antibody, in contrast to negative findings with serum from 40 individuals with illnesses not malignancies.[99] Someday research will reveal the interplay between this blocking antibody of Villequez and the macroglobulin in cancer serum demonstrated by Wallace et al.[91]

WHAT CAN BE DONE NOW?

How can any investigator test the pros and cons of bacterial association in cancer? If the forms seen by White[95] in ascitic fluid of malignancies are microbial, they should grow in media suitable for CWD forms. Their microbial nature can be confirmed or denied by staining with fluorescent muramidase.

Malignant cells cultured 24 to 48 h in broth, according to Glover's record, become heavily populated with "the organisms". Such infected malignant cells can serve as tools for staining and electron microscopy.

A careful study in Thailand[92] and one in Florida[83] suggest that malignant cells host intranuclear forms with the morphology and staining reactions of bacteria. The study in Thailand may relate the organisms to CWD forms since they neither resist common fixatives nor grow on standard culture media. Attempts to identify these and to improve fixation for electron microscopy are needed.

SUMMARY

Thanks to the work of Pollard with germ-free rats, there is little doubt that sterile chemical agents can initiate malignancies without microbial assistance.[77] Injection of methylcholanthrene induces fibrosarcomas in animals which appear to be free of microbes when autopsied. Likewise, viruses alone without bacterial accompaniment can initiate malignancies. For example, Rous sarcoma virus produces metastasizing fibrosarcomas in rats which seem free of all other microbes.

At the other extreme, it is clear that oncogenic bacteria and fungi exist. The grey area is whether any of the *common* malignancies of vertebrates are fostered by bacteria, perhaps stimulating an associated virus. Current methods for culturing[60] the suspected bacterium do not sound complex. The yeast-like forms which White finds omnipresent in cancerous tissue could well be the fungoidal stages of a wall-deficient bacterium. He believes these forms, often abundant in the transudates of malignancies, are confused with fat droplets.[95] When organisms are found, the task remains to show that they are not harmless flora. A role for bacteria in malignancies in either the classic or wall-deficient stage remains to be elucidated. Proponents of the bacterial association theory have different hypotheses to relate bacteria to cancer. Some have suggested that the bacteria carry DNA which directly or indirectly activates oncogenes, as is true for viruses.

And is the DNA inherent in the bacterial strain or acquired by growth in malignant tissue? Is it possible that the malignancy-causing tumors carry a pertinent plasmid? *A. tumefaciens* strains cause plant tumors when they carry the T plasmid, not when they are plasmid-free.[50] These are mysteries which must be explored at the molecular level.[37] However, sufficient data have been amassed to warrant reinvestigations of associated bacteria. There is no subject generally viewed with greater skepticism than an association between bacteria and human cancer. However, the medical profession may look back with irony at the stony reception given by his colleagues to Koch's paper elucidating the etiology of tuberculosis. Similarly, medical students were once taught that whooping cough vaccination was an unrealistic dream reported only by two women at the Michigan Public Health Laboratories and by a deranged pediatrician named Sauer.

REFERENCES

1. **Adamson, C. A.,** Bacteriological study of lymph nodes; analysis of postmortem specimens with particular reference to clinical, serological and histopathological findings, *Acta Med. Scand. Suppl.,* 227, 1–21, 1949.
2. **Alexander-Jackson, E.,** A specific type of microorganism isolated from animal and human cancer; bacteriology of the organism, *Growth,* 18, 37–51, 1954.
3. **Bayreuther, K. E.,** Polyoma virus: production in *Bacillus subtilis, Science,* 146, 778-779, 1964.
4. **Berliner, A. D. and Distenfeld, A.,** Hodgkin's disease in a married couple, *JAMA,* 221, 703–704, 1972.
5. **Biggs, P. M., Payne, L. N., Milne, G. S., Churchill, A. E., Chubb, R. C., Powell, D. G., and Harris, A. H.,** Field trials with an attenuated cell associated vaccine for Marek's disease, *Vet. Rec.,* 87, 704–709, 1970.
6. **Bittner, J. J.,** Some possible effects of nursing on the mammary gland tumor incidence in mice, *Science,* 84, 162, 1936.
7. **Blank, F., Chin, O., Just, G., Meranze, D. R., Shimkin, M. B., and Wieder, R.,** Carcinogens from fungi pathogenic for man, *Cancer Res.,* 28, 2276–2281, 1968.
8. **Blank, H.,** Virus-induced tumors of human skin (warts, molluscum contagiosum), *Ann. N.Y. Acad. Sci.,* 54, 1226–1231, 1952.
9. **Bloomfield, A. L.,** The bacterial flora of lymphatic glands, *Arch. Intern. Med.,* 16, 197–204, 1915.
10. **Bordet, J.,** Le problème de l'autolyse microbienne transmissible ou du bactériophage, *Ann. Inst. Pasteur (Paris),* 39, 717–763, 1925.
11. **Bordet, J.,** Pouvoir lysogène actif ou spontané et pouvoir lysogène passif ou provoqué, *C. R. Soc. Biol. (Paris),* 93, 1054–1056, 1925.
12. **Bostick, W. L.,** The status of the search for a virus in Hodgkin's disease, *Ann. N.Y. Acad. Sci.,* 54, 1162–1176, 1952.
13. **Boyland, E.,** The causes of cancer, *Ist. Tip. Fak. Mec.,* 28, 208–212, 1965.
14. **Brehmer, W.,** ''Siphonospora polymorpha'' n. sp., ein neuer Mikroorganismus des Blutes und seine Beziehung zur Turmorgenese, *Med. Welt,* 8, 1179–1185, 1934.
15. **Brown, R. J., Benedict, A. A., and Armstrong, N.,** Adsorption of Influenza A virus by *Aerobacter aerogenes* spheroplasts, *J. Bacteriol.,* 83, 1124–1130, 1962.
16. **Bunting, C. H.,** The blood-picture in Hodgkin's disease, second paper, *Bull. Johns Hopkins Hosp.,* 25, 173–184, 1914.
17. **Burkitt, D.,** The African lymphoma — epidemiological and therapeutic aspects, in *Proc. Int. Conf. Leukemia-Lymphoma,* Zarafonetis, C. J. D., Ed., Lea & Febiger, Philadelphia, 1968, 321–330.

18. **Burmester, B. R.,** Studies on fowl lymphomatosis, *Ann. N.Y. Acad. Sci.,* 54, 992–1003, 1952.

19. **Cantwell, A. R., Jr.,** Histologic observations of variably acid fast coccoid forms suggestive of CWD bacteria in Hodgkins disease, 4 cases, *Growth,* 45, 168–187, 1981.

20. **Cantwell, A. R., Jr. and Kelso, D. W.,** Microbial findings in cancers of the breast and in their metastases to the skin, *J. Dermatol. Surg. Oncol.,* 7, 483–491, 1981.

21. **Cantwell, A. R., Jr. and Kelso, D. W.,** Variably acid-fast bacteria in a fatal case of Hodgkin's disease, *Arch. Dermatol.,* 120, 401–402, 1984.

22. **Carpenter, C. M., Nelson, E. L., Lehman, E. L., Howard, D. H., and Primbs, G.,** The isolation of unidentified pleomorphic bacteria from the blood of patients with chronic illness, *J. Chron. Dis.,* 2, 156–161, 1955.

23. **Chang, J. C., Appleby, J., and Bennett, J. M.,** Nitroblue tetrazolium test in Hodgkin's disease and other malignant lymphomas, *Arch. Intern. Med.,* 133, 401–403, 1974.

24. **Clark, G. A.,** Successful culturing of Glover's cancer organism and development of metastasizing tumors in animals produced by cultures from human malignancy, *Proc. VI Int. Congr. Microbiol. (Rome),* 6, 41–49, 1953.

25. **Crum, C. P., Barber, S., and Roche, J. K.,** Pathobiology of papillomavirus-related cervical diseases, prospects for immunodiagnosis, *Clin. Microbiol. Rev.,* pp. 270–285, 1991.

26. **Devore, J. W. and Doan, C. A.,** Studies on Hodgkin's syndrome. XII. Hereditary and epidemiologic aspects, *Ann. Intern. Med.,* 47, 300–316, 1957.

27. **Diller, I. C. and Medes, G.,** Isolation of a pleomorphic acid-fast organism from liver and blood of carcinogen-fed rats, *Am. Rev. Respir. Dis.,* 90, 126–128, 1964.

28. **Diller, I. C., Donnelly, A. J., and Fisher, M. E.,** Isolation of pleomorphic, acid-fast organisms from several strains of mice, *Cancer Res.,* 27, 1402–1408, 1967.

29. **Diller, I. C. and Donnelly, A. J.,** Experiments with mammalian tumor isolates, *Ann. N.Y. Acad. Sci.,* 174, 655–674, 1970.

30. **Dmochowski, L.,** Viral studies in human leukemia and lymphoma, in *Proc. Int. Conf. Leukemia-Lymphoma,* Zarafonetis, C. J. D., Ed., Lea & Febiger, Philadelphia, 1968, 97–113.

31. **Domingue, G. J., Acevedo, H. F., Powell, J. E., and Stevens, V. C.,** Antibodies to bacterial vaccines demonstrating specificity for human choriogonadotropin (hCG) and immunochemical detection of hCG-like factor in subcellular bacterial fractions, *Infect. Immun.,* 53, 95–98, 1986.

32. **Ebbesen, P.,** A subcellular agent inducing plasma cell leukemias in mice. The relation of this agent to amyloidosis and chronic murine pneumonia, *Acta Pathol. Microbiol. Scand. Suppl.,* 197, 1968.

33. **Fleisher, M. S.,** Significance of diphtheroid microorganisms in blood cultures from human beings, *Am. J. Med. Sci.,* 224, 548–553, 1952.

34. **Fraenkel, E. and Much, H.,** Über die Hodgkinische Krankheit (Lymphomatosis granulomatosa), insbesondire deren Ätiologie, *Z. Hyg.,* 67, 159–200, 1910.

35. **Fraenkel, E.,** Über die sogen. Hodgkinische Krankheit (Lymphomatosis granulomatosa), *Dtsch. Med. Wochenschr. 1912,* pp. 637–642, 1912.

36. **Frey-Wettstein, M. and Hays, E. F.,** Immune response in preleukemic mice, *Infect. Immun.,* 2, 398–403, 1970.

37. **Gallo, R. C. and Nerurkar, L. S.,** Human retroviruses: their role in neoplasia and immunodeficiency, *Ann. N.Y. Acad. Sci.,* 567, 82–94, 1989.

38. **Gilbey, J. C. and Pollard, M.,** Search for Mycoplasma in germfree leukemic mice, *J. Natl. Cancer Inst.,* 38, 113–116, 1967.

39. **Gilmore, H. R. and Zelesnick, G.,** Environmental Hodgkin's disease and leukemia, *Penn. Med.,* 65, 1047–1049, 1962.

40. **Glaser, S. L.,** Spatial clustering of Hodgkin's disease in the San Francisco Bay Area, *Am. J. Epidemiol.,* 132, S167–S177, 1990.

41. **Glover, T. J.,** The bacteriology of cancer, *Can. Lancet Pract.,* 75, 92–111, 1930.

42. **Gordon, M. H., Gow., A. E., Levitt, W. M., et al.,** Recent advances in the pathology and treatment of lymphadenoma, *Proc. R. Soc. Med.,* 27, 1035, 1934.

43. **Gregory, J. E.,** Pathogenesis of cancer, 3rd ed., Fremont Foundation Publishers, Pasadena, CA, 1955.

44. **Gross, L.,** Mouse leukemia, *Ann. N.Y. Acad. Sci.,* 54, 1184–1196, 1952.

45. **Hayflick, L.,** Mycoplasmas from malignant tissue, in *The Mycoplasmatales and the L-Phase of Bacteria,* Hayflick, L., Ed., Appleton-Century-Crofts, New York, 1969, 38–41.

46. **Heberlein, G. T. and Lippincott, J. A.,** Photoreversible UV enhancement of infectivity in *Agrobacterium tumefaciens, J. Bacteriol.,* 89, 1511–1514, 1965.

47. **Heberlein, G. T. and Lippincott, J. A.,** Ultraviolet-induced changes in the infectivity of *Agrobacterium tumefaciens, J. Bacteriol.,* 93, 1246–1253, 1967.

48. **Hoster, H. A., Dratman, M. B., Craver, L. F., and Rolnick, H. A.,** Hodgkin's disease, 1832–1947, *Cancer Res.,* 8, 1–78, 1948.
49. **Huebner, R. J., Hartley, J. V., Rowe, W. P., Lane, W. T., and Capps, W. I.,** Rescue of the defective genome of Moloney sarcoma virus from a noninfectious hamster tumor and the production of pseudotype sarcoma viruses with various murine leukemia viruses, *Proc. Natl. Acad. Sci.,* 56, 1164–1169, 1966.
50. **Hynes, M. F., Simon, R., and Pühler, A.,** The development of plasmid-free strains of *Agrobacterium tumefaciens* by using incompatibility with a *Rhizobium meliloti* plasmid to eliminate pAtC58, *Plasmid,* 13, 99–105, 1985.
51. **Inoue, S. and Singer, M.,** Experiments on a spontaneously originated visceral tumor in the newt, *Triturus pyrrhogaster, Ann. N.Y. Acad. Sci.,* 174, 729–764, 1970.
52. **Jaffe, E. S.,** The elusive Reed-Sternberg cell, *N. Engl. J. Med.,* 320, 529–531, 1989.
53. **Kagan, G. Ya., Postinikova, S. A., Morgunova, T. D., Rakovskaya, I. V., Neustroeva, V. V., and Smirnova, T. D.,** Mycoplasma-virus infection in experiments in vitro and in vivo, *Proc. 10th Int. Congr. Microbiol. (Mexico City),* 1970, 92.
54. **Kagan, G. Ya., Golosova, T. V., Martynova, V. A., Chumakova, L. P., Koptelova, E. I., and Raskova, T. M.,** Isolation and identification of microbial agents from the bone marrow and blood of patients suffering from acute leukemia, *Zh. Microbial. Epidemiol. Immunobiol.,* 9, 72–77, 1971.
55. **Kassel, R. and Rottino, A.,** Significance of diphtheroids in malignant disease studied by germ-free techniques, *Arch. Intern., Med.,* 96, 804–808, 1955.
56. **Kouttab, N. M. and Jutila, J. W.,** Friend leukemia virus infection in germfree mice following antigen stimulation, *J. Immunol.,* 108, 591–595, 1972.
57. **Lancaster, W. D. and Jenson, A. B.,** Human papillomavirus infection and anogenital neoplasia, speculations for the future, *Obstet. Gynecol. Clins. N. Amer.*14, 601, 1987.
58. **Laqueur, G. L., McDaniel, E. G., and Matsumoto, H.,** Tumor induction in germfree rats with methyl-azoxymethanol (MAM) and synthetic MAM acetate, *J. Natl. Cancer Inst.,* 39, 355–371, 1967.
59. **L'Esperance, E. A.,** Studies in Hodgkin's disease, *Ann. Surg.,* 93, 162–168, 1931.
60. **Livingston, V. W. C. and Alexander-Jackson, E.,** A specific type of organism cultivated from malignancy, bacteriology and proposed classification, *Ann. N.Y. Acad. Sci.,* 174, 636–654, 1970.
61. **Livingston, V. W. C. and Livingston, A. M.,** Demonstration of *Progenitor cryptocides* in the blood of patients with collagen and neoplastic diseases, *Trans. N.Y. Acad. Sci.,* 34, 433–453, 1972.
62. **Lucké, B.,** Kidney carcinoma in the leopard frog, a virus tumor, *Ann. N.Y. Acad. Sci.,* 54, 1093–1109, 1952.
63. **MacMahon, B.,** Epidemiology of Hodgkin's disease, *Cancer Res.,* 26, 1189–1200, 1966.
64. **Macomber, P. B.,** Cancer and cell wall deficient bacteria, *Med. Hypotheses,* 32, 1–9, 1990.
65. **Mankiewicz, E.,** Bacteriophages that lyse Mycobacteria and Corynebacteria and show cytopathogenic effect on tissue cultures of renal cells of *Cercopithecus aethiops,* a preliminary communication, *Can. Med. Assoc. J.,* 92, 31–33, 1965.
66. **Marasca, R., Luppi, M., Montorsi, M., et al.,** Identification of sequences of human herpes virus six (HHV-6) in a case of Hodgkin's disease by polymerase chain reaction, *Med. Riv. Encicl. Med. Ital.,* 10, 43–45, 1990.
67. **Mazar, S. A. and Straus, E.,** Marital Hodgkin's disease, a review of the familial incidence and of etiological factors, *Arch. Intern. Med.,* 88, 819–830, 1951.
68. **Mazet, G.,** Étude bactériologique sur la maladie d'Hodgkin, *Montpellier Med. 1941,* 316–328, 1941.
69. **McKay, K. A., Neil, D. H., and Corner, A. H.,** The demonstration of a single species of an unclassified bacterium in five cases of bovine lymphosarcoma, *Growth,* 31, 357–368, 1967.
70. **Miller, R. W.,** Mortality in childhood Hodgkin's disease. An etiologic clue, *JAMA,* 198, 1216–1217, 1966.
71. **Murphy, E. D.,** Replica of Hodgkin's disease in SJL/J mice, *34th Annu. Rep.,* Jackson Laboratory, Bar Harbor, ME, p. 26, 1963.
72. **de Negri, E. and Mieremet, C. W. G.,** Zur Aetiologie des Malignen Granuloms, *Zentralbl. Bakteriol. Parasitenkd. Infektionskr. Hyg. Abt. 1 Orig.,* 68, 292–308, 1913.
73. **Newell, G. R.,** Etiology of multiple sclerosis and Hodgkin's disease, *Am. J. Epidemiol.,* 91, 119–122, 1970.
74. **Nuzum, J. W.,** The experimental production of metastasizing carcinoma in the breast of the dog and primary epithelioma in man by repeated inoculation of a Micrococcus isolated from human breast cancer, *Surg. Gynecol. Obstet.,* 11, 343–352, 1925.
75. **Oberling, C. and Guerin, M.,** Sur la production de tumeurs par inoculation intracutanu de virus leucémique de la poule, *C. R. Soc. Biol.,* 129, 1059–1060, 1938.
76. **Parsonnet, J., Friedman, G. D., Vandersteen, D. P., Chang, Y., Vogelman, J. H., et al.,** *Helicobacter pylori* infection and the risk of gastric carcinoma, *N. Engl. J. Med.,* 325, 1127–1131, 1991.
77. **Pollard, M.,** Leukemoid changes induced in germfree and conventional rats by Rous sarcoma virus, *J. Reticuloendothel. Soc.,* 7, 254–263, 1970.

78. **Price, J. T., and Bulmer, G. S.,** Tumor induction by *Cryptococcus neoformans, Infect. Immun.,* 6, 199–205, 1972.

79. **Purchase, H. G., Okazaki, W., and Burmester, B. R.,** Field trials with the Herpes virus of turkeys (HVT) Strain FC126 as a vaccine against Marek's disease, *Poult. Sci.,* L, 775–783, 1971.

80. **Razis, D. V., Diamond, H. D., and Craver, L. F.,** Familial Hodgkin's disease, its significance and implications, *Ann. Intern. Med.,* 51, 933–971, 1959.

81. **Rowe, W. P., Hartley, J. W., and Huebner, R. J.,** Polyoma and other indigenous mouse viruses, in *The Problem of Laboratory Animal Disease,* Harris, R. J. C., Ed., Academic Press, New York, 1962, 131–142.

82. **Seibert, F. B., Farrelly, F. K., and Shepherd, C. C.,** DMSO and other combatants against bacteria isolated from leukemia and cancer patients, *Ann. N.Y. Acad. Sci.,* 141, 175–201, 1967.

83. **Seibert, F. B., Feldmann, F. M., Davis, R. L., and Richmond, I. S.,** Morphological, biological and immunological studies on isolates from tumors and leukemic bloods, *Ann. N.Y. Acad. Sci.,* 174, 690–728, 1970.

84. **Spiegelman, A., Burny, A., Das, M. R., Keydar, J., Scholm, J., Travnicek, M., and Watson, K.,** Characterization of the products of RNA-directed DNA polymerases in oncogenic RNA viruses, *Nature (London),* 227, 563–567, 1970.

85. **Spiegelman, A., Burny, A., Das, M. R., Keydar, J., Scholm, J., Travnicek, M., and Watson, K.,** DNA-directed DNA polymerase activity in oncogenic RNA viruses, *Nature (London),* 277, 1029–1031, 1970.

86. **Stearn, E. W., Sturdivant, B. F., and Stearn, A. E.,** The ontogeny of an organism isolated from malignant tumors, *J. Bacteriol.,* 28, 227–245, 1929.

87. **Stewart, S. E., Mitchell, E. Z., Whang, J. J., Dunlop, W. R., Ben, T., and Nomura, S.,** Viruses in human tumors, *J. Natl. Cancer Inst.* 43, 1–14, 1969.

88. **Torrey, J. C.,** Bacteria associated with certain types of abnormal lymph glands, *J. Med. Res.,* 34, 65–80, 1916.

89. **Vianna, N. J., Greenwald, P., and Davies, J. N. P.,** Extended epidemic of Hodgkin's disease in high school students, *Lancet,* 1, 1209–1211, 1971.

90. **Vianna, N. J., Greenwald, P., and Davies, J. N. P.,** Tonsillectomy and Hodgkin's disease, the lymphoid tissue barrier, *Lancet,* 1, 431–432, 1971.

91. **Wallace, R., Diena, B. B., Stoddart, T. G., Catton, G. E., and Greenberg, L.,** A bentonite flocculating factor in cancer sera, *Rev. Can. Biol.,* 29, 203–205, 1970.

92. **Weinman, D., Johnston, E., Saeng-Udom, C., Whitaker, J. A., Tamasatit, P., Panas-Ampol, K., and Fort, E.,** Lymphoma, intranuclear bacilliform structures in a patient with febrile anemia, *Am. J. Pathol.,* 52, 1129–1143, 1968.

93. **Weiss, L. M., Movahed, L. A., Warnke, R. A, and Sklar, J.,** Detection of Epstein-Barr viral genomes in Reed-Sternberg cells of Hodgkin's disease, *N. Engl. J. Med.,* 320, 502–6, 1989.

94. **White, J. E.,** Report on one hundred proven cases of malignancy treated by a specific antiserum, *Proc. VI Int. Congr. Microbiol. (Rome),* 6, 29–40, 1953.

95. **White, M. W.,** Etiology of malignancies, a new concept, *J. Int. Coll. Surg.,* 43, 593–602, 1965.

96. **Wogan, G. N. and Pong, R. S.,** Aflatoxins, *Ann. N.Y. Acad. Sci.,* 174, 623–634, 1970.

97. **Yaoi, H., Nagata, A., and Saito, K.,** Isolation of the virus of equine infectious anemia by serial transmission in rabbits, *Yokohama Med. Bull.,* 11, 1–20, 1960.

98. **zur Hausen, H.,** Papillomaviruses in human cancers, in *Common Mechanisms of Transformation by Small DNA Tumor Viruses,* Villarreal, L. P., Ed., *American Society of Microbiology,* Washington, D.C., 1989, 75–80.

99. **Villequez, E.,** La parasitisme latent des cellules du sang chez l'homme, en particulier le sand du cancereux, Librarie Maloine, Paris, 1955.

Chapter 36

ARTIFACTS AND CONTAMINANTS

Research on CWD forms would progress faster if artifacts did not so exactly duplicate the morphology of variant growth. Examples of false colonies are given in the accompanying micrographs (Figures 1 and 2). Lipids in chick embryo and in tissue can easily be confused with the fungoidal type of wall-deficient bacteria,[11] thus making microscopic studies with these preparations difficult. The artifacts are carried over to egg-yolk-containing media such as Petragnani's and Loewenstein's, used for growth of Mycobacteria. The need to employ media free of egg components becomes obvious.

Relatively early studies by Partridge and Klieneberger[10] showed that cholesterol could form globules in cultures of Mycoplasma or wall-deficient variants. Ørskov[9] found that large spheres which resembled microbial L-bodies might prove to be mere oil droplets. Heating plasma produces strands which stain moderately with acridine orange, causing misinterpretation. Cinatl[4] found that nuclei of dead cells cause precipitation of material easily confused with microbial growth. In early Mycoplasma work, it was found that a microbe-like precipitate develops when stain is added to serum-containing medium.[8] It is unfortunate that basic fuchsin, one of the best stains to accent morphology for photography, may spontaneously form a growth-mimicking precipitate. To be sure that structures in stained material are not produced by the stain, they also must be observed in phase or dark-field microscopy of unstained preparations.

Frobisher noted that granules apparently increase on incubation of protein-containing fluid,[6] imitating a basic morphological feature of stable L-colonies.

A troublesome whorl of precipitate which mimics a Mycoplasma or L-Form colony is commonly found after prolonged incubation of serum-containing medium (Figure 3 A and B).[3] Köhler and Otto[7] avoided such artifacts by pretreating the serum with dextran sulfate to remove lipid. The precipitate can also be inhibited by using agamma rather than unfractionated serum.

However, in an extensive study of serum artifacts, Buchanan concluded that often the whorl-colonies actually are L Forms found in many samples of serum. Not only could the colonies be successfully transferred, but their growth was prevented by vapor of formaldehyde or phenol.[2] Unfortunately these contaminants survived heat applied to remove serum complement. It would be expected that such colonies, which often resist staining by ordinary exposure times, would adsorb acridine orange or fluorochrome conjugated lysozyme applied for 24 h. Dienes early noted that L-Form colonies may adsorb stains slowly. Most significantly, such colonies of whorls have in some cases represented pathogens from disease such as the kidney tissue studied by Fernandes and Panos,[5] or the nocardia cultured by Beaman et al.[1]

Certain media are found to be heavily populated with dead wall-deficient colonies. The morphology of these granular colonies remains surprisingly intact through autoclaving.

Another source of error is contributed by organisms contaminating staining reagents. There seems to be no stain or buffer in which some microorganisms, in the classic or wall-deficient stage, cannot propagate after an adaptation period of several days or weeks. Refrigeration does not deter such contaminating growth. This problem is lessened by freezing fresh solutions of the stain and buffer in aliquots for a day's use. Even fluorescent antibody, freshly diluted from the lyophilized state, may contain microcolonies of wall-deficient bacteria.

A surprising source of contamination was noted by Judge in an unpublished study.

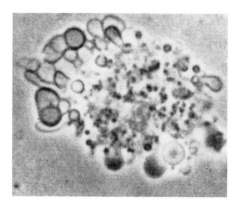

FIGURE 1. Uninoculated chick embryo material streaked on agar plate. (From Prittwitz und Gaffron, J. von, *Naturwissenschaften,* 42, 113–115, 1955. With permission.)

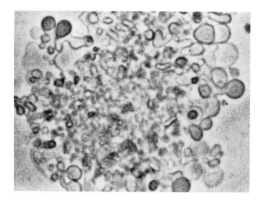

FIGURE 2. L-Form colony of *Corynebacterium diphtheriae.* (From Prittwitz und Gaffron, J. von, *Naturwissenschaften,* 42, 113–115, 1955. With permission.)

Lyophilized antibiotics often contained CWD bacilli which rapidly reverted. If the reconstituted antibiotics were allowed to stand undiluted for a few minutes, the contaminants were killed, but the organisms survived if dilutions were made immediately. It remains to be determined whether organisms are not removed by filtering the antibiotic solution before freeze-drying, or whether the contaminants enter during lyophilization.

SUMMARY

When working with the pleomorphic CWD forms, one becomes acutely aware of artifacts which complicate investigation. These include microbes which grow in common dyes, buffers, and fluorescent antibody. Microcolonies are imitated by precipitated stains, of which carbol fuchsin is the most deceiving. Dead colonies in media preserve their morphology through autoclaving. Serum added to medium when it is too warm precipitates in fine spheroplast-sized granules.

Precautions to avoid misinterpretation of artifacts include examining (1) incubated un-

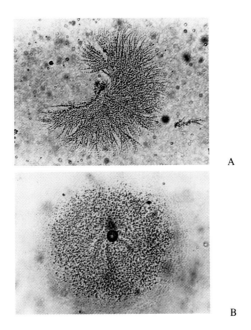

A

B

FIGURE 3   (A) Uninoculated PPLO agar with 20% calf serum after 14-d incubation at 37°C. Fan-shaped structure may be artifact or contaminant colony. Size: 275 μm. (B) Another plate (as in A) with structure suggesting a colony of L-organisms. Size: 330 μm. (From Cinatl, J. and Marhoul, Z., *Pathol. Microbiol.*, 29, 459–469, 1966. With permission.)

inoculated medium and (2) inoculated refrigerated medium. Any structures delineated by stains or antibody must also be seen before adding stain or other reagent.

REFERENCES

1. **Beaman, B. L. and Scates, S. M.,** Role of L Forms of *Nocardia caviae* in the development of chronic mycetomas in normal and immunodeficient murine models, *Infect. Immun.*, 33, 893–907, 1981.
2. **Buchanan, A. M.,** Atypical colony-like structures developing in control media and in clinical L Form cultures containing serum, *Vet. Microbiol.*, 7, 1–18, 1982.
3. **Cinatl, J. and Marhoul, Z.,** Structural changes in aging serum, I. *Pathol. Microbiol.*, 29, 459–469, 1966.
4. **Cinatl, J.,** Initiation of structures by nuclei of dead cells, *Exp. Cell Res.*, 47, 123–131, 1967.
5. **Fernandes, P. B. and Panos, C.,** Wall-less microbial isolate from a human renal biopsy, *J. Clin. Microbiol.*, 5, 106–107, 1977.
6. **Frobisher, M.,** II. Some pitfalls in bacteriology, *J. Bacteriol.*, 25, 565–571, 1933.
7. **Köhler, W. and Otto, R.,** Hemmung der Stegasmenbildung durch Eliminierung der Serumlipide mit Dextransulfat, *Zentralbl. Bakteriol. Parasitenkd. Infektionskr. Hyg. Abt. 1 Orig.*, 202, 212–214, 1967.
8. **Nowak, J. and Lominski, I.,** Morphologie et evolution du microbe de l'agalaxie contagieuse des ovins et des caprins dans les milieux nutritifs, *Ann. Inst. Pasteur* (Paris), 53, 438–452, 1934.
9. **Ørskov, J.,** On the morphology of peripneumonia-virus, agalactis-virus and Seifferts microbes, *Acta Pathol. Microbiol. Scand.*, 19, 586–590, 1942.
10. **Partridge, S. M. and Klieneberger, E.,** Isolation of cholesterol from the oily droplets found in association with the L organism separated from *Streptobacillus moniliformis*, *J. Pathol.*, 52, 219–223, 1941.
11. **Prittwitz und Gaffron, J. von,** Fehlerquellen bei der Begutachtung von Kulturen und mikroskopischen Präparaten pleuropneumonieähnlicher organismen, *Naturwissenschaften*, 42, 113–115, 1955.

Chapter 37

THE XX CHROMOSOME

In the first edition of this book, the author noted that women had contributed greatly to knowledge of CWD organisms, perhaps resulting from the patience for which the sex is famous. Photographs of the following females ended the chapter: Eleanor Alexander-Jackson, Joan Apgar, Lieselotte Bloss-Bender, Antonina Brem, Ann Burgess, Eufalia Cabezas de Herrera, Judith Carleton, Sylvia Coleman, Anna Csillag, Mary Elizabeth Farkas, Laura Gutman, Betty Hatten, Gitta Iakovievna Kagan, Gertraud Kandler, Elena I. Koptelova, Mary Medill-Brown, Phyllis Pease, Jutta Preuner von Prittwitz, Linea Rydstedt, Antoinette Ryter, Edith Schuhmann, Florence Seibert, Lucille Tunstall, and Ruth Wittler. In the ensuing years women have continued to make significant findings as noted in the preceding chapters.

Sadly we note that Alexander-Jackson, though not forgotten, is gone; likewise, a leading Russian pioneer in "L-Form" characteristics and pathogenesis, Gitta Kagan, has expired. Her work is being advanced energetically by her many associates.

However, it must be humbly noted that the male sex has also shown remarkable patience. Examples include: Rodrick Chiodini, following the growth of CWD Mycobacterium paratuberculosis isolates until they finally reverted after 10 months; Arthur Kendall, who lectured unceasingly on bacterial filtrability which can now be demonstrated in a high school biology class; Gerald Domingue, who recently successfully treated idiopathic hematuria after testing the patient's dual CWD forms with 15 antimicrobials; Louis Dienes, who at age 80, manipulated the electron microscope over preparations of CWD microorganisms.

Chapter 38

MEDIA, METHODS, AND STAINS

MEDIA FOR CLINICAL USE

MEDIUM M 70

This medium is excellent for growing CWD organisms of most species, including microaerophiles, from patients' blood. Its suitability for anaerobes in an anaerobic incubator has not been evaluated.

Veal infusion broth Difco 25 g
Soluble starch Difco 0.5 g
Noble's agar 0.5 g
Tap water 900 ml

After dissolving the components the pH is adjusted to 5.5 with HCl and the medium tubed in 9-ml amounts in 15-mm-diameter tubes. These are autoclaved 15 min at 121°C and stored at 4°C for 4 or more days. Just prior to inoculation 1.0 ml of 10% solution of autoclaved yeast extract (Difco 0127) is added to each tube. The yeast extract is refrigerated at least 1 d before use. If particles form in the yeast extract, these are avoided. Sometimes yeast extract contains dead CWD colonies or other artifacts. These are removed by filtering through a 0.22 μm filter or a syringe before autoclaving.

The inoculum is 0.1 ml of the patient's anticoagulated blood. Incubation is with vigorous rotation on an automatic shaker at 35 to 37°C. Good growth of CWD forms is expected in 2 to 3 d.[53] For subculture of genera other than Mycobacteria, this medium may need 10% w/v sucrose.

When cultures are made to harvest DNA for analysis agar should be omitted. Our laboratory uses M 70 but substitutes 0.05% silica gel for agar.

ROSNER'S BLOOD CULTURE MEDIUM

- Brucella broth (Pfizer®)-45 ml per bottle
- Liquoid (sodium polyanetholesulfonate) in a final concentration of 0.05%
- Sucrose in a final concentration of 30%

When Rosner's blood culture medium was compared with Brucella broth, without Liquoid and without sucrose, there was a strong indication that, in blood, the pathogen often exists in an osmotically fragile stage which needs hypertonic medium for its first *in vitro* growth. Table 1 compares isolations of organisms in Brucella broth with and without modifications.

Similarly, classics from wounds and pleural fluids have grown only in hypertonic medium containing 5% horse serum.[109]

DOMINGUE'S MEDIA FOR ISOLATION OF CWD FORMS FROM BLOOD OR URINE[30]

Variant broth
    Brain heart infusion (BHI; Difco), 37 g

TABLE 1
The Frequency with Which Various Species were
Recovered from a Total of 1000 Patients

	Brucella broth, unmodified	Brucella broth with liquoid and sucrose[a]
Alpha-hemolytic streptococci	37	41
Pneumococcus	24	34
β-Hemolytic streptococci	10	12
Bacteroides species	—	11
Anaerobic streptococci	—	9
Neisseria meningitidis	—	5
Haemophilus species	—	4
Staphylococcus		
Mannite + Coagulase +	3	3
Pasteurella-like organisms	—	2

[a] The strains recorded as growing only in this medium did not need hypertonicity in subculture. This study did not include examination for nonreverting organisms. (The *Enterobacteriaceae* are omitted.)

    Yeast extract (Difco), 5 g
    Sucrose (Baker), 100 g
    Distilled H$_2$O, 1000 ml
Variant agar
    Variant broth, plus: Agar (Difco), 12 g/l
B9 broth
    Tryptic soy broth (Difco), 30 g
    Sodium chloride (Baker), 2 g
    Yeast extract (Difco), 10 g
    Distilled H$_2$O, 960 ml

    For solid medium (B9 agar) 10 g of agar is added.

HORSE MUSCLE INFUSION MEDIUM

    This medium, described in Section "Induction of Variants", is often, without antibiotic, used to culture L Forms from clinical materials in Russia.

GALACTOSE-CONTAINING MEDIUM

    Zafar[122,123] found 1.% D-galactose superior to sucrose in growing CWD forms of bacteria and Candida from pretherapy blood cultures. The variants grew in subculture when diluted 10^{12} in the presence of galactose, and only from dilutions through 10^4 in the absence of this carbohydrate. The medium also contained Casman's broth base (Difco), purified agar, filter-sterilized swine serum, filtered erythrocyte lysate, and separately autoclaved yeast extract. Because sterilizing serum with filtration often fails, treatment with β-propiolactone should be considered. The erythrocyte lysate of human cells was filter sterilized and used in a final concentration of 0.04%. Before autoclaving the yeast extract was filtered through a 0.22-μm filter to remove artifacts.

INDUCTION OF VARIANTS

PENICILLIN INDUCTION AND SUBCULTURE OF L FORMS OF STAPHYLOCOCCI[102]

This medium is suitable for induction of L Forms of any Staphylococcus species, for research or classroom study. L-Form colonies are anticipated in 4 to 7 d.

BHI (Difco)	3.7%
Agar, Noble's	1.3%
NaCl	3.0%
Ascitic fluid (Difco)	20% (preferable to 10% horse serum)
Penicillin G	100 U/ml to 1000 U depending on strain of Staphylococcus

The broth, agar, and NaCl are autoclaved at 120°C for 20 min. The ascitic fluid, which has been rapidly warmed to room temperature, is added to the agar which has been cooled to 55°C. The antibiotic is then added and the medium poured into petri dishes. An alternate method is to use this or another antimicrobial agent on a gradient as described below.

The medium is inoculated by spreading several drops of a dense 24-h culture on the surface. Aerobic and anaerobic incubations are equally effective. Cultures are sealed to prevent exsiccation.

GRADIENT PLATES FOR INDUCTION OR TO TEST VARIABLES

Information on a great number of variables may be obtained by gradient plates, incorporating any inhibitor or growth factor of interest or modifying the pH. It is convenient to employ a square petri dish and place the antibiotic, e.g., at one edge of the plate in a ditch or on filter paper.

Madoff and Dienes[69] used a gradient plate to obtain L Forms of *Neisseria meningitidis* in the following manner: a broth culture is streaked on a plate containing BHI broth, 1.2% agar, 10% sucrose, yeast extract 0.5%, and heat inactivated horse serum 10%. After distributing the inoculum over the agar surface, 0.25 ml of benzylpenicillin (100,000 U/ml) is placed in a trough or on a strip of filter paper. Plates are incubated at 37°C in candle jars containing moist gauze and are examined each day for 10 d under a dissecting microscope. When L-Form growth is present, blocks of agar containing L-Form colonies are cut out with a small metal spatula, inverted onto fresh serum-agar plates, and streaked with a spatula.

SEMIDEFINED MEDIUM FOR INDUCTION OF *STAPHYLOCOCCUS AUREUS* L FORMS[7]

Component	Amount	
Glucose	1.0	g
NaCl	5.0	g
Sodium lactate	2.0	g
Noble agar	750	mg
Vitamin free Casamino acids (Difco)	500	mg
L-Cystine	10	mg
L-Tryptophan	10	mg
K_2HPO_4	500	mg

Component	Amount	
$MgSO_4$-$7H_2O$	40	mg
$FeSO_4$-$4H_2O$	2	mg
$MnCl_2$-$4H_2O$	5	mg
Biotin	0.1	μg
Nicotinic acid	500	μg
Thiamine hydrochloride	100	μg
Penicillin	100,000	U
Distilled water	100	ml

Inocula

Growth of each strain in 18-h tryptose phosphate broth is packed by centrifugation and washed three times with 0.85% saline. The washed cells are diluted with saline to 80% transmittance at 530 μm in a Bausch & Lomb Spectronic-20 colorimeter. A 0.1-ml portion of this suspension is spread over the surface of each agar plate with a glass rod. Plates are incubated at 37°C. Aerobic cultures are sealed to prevent moisture loss. Anaerobic incubation is in Brewer anaerobic jars.

The optimum pH for L-colony formation is 6.5. More colonies develop using 0.75% agar than at higher concentrations. More L-colonies develop in pour plates than on streak plates and sometimes with anaerobic incubation.[7]

TO INDUCE L FORMS FROM STREPTOCOCCI[124]

Liquid tryptic digest of beef cardiac muscle
  Horse serum 10%
  NaCl 2.6%
  Penicillin 1000 U/ml
  Agar for semiliquid 0.3%
  Agar for solid medium 1.3%

TRYPTIC DIGEST OF CARDIAC MUSCLE[130]

Water is added to chopped fresh beef heart muscle for a ratio of water to meat of 1.5:1. This is boiled for 10 min and cooled. The meat is removed and ground. The pH of the broth is adjusted to 8.0. To 10 l of broth is added 6 kg of ground muscle, 1.5 kg of fresh pancreas, and 300 ml of chloroform. This mixture is hermetically sealed and incubated at 45 to 48°C for 10 d with regular mixing during the first 3 d. After 10 d the mixture is filtered and, if necessary, the pH of the filtrate adjusted to pH 8.0. The amino nitrogen content should be 600 mg%.

For nutrient medium the following are combined:

Muscle hydrolysate as above 200 ml
Yeast extract 100 ml
NaCl 5 g (0.5%)
Beef infusion 400 ml
Tap water 300 ml

Beef infusion is prepared by refrigerating ground beef with tap water for 24 h. The liquid is separated, boiled 5 min, filtered, and restored to twice its original volume.

TO INDUCE L FORMS FROM SALMONELLA[124]

Basic medium and horse serum are the same as for Streptococci but penicillin is 250 to 500 U/ml.

FOR INDUCING L FORMS OF MYCOBACTERIA[124]

KH_2PO_4 1.5 g	Na_2HPO_4 2.5 g
$MgSo_4 \cdot 7H_2O$ 0.5 g	Na citrate 1.5 g
Ferric ammonium citrate 0.05 g	L-Asparagine 1.0 g
Glycerol 30 ml	Agar 0.3%
Saccharose 20%	Beef serum 10%
Penicillin 5 to 10 U/ml	Streptomycin 5 to 10 U/ml
Distilled water to make 1000 ml	

Russian laboratories also use a membrane method to obtain pure cultures of induced L Forms of Mycobacteria. A sterile, semi-permeable membrane disk is placed on a petri dish of the antibiotic-containing medium and the Mycobacteria inoculated on the surface. Growth of the variants is expected in 2 to 3 weeks.[13] A profitable gradient effect can result from placing concentrated antibiotic in the center or at the edge of the plate when the medium is made semisolid.

Another method of inducing the L phase of Mycobacterium tuberculosis consists of culture on 10-ml slants of Kirschner's medium containing 50 μg/ml of cycloserine with an overlay of 10 ml of Kirscher's broth containing cycloserine.[75] CWD colonies are usually present by 72 h. For all culture of CWD forms large inocula are necessary.

INDUCTION OF CWD FORMS FROM *N. ASTEROIDES*[8]

Log phase (16 h) culture
Ingredients, w/v

BHI powder 4%
Yeast extract 0.4%
Sucrose 15%
NaCl 3%
Glycine 4%
Horse serum 10%
Lysozyme 2 mg/ml
Double glass-distilled water

The medium is sterilized by filtration. Bacteria are added to give an absorbency of 0.1 (λ = 580 nm). Two hundred milliliters of this suspension is placed in a sterile 2800-ml Fernbach flask and incubated with rotational agitation (150 rpm) for 6 h, followed by stationary incubation in a CO_2 incubator (5% CO_2 in air) for one week.

The growth is concentrated and transferred to fresh induction medium which differs from the above in substituting D-cycloserine (0.2 mg/ml) for the lysozyme. This culture is again incubated for 1 week in CO_2.

A third week of incubation is done after the growth is transferred to the new medium, omitting both inducing agents, lysozyme, and D-cycloserine.

FIGURE 1.   Stable L Forms of *Clostridium perfringens* and their growth on glass surfaces. (From Mahoney, D. E. and Moore, T. I., *Can. J. Microbiol.*, 22, 953–959, 1976. With permission.)

Additional subcultures are incubated in the BYE broth without glycine. All incubation is at 37°C.*

MEDIUM TO PRODUCE STABLE L FORMS OF *CLOSTRIDIUM PERFRINGENS*[71]

10 ml BHI broth (Difco)
10% Sucrose
2 U of penicillin G added to cooled mixture

The previously autoclaved BHI-sucrose medium was boiled and cooled just before inoculating the penicillin and 1 ml of overnight culture grown in cooked meat medium. Most experimental work was done with subcultures in this medium supplemented with sodium thioglycolate (0.1% w/v) and agar 0.05%. Penicillin could be omitted. The growth could be harvested from glass surfaces such as Leighton tubes or from bottles (Figure 1).[71]

MEDIUM TO OBTAIN L FORMS OF *N. MENINGITIDIS*[23]

BHI
Agar 1%
Polyvinylpyrrolidone (PVP) 7% Sigma
Horse serum 10%
Penicillin 10 U/ml

As a stabilizing agent PVP was superior to sucrose but was sometimes toxic if not cleared by dialysis of low-molecular-weight contaminating molecules. A CO_2 concentration of 10% or higher was required.

---

* The investigators explored several hundred variations of induction media and conditions. Significant factors were: age of culture; lot number of BHI, yeast extract and serum; temperature of incubation; CO_2 content of atmosphere.

MEDIUM FOR CONTINUED PROPAGATION
OF CWD FORMS

Nimmo and Blazic[88] studied optimum conditions for maintenance in the L phase of six common pathogens: *Escherichia coli; Proteus mirabilis; Klebsiella*, coagulase positive and negative Staphylococci; Enterococci; and β-hemolytic Streptococci. Important were highly purified agar, magnesium, and sucrose. Soluble starch in a concentration of 0.5% is well known to neutralize toxic components of culture media.

A chemically defined medium revealed that requirements for a lysozyme-induced CWD variant of a streptococcus did not differ from those of its parent form.[43] The functional medium is given in the following table:

Composition of Medium for Growth of
Streptococci faecium F24 and a Derived L-Phase Variant[43]

Component	Amount (μg/ml)
Amino acids[a]	
Arginine-HCl	200
Asparagine	5
Aspartic acid	100
Cystine	200
Glutamic acid	300
Glycine, histidine-HCl, hydroxyproline	200
Isoleucine, leucine, lysine-HCl, methionine, and	
phenylalanine	100
Proline	200
Serine, threonine	100
Tryptophan	200
Tyrosine and valine each	100
Glucose	2.0×10^4
Purines and pyrimidine	
Adenine-H_2SO_4-$2H_2O$, guanine-HCl-H_2O, and uracil each	30
Salts	
$(NH_4)_2SO_4$	600
$FeSO_4$-$7H_2O$	10
$MgSO_4$-$7H_2O$	200
$MnSO_4$-$4H_2O$ and NaCl each	10
Na_2HPO_4	1.6×10^4
KH_2PO_4	442
K_2HPO_4	305
NH_4Cl	1.5×10^4
Vitamins	
para-Aminobenzoic acid	0.04
Biotin	0.005
Folic acid	0.05
Pantothenic acid (Ca salt), pyridoxamine-2HCl	0.4

Component	Amount (μg/ml)
Nicotinamide	1.0
Riboflavin, thiamin-HCl	0.2

[a] L-Isomers, except alanine, were used.

MEDIUM FOR MAINTAINING L FORM OF *STAPHYLOCOCCUS AUREUS*

The semisynthetic medium described below which permitted continued propagation of *S. aureus* L Forms was devised by Takahashi and associates.[111]

Component	Amount (μg/ml)
Amino acids	
L-Arginine-HCl	1530
L-Valine	85
L-Threonine	87
L-Lysine-HCl	265
L-Methionine	33
Salts	
NaCl	5.5×10^4
KCl	1.6×10^3
Na$_2$HPO$_4$ (anhydrous)	1.0×10^3
CuSO$_4$-5H$_2$O	1.8×10^{-2}
ZnSO$_4$-7H$_2$O	6.3
FeSO$_4$-7H$_2$O	6.1
MgCl$_2$ (anhydrous)	420
CaCl$_2$ (anhydrous)	240
MnCl$_2$-4H$_2$O	5.0×10^1
Also needed	
Hypoxanthine	30
Sodium lactate (70% solution)	2.0×10^4
Sodium pyruvate	800
HEPES[a]	6.0×10^4
Nicotinic acid	1.0×10^1
BME vitamin solution (GIBCO Labs, Grand Island, NY)	

        Biotin, D-Ca pantothenate, choline chloride, folic acid, nicotinamide, pyridoxal-HCl, thiamine-HCl, 100 mg/l; *i*-inositol 180.0 mg/l; riboflavin 10.0 mg/l.

[a] *N*-2-Hydroxyethylpiperazine-*N'*-2-ethanesulfonic acid.

MEDIUM FOR MAINTENANCE OF CWD MYCOBACTERIA

CWD Forms of Mycobacteria may be maintained on the inducing medium or on the following agars. When the inducing agent is removed, they may or may not revert. Ratnam and Chandrasekhar have employed Murohashi and Yoshida's medium[85] in culture of *Mycobacterium leprae* and Kirchner's medium prepared as a semisolid. Murohashi and Yoshida's medium:

Monopotassium orthophosphate	2 g
Disodium orthophosphate (12 H$_2$O)	1.5 g

Asparagine	3 g
Magnesium sulfate	0.02 g
Sodium citrate	2 g
Calcium chloride	0.0025 g
Yeast RNA	100 μg
Glycerol	30 ml
Distilled water to	1000 ml

The pH of the solution is adjusted to 6.6 to 6.8 with sodium hydroxide, 1 g of powdered agar is added, and the mixture is autoclaved. This agar medium is stored at 5°C.

Separately, a 5% solution of bovine serum albumin Fraction V[1] and a 20% solution of glucose are sterilized by filtration.

The agar medium is dispensed in 9-ml quantities and to each tube, 1 ml of bovine serum albumin solution and 0.5 ml of glucose solution is added. The tubes are then heated to 40 to 45°C and kept at this temperature until inoculated.

CAPRICE OF CULTURE MEDIA
AND OF CWD MICROBES

Surprising differences exist between culture of CWD forms and the orthodox stage studied for 100 years. Even rules sufficient for Mycoplasma are inadequate here.

AGE OF MEDIA

For many classic organisms it is imperative to use medium on the day of preparation. For CWD forms the converse is usually true. Farrell[127] found that freshly prepared medium may permit no growth in pour plates, whereas several thousand colonies develop in the same medium aged 5 d before supplements are added and pour plates made. Farrell used trypticase soy agar plus 0.5% additional agar. Horse serum and filtered yeast extract were added to the autoclaved medium to give a concentration of 1% of each supplement.

Altenbern found that aging media lessened its toxic properties, determining that molecules resembling formate might be the inhibitors.[4]

WATER

When O'Beirne and Eveland[125] set out to extend the laboratory's studies of Listeria L Forms, induction of L-variants failed to occur. Water in the medium was found to be the negating factor. With glass-distilled water, variation occurred as before.

Makemson and Darwish noted that successful culturing of *E. coli* L Forms was associated with cleaning the distilled water system. Unless the still was freshly cleaned, calcium in the water was deposited along the pipes. Growth was successful when they supplemented their heart infusion broth-sucrose-agar-penicillin with finely titrated amounts of calcium chloride and magnesium sulfate. It was not satisfactory to use both salts as chlorides.[72] It will be noted that the medium used at Gamalaya Institute is made with tap water. We agree that tap water usually contains necessary salts.

AGAR SOURCE AND CONCENTRATION

The most carefully formulated medium will not permit growth of wall-deficient bacteria if it contains certain agar. Agar toxic for Mycoplasma was noted by Lynn and Morton[67] among agars satisfactory for classic bacteria. The situation is comparable for wall-deficient bacteria, whether the strain originates *in vitro* or *in vivo*. Schuhmann and Taubeneck found that agar had to be carefully selected from lots on the international market.[104] In independent studies in our laboratory, Marilyn Williams and King Adamson[128,129] evaluated agars in culturing 25 CWD isolates from bacteremia. Agarose, samples especially purified for elec-trophoresis, tissue culture agar, and agars for routine bacteriological media were compared. Some standard agar-agar was entirely satisfactory. On the other hand, toxic lots were en-countered in both standard and purified samples. In general, purified agars were superior. Agarose was not superior and was inadequate for some fungi in the classic stage.

Most CWD forms require a trace of agar, e.g., 0.07%, if they are to grow in broth. For some species gelatin or silica gel can substitute for agar. Pour plates show better growth if the consistency is soft. Fastidious species will not grow on media surfaces but grow in pour plates.

In our laboratory benefits occur from (1) using a total of 1% yeast extract, autoclaved separately, (2) adding purified agar to broth to make a semiliquid, (3) aging the medium 5 d, and (4) making pour plates rather than surface inoculation. The source of water can be critical. For all autoclaved medium we use tap water. Someone has remarked "There is something good about Detroit!"

MEAT INFUSIONS

Meat infusion, the common ingredient for culture of classic bacteria, can negate growth of CWD forms,[104] both of the *in vitro*-induced and *in vivo*-occurring organisms. Casamino acids rather than tissue infusions are needed for some isolates.

SERA

Sera of man and other sources need heat inactivation, as L-phase variants of staphylococci and presumably other species are killed by normal serum.[17] Agamma swine serum is a preferable enrichment.

It cannot be emphasized too strongly that serum, even after filtration, usually contains microorganisms. Filtering twice through 0.22-μm filters is recommended with a day at room temperature between filtrations. Obviously, minimally contaminated serum causes less prob-lem in solid than in other media. Contaminants are especially troublesome in CWD studies because the cultures are scrutinized closely.

Components in serum toxic to CWD forms were investigated by Godzeski et al.[37] They found that inhibition occurred only from the combination of serum fractions soluble and insoluble in 50% acetone. Lorkiewicz[64] concluded that serum does nothing in culture of CWD Proteus except neutralize inhibitors in the medium. He found that 0.5% charcoal could substitute for serum. Thus, although serum may be inhibitive, usually it neutralizes more toxicity than it contributes.

CHANGES IN THE SAME MEDIUM

The literature shows that it is not rare to culture forms on a certain medium at one interval and later find this medium unsatisfactory. One experience concerns thiamine assay medium (Difco). A long-stored sample was an excellent medium. New lots did not support L-phase growth.

eH REQUIREMENTS

Some aerobic bacteria are no longer O_2-tolerant in the CWD state. The penicillin-induced variant of *Salmonella typhimurium* may develop only under anaerobic conditions.[105] It was early noted that unless medium is ideal, growth of Salmonella CWD forms is limited to an anaerobic milieu.[29] Aged cultures of *P. mirabilis* often grow only as CWD anaerobes. This anaerobic requirement of many isolates is significant in clinical work. The variant of *Staph-*

ylococcus aureus in the spinal fluid of an untreated patient initially grew only anaerobically, as described in the Meningitis chapter.

pH REQUIREMENTS

The optimum pH for variant growth may not resemble the pH optimum of the parent. In agitated cultures, in medium containing a trace of agar, CWD forms of most species show greatly concentrated growth at pH 5.5 to 6.0. This includes the Mycobacteria.[53]

Without agitation, many variants prefer neutrality.[72] Very commonly a pH of 7.8 is preferred, as noted in studies of *Listeria monocytogenes*.[32]

ERYTHROCYTE LYSATE

As true for Mycoplasma, growth of wall-deficient variants is often improved when the serum employed for enrichment is from hemolyzed blood.[83] The entire content of the red cell is needed, as dried hemoglobin or hemin, while improving media, will not foster as small an inoculum as will erythrocyte lysate.[122]

OSMOTIC STABILIZERS AND ADJUSTING TO NORMAL OSMOLALITY

For many years, attempts to propagate CWD forms met with minimal success. This situation changed after two reports. One, by Lederberg and St. Clair,[63] noted that sucrose stabilizes penicillin-induced wall-deficient bacteria in accord with Weibull's data that protoplasts of Bacilli are osmotically stabilized by this carbohydrate.[119] The second publication, by Dienes and Sharp,[28] stated that sodium chloride could supply the hypertonicity needed for maintenance of streptococcal CWD forms. Awareness of the necessity for osmotic balance brought CWD-form investigations into many laboratories.

NaCl and sucrose remain the most commonly used agents to create hypertonicity. Sodium chloride, a favorite in induction and propagation of CWD forms of Staphylococci, often causes extracellular masses of DNA to form. Sometimes one must let the organism make the choice. For instance, with Streptococcus MG, Madoff[68] found that one strain could only be induced in the presence of 3% sodium chloride, whereas two other strains required sucrose in a concentration of 10% or greater. Sodium chloride brings almost universal success with Staphylococci, but results vary with streptococci.[79]

Except for induction of Staphylococci, sucrose is usually superior to NaCl. For example, Watanakunakorn and associates used high salt for L-Form induction of Staphylococci[117] but employed 10% sucrose to induce L Forms from Pseudomonas.[118] Table 2 shows that a combination of NaCl and sucrose may be ideal.[34]

SOLID VS. LIQUID MEDIUM

Stable L Forms of *S. aureus* which can grow on solid medium containing 0.85% NaCl may require 2.0% NaCl for propagation in a liquid medium. This augmentation generally applies to other stabilizing agents also.

STABILIZING AMINES

As spermine stabilizes the CWD strains, viability may be retained or lost. Organisms retaining viability have been *E. coli*,[45,110] Erwinia,[47] Hemophilus influenzae,[62] and an

TABLE 2
Optimum Concentrations of NaCl Plus Sucrose in
Conversion of *S. aureus* to CWD Form

NaCl (%)	Sucrose conc. (%)	Number of colonies per plate			
		1st day	3rd day	4th day	10th day
5.0	—	—	—	80	87
3.0	—	—	—	76	93
2.5	—	—	—	3	3
0.0	—	—	—	—	—
5.0	5.0	138	518	518	518
3.0	5.0	270	1120	1120	1120
2.5	5.0	297	690	690	690
2.0	5.0	40	263	263	263
1.5	5.0	—	10	92	92
0.0	5.0	—	—	—	—

Note: Incubation at 37°C on L-medium. Numbers were similar with aerobic and anaerobic incubation.

From Fodor, M., *Z. Allg. Mikrobiol.*, 9, 95–101, 1969. With permission.

unidentified Gram-negative marine bacterium.[18] Other organisms were protected from lysis but lost viability, namely, Streptococcus[49] and Micrococcus.[18]

Harold[49] concluded that amines stabilize by cation binding to cell membrane, a theory supported by the finding that acidic polymers such as RNA and heparin nullify the protective effect of the polyamines.[70] The action of spermine is not limited to the membrane, as shown by a study of Groman and Suzuki.[45] Phage maturation in *E. coli* is effected: phages Δ and f_2 are inhibited by spermine, whereas reproduction of ΦX 174 is enhanced.

INORGANIC SALTS IN STABILIZING

Na^+, Ca^{2+}, Mg^{2+}, and K^+ appear to be interchangeably adequate in stabilizing some penicillin-induced CWD forms. However, requirements for many genera are unique. There may even be differences for species within a genus as in Nocardia.[8] More than one cation may be needed in a finely titrated level.[72] Some of the salts employed are listed in Table 2. It is enigmatic that EDTA can stabilize some wall-deficient forms of *Clostridium sporogenes*,[35] whereas calcium ions are needed for induction and propagation of *E. coli* spheroplasts.[72]

Protoplasts of *Bacillus megaterium* and *B. cereus,* prepared by autolytic enzymes, are best stabilized at pH 7 to 7.5 in the presence of 0.2 *M* sucrose, plus cobalt ions.[25] KCl is superior to 20% sucrose to prevent lysis of *Neurospora crassa*.[116] This is an area where more work is needed. In many instances, available data do not denote whether multiplication occurs in the salt studied.

HYPERTONICITY FROM CARBOHYDRATES

As discussed above, sucrose has been used in induction in widely varying concentrations. Therefore, a titration is desirable for each organism to determine the optimum to prevent lysis, yet allow growth. Galli and Hughes found that 34% sucrose was required to prevent protein leakage from *C. sporogenes*. Sixty percent sucrose, which approaches saturation, was employed to stabilize spheroplasts of *E. coli*, but their viability was not recorded.[35] A

TABLE 3
Osmotic Stabilizers of CWD Forms including
Protoplasts or Spheroplasts

Carbohydrates	Salts or ions[a]
Cellobiose(1)[b]	Na$^+$(16, 19, 27, 48, 65, 95, 103, 121)
Lactose (1)	KCl, 1 M (28, 93, 116)
Maltose 17% (46)	Cobalt sulfate (25)
Mannitol (31, 51)	NH$_4$Cl (98, 12)
Melibiose (1)	Ca^{2+} (28, 72)
Raffinose (1)	Na$_2$HPO$_4$ (28, 61)
Rhamnose (31)	Mg^{2+} (14, 28, 88)
Sucrose	
15% for *V. comma*(21)	
20% for *N. crassa* (73)	
Disodium succinate (4, 33, 59)	

Polyamines and peptides	Amino acids and their methyl esters
Spermine (18, 46, 49, 62, 110)	Cysteine (35)
Spermidine (49, 110)	Methyl esters of, (49)
Streptomycin (49, 93, 110)	Lysine
Protamine (49)	Arginine
Polylysine (49)	Histidine
Cadaverine (49)	0.5 M L-proline for
Putrescine (49)	Cyanobacteria (26)
Polymyxin (87)	

Miscellaneous

EDTA(35)
Polyethylene glycol 17% (81)
Polyvinylpyrollidone 7% (6)

[a] Commonly a salt and larger molecule are cooperative.
[b] Numbers in parentheses indicate Reference numbers.

great many small carbohydrate molecules may substitute for sucrose, as shown in Table 3. One commonly employed is succinate.

ADAPTATION OF CWD *S. AUREUS* TO NORMAL OSMOLALITY

NaCl added (%)	Osmolality of medium	No. of serial transfers
4.5	1950	3
4.0	1775	3
3.5	1600	3
3.0	1425	5
2.5	1235	5
2.0	1075	5

NaCl added (%)	Osmolality of medium	No. of serial transfers
1.5	900	49
1.0	725	58
0.5	550	72
0.0	360	120

Tubes containing 3.5 ml of BHI broth supplemented with NaCl were inoculated with 0.01 ml of 3- to 4-d-old cultures.[113]

ESSENTIAL VITAMINS

Just as growth requirements assist in taxonomy of classic microorganisms, vitamin requirements may help determine the identities of all the look-alike CWD microbes. The few facts known about their nutrition indicate that although there are remarkable differences in requirements, there are enough similarities between parent and CWD offspring to show relationship. The most studied organism in this respect is *P. mirabilis,* which as a classic bacterium and as an L-phase variant needs only one vitamin — nicotinamide. The fundamental study of Medill and O'Kane[78] showed that *P. mirabilis* variants grow at their maximum rate when all vitamins except nicotinamide are excluded from the medium. This same monovitamin requirement applies to the L-phase of other Proteus species: *P. vulgaris,*[76] *P. morgani,*[76] and *P. rettgeri.*[58,76]

Banville showed that the wall-deficient stage of *S. aureus* differs from that of Proteus in requiring three vitamins: thiamin, niacin, and biotin.[7] Brief multiplication of staphylococcal CWD forms is possible with only one or two vitamins.[77] However, for sustained growth the staphylococcal variants need the same additives as the parent Staphylococcus.

Similarly, CWD colonies of Streptococci[115] may bypass the special growth requirements of the parent organism for one subculture, but such lack of fastidiousness does not persist through repeated subcultures. It seems that required growth supplements are carried over in adequate amounts for one subculture. It can be concluded that wall-deficient variants usually have the same basic requirements as the parents in a qualitative sense, but the quantity of the vitamin required may be minute. This could mean that the classic wall nonspecifically absorbs nutrilites.

In general, growth factors needed by the classic bacteria are needed for continuous propagation of the wall-deficient stages. Thus, Pseudomonas[39,56,107] and Providencia[20] spheroplasts can be expected to have no vitamin or nucleotide requirements. A more demanding organism such as *Vibrio cholerae*[40] may need multiple nucleotides for the variant as well as the parent. When serum is required by the variant only, its function is assumed to be adsorbing or competing with toxic molecules which otherwise damage the relatively exposed membrane of the variant.

YEAST EXTRACTS: SOURCES, "STERILIZING", AND CONTENT

One commercial company offers seven yeast extracts. Which ones are preferable for wall deficient bacteria? Godzeski[37] noted that not only the source, but also the lot number is important. The brand of the yeast is as important as its concentration.

Since part of the vitamin content of yeast is thermolabile, filtration of preparations is preferable to autoclaving. Pistole,[126] studying growth factors for Peptococci, found Difco's

0127 yeast extract stimulated growth very little if autoclaved with the medium, gave improvement if autoclaved separately, and was maximally effective if filter-sterilized. A concentration of 1% was optimum and greater concentrations inhibitory. These facts also apply to enriching medium for CWD forms. Sterility is not approached unless twice filtered through a 0.22-μm filter. Many commercial preparations have a few organisms which cause no confusion in solid medium but are unsuitable for liquid culture.

The essential ingredients in yeast extract need to be identified. It is usually assumed that nucleotides are contributed as well as vitamins. The requirements for CWD variants are more complex than the nicotinamide nucleoside needed for *H. influenzae*.[66] Although they occasionally grow on chocolate agar supplemented with yeast extract, adding nucleotides and hemin to a base medium does not usually result in adequate growth. Testing yeast components for growth-fostering ability is complicated because, occasionally, yeast extract merely buffers toxic molecules in culture medium.

AMINO ACID REQUIREMENTS FOR NUTRITION

DL-amino acids usually are readily assimilated by both parent and aberrant bacteria, enhancing their growth. However, inherent differences in nitrogen requirements exist between genera. Vitamin-free casamino acids in a final concentration of 0.5% support growth of classic and variant *P. mirabilis* in synthetic solid medium.[78] Similarly, 1% casamino acids is adequate for culturing L-phase colonies of *P. vulgaris* in liquid medium.[2] Salton and Shafa, however, were unable to grow spheroplasts of *Salmonella gallinarum* until they increased the concentration of casamino acids from 0.5 to 2.0%.[100]

Atmospheric environment affects the form in which some variants can accept nitrogen. Inclusion of a series of amino acids in solid medium for the cultivation of Proteus L Forms actually inhibits L-growth under aerobic conditions, but does not under reduced oxygen tension. The reverse is true when NH_4SO_4 replaces the amino acids as a nitrogen source.[80]

MEDIA COMPONENTS WHICH INHIBIT CWD VARIANTS

As described above, most media contain bactericidal or bacteriostatic compounds inhibitive to CWD organisms.

Medill and O'Kane[78] convincingly demonstrated that for many species, serum is needed for CWD growth only if the medium is toxic, as is true of any meat infusion product.

However, the parent, as well as its CWD stage, usually has complex needs for nutritive supplements. These vary tremendously; e.g., they are not identical even for different species of Bacteroides. The common solution is to add serum to a concentration of 30% or whole blood. Here it becomes apparent that there are two facets to additives. There are few growth stimulants which do not inhibit some microbial species. Other growth factors which are known inhibitors for some microorganisms include the following: folic acid (as little as 0.001 μg/ml),[97] biotin,[24,86] thiamin,[94] heme in the same concentration needed as the X factor for Hemophilis[36] (heme, which is in meat infusion, can be neutralized if a reducing substance is added to the medium), nicotinic acid (inhibition due to an excess can be reversed by yeast extract),[57] and riboflavin (at least for one microorganism, riboflavin inhibition is specifically reversed by excess leucine).[106] It seems reasonable that the CWD stages of bacteria may be hypersensitive to these additives. The basic Proteus model shows that the classic bacterium tolerates inhibitors in natural media which are completely bacteriostatic for the L-phase.

REVERSION METHODS

Many, including Nimmo and Blazevic,[88] have noted that reversion improves by increasing agar concentration and by omitting serum and antibiotic. Reversion in thiogycolate agar is improved by inoculating a tube of heat-liquified medium at 50°C.

REVERSION A LA DOMINGUE

Domingue found "In most instances reversion occurred after one or two passages in tubes containing equal volumes of variant broth and trypticase soy broth with 5% horse serum or tubes devoid of horse serum. In some instances reversion occurred after variant broth cultures were streaked directly onto sheep blood agar plates."[30]

OGAWA MEDIUM FOR MYCOBACTERIA

To our knowledge this medium has not been used in studies of CWD Mycobacteria. However, its reported superiority to the more commonly employed Lowenstein-Jensen's medium suggests that it can have a place in reversion studies. It is thought the improved growth rate relates to its acid pH.[92]

Composition of Ogawa Egg Yolk Medium
for Cultivation of Mycobacteria

Basal medium: KH_2PO_4 1 g
            Na glutamate   1 g
            Glycerol   6 ml
            Water   100 ml
Autoclave: 121°C, 15 min.
Add: Egg yolk   200 ml
      Malachite green 2%   5 ml
pH = 5.9–6.1
Inspissate: On two consecutive days at 85 and 75°C for 40 min and 30 min, respectively[90]

STORAGE OF THE VARIANTS

The variants have retained viability and characteristics when stored in glycerol or 10% dimethylsulfoxide at −80°C.[112] CWD forms of many genera, including mycobacteria, retain viability at 4°C for several years. The organisms of tuberculosis and sarcoidosis in blood cultures usually live for 10 years or more at 4°C.

POUR PLATES OF DISTILLED OR DEIONIZED WATER

This makes a simple, interesting experiment for a high school biology class; the more days the water stands at room temperature the better. Fifteen milliliters of distilled or deionized water is centrifuged at 5000 rpm for 30 min, the supernate aseptically removed, and the sediment in 0.3 ml of the water transferred to sterile petri dishes.

Pour plates are made in triplicate using any relatively artifact-free agar medium. Artifacts can be removed before autoclaving by filtering the hot medium through Kim Wipes®. It usually is not necessary to add serum or ascitic fluid. The plates are incubated under refrigeration, at room temperature, and at 37°C, taking care to prevent evaporation. Different colonies will appear with temperature variation. Growth may be best at or near the bottom of the plate. Colonies are expected within 7 d. The plates are examined at a magnification of 100 or 200 ×.

PREPARATIONS FOR ELECTRON MICROSCOPY

Lysis of fragile CWD forms often occurs in preparation for electron microscopy. Obviously, lysis indicates a need for osmotic stabilizers during fixation preceding electron microscopy.

Domingue obtained excellent electron micrographs using the following procedure with hypertonic buffer. A 35-ml sample of variant broth culture is centrifuged at 10,000 rpm for 30 min at 20°C. The resulting pellet is fixed at 4°C for 2 h. in 3% glutaraldehyde in 0.1 *M* sodium cacodylate buffer, pH 7.2, with 7.5% sucrose as osmotic stabilizer. After buffer wash (3 × 5 min) the pellet is postfixed 2 h at 4°C in 1% osmium tetroxide. Following another buffer wash the pellet is stained for 1.5 h at 4°C with 1% aqueous uranyl acetate. A final buffer wash precedes dehydration through ethanol, epoxypropane (10 min) and an overnight treatment in epoxypropane-Epon 812 (1:1) at 4°C. The pellet is then broken into small fragments for embedding into Epon 812. Thin sections (60 to 90 μm) are cut on a Reichert ultramicrotome and grid-stained with lead citrate. Electron photomicrographs are taken using a Siemens Elmiskop 102 electron microscope.[30]

Susan[108] found that CWD *Pseudomonas aeruginosa* disappeared when using buffers of normal osmolality in either the OsO_4-uranyl acetate or the glutaraldehyde OsO_4 method. However, the 0.05 μm bodies were numerous when shadowed preparations were made. For this purpose a small drop of culture medium (control) or diluted suspension of organisms was placed on a 200 mesh grid coated with Formvar and stabilized with carbon. The grid was allowed to stand 1 to 5 min before the drop was washed off with distilled, filtered, water. The grids were shadowed at 20°C at a distance of 5 in. from a platinum-carbon pellet in an Illini vacuum evaporator.[108]

Helpful modifications for electron microscopy include treatment of bacteria with 0.5% uranyl acetate for 30 min before dehydration and use of Lowicryl HM20 rather than the commonly used Lowicryl K4M. Preembedding in gelatin rather than agar improves sectioning of some species.[9]

STAINS

FIXATION AND WASHING SMEARS

Since most CWD organisms cannot be heat fixed without destroying their morphology, attaching them to a glass slide becomes a problem. After chemical fixation the majority of microcolonies may wash down the sink. It is advisable to check a wet, unstained drop of the culture microscopically to learn what should be seen in the fixed, stained preparation. It is helpful to add a loop of Fluorochrome B solution (Polysciences Inc., Warrington, PA) to the wet preparation to color the colonies. As a fixative 10% formalin may be superior to methanol. Dilute protein added to the slide increases retention of the slippery organisms. Acetone added as a drop and allowed to evaporate gives excellent preservation of organisms' morphology, but destroys blood cells. Washing of stained slides must be done gently with cold water from a beaker. Centrifuging a small quantity of culture and staining the sediment is a certain way to have good results when the culture broth is semiliquid, i.e., contains a little agar.

ACRIDINE ORANGE STAINING[22,50]

Acridine orange (AO) was early used as a cytological stain by Von Bertalanffy and co-workers[11] and by Armstrong.[5]

A stable preparation is available from Difco. Or the stain can be prepared in the laboratory as follows:

1. Stock Stain:

Acridine orange (G. T. Gurr, Ltd., London)	1.0	g
Tween 80	2.0	ml
Tap water	1000.0	ml

2. McIlvaine's Buffer (pH 3.8)

Na$_2$HPO$_4$	10.081	g
Citric acid monohydrate	13.554	g
Tap water	1000.0	ml

3. Working stain: Dilute one part stock stain with one part McIlvaine's buffer. Store the stock stain and the buffer in aliquots at $-10°C$ and mix the day of use. Both stain and buffer are stable for at least a year if kept frozen. Interesting microorganisms can grow in the solutions if stored unfrozen (see Frontispiece).

PROCEDURE FOR ACRIDINE ORANGE STAIN

Glass slides of a type recommended for fluorescent microscopy are used. The procedure can be simplified by adding a loop of liquid culture or a few colonies from solid medium to a drop of AO on the slide and placing a coverslip on this.

When examined with a fluorescence microscope classic bacteria and CWD forms stain bright orange, except aged growth, which appears green. Five min is usually adequate for staining.

The routine methods used by most clinical laboratories, with alcohol-fixed smears, reveal many CWD microcolonies accompanying any classic organisms. The most interesting slides are ones showing only CWD forms. Since these variants do not grow on the surface of plates in macroscopic colonies, the CWD variants are usually ignored as artifacts. They often grow as minute colonies on the surface of blood agar plates and better on chocolate agar, especially if enriched with separately autoclaved yeast extract.

As noted by Dienes, some CWD organisms absorb stains slowly, and this is true of AO. Incubation overnight in a Coplin jar is necessary for some species.

ACRIDINE ORANGE ACID-FAST STAIN

AO was used in the detection of acid-fast organisms in the diagnosis of tuberculosis by Bravo Oliva and Garcia Rodreguez.[15] Our laboratory modified their method to reveal acid-fast characteristics of the CWD forms of Mycobacteria.

Prepared smears are heat-fixed. Mild heat fixation is not as damaging to the morphology of CWD mycobacteria as to most species in this phase. The AO stain in distilled water is applied 24 to 36 h in a staining jar, then washed briefly and gently with cold water. Better staining results when buffer is not added to the AO. Acid alcohol (3% HCl in 95% ethanol) is applied for decolorizing. The smear is washed, then counterstained for only a few seconds with a mixed solution of methylene blue and crystal violet. After the final washing, the smears are mounted and protected with a cover slip before examination. The buffer of the standard AO stain (pH 3.8) may be employed as a mounting fluid. Acid-fast organisms appear an intense orange on a dark background.

Known positive and negative smears should be stained as controls to gauge decolorizing time. The most satisfactory specimens for positive controls are sputum samples from untreated cases of pulmonary tuberculosis or 3-d blood cultures from febrile patients with mycobacteriosis.

FLUORESCENT MURAMIDASE

Fluorescent stains are needed which are specific for microbes rather than tissue cells. Fluorescein or rhodamine-labeled muramidase is a valuable tool in this respect. A lysozyme sensitive strain of *B. megaterium* is used as the positive control for the activity of the labeled muramidase. For this purpose an 18-h culture of *B. megaterium* is suspended in sterile 6% sucrose (see Frontispiece). Other species of Bacilli may be substituted. A smear of gum scraping from a human mouth is also suitable.

For staining, equal volumes of conjugate and specimen are diluted on a slide and a cover slip is applied. This is examined after 5 min. If cells are very numerous, the specimen is preliminarily diluted with an equal volume of broth.

RHODAMINE B-LABELED MURAMIDASE[41,91]

A. Carbonate bicarbonate buffer
1. Na_2CO_3 26.5 gm
H_2O q.s.   500.0 ml
2. $NaHCO_3$ 21.0 gm
H_2O q.s.   500.0 ml

One part of (1) was mixed with four parts of (2). If necessary, the mixture is adjusted to pH 9.0 using either HCl or NaOH.

B. Buffered saline for dialysis of rhodamine B-labeled muramidase
NaCl   120.0 g
NaH_2PO_4 20.7 g
H_2O q.s.   15.0 l

If necessary, the mixture is adjusted to pH 7.0 with 40% NaOH.

C. Conjugation of rhodamine B to muramidase

Five-hundredths milligram of Rhodamine B (CI 45170, Matheson, Coleman & Bell) is used for every milligram of muramidase (salt-free, Nutritional Biochemical Corporation). Ten milliliter buffered saline, 3.0 ml carbonate bicarbonate buffer, and 2.0 ml water-free acetone are combined in a 50.0-ml Erlenmeyer flask. This mixture is cooled in an acetone dry-ice bath until crystals of ice form in the flask. The mixture is then placed on a magnetic stirrer and 250.0 mg of muramidase, suspended in 10.0 ml of buffered saline is added slowly in a fine stream. When the addition is complete, 12.5 mg of rhodamine B, dissolved in 1.5 ml acetone, is added dropwise to the cold, stirred protein solution. The mixture is transferred to a refrigerator at 4°C and agitated with a magnetic stirrer for 18 h.

The conjugate is transferred to a mechanical dialyzer and dialyzed against the phosphate buffered saline with changes as necessary until the dialysate shows no fluorescence when viewed under ultraviolet light. The conjugate is centrifuged for 15 min at 10,000 rpm. The supernate is removed and filtered through a sterile 0.22-μm Millipore-membrane filter. For use 0.03 ml of the specimen or culture is mixed on a slide with 0.03 ml of labeled muramidase and examined with the fluorescent microscope. The conjugate is stored in 2.0-ml ampules at 4°C with a final concentration of 10.0 mg muramidase per milliliter. Unfortunately, in our experience, freezing or lyophilizing causes the stain to become insoluble.

Previously Gould et al.[41] found that a minimal concentration of fluorescent muramidase would highlight microbes without causing their lysis. Fortunately, much CWD growth contains sufficient murein to adsorb the enzyme (see Frontispiece).

DIENES' STAIN FOR IMPRESSION SMEARS

Methylene blue	2.5 g
Azur II	1.25 g
Maltose	10.0 g
Na_2CO_3	0.25 g
Benzoic acid	0.2 g
Water	100.0 ml

A clean cover slip is flamed and pressed firmly on agar surface containing colonies. With forceps the cover slip is removed, placed with smear up on a clean slide, and secured with drops of rubber cement at the corners. The smear is fixed with a drop of methanol which is allowed to evaporate. Dienes' stain is applied for 3 min and the preparation washed gently with cold water.

Colonies of mycoplasma and L Forms stain bright blue, as do young viable bacterial colonies. Autolyzed bacteria, cellular material, and debris take on a pink hue. The agar stains faintly blue or violet. If the agar stains too deeply, the stain may be diluted.

CWD colonies of either the "fried egg" or granular type can be stained on the agar surface if the medium is clear, i.e. contains only serum, ascitic fluid or erythrocyte lysate. The plate is flooded with Dienes' stain and immediately tilted to prevent deep penetration of the stain into the agar. Dienes' stain is usually too concentrated for this purpose until diluted approximately 1:4. The colonies are then viewed with a magnification of 100 or 200 × or on very thin medium, with 400 ×. Methylene blue or dilute basic fuchsin can also be used for staining surface growth. Any stain should be filtered through Whatman No.1 filter paper just before use.

VITAL STAINING OF L FORMS WITH CHLORAZOL BLACK E[10]

Chlorazol black E (CBE) is an amphoteric tris-azo dye, also known as Pontamine Black E, Erie Black, Direct Black, and Renol Black. It stains colonies during growth without inhibition; in some cases it actually stimulates growth.

The dye may be incorporated in pour plates in final concentrations varying from 0.1 to 0.001%. The optimum concentration varies with the medium and organism.

KINYOUN'S STAIN,[54] INTENSIFIED FOR DETECTION OF ACID-FAST CWD FORMS

Commercial Kinyoun's stain may be used. Different lots from the same company vary in effectiveness. The stain may be prepared as follows:

Basic fuchsin	4.0 g
Phenol, melted	8.0 ml
95% Methyl alcohol	20.0 ml

Distilled water	100.0 ml
Tween 80	0.1 ml

Add fuchsin to the alcohol, then add distilled water, phenol, and, last, the Tween 80. No heating is required. Stain usually keeps 2 months in the dark but must be checked with known positive smears. The stain is filtered through Whatman No. 1 filter paper on the day of use.

The slides are fixed by flooding with methanol, draining, and allowing the alcohol to evaporate. Blue wax pencil lines are made at each end of the slide to contain the fuchsin.

Immediately before use, 10 ml of the stain is usually buffered with 0.1 or 0.2 ml of 5% $NaHCO_3$. Pretesting with each batch of Kinyoun's should be done, as color is sometimes best with more buffer. Staining proceeds 2 to 5 min at room temperature. Before the stain is washed away, the decolorizer is applied and the smear rocked to mix evenly. The decolorizer is standard for Kinyoun's: 3% HCl v/v in 95% ethanol. After 5-s decolorizing, the smear is washed gently with cold water. The slide is drained and metanil yellow applied for 2 min to counterstain. The counterstain is not washed off; the slide is drained and allowed to dry. It is better to have thinner slides than to apply the decolorizer two times.

Sputum from untreated tuberculosis cases will usually demonstrate colonies of acid-fast CWD forms, as will blood cultures and pleural fluids. Much less decolorizing is required for materials other than sputum. Blood cultures are best taken when the patient is febrile.

In subcultures lacking much protein, it is helpful to mix the sample with an equal volume of 25% serum or ascitic fluid. Positive controls consist of 3-d blood cultures from patients with Mycobacteriosis. Negative smears can be blood cultures containing classic *Staphylococcus aureus* or *Propionibacterium acnes*. Fungi and some Bacilli tend to be acid fast but are recognizable by their morphology.

Why some batches of basic fuchsin are so superior to others remains enigmatic. In a cooperative study with Allied Chemicals Company we learned that relative concentration of rosaniline to pararosaniline is not pertinent.

The amount of $NaHCO_3$ added and the optimum staining time vary with dye lot. Too much alkalinity or excess staining results in precipitation. Unexpectedly, Bacilli and fungi can grow in Kinyoun's stain or metanil yellow. Therefore, we filter these dyes onto the slides, through a 0.22-μm filter.

METANIL YELLOW COUNTER STAIN[3]

Metanil yellow (Harleco)	0.5 g
Water	1000.0 cc

Methylene blue and most other counterstains tend to remove fuchsin.

PERIODIC ACID PRETREATMENT

With some batches of fuchsin it is necessary to pretreat the smears with periodic acid as tested in studies by Anthony Atkins and by Carolyn Steffen. For this purpose methanol-fixed slides are flooded with 10% aqueous periodic acid, steamed for 10 min, cooled, rinsed with cold water, and drained. The Kinyoun's staining procedure is then followed.

AURAMINE-RHODAMINE FLUORESCENT STAIN FOR MYCOBACTERIA

This stain has a place in presumptively ruling out the presence of *Mycobacterium tuberculosis* in sputum. The method is that of Truant et al.,[114] who modified the procedure of Gray.[42] A satisfactory preparation is available from Difco. The supplier's directions are followed with the exception that decolorizing is for only 5 s when the aim is to see staining of CWD forms.

When only classic bacteria are considered, auramine-rhodamine (AR) staining correlates to an excellent degree with acid-fast properties. However, CWD forms of many genera are AR positive in sputum smears when decolorizing is brief.

Oddly, CWD colonies of Mycobacteria in blood cultures are usually not AR positive, as shown by Garvin and by Steffen.

VICTORIA BLUE ACID-FAST STAIN FOR STAINING CLASSIC AND CWD MYCOBACTERIA

Some laboratories find the Victoria blue stain more functional than basic fuchsin for detecting Mycobacteria.[84] Examples showing that it may be interchangeable with Kinyoun's are shown in the Frontispiece. It requires experience to judge the degree of decolorization, and at first it seems strange to have the insignificant objects stain red.

DIRECTIONS FOR STAINING

1. Flood entire slide with methanol, shake excess off immediately, and let air dry at a 45° angle.
2. Place slides in Coplin jars (one jar for each patient). Add enough Victoria blue to cover smears.
3. Incubate at 37°C for approximately 21 h. For some batches of Victoria blue, better staining results from steaming for 5 min.
4. Decolorize as follows: Drain dye from slide and flood with 0.1% nitric acid in 70% ethanol and let stand for 15 s. Rinse with cold water for 5 s, holding slide at a 45° angle. Place slides at a 45° angle and let air dry.

If the control negative slide (blood culture containing *S. aureus* with their omnipresent CWD microcolonies) is not almost completely decolorized by macroscopic appearance, repeat the decolorization as necessary; then apply the decolorizer to the test slides according to this guideline.

COMPOSITION OF STAIN, DECOLORIZER, AND COUNTERSTAIN

Potassium acetate, 1.5 g, and Victoria blue dye, 5.0 g, are dissolved in 100 ml of 95% ethanol and allowed to stand overnight for the precipitate to settle. The stain is diluted 1:10 in distilled water on the day of use and filtered through Whatman No. 1 filter paper. The counterstain consists of Kinyoun's carbol-fuchsin diluted 1:10 applied 10 s.

Positive controls: 3-d blood culture from known case of tuberculosis, stored refrigerated.

THE PHENOL PHENOMENON

A characteristic of some Mycobacteria has been noticed so recently that the significance and applications have not been fully evaluated. When a drop of patient's blood is smeared on a slide as for a leukocyte differential, growth occurs if the slide is placed in buffered AR stain. The remarkable aspect is that the dye preparation contains a final concentration of approximately 5.0% phenol. The growth consists of L-bodies which hemolyze the surrounding erythrocytes (see Frontispiece). The result is small, cleared areas which permit macroscopic estimate of the bacteria in circulating blood. The phenomenon occurs with tuberculosis, sarcoidosis, Crohn's disease, and *Mycobacterium avium* infection. In Crohn's disease the identity as a Mycobacterium is assumed from survival in phenol and because of the literature involving Mycobacteria in this syndrome. The essential collaborator in this study is Alva Johnson of the Eastern Virginia Medical School.

Blood from a finger prick is satisfactory. Prior to staining, the smear is fixed by flooding with methyl alcohol left on 10 s before draining. The AR stain, which may be purchased from Difco Corp., is buffered with 10% $NaHCO_3$ in a ratio of 3:1. Incubation is at 35 to 37°C in a Coplin jar for 3 to 4 days. Plastic Coplin jars are autoclaved before use. The stained slide is clarified with methyl alcohol, or if necessary, HCl in a final concentration of 1.0% in alcohol.

The long-ago-published slide method of culturing classic *M. tuberculosis* from sputum is well known. However, to our knowledge, this is the first report of slide culture of Mycobacteria from blood.

SUMMARY

Without addition of serum or blood, perhaps because the variants have lost some of their insulating wall, most culture media are toxic to these variants. The toxic molecules, found in meat infusions and most samples of agar, remain unidentified. Most media will not support growth of wall-deficient microbes without the addition of serum. However, when the essential vitamins and nucleotides are supplied, serum may not be required. It is thought that serum plays a dual role, furnishing growth factors and simultaneously neutralizing the toxicity of meat infusion agar. Omission of serum may be fortuitous, as it often contains antibody to which the forms are exquisitely sensitive.

REFERENCES

1. **Abrams, A.,** Reversible metabolic swelling of bacterial protoplasts, *J. Biol. Chem.,* 234, 383–388, 1959.
2. **Abrams, R. Y.,** A method for the cultivation of L Forms in liquid media, *J. Bacteriol.,* 70, 251, 1955.
3. **Alexander-Jackson, E.,** A differential triple stain for demonstrating and studying non-acid-fast forms of the tubercle bacillus in sputum, tissue and body fluids, *Science,* 99, 307–308, 1944.
4. **Altenbern, R. A.,** Critical factors influencing growth of L Forms of *Proteus mirabilis, J. Bacteriol.,* 81, 586–594, 1961.
5. **Armstrong, J. A.,** Histochemical differentiation of nucleic acid by means of induced fluorescence, *Exp. Cell. Res.,* 11, 640–643, 1956.
6. **Bacigalupi, B. A. and Lawson, J. W.,** Defined physiological conditions for the induction of the L-form of *Neisseria gonorrhoeae, J. Bacteriol.,* 116, 778–784, 1973.
7. **Banville, R. R.,** Factors affecting growth of *Staphylococcus aureus* L Forms on semidefined medium, *J. Bacteriol.,* 87, 1192–1197, 1964.
8. **Beaman, B. L.,** Biology of cell wall defective forms of *Nocardia,* in *The Bacterial L-Forms,* Madoff, S., Ed., Marcel Dekker, New York, 1986, 203–227.
9. **Benichou, J. C., Frehel, C., and Ryter, A.,** Improved sectioning and ultrastructure of bacteria and animal cells embedded in Lowicryl, *J. Electron Microsc. Tech.,* 14(4), 289–297, 1990.
10. **Berliner, M. D., Kundsin, R. B., and Allred, E. N.,** Vital staining of Mycoplasma and L Forms with chlorazol black E, *J. Bacteriol.,* 99, 1–7, 1969.
11. **Bertalanffy, L. von, Masin, M., and Masin, M.,** Use of acridine orange fluorescence in exfoliative cytology, *Science,* 124, 1024–1025, 1956.
12. **Bibel, D. J. and Lawson, J. W.,** Morphology and viability of large bodies of streptococcal L-forms, *Infect. Immun.,* 12, 919–930, 1975.

13. **Bilko, I. P.,** Cultivation of *M. tuberculosis* L Forms on semipermeable cellophane membranes, *Probl. Tuberk.,* 8, 58–61, 1986.

14. **Bleiweis, A. S. and Zimmerman, L. N.,** Formation of two types of osmotically fragile bodies from *Streptococcus faecalis var. liquefaciens, Can. J. Microbiol.,* 7, 363–373, 1961.

15. **Bravo Oliva, J. and Garcia Rodriquez, J. A.,** El colorante naranja de acridina en el diagnostico bacteriologico de la tuberculosis, *Med. Trop.,* 44, 276–281, 1968.

16. **Brorson, J. E., Lundholm, M., and Seeberg, S.,** L-colony production by different strains of *Staphylococcus aureus* in the absence of antibiotics, *J. Infect. Dis.,* 128, 335–339, 1973.

17. **Brorson, J. E.,** Influence of normal serum on L phase variants of *Staphylococcus aureus* on solid medium, *Scand. J. Infect. Dis.,* 6, 45–47, 1974.

18. **Brown, A. D.,** Inhibition by spermine of the action of a bacterial cell wall lytic enzyme, *Biochim. Biophys. Acta,* 44, 178–179, 1960.

19. **Brown, G. W., Jr., King, G., and Sugiyama, H.,** Penicillin-lysozyme conversion of *Clostridium botulinum* types A and E into protoplasts and their stabilization as L-form cultures, *J. Bacteriol.,* 104, 1325–1331, 1970.

20. **Buttiaux, R., Osteux, R., Fresnoy, R., and Moriamez, J.,** Les proprietes biochimiques caracteristiques du genre Proteus; inclusion souhaitable des Providencia dans celui-ci, *Ann. Inst. Pasteur (Paris),* 87, 375–386, 1954.

21. **Chattergee, B. R. and Williams, R. P.,** Preparation of spheroplasts from *Vibrio comma, J. Bacteriol.,* 85, 838–841, 1963.

22. **Chattman, M. S., Mattman, L. H., and Mattman, P. E.,** L Forms in blood cultures demonstrated by nucleic acid fluorescence, *Am. J. Clin. Pathol.,* 51, 41–50, 1969.

23. **Chin, W. L. and Lawson, J. W.,** Effect of antibiotics on L Form induction of *Neisseria meningitidis, Antimicrob. Agents Chemother.,* 9(6), 1056–1065, 1976.

24. **Clark, F. M. and Mitchell, W. R.,** Studies on the nutritional requirements of *Clostridium thermosaccharolyticum, Arch. Biochem.,* 3, 459–466, 1944.

25. **Dark, F. A. and Strange, R. E.,** Bacterial protoplasts from Bacillus species by the action of autolytic enzymes, *Nature (London),* 180, 759–760, 1957.

26. **Delaney, S. F.,** Spheroplasts of *Synechococcus* PCC 6301, *J. Gen. Microbiol.,* 130, 2771–2773, 1984.

27. **Dienes, L.,** Alterations of the L-forms of a sporebearing bacillus, *J. Bacteriol.,* 104, 1378–1385, 1970.

28. **Dienes, L. and Sharp, J. T.,** The role of high electrolyte concentration in the production and growth of L Forms of bacteria, *J. Bacteriol.,* 71, 208–213, 1956.

29. **Dienes, L., Weinberger, H. J., and Madoff, S.,** The transformation of typhoid bacilli into L Forms under various conditions, *J. Bacteriol.,* 59, 755–764, 1950.

30. **Domingue, G. J., Ed.,** *Cell Wall Deficient Bacteria,* Addison-Wesley, Reading, MA, 1982.

31. **Eddy, A. A. and Williamson, D. H.,** A method for isolating protoplasts from yeast, *Nature (London),* 179, 1252–1253, 1957.

32. **Edman, D. C., Pollock, M. B., and Hall, E. R.,** *Listeria monocytogenes* L Forms. I. Induction, maintenance, and biological characteristics, *J. Bacteriol.,* 96, 352–357, 1968.

33. **Fitz-James, P. C.,** Cytological and chemical studies of the growth of protoplasts of *Bacillus megaterium, J. Biophys. Biochem. Cytol.,* 4, 257–265, 1958.

34. **Fodor, M.,** Untersuchungen zur osmotischen Empfindlichkeit der L Formen von *Staphylococcus aureus, Z. Allg. Mikrobiol.,* 9, 95–101, 1969.

35. **Galli, E. and Hughes, D. E.,** The autolysis of *Clostridium sporogenes, J. Gen. Microbiol.,* 39, 345–353, 1965.

36. **Glass, V.,** Effect of blood digest and haem on the growth of *C. diphtheriae, J. Pathol. Bacteriol.,* 49, 549–561, 1939.

37. **Godzeski, C. W., Hisker, R. M., and Brier, G.,** Predominant chaacteristics of the growth of some antibiotic-induced L phase bacteria, *Antimicrob. Agents Chemother.,* pp. 507–516, 1963.

38. **Gomori, G.,** A new histochemical test for glycogen and mucin, *Am. J. Clin. Pathol.,* 10, 177–179, 1946.

39. **Gordon, M. H.,** On the nitrogenous food requirements of some of the commoner pathogenic bacteria, *R. Army Med. Corps,* 28, 371–376, 1917.

40. **Gota, S., Ota, K., and Kumahara, F.,** Studies on the purine requirements of Vibrio El Tor, *Jap. J. Microbiol.,* 3, 223–230, 1959.

41. **Gould, G. W., Georgala, D. L., and Hitchens, A. D.,** Fluorochrome labeled lysozyme: reagent for the detection of lysozyme substrate in cells, *Nature (London),* 200, 385–386, 1963.

42. **Gray, D. F.,** Detection of small numbers of Mycobacteria in sections by fluorescence microscopy, *Am. Rev. Tuberc.,* 69, 82–85, 1953.

43. **Gregory, W. W. and Gooder, H.,** Identical nutritional requirements of *Streptococcus faecium* F24 and a derived stable L-phase variant, *J. Bacteriol.,* 129, 1151–1153, 1977.

44. **Grocott, R. G.,** A stain for fungi in tissue secretions and smears using Gomori's methenamine-silver nitrate technique, *Am. J. Clin. Pathol.,* 25, 975–979, 1955.

45. **Groman, N. B. and Suzuki, G.,** Effect of spermine on lysis and reproduction by bacteriophages $\phi174$, λ, and f_2, *J. Bacteriol.,* 92, 1735–1740, 1966.

46. **Grula, E. A.,** Cell division in a species of Erwinia. I. Inhibition of division by D-amino acids, *J. Bacteriol.,* 80, 375–385, 1960.

47. **Grula, E. A., Smith, G. L., and Grula, M. M.,** Cell division in Erwinia, inhibition of nuclear body division in filaments grown in penicillin or mitomycin C, *Science,* 61, 164, 1968.

48. **Guze, L. B., Harwick, H. J., and Kalmanson, G. M.,** Klebsiella L-forms: effect of growth as L-form on virulence of reverted *Klebsiella pneumoniae, J. Infect. Dis.,* 133, 245–252, 1976.

49. **Harold, F. M.,** Stabilization of *Streptococcus faecalis* protoplasts by spermine, *J. Bacteriol.,* 88, 1416–1420, 1964.

50. **Hüi, L., Tien-Lin, L., and Ai-Te, L.,** The L Forms of B. *Proteus vulgaris.* II. Cytochemical study of the L Form B. Proteus, using Feulgen's staining methods and fluorescence microscopy with acridine orange, *Sci. Sinica* 13, 1829–1834, 1964.

51. **Indge, K. J.,** Metabolic lysis of yeast protoplasts, *J. Gen.Microbiol.,* 51, 433–440, 1968.

52. **Inoue, A., Hasegawa, D., and Takagi, A.,** Spheroplast formation in vibrios, *Jpn. J. Bacteriol.,* 24, 533–534, 1969.

53. **Judge, M. S.,** Evidence Implicating a Mycobacterium as the Causative Agent of Sarcoidosis, and Comparison of This Organism with the Blood-Borne Mycobacterium of Tuberculosis, Dissertation, Wayne State University, Detroit, 1979.

54. **Kinyoun, J. J.,** A note on Uhlenberth's method for sputum examination, *Am. J. Public Health,* 5, 867–870, 1915.

55. **Klodnitskaia, N.,** Isolation of L Forms of streptococci from the penicillin-treated scarlet fever patients and the experience of obtaining ovohemocultures, *Zh. Mikrobiol. Epidemiol. Immunobiol.,* 33, 31–35, 1962.

56. **Koser, S. A. and Rettger, L. F.,** Studies of bacterial nutrition. The utilization of nitrogenous compounds of definite chemical composition, *J. Infect. Dis.,* 24, 301–321, 1919.

57. **Koser, S. A. and Kasai, G. J.,** The effect of large amounts of nicotinic acid and nicotinamide on bacterial growth, *J. Bacteriol.,* 53, 743–753, 1947.

58. **Koser, S. A.,** *Vitamin Requirements of Bacteria and Yeasts,* Charles C. Thomas, Springfield, IL, 1968.

59. **Kusaka, I.,** Growth and division of protoplasts of *Bacillus megaterium* and inhibition of division by penicillin, *J. Bacteriol.,* 94, 884–888, 1967.

60. **Landman, O. E. and Forman, A.,** Gelatin-induced reversion of protoplasts of *Bacillus subtilis* to the bacillary form, Biosynthesis of macromolecules and wall during successive steps, *J. Bacteriol.,* 99, 576–589, 1969.

61. **Landman, O. E. and Speigelman, S.,** Enzyme formation in protoplasts of *Bacillus megaterium, Proc. Natl. Acad. Sci. (USA),* 41, 698–704, 1955.

62. **Lapinski, E. M. and Flakas, E. D.,** Reversal of penicillin-induced L phase growth of *Haemophilus influenzae* by spermine and its effects on antibiotic susceptibility, *Infect. Immun.,* 1, 474–478, 1970.

63. **Lederberg, J. and St. Clair, J.,** Protoplasts and L-type growth of *Escherichia coli, J. Bacteriol.,* 75, 143–160, 1958.

64. **Lorkiewicz, Z.,** Growth of stabilized L Forms of *Proteus vulgaris* without the addition of serum and penicillin, *Acta Microbiol. Pol.,* 6, 3–8, 1957.

65. **Lovett, P. S.,** Spontaneous auxotrophic and pigmented mutants occurring at higher frequency in *Bacillus pumilus* NRRLB-3275, *J. Bacteriol.,* 112, 977–985, 1972.

66. **Lwoff, A. and Lwoff, M.,** Studies of codehydrogenases. II. Physiological function of growth factor ''V'', *Proc. R. Soc. London (Biol.),* 122, 360–373, 1937.

67. **Lynn, R. J. and Morton, H. E.,** The inhibitory action of agar on certain strains of pleuropneumonia-like organisms, *Appl. Microbiol.,* 4, 339–341, 1956.

68. **Madoff, S.,** L-forms from *Streptococcus* MG, induction and characterization, *Ann. N.Y. Acad. Sci.,* 174, 912–921, 1970.

69. **Madoff, S. and Dienes, L.,** L Forms from pneumococci, *J. Bacteriol.,* 76, 245, 1958.

70. **Mager, J.,** The stabilizing effect of spermine and related polyamines and bacterial protoplasts, *Biochim. Biophys. Acta,* 36, 529–531, 1959.

71. **Mahony, D. E. and Moore, T. I.,** Stable L Forms of *Clostridium perfringens* and their growth on glass surfaces, *Can. J. Microbiol.,* 22, 953–959, 1976.

72. **Makemson, J. C. and Darwish, R. Z.,** Calcium requirement and magnesium stimulation of *E. coli* L Form induction, *Infect. Immun.,* 6, 880–882, 1972.

73. **Manocha, M. S.,** Electron microscopy of the conidial protoplasts of *Neurospora crassa, Can. J. Bot.,* 46, 1561–1564, 1968.

74. **Mattman, L. H.,** Cell wall deficient forms of Mycobacteria, *Ann. N.Y. Acad. Sci.,* 174, 852–861, 1970.

75. **Mattman, L. H.,** L Forms isolated from infections, in *Spheroplasts, Protoplasts and L Forms of Bacteria,* Roux, J., Ed., Inserm, Paris, 1977, 472–483.

76. **Mattman, L. H., Burgess, A. R., and Farkas, M. E.,** Evaluation of antibiotic diffusion in L variant production by *Proteus* species, *J. Bacteriol.,* 76, 333, 1958.

77. **Mattman, L. H., Tunstall, L. H., and Rossmoore, H. W.,** Induction and characteristics of staphylococcal L Forms, *Can. J. Microbiol.,* 7, 705–713, 1961.

78. **Medill, M. A. and O'Kane,** A synthetic medium for the L type colonies of Proteus, *J. Bacteriol.,* 68, 530–533, 1954.

79. **Michel, M. P. and Hijmans, W.,** The additive effect of glycine and other amino acids on the induction of the L-phase of Group A β-hemolytic streptococci by penicillin and D-cycloserine, *J. Gen. Microbiol.,* 23, 35–46, 1960.

80. **Minck, R., Kirn, A., and Galleron,** Recherches dur la transformation L (formes naines) en milieux synthetiques d'une souche de Proteus, *Ann. Inst. Pasteur (Paris),* 92, 138–141, 1957.

81. **Mohan, R. R., Kronish, D. P., Pianotti, R. S., Epstein, R. L., and Schwartz, B. S.,** Autolytic mechanism for spheroplast formation in *Bacillus cereus* and *Escherichia coli, J. Bacteriol.,* 90, 1355–1364, 1965.

82. **Montgomerie, J. Z., Kalmanson, G. M., and Guze, L. B.,** Fatty acid composition of L-forms of *Streptococcus faecalis* cultures at different osmolalities, *J. Bacteriol.,* 115, 73–75, 1973.

83. **Moustardier, G., Brisou, J., and Perrey, M.,** Milieu de culture pour l'isolement des organismes L dans Les urethrites amicrobiennes a inclusions, *Ann. Inst. Pasteur (Paris),* 85, 515–520, 1953.

84. **Murosashi, T. and Yoshida, K.,** Relation between the loss of acid-fastness due to UV irradiation and the pathogenicity of tubercle bacilli, *Am. Rev. Respir. Dis.,* 94, 86–93, 1966.

85. **Murohashi, T. and Yoshida, K.,** An attempt to culture *Mycobacterium leprae* in cell-free, semisynthetic, soft agar media, *Bull WHO,* 47, 195–210, 1972.

86. **Murphy, S. G. and Elkan, C. H.,** Growth inhibition by biotin in a strain of *Rhizobium japonicum, J. Bacteriol.,* 86, 884–885, 1963.

87. **Newton, B. A.,** A fluorescent derivative of polymyxin: its preparation and use in studying the site of action of the antibiotic, *J. Gen. Microbiol.,* 12, 226–236, 1955.

88. **Nimmo, L. N. and Blazevic, D. J.,** Selection of media for the isolation of common bacterial L-phase organisms from a clinical specimen, *Appl. Microbiol.,* 18, 535–541, 1969.

89. **Panos, C. and Barkulis, S. S.,** Streptococcal L-forms. I. Effect of osmotic change on viability, *J. Bacteriol.,* 78, 247–252, 1959.

90. **Pattyn, S. R. and Portaels, F.,** In vitro cultivation and characterization of *Mycobacterium lepraemurium, Int. J. Leprosy,* 48, 7–14, 1980.

91. **Pohlod, D. J., Mattman, L. H., and Tunstall, L.,** Structures suggesting CWD forms detected in circulating erythrocytes by fluorchrome staining, *Am. Soc. Microbiol.,* 23, 262–267, 1972.

92. **Portaels, F. and Pattyn, S. R.,** Growth of Mycobacteria in relation to the pH of the medium, *Ann. Microbiol. (Inst. Pasteur),* 133B, 213–221, 1982.

93. **Rensburg, A. J. van,** Properties of *Proteus mirabilis* and Providence spheroplasts, *J. Gen. Microbiol.,* 56, 257–264, 1969.

94. **Robbins, W. J. and Kavanaugh, F.,** Vitamin B_1 or its intermediates and growth of certain fungi, *Am. J. Bot.,* 25, 229–236, 1938.

95. **Roberts, R. B. and Wittler, R. G.,** The L-form of *Neisseria meningitidis, J. Gen. Microbiol.,* 44, 139–148, 1966.

96. **Roberts, D. H. and Little, T. W. A.,** A note on spheroplasts and Mycoplasma in synovial fluid, *Br. Vet. J.,* 127, 143–147, 1971.

97. **Rogosa, M., Franklin, J. G., and Perry, K. D.,** Correlation of the vitamin requirements with cultural and biochemical characteristics of *Lactobacillus* spp., *J. Gen. Microbiol.,* 25, 473–482, 1961.

98. **Roth, G. S., Shockman, G. D., and Daneo-Moore, L.,** Balanced macromolecular biosynthesis in "protoplasts" of *Streptococcus faecalis, J. Bacteriol.,* 105, 710–717, 1971.

99. **Sacks, L. E.,** A pH gradient agar plate, *Nature (London),* 178, 269–270, 1956.

100. **Salton, M. R. J. and Shafa, F.,** Some changes in the surface structure of Gram-negative bacteria induced by penicillin action, *Nature (London),* 181, 1321–1324, 1958.

101. **Schell, G. H.,** An Artifact-Free Medium for the Isolation of L Forms from Clinical Blood Cultures, M.S. thesis, Wayne State University, Detroit, 1968.

102. **Schönfeld, J. K.,** L Forms of staphylococci; their reversibility; changes in the sensitivity pattern after intermediary passages in the L phase, *Antonie van Leeuwenhoek,* 25, 325–331, 1959.

103. **Schuhardt, V. T. and Klesius, P. H.,** Osmotic fragility and viability of lysostaphin-induced staphylococcal spheroplasts, *J. Bacteriol.,* 96, 734–737, 1968.

104. **Schuhmann, E. and Taubeneck, U.**, Stabile L Formen verschiedener *E. coli*-Stamme, *Z. Allg. Mikrobiol.*, 9, 297–313, 1969.

105. **Silberstein, J. K.**, Observations on the L Forms of Proteus and Salmonella, *Z. Allg. Pathol. Bakteriol.*, 16, 739–755, 1953.

106. **Slotnick, I. J. and Dougherty, M.**, Unusual toxicity of riboflavin and flavin mononucleotide for *Cardiobacterium hominis*, *Antonie van Leeuwenhoek*, 31, 355–360, 1965.

107. **Sullivan, M. X.**, Synthetic culture media and the biochemistry of bacterial pigments, *J. Med. Res.*, 14, 109–160, 1905–1906.

108. **Susan, S. R.**, A Fine Structure Study of *Pseudomonas aeruginosa* in Media and Distilled Water, M.S. thesis, Wayne State University, Detroit, 1972.

109. **Swieczkowski, D. M.**, A Comparison of the Efficacy of Two Media Used for the Detection of CWD Forms of Microorganisms from Clinical Specimens, M.S. thesis, Wayne State University, Detroit, 1970.

110. **Tabor, C. W.**, Stabilization of protoplasts and spheroplasts by spermine and other polyamines, *J. Bacteriol.*, 83, 1101–1111, 1962.

111. **Takahashi, T., Ichikawa, S., and Tadokoro, I.**, A chemically defined medium for the stable L Forms of *Staphylococcus aureus*, *Microbiol. Immunol.*, 25(12), 1357–1362, 1981.

112. **Takahashi, T. and Tadokoro, I.**, Significance of bacterial L-forms in diseases, Isolation of L-forms from clinical specimens, *Yokohama Med. Bull.*, 32(3–6), 141–163, 1981.

113. **Takahashi, T., Tadokoro, I., and Adachi, S.**, An L-form of *Staphylococcus aureus* adapted to a brain heart infusion medium without osmotic stabilizers, *Microbiol. Immunol.*, 25(9), 871–886, 1981.

114. **Truant, J. P., Brett, W. A., and Thomas, W. J.**, Fluorescence microscopy of tubercle bacilli stained with auramine and rhodamine, *Henry Ford Hosp. Med. Bull.*, 10, 287–296, 1962.

115. **Tunstall, L. H. and Mattman, L. H.**, Growth of hemolytic streptococci in vitamin free medium, *Experientia*, 17, 190–191, 1961.

116. **Vallee, M. and Segel, I. H.**, Sulphate transport by protoplasts of *Neurospora crassa*, *Microbios*, 4, 21–31, 1971.

117. **Watanakunakorn, C., Goldberg, L. M., Carleton, J., and Hamburger, M.**, Staphylococcal spheroplasts and L-colonies. III. Induction by lysostaphin, *J. Infect. Dis.*, 119, 67–74, 1969.

118. **Watanakunakorn, C. and Hamburger, M.**, Induction of spheroplasts of *Pseudomonas aeruginosa* by carbenicillin, *Appl. Microbiol.*, 17, 935–937, 1969.

119. **Weibull, C.**, The isolation of protoplasts from *Bacillus megaterium* by controlled treatment with lysozyme, *J. Bacteriol.*, 66, 688–695, 1953.

120. **Wyrick, P. B. and Rogers, H. J.**, Isolation and characterization of cell wall-defective variants of *Bacillus subtilis* and *Bacillus licheniformis*, *J. Bacteriol.*, 116, 456–465, 1973.

121. **Young, F. E., Haywood, P., and Pollock, M.**, Isolation of L-forms of *Bacillus subtilis* which grow in liquid medium, *J. Bacteriol.*, 102, 867–870, 1970.

122. **Zafar, R. S.**, Quantitation and Identification of Cell Wall Deficient Microorganisms Occurring in Pre-Therapy Blood Cultures, M.S. thesis, Wayne State University, Detroit, 1975.

123. **Zafar, R. S. and Mattman, L. H.**, Characteristics of L phase variants in pretherapy blood cultures, *ASM Abst. Annu. Meet.*, 1977.

124. **Prozorovsky, S., Dorozhkova, I., Vulfotch, Yu., et al.**, Gamaleya Institute, Moscow.

125. **O'Beirne, A. and Eveland, W.**, personal communication.

126. **Pistole, T.**, unpublished study.

127. **Farrell, J.**, unpublished study.

128. **Williams, M.**, unpublished study.

129. **Adamson, K.**, unpublished study.

130. **Prozorovski, S.**, personal communication.

Chapter 39

IDENTIFICATION

STAINING

The Gram stain has a place in identifying CWD variants. Although Gram-positive species tend to be Gram negative when highly pleomorphic, the edge of a microcolony may reveal the true reaction and even suggest the morphology. Often Gram-positive cocci are evident without differentiating streptococci from staphylococci.

The formulas of other useful stains are given in the chapter on Media, Methods, and Stains. Examination of unstained preparations by phase microscopy remains an almost mandatory tool to confirm and check staining results, as stains may contain contaminants or precipitate into interesting artifacts. Identification as growth but not species can be made with dilute fuchsin applied to agar in a pour plate (Figure 1).[46,47]

TEMPERATURE REQUIREMENTS

The mesophilic characteristic of a species is evident in the CWD stage and will rule out the species more fastidious in this respect. An example is the rapid growth of CWD *Pseudomonas aeruginosa* at room temperature.

BIOCHEMICAL REACTIONS AND UNIQUE COMPONENTS

Biochemical reactions of 51 clinical strains of CWD Staphylococci were studied by Whalen.[44] He found major biochemical activities usually still present, namely, reactive colonies with tellurite and tetrazolium chloride, the production of catalase, DNAase, lipase, lecithinase, and hemolysin. He modified the method of Willis by using organisms concentrated by centrifugation as the pour plate inoculum. This greatly intensified reactions. Coagulase was negative in the clinical CWD strains in contrast to its presence in Staphylococcal variants made *in vitro*.

Cohen et al.[8] tested the applicability of diagnostic laboratory tests in identifying *in vitro* produced CWD variants. Thirteen of 29 procedures were informative: tests for oxidase, catalase, phosphatase, glucose utilization, phenylalanine deamination, esculin hydrolysis, and nitrate, tetrazolium, and tellurite reduction. *Proteus mirabilis* was identified by catalase, phosphatase, phenylalanine deamination, urease, nitrate reduction, and breakdown of glucose. It has long been known that enzymatic activity is usually diminished with loss of wall components. This is ameliorated by using the sediment from a centrifuged culture as inoculum.

In general, biochemical reactions of CWD variants mimic those of the classic but may be delayed and less marked. In some cases enzymes are lost, as found by Takahashi et al.,[41] noting that variants of staphylococci and streptococci failed to ferment trehalose.

REACTIONS WITH ANALYTAB SUBSTRATES

The strips of substrates commonly used in clinical laboratories can give clues to identity

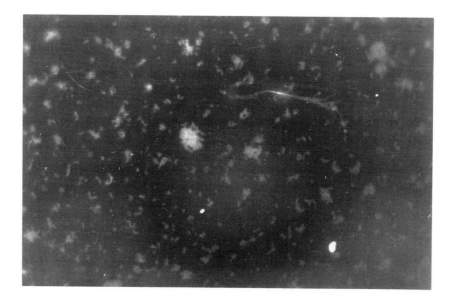

FIGURE 1. Fuchsin stains CWD colonies of *Klebsiella pneumoniae* in pour plate of blood culture. (Original magnification × 1000.)(From Zafar, R. S., M.S. thesis, Wayne State University, Detroit, 1975.)

of many amorphous growths. The procedure for inoculation of API strips (see Frontispiece) with blood cultures from tuberculosis patients follows:

1. Incubate blood culture 3 to 4 d. Standard blood culture media is suitable.
2. Gently tip blood culture to mix.
3. Allow to stand only long enough for red cells to settle.
4. Remove 3 ml of supernate and adjust pH to 7.0 with sterile NaOH.
5. Inoculate each cupule with supernate.
6. Cover cupules with oil as per API directions.
7. Incubate 24 h; if negative, 48 h.

In a short series, pleural fluids from tuberculosis cases gave API reactions with ADH, LDC, and urea. The panel of reactions differed from that of growth of CWD *Mycobacterium tuberculosis* in blood cultures only in fermentation of glucose without gas.

CYTOCHROME OXIDASE

N, N-dimethyl-*p*-phenylenediamine monohydrochloride (Eastman Kodak Co.) at a 0.1% concentration detects cytochrome oxidase in subsurface growth of CWD gonococci in pour plates (Figure 2). Using bright-field illumination, tests are examined microscopically for color changes at intervals through 1 h. Control preparations might contain classic *Neisseria gonorrhoeae, P. aeruginosa,* and *Staphylococcus aureus.*[13] Twenty years ago, cytochrome oxidase marked CWD *N. gonorrhoeae* colonies in a study of synovial fluid by Holmes et al.[16]

Besserer[4] found cytochrome oxidase in blood cultures containing only CWD *P. aeruginosa.* With Pseudomonas the enzyme usually was demonstrable by applying a drop of culture to a Pathotec® cytochrome oxidase strip.

LIPASE[37]

Cream is sterilized at 115°C for 10 min and added to the basic medium which may

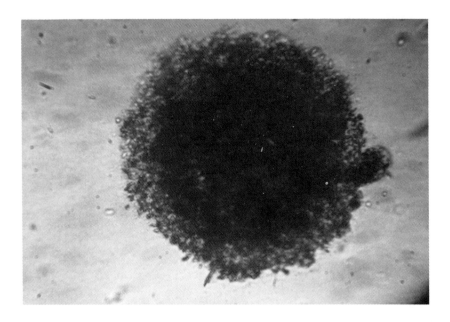

FIGURE 2. Cytochrome oxidase reaction in CWD *Neisseria gonorrhoeae* colony from blood of a patient with genital infection. (Original magnification × 450.)(From Gray, J. M., Ph.D. dissertation, Wayne State University, Detroit, 1980.)

contain horse serum. Since many CWD species will not grow on this medium, readings are obtained by inverting an agar block of growth onto the medium and incubating. Lipase activity is detected around the block.

Lipase may also be detected on tributyrin agar. The basic medium is prepared with addition of 0.2% phenol. Glycerol tributyrate in a final concentration of 0.5% is emulsified in the hot medium using a mixer. No serum is added. Inoculation again is by inverting a growth-containing agar block onto the medium surface. Preformed lipase is expected to produce clearing of the emulsion in 18 h.

CATALASE[37,44]

The medium is pretested to assure absence of catalase. To test for the enzyme a loopful of hydrogen peroxide (10% v/v) is placed on the CWD growth. Immediate effervescence indicates catalase formation. For clinical strains where pour plates are necessary, the H_2O_2 should be applied to the subsurface growth.

DNAase[11,37]

DNA (sodium salt of thymus gland DNA, 0.2% BDH) is added to a basic L-Form medium before autoclaving. Deoxyribonuclease activity is detected after growth has occurred by flooding the plate with 1.5 N HCl. Unattached DNA precipitates in the medium, leaving a clear zone in the areas where DNAase has been produced. Again, for testing for DNAase and all other enzymes, it is advisable to inoculate pour plates with concentrated organisms.[44]

Smith and Willis[37] note that when horse serum is part of the basic medium used for detecting DNAase and other biochemical reactions, it should be heated at 60°C for 1 h, not only to inactivate complement, but also to destroy enzymes complicating biochemical reactions. It will be interesting to learn whether pretreating the organisms with polymyxin B increases sensitivity of this reaction. Such treatment is recommended in studying classic bacteria.[15]

AGAR HYDROLYSIS

Clostridium perfringens and *S. aureus* are distinctive in lysing agar when in their CWD state.[19,24] Logically, this is most noticeable with soft agar. At least for Staphylococci, the lysis is more rapid at an acid pH. With current methods lysis is slow.

FREE COAGULASE

To detect free coagulase, a 1-cm² block of agar culture is added to 2 ml of 2% solution of human fibrinogen in sterile physiological saline. Free coagulase is demonstrated by solid clot formation after overnight incubation at 37°C. Control tubes consist of medium with no inoculum and medium inoculated with the sterile medium. Whole plasma is unsatisfactory for the test.[37] In contrast to reactions with laboratory produced CWD staphylococci, clinical strains usually fail to make this enzyme.[44]

HIPPURICASE

In vitro-produced CWD Group β-hemolytic streptococci retain this enzyme. Thus, it will be interesting to test for its production by naturally occurring variants.[9]

DYE REDUCTIONS

TRIPHENYLTETRAZOLIUM CHLORIDE REACTIONS

Triphenyltetrazolium chloride (TTC) reduction by CWD staphylococci can be much brighter than by the classic form, especially when the medium contains 0.04% erythrocyte lysate[46,47] (see Frontispiece). Zafar and Mattman noted that the largest variant colonies grew tenaciously against the plastic of the petri dish bottom.

Whalen, who tested CWD staphylococci in cultures of blood and synovial fluids, found TTC reduction by all 42 strains examined.[44] Most interesting, TTC reduction in colonies with morphology typical of CWD staphylococci was found in cultures of four patients, still ill, whose classic *S. aureus* had disappeared from their cultures. TTC reduction in all 9 strains of CWD *S. epidermidis* resembled the reaction with *S. aureus*. TTC was incorporated in medium in a final concentration of 0.035%.

CWD *P. mirabilis* in pour plates containing TTC gave unique surface growth of red filaments[46,47] (Figure 3).

Early studies by Holmes and colleagues[16] employed TTC to indicate growth of CWD gonococci by color appearing throughout the plates as well as in colonies.

TELLURITE REDUCTION MEDIUM[44]

Veal Infusion Broth (Difco 0344)	20 g
Soluble Starch (Difco 0178)	1 g
NaCl	30 g
LiCl₂	2 g
Noble's agar (Difco 0142)	18 g
Mannitol	5 g
MgSO₄; 1H₂O	2.5 g
Tap water	100 ml

Before making pour plates, 100 ml of the basal medium is supplemented with 10 ml of separately autoclaved yeast extract (Difco 0127) and 4 ml of tellurite enrichment. The tellurite enrichment contains: 5 ml of Chapman tellurite solution; 40 ml of inactivated swine serum;

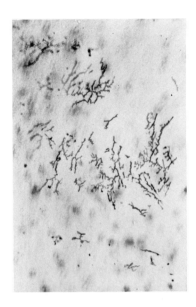

FIGURE 3. CWD *Proteus mirabilis* in pour plate grows on surface as filaments reducing triphenyl tetrazolium chloride. Inoculum was a pretherapy blood culture. (Original magnification × 100.)(From Zafar, R. S., M.S. thesis, Wayne State University, Detroit, 1975).

5 mg of gentamicin sulfate. Chapman tellurite solution consists of a 1.0% solution of potassium tellurite. Reduction of tellurite by *S. aureus* from a blood culture is shown in Figure 4.[44]

HEMOLYSINS

Hemolysins of CWD bacteria are best detected by making pour plates of concentrated organisms and then incubating an agar block of the growth on 5% blood agar.[44] *Streptococcus agalactiae*, penicillin-induced into the L-phase, makes β-hemolysin in both liquid and agar medium.[9]

CAMP FACTOR

The CAMP Factor was produced in equal amounts by penicillin-induced Group β-hemolytic streptococci and by the parent streptococcus.[9]

PIGMENTATION

Galactose in a final concentration of 1.5% autoclaved in the basic medium causes pigmentation in staphylococcal CWD variants.[37] The variant colonies can also be pigmented when grown on sartorius membranes of 100 and 200 μm sterilized with ultraviolet light. More work is needed concerning characteristics of CWD staphylococci occurring in infections.

M PROTEIN AND HYALURONIC ACID FROM β-HEMOLYTIC STREPTOCOCCUS

The type M protein produced by the CWD variants of some β-hemolytic streptococci[11] is secreted into the medium where it causes opacity in horse-serum-containing media.[45] Thus, this opacity might be a simple visual aid assisting identification.

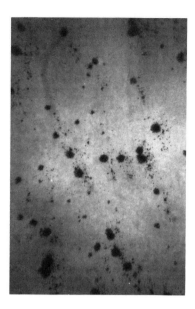

FIGURE 4. *Staphylococcus aureus* reduces potassium tellurite in a pour plate of a blood culture. (Original magnification × 1000.)(From Whalen, M. A., M.S. thesis, Wayne State University, Detroit, 1975.)

The well-known hyaluronic acid capsule of β-hemolytic streptococci is retained by some CWD variants. This mucopolysaccharide is demonstrated in a unique way by showing diminution of the colonies grown in pour plates containing hyaluronidase.[27]

AGGLUTINATION AND HEMAGGLUTINATION

Simple agglutination using antibody for the variants or classic stage can presumptively identify the variants. This has been profitable in examining variants in blood cultures in tuberculosis (Figure 5 A and B) and pneumococcal meningitis and in showing that variants in the blood of pneumonia patients are of the same pneumococcus type as the classic forms in sputum. Often it is more feasible when working with blood cultures to agglutinate the parasitized erythrocytes than the organism.

Specific immune serum agglutinated stable L variants of *N. meningitidis*.[6] Passive hemagglutination identified and typed L Forms of *Vibrio cholerae*.[5]

FLUORESCENT ANTIBODY

Fluorescent antibody (FA) reactions are invaluable in identification of CWD variants in clinical problems and research (see Frontispiece). Barile and DelGiudice[2] used fluorescent antibody to identify Mycoplasma on agar surfaces using epifluorescence. This technique is applicable to CWD colonies cultivable on the surface of clear medium such as serum agar. Necessarily, clues must have been provided by physiochemical properties shown by the variant or in clinical work by the patient's symptoms.

A useful tool is antibody made vs. the classic form which will also react with the variant. Antibody to variants has been made in many laboratories, but more time is required to obtain a good immunoglobulin response. Fortunately, identifying reactions may be obtained with organisms concentrated from blood, in blood cultures, in urine, spinal fluid or exudates.

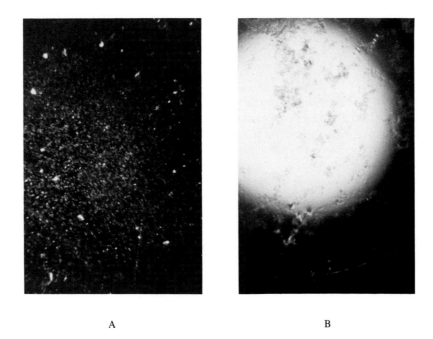

A   B

FIGURE 5. (A) Negative test with normal rabbit serum. (B) Agglutination of CWD growth in blood culture by antibody vs. H37Rv diluted 1:1000. Red cells in culture were allowed to settle before using supernatant.

Both direct or indirect staining methods are applicable with incubation times as brief as possible to avoid loosening the microcolonies with exposure to liquid. Working with wall-deficient variants requires a step unnecessary with classic bacteria and fungi. Before applying the diluted fluorescent antibody it must be centrifuged to remove wall-deficient colonies which grow in sera.

Gamaleya Institute prepares hyperimmune sera to L Forms of Streptococci and Salmonella by injecting vaccines of the L Forms into rabbits with this schedule:

1. First injection intravenously and into foot pads
2. A month later, injections intravenously and intracutaneously
3. Two weeks later, intravenous injection
4. Two weeks later, intravenous and intramuscular injections

If titer is low, the series is repeated. A titer of 1:100,000 is expected. As is well known, antibody appears earlier if an adjuvant such as Freundt's (Difco) or Ribi's (Hamilton, MT) is used.

FA will reveal L Forms of *Salmonella typhi* in the bone marrow of cases and carriers, as shown by Prozorovsky and associates.[30]

FA supplied rapid identification of CWD pseudomonas in blood cultures studied by Besserer.[4] CWD forms identified by FA in blood cultures in a single series included *Candida albicans, P. mirabilis, Streptococcus pneumoniae, Escherichia coli, Staphylococcus aureus, Klebsiella pneumoniae,* alpha hemolytic streptococcus, and *Corynebacterium xerosis.*[46,47]

FA to *M. tuberculosis* has identified CWD variants of this species in isolator tube concentrates of venous blood, in sputum and in cultures of blood. Alva Johnson and Peter Almenoff[49,50] found that a mouse monoclonal antibody to the tubercle bacillus showed complete cross-reactivity to the sarcoid organism growing in patients blood. A Russian laboratory noted that the indirect FA method is functional in recognizing both classic and CWD forms of *M. tuberculosis* in pleural and bronchial exudate.[21,40]

FLUORESCENT LYSOSTAPHIN

Lysostaphin labeled with fluorescein or rhodamine is useful for rapid identification of the common pathogen, *S. aureus*. The stain can be applied directly to smears of the blood culture or to concentrates of bacteria from the blood itself (see Frontispiece).

The procedure is based on the methods employed for the conjugation of immunoglobulins with FITC as outlined in Nowotny's manual.[29] All the work is done at approximately 4°C. Thirty milligrams of lysostaphin is mixed with 12.5 ml of 0.2 *M* sodium phosphate (dibasic), pH 8.9, and stirred overnight with a magnetic stirrer. The lysostaphin solution is then dialyzed for 5 d with daily changes of 2% NaCl. The dialyzed solution is restored to its original volume of 12.5 ml by adding 0.2 *M* dibasic sodium phosphate solution pH 8.9.

A fluorescein isothiocyanate solution is made by dissolving 1.4 mg of FITC in 2.5 ml of 0.2 *M* dibasic sodium phosphate, pH 8.9, and 2.5 ml of distilled water.

The lysostaphin and FITC solutions are slowly mixed with constant stirring for 20 min after which the pH is adjusted to 9.0 with 0.1 *M* sodium phosphate (tribasic). Four milliliters of physiological saline is added to bring the volume to 20 ml. The solution is covered with foil and stirred overnight with a magnetic stirrer.

The mixture is then placed in a dialysis bag and dialyzed against 1000 ml of phosphate-buffered saline, pH 7.5 for 6 d with daily changes of dialysate. On the sixth day, the fluorochrome-lysostaphin solution is dispensed in 1.0-ml aliquots and stored at 4°C. If during storage a sediment develops, only the supernatant is used in staining.[17] Applications of this fluorochrome have been made especially by Jubinski[17] and Chopra[7] (see Frontispiece).

GEL ELECTROPHORESIS

Polyacrylamide gel (PAG) studies show that *in vitro*-produced variants retain most but not all of the cellular proteins of the parent strains.[20,31,42] Densitometric scan of proteins from parent and variant of *Mycobacterium phlei* resolved by gel electrophoresis are shown in Figure 6.[38,39]

The most extensive examination of proteins in naturally occurring CWD organisms was done by Motwani,[28] utilizing skip culture as sources of the variants. A skip culture is one negative for classics which occur between cultures yielding classics. The retention of proteins varied between CWD cultures of the same strain, but sufficient pertinent proteins were retained to suggest the identity of *C. albicans,* Group A and Group D streptococci, *Staphylococcus pneumoniae, Pseudomonas aeruginosa, Staphylococcus aureus, Streptococcus pneumoniae,* and *Proteus mirabilis* (Figure 7 A, B, C).[28] Some of the gel patterns of the *in vivo* occurring variants definitely resemble patterns from L variants made in the laboratory of Theodore et al.[42] A marked resemblance between the sarcoid organism and *M. tuberculosis* was shown by the gel electrophoresis in studies by Steffen,[38] Steffen and Mattman,[39] and D. Garvin.

It can be expected that the membrane-bound enzymes of organisms can be recognized from CWD variants as they are from the classic stage. Examples from *E. coli* are nitrate reductase, formic dehydrogenase, cytochrome b, and succinic dehydrogenase.[23]

GROWTH READINGS SUGGEST CWD ORGANISMS

Isotope-labeled substrates in the Bactec Instrument can indicate multiplication of wall-

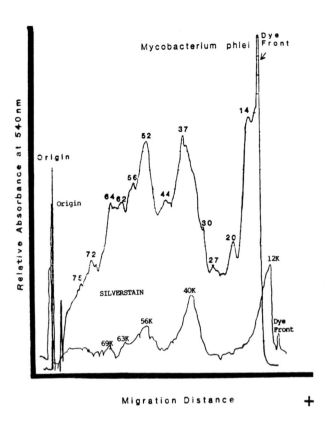

FIGURE 6. Densitometer curves from polyacrylamide gel profiles of *Mycobacterium phlei*. Upper curve from classical organisms, lower curve from cell wall deficient variants. (From Steffen, C. M., Ph.D. dissertation, Wayne State University, Detroit, 1985.)

deficient organisms just as they detect classics. The findings of Motwani are given in Figure 8.[28]

BACTERIOCINS AND BACTERIOPHAGE

BACTERIOCIN PRODUCTION

Pyocins are excreted by CWD *Pseudomonas aeruginosa* on suitable media. Besserer[4] showed that pyocins for all indicator strains were made when the CWD *P. aeruginosa* was grown on medium composed of brain heart infusion broth, sucrose, purified agar, MgSO$_4$, and swine serum. No pyocin production was evident when polyvinylpyrrolidone was substituted for the sucrose.

PHAGE LYSIS

Cholera bacteriophage identified and typed L Forms of cholera vibrios, thus distinguishing them from other bacterial species and from Mycoplasma.[5] It would be expected that most naturally occurring CWD variants are identifiable with phages or bacteriocins.

PATHOGENESIS

As described in the chapters dealing with specific organisms, virulence of the CWD stage can often be demonstrated in mice, rabbits, monkeys, goats, man, and tissue culture.

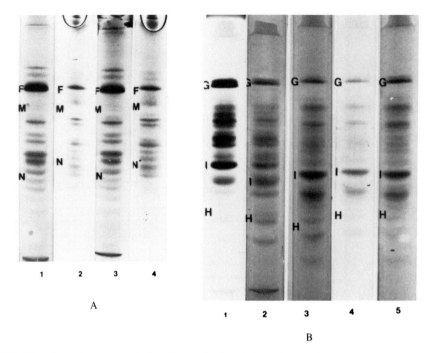

A

B

FIGURE 7A. Group A streptococci. 1 and 2 = classic and skip culture variant from patient X. 3 and 4 = classic and skip culture variant from patient Y.

FIGURE 7B. Pattern No. 1 made by proteins from a classic *Pseudomonas aeruginosa*. Gels 2–5 are from CWD *P. aeruginosa* in skip blood cultures.

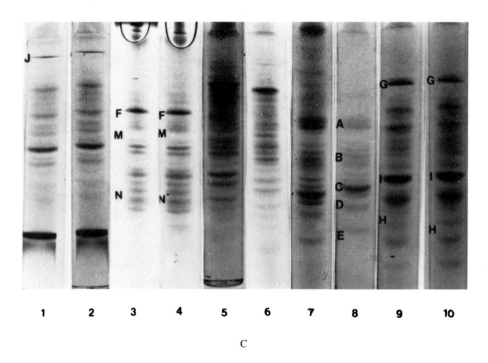

C

FIGURE 7C. Profiles from CWD organisms found in pure culture in patients' bloods. 1 and 2 = pneumococci. 3 and 4 = Group A streptococci. 5 and 6 = Group D streptococci. 7 and 8 = *Staphylococcus aureus*. 9 and 10 = *P. aeruginosa*. (From Motwani, N., Ph.D. dissertation, Wayne State University, Detroit, 1976.)

	Growth Index Readings on Blood Cultures Lacking Classical Organisms			
	30–40	40–50	50–60	Above 60
CWD variants isolated	27%(17/63)	72%(54/75)	85%(35/41)	100%(2/2)
No growth	73%(46/63)	28%(21/75)	15%(6/41)	0%(0/2)

FIGURE 8. Correlation of growth index readings with CWD form isolations. (From Motwani, N., Ph.D. dissertation, Wayne State University, Detroit, 1976.)

Experimentally, success is greater with fresh isolates. Success is often greater when the resistance of the animal is lowered artificially. Judge[18] showed that the sarcoid mycobacterium causes disease in experimental animals earlier if they are pretreated with cortisone. As reported by Judge, Golden found that characteristics and locations of the granulomas distinguish the lesions of sarcoidosis from those caused by the CWD stages of *M. tuberculosis*.[18,25]

Godzeski showed that latency with the CWD stages of β-hemolytic streptococci can change to fulminate infection in immunosuppressed laboratory animals.[48] Fourteen CWD strains from uveitis caused pathology in cortisone treated mice without reversion of the organisms.[3] Pathogenic potential was shown without indicating genus of the organisms.

IMMUNOASSAY AND RADIOIMMUNOASSAY

Enzyme immunoassay (EIA) was found to be a sensitive method to detect *Salmonella typhi* in blood and urine. As little as 20 ng/ml of antigen was recognized.[12]

Radioimmunoassay of *S. typhi* antigen applied to tissues of animals revealed that specific antigen was present for 6 months after living L Forms were injected. Wide dissemination as well as prolonged persistence was evident. If dead L Forms were injected, they disappeared by 19 d.[22]

A commercially available kit (Gonozyme™, Abbott Laboratories, Chicago, IL), serves to identify L Forms of *N. gonorrhoea* in pure and mixed culture.[1] Although the investigation concerned forms produced *in vitro*, the system can be expected to identify the clinical variants. The L Form of the gonococcus may be the only form seen in 25% of cervical exudates, and its abundance in circulating leukocytes can constitute the most sensitive clue for diagnosis.[13,14]

NUCLEIC ACID PROBES AND HYBRIDIZATION

Nucleic acid homology and similar approaches have to date rarely been applied to identification of CWD microbes. This is surprising since in 1963 DNA homology was used to relate *in vitro*-made L Forms to their parents.[26] However, DNA homology is expected to be useful, as it has been in the analysis of the noncultivable agent of bacillary angiomatosis by Relman and associates.[32]

Chemiluminescent DNA probes, applied to recognizing Group β-streptococci, can be notably rapid.[33] Most interesting is that *S. typhi* infections, still common in developing countries, can be detected in culture-negative blood by using a DNA probe.[34]

Mating of DNA between classic and L-Form *Brucella abortus* can identify the variant.[43]

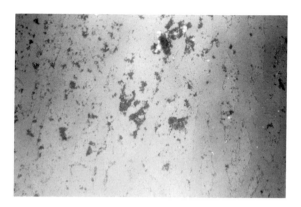

FIGURE 9.   Monoclonal antibody to *Mycobacterium tuberculosis* reacts in an immunogold technique with CWD growth from blood of a pulmonary tuberculosis patient. (Photo submitted by Alva Johnson.)

Control variants included L Forms of *S. typhimurium*. Such methods will certainly be useful for identification of nonreverting variants.

GAS CHROMATOGRAPHY-MASS SPECTROSCOPY

Gas chromatography combined with spectroscopy may well be applied to microbial variants. Fox and research team[10] found that analyzing the carbohydrate content of Legionella was informative for species identification. Senatore's studies indicated that a reference library of profiles of CWD variants may be required as the added lipids in the variants cause marked changes.[35]

IMMUNOGOLD REACTIONS

The Ted Pella Kit (Ted Pella Inc. Redding, CA) can be used to obtain immunogold reactions as shown in Figure 9.

ANTIBIOTIC SENSITIVITY TESTING

For a few genera antibiotic susceptibility can give a clue regarding genus. An example is the response of Mycobacteria to Isoniazid. Many isolates of CWD Mycobacteria will grow as minute colonies on chocolate agar or 10% horse serum agar if the media are enriched with 1% separately autoclaved yeast extract. Impression smears are helpful in interpreting the results (Figure 10).[18,25]

ELECTRON MICROSCOPY WITH ANTIBODY

Ferritin-labeled antibody has been used to identify the L Form of *S. typhi* (Figure 11).

LECTINS

Lectins from snail and from several plants specifically agglutinate Group C β-hemolytic streptococci; numerous other lectins are useful in rapid identification of pathogens as re-

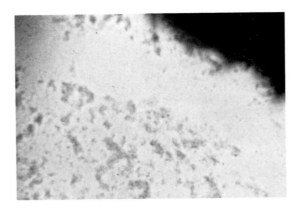

FIGURE 10. Isoniazid disk (upper right) inhibits growth of CWD *M. tuberculosis* from blood culture. (From Judge, M. S., Ph.D. dissertation, Wayne State University, Detroit, 1979.)

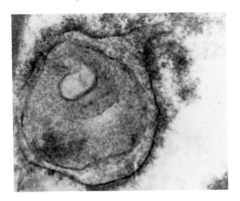

FIGURE 11. Ferritin-labeled antibody to L Form of *Salmonella typhi* is on surface of spheroplast in blood of typhoid convalescent. (Magnification × 108,000.)(Photo submitted by Didenko, L. V., Konstantinova, L. V., and Konstantinova, N. D.)

viewed by Slifkin and Doyle.[36] However, to our knowledge, the field of lectin interaction with CWD microbes has not been explored.

SUMMARY

Broad classification results from (1) the Gram stain, especially if a Gram-positive organism makes reversion steps toward the classic state, (2) growth present or absent at room temperature.

More specific classification results from the following: (1) tests for TTC and tellurite reduction, (2) a search for common enzymes with API panels or other reagents, (3) agar lysis and pigmentation when they occur, (4) phages and bacteriocins identify many species.

Mycobacterial CWD forms can be identified not only by specific staining, but by inhibition by Isoniazid.

Specific identification depends on antibody reactions or, for staphylococci, staining with fluorescent lysostaphin.

REFERENCES

1. **Bae, B. H. C., Amin, R. M., and Korzis, J.,** Detection of L Forms of *Neisseria gonorrhoeae* in pure and mixed culture suspensions by an enzyme immunoassay, *Am. J. Clin. Pathol.*, 85, 618–622, 1986.

2. **Barile, M. F. and DelGiudice, R. A.,** Isolations of mycoplasmas and their rapid identification by plate epi-immunofluorescence, in *Pathogenic Mycoplasmas*, Ciba Symposium, 1972, Elsevier, Excerpta Medica.

3. **Barth, C. L. and Hessburg, P. C.,** Role of cell wall deficient bacteria in uveitis, in *Cell Wall Deficient Bacteria*, Domingue, G. J., Addison-Wesley, 1982, 465–488.

4. **Besserer, J. A.,** Identification of Wall Deficient *Pseudomonas aeruginosa* by Immunological, Biochemical and Biological Methods, M.S. thesis, Wayne State University, Detroit, 1978.

5. **Bichul, K. G., Lomov, Y. M., Prozorovsky, S. V., Kopylov, V. A., Mazrukho, B. L., and Polkovnikova, O. V.,** Concerning the method of identification of the L Forms of *Cholera vibrios*, *Zh. Mikrobiol., Epidemiol., Immunol.*, 11, 86–89, 1977.

6. **Bohnhoff, M. and Page, M. I.,** Experimental infection with parent and L phase variants of *Neisseria meningitidis*, *J. Bacteriol.*, 95, 2070–2077, 1968.

7. **Chopra, C. S.,** Rapid Methods for Identification of Common Pathogenic Cocci in Blood Cultures, Ph.D. dissertation, Wayne State University, Detroit, 1981.

8. **Cohen, R. L., Wittler, R. G., and Faber, J. E.,** Modified biochemical tests for characterization of L phase variants of bacteria, *Appl. Microbiol.*, 16, 1655–1662, 1968.

9. **Flores, A. E. and Ferrieri, P.,** Biochemical markers of the penicillin-induced L phase of a Group B, Type III *Streptococcus* sp., *J. Clin. Microbiol.*, 18, 961–967, 1983.

10. **Fox, A., Rogers, J. C., Fox, K. F., Schnitzer, G., Morgan, S. L., Brown, A., and Aono, R.,** Chemotaxonomic differentiation of Legionellae by detection and characterization of aminodideoxyhexoses and other unique sugars using gas chromatography-mass spectrometry, *J. Clin. Microbiol.*, 28, 546–552, 1990.

11. **Freimer, E. H., Krause, R. M., and McCarty, M.,** Studies of L Forms and protoplasts of Group A streptococci. I. Isolation,growth and bacteriologic characteristics, *J. Exp. Med.*, 110, 853–874, 1959.

12. **Gorelov, A. L., Levina, G. A., and Prozorovski, S. V.,** Study of the possibility of using different variants of the enzyme immunoassay for the detection of *Salmonella typhi* L Forms in biological fluids, *Zh. Mikrobiol. Epidemiol. Immunobiol.*, 8, 60–64, 1989.

13. **Gray, J. M.,** The *In Vitro* Stability of Betalactamase Production by *N. gonorrhoeae* Strains, and Improved Diagnosis by Detection of Cell Wall Deficient Forms, Ph.D. dissertation, Wayne State University, Detroit, 1980.

14. **Gray, J. M. and Mattman, L. H.,** Cell wall deficient *Neisseria gonorrhoeae*, in *Cell Wall Deficient Bacteria*, Domingue, G. J., Ed., Addison-Wesley, Reading, MA, 1982, 409–451.

15. **Hänninen, M. L.,** Rapid method for the detection of DNase of Campylobacters, *J. Clin. Microbiol.*, 27, 2118–2119, 1989.

16. **Holmes, K. K., Gutman, L. T., Belding, M. E., and Turck, M.,** Recovery of *Neisseria gonorrhoeae* from "sterile" synovial fluid in gonococcal arthritis., *N. Engl. J. Med.*, 284, 318–320, 1971.

17. **Jubinski, J. C.,** Fluorochrome Labeling of Lysostaphin and Its Application to Identification of Classical and L Phase Staphylococci, M.S. thesis, Wayne State University, Detroit, 1977.

18. **Judge, M. S.,** Evidence implicating a Mycobacterium as the causative agent of sarcoidosis, and comparison of this organism with the blood-borne Mycobacterium of tuberculosis, Ph.D. dissertation, Wayne State University, 1979.

19. **Kawatomari, T.,** Studies of the L Forms of *Clostridium perfringens*, *J. Bacteriol.*, 76, 227–232, 1958.

20. **King, J. R., Theodore, T. S., and Cole, R. M.,** Generic identification of L Forms by polyacrylamide gel electrophoretic comparison of extracts from parent strains and their derived L Forms, *J. Bacteriol.*, 100, 71–77, 1969.

21. **Kovalenko, I. V. and Dorozhkova, I. R.,** Substantiation and development of immunofluorescent techniques for the detection and identification of the L Forms of Mycobacteria, *Zh. Mikrobiol. Epidemiol. Immunol.*, 6, 35–38, 1986.

22. **Levina, G. A., Gorelov, A. L., Knyazeva, E. N., Alekseeva, N. V., and Prozorovski, S. V.,** Immuno-radiometric study of the persistence of *Salmonella typhi* L Forms in the body of experimental animals, *Zh. Mikrobiol. Epidemiol. Immunobiol.*, 3, 85–90, 1988.

23. **MacGregor, C. H. and Schnaitman, C. A.,** Alterations in the cytoplasmic membrane proteins of various chlorate-resistant mutants of *E. coli*, *J. Bacteriol.*, 108, 564–570, 1971.

24. **Mattman, L. H., Tunstall, L. H., and Rossmoore, H. W.,** Induction and characteristics of staphylococcal L Forms, *Can. J. Microbiol.*, 7, 705–713, 1961.

25. **Mattman, L. H. and Judge, M. S.,** Septicemia and some associated infections, demonstration of CWD bacteria, in *Cell Wall Deficient Bacteria*, Domingue, G. J., Ed., Addison-Wesley, Reading, MA, 1982, 427–451.

26. **McCarthy, B. J. and Bolton, E. T.,** A general method for the isolation of RNA complementary to DNA, *Proc. Natl. Acad. Sci. USA,* 48, 1390–1397, 1963.

27. **Mortimer, E. A. and Vastine, E. L.,** Production of capsular polysaccharide (hyaluronic acid) by L colonies of Group A streptococci, *J. Bacteriol.,* 94, 268–271, 1967.

28. **Motwani, N.,** Identification of Cell Wall Deficient Bacteria by Immunofluorescence and Polyacrylamide Gel Electrophoresis, Ph.D. dissertation, Wayne State University, Detroit, 1976.

29. **Nowotny, A.,** *Basic Exercises in Immuno Chemistry, A Laboratory Manual,* Springer-Verlag, Berlin, 1969.

30. **Prozorovsky, S. V., Vulfovich, Y. V., Pogorelskaya, L. V., Levina, G. A., Ilinskii, Y. A., and Gorlov, A. L.,** Detection of the microbial variants of *Salmonella typhi* in the bone marrow of typhoid patients and carriers, *Zh. Mikrobiol. Epidemiol. Immunobiol.,* 12, 15–17, 1986.

31. **Razin, S. and Shafer, Z.,** Incorporation of cholesterol by membranes of bacterial L phase variants with an appendix on the determination of the L phase parentage by the electrophoretic patterns of cell proteins, *J. Gen. Microbiol.,* 58, 327–339, 1969.

32. **Relman, D. A., Loutit, J. S., Schmidt, T. M., Falkow, S., and Tompkins, L. S.,** The agent of bacillary angiomatosis. An approach to the identification of uncultured pathogens, *N. Engl. J. Med.,* 323, 1573–1580, 1990.

33. **Rosen, I. G., Alden, M. J., Milliman, C. L., and Rubin, S. M.,** A chemiluminescent DNA probe assay for the identification of Group B streptococci from culture plates, in *Abstr. 29th Interscience Conf. Antimicrob. Agents Chemother.,* 867, 246, 1989.

34. **Rubin, F. A., McWhirter, P. D., Punjabi, N. H., Lane, E., Sudarmono, P., Pulungsih, S. P., Lesmana, M., Kumala, S., Kopecko, D. J., and Hoffman, S. L.,** Use of a DNA probe to detect *Salmonella typhi* in the blood of patients with typhoid fever, *J. Clin. Microbiol.,* 27, 1112–1114, 1989.

35. **Senatore, G.,** Presumptive Identification of Classical and Cell Wall Deficient Microorganisms by Gas-Liquid Chromatography, M.S. thesis, Wayne State University, Detroit, 1983.

36. **Slifkin, M. and Doyle, R. J.,** Lectins and their application to clinical microbiology, *Clin. Microbiol. Rev.,* 3, 197–218, 1990.

37. **Smith, J. A. and Willis, A. T.,** Some physiological characters of L Forms of *Staphylococcus aureus, J. Pathol. Bacteriol.,* 94, 359–365, 1967.

38. **Steffen, C. M.,** Identification of Cell Wall Deficient Mycobacteria with Auramine Rhodamine Staining and Polyacrylamide Gel Electrophoresis, Ph.D. dissertation, Wayne State University, Detroit, 1985.

39. **Steffen, C. M. and Mattman, L. H.,** Identification of cell wall deficient Mycobacteria with polyacrylamide gel electrophoresis, *Mich. Acad. Papers of Mich. Acad. Sci. Arts, Lett.,* 17, 303–313, 1986.

40. **Surkova, L. K., Kovalenko, I. V., Egorova, G. V., Lavor, Z. V., and Kazakov, A. F.,** Immunofluorescence in differential diagnosis of tuberculosis, *Probl. Tuberk.,* 8, 24–27, 1985.

41. **Takahashi, T., Nakatsuka, S., Okuda, K., and Tadokoro, I.,** Preservation of bacterial L Forms by freezing, *Jpn. J. Exp. Med.,* 50(4), 313–318, 1980.

42. **Theodore, T. S., Tully, J. G., and Cole, R. M.,** Polyacrylamide gel identification of bacterial L Forms and Mycoplasma species of human origin, *Appl. Microbiol.,* 21, 272–277, 1971.

43. **Tolmacheva, M.,** Molecular DNA-DNA hybridization reaction as a method for identifying L Form Brucella, *Zh. Mikrobiol. Epidemiol. Immunobiol.,* 10, 110, 1988.

44. **Whalen, M. A.,** Biochemical Reactivity of Staphylococci in the Cell Wall Deficient Stage, M.S. thesis, Wayne State University, Detroit, 1975.

45. **Widdowson, J. P., Maxted, W. R., and Grant, D. L.,** The production of opacity in serum by Group A streptococci and its relationship with the presence of M antigen, *J. Gen. Microbiol.,* 61, 343–353, 1970.

46. **Zafar, R. S.,** Quantitation and Identification of CWD Microorganisms Occurring in Pre-Therapy Blood Cultures, M.S. thesis, Wayne State University, Detroit, 1975.

47. **Zafar, R. S. and Mattman, L. H.,** Characteristics of L phase variants in pre-therapy blood cultures, *ASM Abstr. Annu. Meet.,* 1977.

48. **Godzeski, C. W.,** personal communication.

49. **Almenoff, P., Johnson, A., and Mattman, L. H.,** Monoclonal antibody to detect cell wall deficient forms of acid fast bacteria for rapid diagnosis of tuberculosis, American College of Chest Physicians, 56th Annual Science Assembly, October 2–16, 1990.

50. **Almenoff, P., Johnson, A., Judge, M. S., and Mattman, L. H.,** Acid fast cell wall deficient forms in Sarcoidosis, Annual Meeting, American Thoracic Soc., Anaheim, CA, May 1991.

APPENDICES

DISEASE STATES FROM WHICH L FORMS HAVE BEEN ISOLATED IN PURE CULTURE

Arthritis
    Acute — *S. aureus*
            Clostridium ramosum
            N. gonorrhoeae
    Rheumatoid
Aster yellows disease
Brain abscesses
Crohn's Disease
Endocarditis (20% of cases)
    Acute — *S. aureus*
            Flavobacterium
    Subacute — Streptococci
            Fungi
Gonorrhea
Loeffler's syndrome
Meningitis
Mycobacteriosis
    M. fortuitum
    M. intracellulaire
    M. leprae
    M. scrofulaceum
    M. tuberculosis
    Sarcoidosis organism
Osteomyelitis
    Typical — *S. aureus*
    Atrophy of marrow — Listeria

Peritonitis and salpingitis
Pleurisy
Postoperative thrombi
Postorgan transplant — fatal Pneumonia. Organism reverted to pneumococcus type 3
Pyelonephritis, cystitis and other urinary tract infections
Rheumatic fever
Septicemia — wide spectrum of organisms
Sterility of male
Stones in urinary bladder
Streptobacillus moniliformis
    Infections in mice (1937)
    Infections in man
Uveitis
Whipple's disease

TABLE OF PATHOGENS
Organisms Shown to Be Invasive or Toxigenic in the L-Stage

Acinetobacter calcoaceticus
A. endocarditis
Actinomyces odontolyticus
Aerococcus viridans
Agrobacterium tumefaciens
Aspergillus parasiticus
    Spheroplasts make aflatoxin
Bordetella pertussis
Brucella suis
Candida albicans
Citrobacter freundii
Clostridium botulinum
C. perfringers
C. rhamosum
C. tetani
Corynebacterium diphtheriae
C. hemolyticum
C. xerose
Erwinia carotovora
Erysipelothrix rhusopathiae
Escherichia coli

M. tuberculosis
    Nephritis in guinea pig after 18 mos. in immunosuppressed G.P., hamsters, gerbils, mice — 21 d, in man in 2 cases, incubation period unknown
Neisseria meningitidis
Nocardia asteroides
N. caviae
Pasteurella hemolytica
P. multocida
P. pseudotuberculosis
Propionibacterium acnes
Proteus OX$_{19}$
    Pathogenic for chick embryo
P. mirabilis
P. vulgaris
Pseudomonas aeruginosa
Salmonella dublin
S. enteritidis
S. typhi
S. typhimurium
    Hemorrhage in chick chorioallantoic membrane

TABLE OF PATHOGENS (continued)
Organisms Shown to Be Invasive or Toxigenic in the L-Stage

Flavobacterium — heart valves
Hemophilus influenzae
H. parainfluenzae
Klebsiella pneumoniae
Listeria monocytogenes
Mycobacterium of sarcoidosis
M. paratuberculosis
M. scrofulaceum

Staphylococcus aureus
S. epidermidis
Streptococcus Group A
S. bovis
S. fecalis
S. pneumoniae
S. salivarius
Vibrio cholerae

AUTHORS CITED BY CHAPTERS

INDEX